National Contexts for Mathematics and Science Education

AN ENCYCLOPEDIA
OF THE EDUCATION SYSTEMS
PARTICIPATING IN TIMSS

$75

National Contexts for Mathematics and Science Education

AN ENCYCLOPEDIA
OF THE EDUCATION SYSTEMS
PARTICIPATING IN TIMSS

Edited by David F. Robitaille

Pacific Educational Press
Vancouver Canada

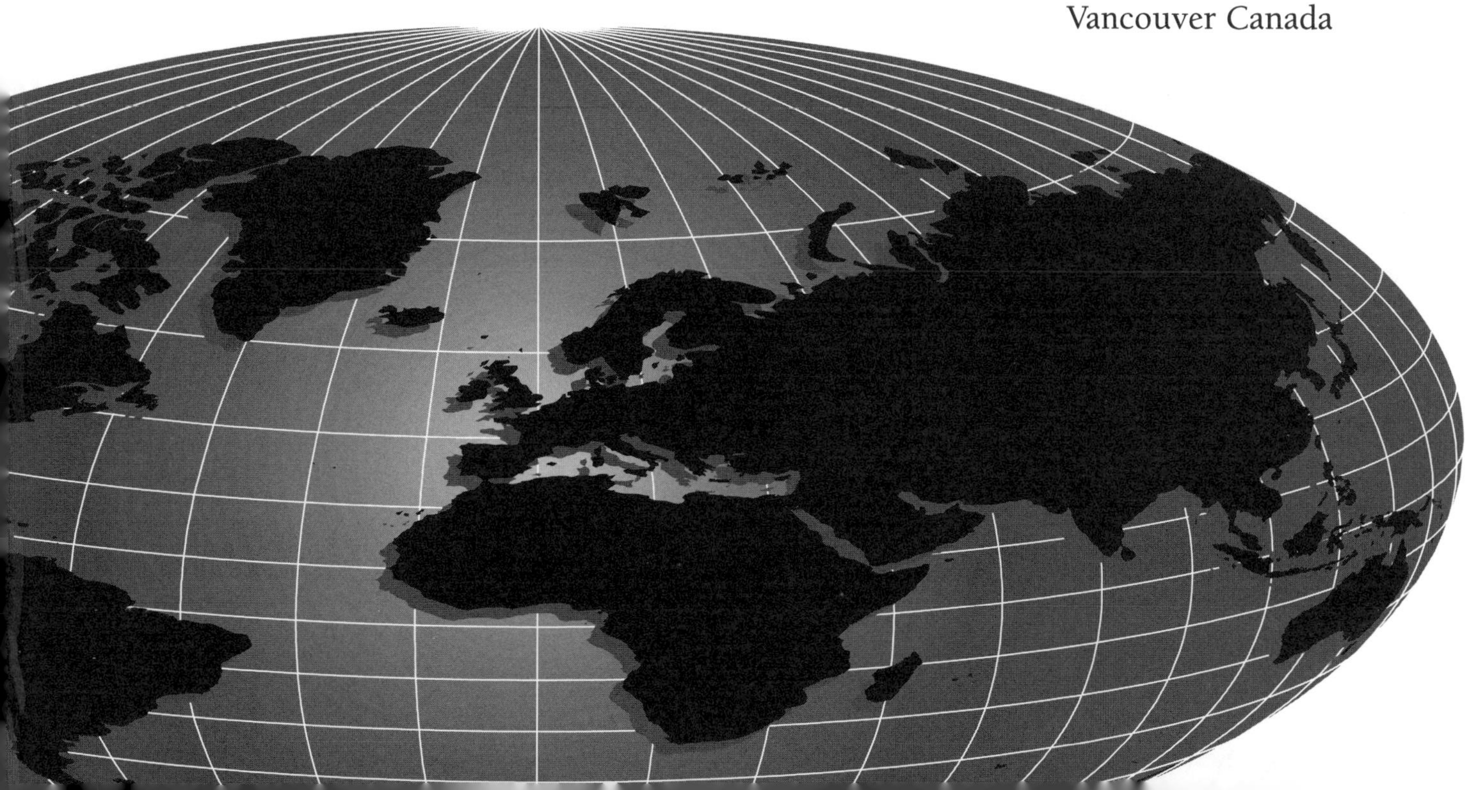

Published by Pacific Educational Press
Faculty of Education
University of British Columbia
Vancouver Canada V6T 1Z4
Telephone: (604) 822-5385
Facsimile: (604) 822-6603
E-mail: cedwards@unixg.ubc.ca

Acknowledgements

The publisher would like to acknowledge the financial contribution made by Canadian Heritage in support of its publishing program. The Third International Mathematics and Science Study wishes to acknowledge that funding for the publication of this book was provided by Employment and Immigration Canada and Industry, Science and Technology Canada.

Canadian Cataloguing in Publication Data

Main entry under title:

National contexts for mathematics & science education

ISBN 1-895766-25-7

1. Third International Mathematics and Science Study. 2. Mathematics--Study and teaching. 3. Science--Study and teaching. 4. Curriculum evaluation. I. Robitaille, David F.
Q181.N37 1997   507'.1   C97-910142-5

Editing: Katy Ellsworth
Design: Warren Clark
Printed and bound in Canada

97 98 99 10 9 8 7 6 5 4 3 2 1

Contents

Foreword

The Third International Mathematics and Science Study (TIMSS) has collected a wealth of data on student performance in over 40 countries around the world. How students in these countries performed was measured by carefully designed and translated tests, tests with questions that required written responses as well as selecting among multiple options. Some of the students were assigned hands-on tasks so their ability to discover solutions could be observed. Along with the test information, the students, their teachers, and their principals filled out detailed questionnaires to supply information about factors that may be related to student performance.

TIMSS's first step has been to establish the similarities and differences among students in different countries on the mathematics and science tests, and these results will no doubt be of great public interest. More important than this, however, is to come to an understanding of these differences and similarities, and why they occur. Although a survey such as TIMSS can never definitively explain student achievement, it can explore the plausibility of many hypothetical reasons for it, perhaps eliminating some and lending support to others.

Insights into the possibilities and limitations of educational practice should lead to the improvement of education around the world.

Although the statistical data collected for TIMSS are wide-ranging, they need to be interpreted in light of the educational systems in which they were collected. For this reason, David Robitaille has edited *National Contexts for Mathematics and Science Education: An Encyclopedia of the Education Systems Participating in TIMSS.* This volume provides an opportunity for each TIMSS country to describe in some detail the workings of its education system. The descriptions have been carefully coordinated so as to facilitate comparisons among the education systems.

This volume should be of great help to those interested in understanding and interpreting the TIMSS data. TIMSS owes Dr. Robitaille thanks for conceiving and bringing this book to fruition. TIMSS is also indebted to the many authors and other people who contributed to this book.

Albert E. Beaton

Preface

This volume is published as part of the Third International Mathematics and Science Study (TIMSS), itself a research initiative of the International Association for the Evaluation of Educational Achievement (IEA).

IEA was founded in 1959 for the purpose of conducting comparative studies focusing on educational policies and practices in various countries and education systems around the world. The membership of IEA has grown over the years from a small number of countries to a group of about fifty-five nations. Its secretariat is located in Amsterdam, the Netherlands. IEA studies have reported on a wide range of topics, each contributing to a deeper understanding of educational processes.

The Third International Mathematics and Science Study (TIMSS) was conceived in 1990 as a response to the need for information about how well children learn mathematics and sciences, and what factors influence the acquisition of these important academic disciplines. TIMSS, with more than 40 countries participating, has become the biggest international comparative study ever conducted. It comprises several components, including a survey of mathematics and science achievement, a curriculum analysis, and a compilation of national profiles of the teaching of mathematics and sciences. This book gathers together the national profiles into a single reference volume.

National Contexts for Mathematics and Science Education: An Encyclopedia of the Education Systems Participating in TIMSS provides readers with information about national education systems, with a specific focus on mathematics and science education, curricula,

textbooks, and assessment. It also provides a background that will help interpret the results of other parts of TIMSS.

IEA is very grateful to the following organizations that have contributed financing to defray the costs of running the International Coordinating Centre (ICC) of the study: the Ministry of Education, Skills, and Training, British Columbia; Human Resource Development Canada; the USA National Center for Educational Statistics; and the USA National Science Foundation.

This book is the result of many individuals' efforts. Special thanks must go to the staff of the TIMSS International Coordinating Centre, particularly to Katy Ellsworth and Beverley Maxwell, who, under the leadership of David Robitaille, did an excellent job in producing this volume.

Readers wishing additional information about this or other IEA studies may correspond directly with IEA at the following address:

IEA Secretariat
Herengracht 487
1017 BT Amsterdam
The Netherlands
Telephone: +31 20 625 3625
Facsimile: +31 20 420 7136
E-mail: Department@IEA.NL

Tjeerd Plomp
Chair of IEA

Acknowledgements

The idea for a TIMSS encyclopedia has been around almost as long as TIMSS has. Skip Kifer of the University of Kentucky may have been the first person to suggest we put such a book together, and I am grateful to him for that suggestion (along with quite a few other things). The structure of the encyclopedia has been modified and refined over time, so it now looks rather different from what was first envisioned. What has not changed from the outset is the recognition that there is a need for a source that could help readers learn more about science and mathematics education in the TIMSS countries, and one they could consult for important contextual information in attempting to interpret the meaning and significance of the achievement results.

One of the decisions we made early in the process was to require all the authors to follow a common, detailed outline. This, we believed, would make it easier for readers to find the information they wanted and ensure, to a large extent, that the same topics were dealt with in pretty much the same way for all of the participating countries. Several people from the TIMSS International Coordinating Centre (ICC) at the University of British Columbia assisted in the development and refinement of that outline, and I am grateful to them for their input. The list includes Barry Anderson, Katy Ellsworth, Bob Garden, Beverley Maxwell, Ed Robeck, and Alan Taylor. Bob, Ed, and Alan also helped by writing three of the national chapters early so that we could send them to other chapter authors to use as exemplars.

Thanks also to the more than sixty authors of the thirty-eight national chapters that are included in this volume. A lot of those authors are either National Research Coordinators or otherwise heavily involved in TIMSS, and this writing had to be done in the midst of one of the most hectic periods of the study. The degree of cooperation we had from all of them was much appreciated and made our job at the ICC a lot easier than it would otherwise have been.

The manuscript was reviewed by three members of the TIMSS Editorial Board—Judith Torney-Purta, Svein Lie, and Al Beaton—and by the chairperson of IEA's Publications and Editorial Committee, Dick Wolf. Their comments were constructive and helpful, and I am grateful to them for taking the time to help us in this way.

A few people did a lot of work to bring this encyclopedia to completion. Ed Robeck and Alan Taylor worked closely with me on the writing of the first two chapters, and were heavily involved in the entire process. Beverley Maxwell coordinated the production of the system diagrams for the national entries, and she was a consistent source of expert advice, suggestions, and assistance throughout the project.

I am particularly grateful to Katy Ellsworth who has shepherded this project to completion. Katy worked closely with each of the authors and with me, keeping us on task and on schedule with patience, tact, and good humor. She now has friends in all of the TIMSS countries, but especially right here in Vancouver.

David F. Robitaille
October 1996

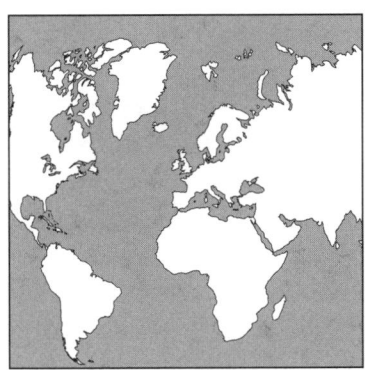

The Importance of Context for International Comparisons

David F. Robitaille, University of British Columbia

Edward C. Robeck, Hastings College

People all over the world are interested in knowing about the status of their education system and how that system compares with those in other countries. The author of the chapter on Belgium in this book, for example, has this to say about the interest in comparative studies: "On a recurrent basis, the IEA surveys have revealed wide variations in pupil performance in Belgian schools in both mathematics and science. This is an important issue for policymakers, particularly when assessing evaluation policies and practices. Although Belgium compared rather well internationally on mathematics in the last comparative study of mathematics and science education (SIMS), a careful analysis of the TIMSS data is necessary. Situated in the heart of Europe, Belgium cannot ignore international comparisons that might help avoid excessive disparities in regard to its partners."

Individual countries typically have a particular interest in the relative results among their important trading partners or competitors, countries that share a common political or cultural heritage with them, or countries that are held up as models of excellence in education. Evidence of this interest is found in the fact that over fifty countries have participated in one or more of segments of IEA's Third International Mathematics and Science Study (TIMSS)—about twice as many as have participated in any previous study of this kind.

Whenever the results of an international study are announced, and especially when those results include scores on some measure of student achievement, a great deal of media attention is focused on the results, albeit usually for only a short time. Taxpayers, parents, educators, educational policymakers, and researchers all use the findings from such studies; and hardly a week goes by in many countries without some comment or criticism about the quality of the national or regional educational system being featured in newspapers, magazines, or broadcast media. Unfortunately such comments and criticisms are frequently not based primarily on fact, but on opinion or hearsay.

Valid international comparisons among systems of education are extremely difficult to make because of the numbers of variables involved, because of our lack of sophisticated ways of accounting for the influence of many of those variables, and because of the fact that some of the variables involved are beyond the control of the educational system. Torsten Husén, the founder of IEA, once referred to international studies in education as exercises in comparing the incomparable. In saying this, his intention was not to indicate that international comparisons and international studies were fruitless activities. On the contrary he was saying that international studies were important precisely because the comparisons were so difficult to make. Education systems, as products of particular cultures, are very different from one another in a number of fundamental ways, and one needs to guard against the temptation to make oversimplified generalizations or comparisons.

There is, therefore, always a danger that unwarranted comparisons will be made or inappropriate conclusions arrived at. However, comparisons based on good data certainly have a greater likelihood of providing useful information for policymakers than comparisons based on no data at all or on ill-informed opinion.

The results of well-designed international comparative studies in education can contribute significantly to the debates that swirl constantly around questions of educational excellence and calls for increased accountability. The findings from an international comparative study of education such as TIMSS can and should be used to inform such debates. The design of TIMSS was structured so as to allow appropriate and valid comparisons to be made among all of the participating countries.

The Development of TIMSS

TIMSS was designed to respond to the wide variety of reasons motivating national education systems to take part in large-scale comparative studies and to adhere to the highest international standards for conducting research into educational practices and outcomes. While the design of TIMSS owes a debt to earlier studies carried out by the IEA since its founding in 1959, it is also different from those earlier studies in important ways.

The first IEA study investigated students' achievement in mathematics, but it was not designed primarily as a study of mathematics. Rather, it was a study of schools and the outcomes of schooling that used achievement in mathematics as a surrogate for the overall outcomes of schooling in general. What became apparent during the course of the study were both the possibilities and the limitations presented by such a study. On one hand, it provided insights into mathematics education that many teachers, administrators, policymakers, and others found useful. On the other, it demonstrated the importance of considering disciplines separately in terms of the concerns that they raise for the policies and practices of education.

The second major IEA study, the Six–Subject Survey, was designed to complement the first by providing information about several other academic subjects in the school curriculum, including science, literature, reading comprehension, English and French as foreign languages, and civic education. Subsequent IEA studies, including the Second International Mathematics Study and the Second International Science Study, have looked at each of these areas in more depth. Between 20 and 30 countries took part in those studies.

The design of TIMSS recognizes that there is nothing inevitable about the way in which formal education is organized. On the contrary, education systems are organized through a series of decisions and compromises that respond to the specific goals, priorities, resources, and historical models that impinge on a particular system. It is important to understand the factors that influence system-level decision making both inside and outside of the settings where actual instruction takes place. These "settings" can be considered at different levels within the organization, from the individual student to the community level—which itself might be thought of more or less broadly—and finally to the level of the general social matrix in which the schools are organized.

A distinction must be made between what is decided in an education system and what is actually taught. For example, countries report that reforms in science and mathematics education have led to the adoption of standards that call for nontraditional problem-solving approaches to instruction. Yet, often the available facilities, supplies, and teacher-education programs influence actual practice in a way that means that traditional approaches are far more likely to be found. There is an important distinction between what, in IEA terms, are called "the intended curriculum" and "the implemented curriculum." By intended curriculum, IEA researchers mean the goals that society has for teaching and learning. The implemented curriculum refers to what is actually taught in schools.[1] Both are important, yet what is intended generally has more to do with social priorities while what is implemented is more likely to be closely associated with conditions within schools.

The importance of distinguishing between intended and implemented curricula is illustrated in systems undergoing major shifts in educational policy. In the Republic of Korea, for instance, policy changes in mathematics education have, as in many other systems, encouraged movement toward problem solving, relevance, active learning, and an appreciation of the importance of helping students develop a positive attitude toward mathematics and science. For these shifts to be implemented, however, teachers' workloads and other working conditions (for example, class size) have to be taken into account. While movement toward the new policies is occurring, to consider either the policies or the practices alone would give a view of the situation that was inadequate to describe the trends accurately and completely.

In addition, although it may be decided that a particular instructional approach should be used and sys-

---

[1] A complete discussion of the intended, implemented, and attained curricula is found under the heading "Conceptual Framework for TIMSS," in this chapter.

tems set in place so that it can be implemented effectively, one cannot assume that students will learn what they were intended to learn. This is the point at which the achievement instruments become important. What the students know and can do (the attained curriculum) needs to be compared to what teachers report teaching (the implemented curriculum) and to the content that is called for in official policies (the intended curriculum). Together, these can be taken as a demonstration of the interconnections among features that are contextually related. For some purposes, student achievement is the central feature of concern, and the intended curriculum becomes an integrally related influence. For other purposes, it is the intended curriculum that is the focus of attention, and current practices in schools—the implemented curriculum—can be seen as the backdrop against which those policy decisions are being made.

Conceptual Framework for TIMSS

TIMSS is the most ambitious study ever undertaken by the members of IEA. About twice as many countries as had ever participated in any IEA study took part in one or more of the components of TIMSS. The study included three populations of students spread across five or more grade levels, the teachers of those students, two subject areas from the school curriculum, an in-depth analysis of curricular materials and opportunity to learn, and achievement testing that involved open-ended items and performance tasks in addition to multiple-choice items.

The conceptual framework for the study was de-

rived from earlier IEA studies, particularly the Second International Mathematics Study. The framework highlights the fact that a given curriculum may be viewed from three perspectives—curriculum as intended, as implemented, and as attained—and that, in order to get a more complete picture of what a curriculum in a particular jurisdiction is like, one must examine it from these three perspectives.

The intended curriculum is a statement of a society's goals for teaching and learning. In TIMSS, the intended curriculum was studied through an analysis of its prescribed textbooks, curriculum guides, examinations, and official policy statements.

The implemented curriculum—what teachers actually teach—was more difficult to study, because it can vary for different teachers or for the same teacher over different years. In order for it to be relevant to the measures of the intended and the attained curricula, for TIMSS's purposes the implemented curriculum was conceptualized as what the teacher taught in the current year to the students in the TIMSS sample. The implemented curriculum, although based on the intended curriculum, was influenced by local contextual variables, including the school's administration and climate, classroom characteristics, resources, and local or community interests and involvement.

The attained curriculum consisted of the outcomes of schooling, or more particularly in the case of TIMSS, the mathematics and the science that students learned in the course of their studies and their attitudes toward these subjects. Although student achievement is greatly affected by both the implemented curriculum and societal contexts, it is also a function of factors under the control of individual students. These factors include the student's effort, attitude, and personal interests. In TIMSS, student achievement was measured using multiple-choice and free-response items, as well as mathematics and science performance tasks. Student attitudes were assessed through a questionnaire administered as part of the survey.

The Intended Curriculum

Four basic research questions (see Robitaille & Garden, 1996) guided the development

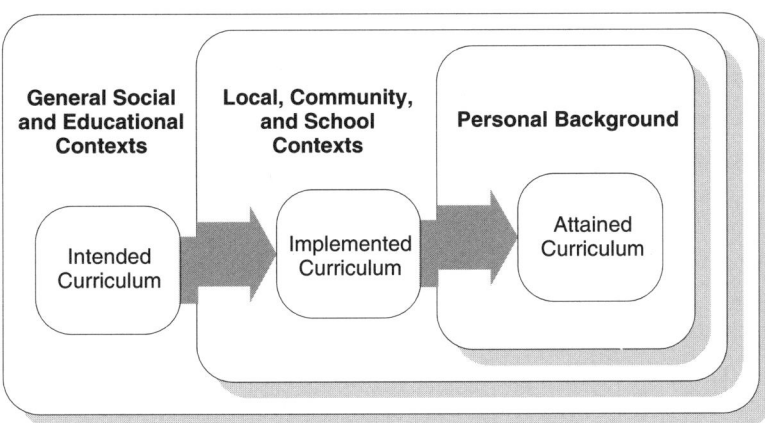

of the study, and the first of the four addressed the intended curriculum:

How do countries vary in the intended learning goals for mathematics and science; and what characteristics of education systems, schools, and students influence the development of those goals?

The intended curriculum consists of the mathematics and science goals defined at the system level. Although the political and administrative structures of educational systems vary, they all include mechanisms for identifying and articulating the educational experiences that are most valued and desired in the system. The greatest diversity between systems regarding the intended curriculum lies in the locus of curriculum decision making within the system. Many countries have established a nationally recognized curriculum, whereas others may have decentralized decision-making structures. For example, each Canadian province determines the curriculum for its students.

The intended curriculum is described in policies, regulations, curriculum guides, and other official statements generated to direct the education system; and reflected in textbooks, resources and examinations. Policies may include decisions to direct additional funding toward the development of specific resources (for example, laboratory equipment or computer hardware and software). Regulations usually direct the number of years students spend in school, the mandatory and optional courses, and the diversity of school programs. Curriculum guides may vary in level of detail, but they are all intended to ensure that students will have similar ranges of educational opportunities within the system. Other documents would include philosophical, pedagogical, and social statements of goals.

Textbooks, resources and examinations are of interest because they serve and reflect the intended curriculum. Textbooks are either written to embody the goals articulated in the curriculum guides or statements, or they are selected by educators because they most closely meet the requirements of the curriculum. Similarly, the relative emphasis on providing teacher support, laboratory materials, computers, teaching aids, films, books, and funding for travel (within the limitations of the system's economics) all reflect the priorities of the intended curriculum. System-wide entrance or graduation examinations further indicate the expected outcomes of the education system, both in their content and their form.

In TIMSS, the intended curriculum, the mathematics and science content that students are expected to study and learn, is described in the form of concepts, processes, skills, and attitudes in mathematics and science. This is consistent with the conceptualization of curriculum in most education systems, and provides the soundest base for making reliable comparisons across systems.

TIMSS and the Place of Mathematics and Science in School Curricula

The fact that education systems in so many countries were willing and able to find the resources needed to participate in one or more components of TIMSS demonstrates the importance that these countries attach to mathematics and science. This point is further illustrated by the fact that other countries expressed interest in participating in the study, but were unable to do so for a number of reasons. A recent OECD (1995) study reported that both science and mathematics are given high priority in most countries, while mathematics was among the top two subject areas in every country surveyed (p. 47). That there is widespread concern about the teaching and learning of mathematics and science is clear from remarks such as the statement by Francisco Ayala (1996), writing in the recently published *World Science Report: 1996*. He says, "In virtually every nation there is dissatisfaction with the scientific education imparted in schools."

The premium placed on ensuring that students are "literate" in mathematics and science, whatever that might mean in the countries concerned, is one of the most prevalent trends that can be identified in the chapters that follow.

While mathematics and science have been important components of every school curriculum for many years, the rationale supporting their importance has changed over time. Thirty years ago, in the United States, the prominence of mathematics and science in school curricula was legitimated as a response to the launching of the first artificial Earth satellite, Sputnik I. At that time, arguments in support of the need for increased support for mathematics and science education were often stated in terms of concerns about threats to national security unless drastic measures were taken in the schools.

Today, in the United States as well as in many other countries, such arguments are much more likely to be framed in terms of an assumed linkage between the caliber of mathematics and science education in a country and the level of economic competitiveness. While fewer than 1 percent of the 25- to 34-year-old members of the work force in an industrialized country are likely to be involved in technical fields (OECD, p. 224), it is widely believed that these professionals play a crucial role in a country's long-term economic vitality. One of the current issues in Singapore, for example, is how to approach secondary science education in a way that encourages students to pursue science and engineering programs at the tertiary level. The rationale for this initiative relates directly to the potential that it would create for technical innovation.

Claims about the link between students' success in mathematics and science and a nation's economic competitiveness in the future are made in a variety of ways. In some systems, there is an emphasis on new technologies, which can be used to open new markets, and therefore provide new jobs. In others, the focus is on preparing workers in a way that is consistent with the global trend toward more technically oriented goods and services. In still others, the focus is on national development in a manner consistent with other systems so that there will be a compatibility among workers, products, and markets. In most countries the arguments advanced would likely involve a combination of these and other factors. The basic point remains, however, that factors related to economic development play a crucial role in the arguments currently given for the importance of mathematics and science education, especially in developing countries (Salomon, Sagasti, & Sachs-Jeantet, 1994).

More generally, science and mathematics are considered fundamental components of comprehensive education. Educating students in ways that will allow them to be critical consumers, discerning advocates of political causes, and informed members of families, work forces, and other social groupings are all reasons given for advocating mathematics and science education for the general populace. In many systems, this has come to include explicit attention to affective factors related to mathematics and science. In England this attention is invigorated by the connection that has been made there, as it has in other places, between low achievement and poor attitudes regarding the subjects in general. By addressing affective factors, the hope is that cognitive abilities in the subjects will be improved as well.

There are also reasons that apply on a more personal level. In virtually every contemporary industrial society, elements of individual standards of living, such as personal income, are highly correlated with educational attainment (OECD, 1995, p. 232–238). This means not just educational attainment in mathematics and science, but in general. Yet both mathematics and science are part of that, and are often considered important indicators of general academic ability and are therefore used as prerequisites for entry into postsecondary education. In these and other respects, arguments can be given for the importance of education in both mathematics and science that relate to an individual's opportunity for social mobility, a satisfying career, and the general fulfillment of personal goals.

For all of these reasons there is growing concern internationally regarding the quality of mathematics and science education and students' levels of attainment. In many systems, current reform efforts include the establishment of standards for the content, instructional practices, and levels of student performance that will be considered sufficient and for which the system is ultimately striving. In some systems, the move to national standards is a fairly new phenomenon. In others it can be seen as part of a sequence of events with antecedents in the years immediately following World War II. In some systems, such as Hungary, the intent to set standards has been in place for some time, but owing to political and social change, the content of such standards has never been settled.

The Plan for This Volume

Whenever the findings from an international comparative study of education are released, interest and attention first focus on the achievement results: "Who won?" Interest in the achievement results is understandable, but those results are like the proverbial iceberg: most of the important and interesting information is hidden from view. The real meaning and the important implications of the results lie in an understanding of and appreciation for the different contexts and variables that give rise to those results. Achievement results at the national level are highly aggregated; and, in the process of carrying out that aggregation, much of the detail that is essential for an adequate understanding of the actual meaning of the achievement results is hidden from view.

One of the hallmarks of every IEA study has been a concerted effort to address more substantive and important questions about what the achievement results might signify, what characteristics of national educational systems might account for differences in achievement, and what relationships might exist among the many variables addressed in the study and student outcomes. In TIMSS, a great deal of care and attention was devoted to ensuring that a large quantity of contextual information was collected from the participating countries and that such information was used extensively in interpreting the findings of the study.

The overall goal of TIMSS is to help improve the teaching and learning of mathematics and science in elementary and secondary schools by providing detailed, valid, and useful information about mathematics and science curricula, the instructional practices utilized by teachers of mathematics and science, the achievement and attitudes of elementary and secondary school students, and the relationships among these variables. This goal requires, however, that contextual information be provided that will help educational stakeholders with the widest variety of interests in education to interpret the findings of TIMSS appropriately; that is, in ways that take advantage of the potential offered by comparative international studies to understand and explain differences in student achievement and other outcomes of schooling, without slipping into the trap of overly simplistic comparisons.

The first data-collection activity concerning the intended curriculum in TIMSS was the administration of two so-called Participation Questionnaires. These questionnaires dealt with system-level characteristics including the grade structure of the system, age-grade distributions of students, course structures in mathematics and science, school year structure, certification requirements for teachers, and many others. At about the same time, two Expert Questionnaires—one for mathematics and one for science—were completed by someone in each participating country who was familiar with the curriculum in that area. The questionnaires, which included many open-ended items, were designed to obtain system-level information about recent trends in mathematics and science curricula, im-

pact of technology, current issues affecting the teaching and learning of mathematics, approaches to evaluation, and several other topics.

This book is intended to serve as a complement to other aspects of the TIMSS investigation of the intended curriculum. Questionnaire responses, particularly to items in a forced-choice format, necessarily restrict the choices respondents have in completing the questionnaire. Open-ended items provide more choice, but respondents frequently do not provide a lot of detail in their responses to such items. What was needed, it was decided, was a thorough description of national systems of education with a particular emphasis on the teaching and learning of mathematics and science. Someone from each country would write a chapter explaining a number of aspects of mathematics and science education that would help flesh out the picture in that country and assist others in interpreting the other findings from the study.

In order to make the process of comparing countries on a given topic somewhat easier, a detailed outline to be used by all of the chapter authors was prepared and agreed upon. A copy of that outline is included in Appendix A.

The national chapters in this volume were written by persons who have expert knowledge about the education system in their country (see Appendix B for a list of the authors). The authors were drawn almost exclusively from the ranks of personnel within ministries of education, research institutes, or institutions of higher education in the countries participating in TIMSS.

Despite the fact they were written to a common outline, all chapters include information that speaks to the uniqueness of each educational setting. This reinforces the importance of a volume such as this. No two systems are alike in the solutions they develop to address educational issues. Even in those cases that appear to be highly similar, the processes or paths by which they arrived at their particular solutions were undoubtedly not at all similar. For instance, many of the chapters report an increasing use of technology in the teaching of mathematics and science. In the case of some countries, this could be interpreted as an attempt to ensure that students are kept abreast of the most current technologies. In other countries, it can be seen as a way of addressing other issues. Colombia provides an illustration of the latter. There, technology—in the form of computer simulations—is being used to compensate for the lack of expensive laboratory equipment in science. In Singapore technology is used not only for a variety of instructional purposes, but also as one way in which explicit national goals aimed at familiarizing students with the interactions of technology and society can be addressed.

A study such as TIMSS provides an unparalleled opportunity to compare and contrast the common and unique features of education systems. The breadth of design and international scope of TIMSS contribute to the opportunities it offers anyone interested in education in mathematics and science. Those opportunities will be realized in the most meaningful ways by appreciating the context-dependent character of the results. It is with the hope of encouraging and facilitating such an appreciation that this volume is provided as an accompaniment to the findings of TIMSS.

References

Ayala F J 1996 Introductory essay: The case for scientific literacy. In H Moore (ed.) *World Science Report.* UNESCO Publishing, London (p. 4)

Elley W B 1992 *How in the World Do Students Read?* International Association for the Evaluation of Educational Achievement, Hamburg

Organisation for Economic Co-operation and Development 1995 *Education at a Glance: OECD Indicators.* Organisation for Economic Co-operation and Development, Paris

Robitaille D F, R A Garden (eds.) 1996 *The Third International Mathematics and Science Study Monograph Number 2: Research Questions & Study Design.* Pacific Educational Press, Vancouver

Travers K J, I Westbury 1989 *The IEA Study of Mathematics I: Analysis of Mathematical Curricula.* Pergamon Press, Oxford

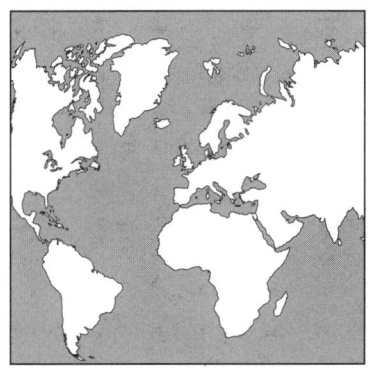

Cross-National Similarities and Differences

David F. Robitaille, University of British Columbia

Alan R. Taylor, University of British Columbia

Even a cursory examination of the country entries that make up the remainder of this volume will show that there are a wide range of topics associated with the teaching and learning of mathematics and science of concern to the TIMSS countries. Yet, despite this range, similar initiatives are underway in a number of countries, pointing to a sharing of common concerns. Certainly there are differences as well; not all countries are interested in or are doing the same things in precisely the same ways or at the same time. But, leaving aside differences in the ways system-wide change might be implemented in different countries, variations in degree of emphasis, or differences in the order in which topics are addressed, there seems to be a widespread consensus that a number of topics are important in mathematics and science education.

The following discussion is structured around the topics in the outline the authors used to write the entries for their respective countries. The five major topics in that outline were: governance and organization of the school system, profiles of teachers, issues and trends in the teaching and learning of mathematics, issues and trends in the teaching and learning of science, and evaluation policies and practices.

GOVERNANCE AND ORGANIZATION

Three main subtopics concerned with how countries organize their education systems and certain aspects of how those systems operate were covered in the first part of the outline. The three topics were locus of decision making, years of compulsory schooling, and the length and timing of the school year.

Locus of Decision Making

Some countries have highly centralized systems of education, and, in such systems, the majority of policy-related decisions are made at the national or regional level. Other countries have much more decentralized systems, with many important educational decisions being made at the local or school level. In most TIMSS countries, decisions about policy, curriculum, and selection of textbooks are centralized; decisions about hiring of staff members, instructional practices, and student evaluation are largely decentralized.

Of the 38 countries, 26 can best be categorized as moderately to highly centralized, with administrative and financial control of the education system largely concentrated at the national level. Among the most highly centralized systems are those belonging to Cyprus, France, Greece, Hong Kong, Iran, Japan, and the Republic of Korea. In the Republic of Korea and the rest of the countries that have a more centralized system, policies and regulations are established centrally on a wide variety of educational matters, leaving little flexibility at the local level. In some cases, centralization extends as far as the requirement that textbooks and other support materials be approved or even developed at the national or regional level, and that student evaluation programs be controlled centrally. In the Republic of Korea, for example, the Ministry of Education sets a compulsory, national curriculum, publishes and approves textbooks, administers all finances, and directs all subordinate agencies in planning and policy implementation. Although there are local offices of education, in practice these bodies are responsible to the Ministry.

Hungary and Denmark are both examples of

moderately centralized systems. In Denmark, the Ministry of Education sets learning objectives and frameworks, and issues curriculum and teaching guidelines. At the local level, however, a school is permitted to prepare its own curriculum and submit it to a local education authority for approval. Similarly, school boards in Hungary are able to set local policy on school operations, curriculum implementation, and teacher hiring, within guidelines established by the Ministry.

In 11 countries, education is organized through autonomous regional authorities. Typically these regions correspond to provinces, states, or other levels of government jurisdiction. In Australia, for example, education is the responsibility of individual states and territories: "State education departments recruit and appoint the teachers in government schools, supply buildings, equipment, and materials, and provide limited discretionary funding for use by schools. . . . Within states the general pattern is that central authorities specify broad curriculum guidelines and schools have considerable autonomy in deciding curriculum detail, textbooks, and teaching methodology."

In a number of countries, although the central government has a significant role in areas such as funding for education, educational policy, and the establishment of national goals for the educational system, quite a high degree of autonomy is devolved to the local level. In these cases, responsibility for decisions concerning matters such as curricular objectives and support materials is exercised by local school districts or school councils. Among the countries reporting this type of organization are England, Iceland, New Zealand, the Russian Federation, the Slovak Republic, the Netherlands, and the United States.

Recent shifts in the levels at which decision-making authority is exercised have been reported in the Czech Republic, Russia, and Sweden. In each case more autonomy has been given to local boards and to schools. Among the responsibilities now assigned at this level are jurisdiction over curriculum, the school calendar, funding, standards, and hiring. In Sweden, for example, schools now develop an individual school plan that describes how goals will be achieved. Local school boards appointed by municipal councils ensure that all schools maintain uniform standards. Formerly, administration of the Swedish education system was centralized and highly bureaucratic.

Whether a system is highly centralized or not, teachers in most countries are responsible for making decisions about instructional methods. Classroom processes and the selection of supplementary materials are typically decided upon either by teachers on their own or in consultation with colleagues, supervisors, and consultants.

Years of Compulsory Schooling

As Figure 1 shows, the number of years of compulsory, formal schooling (that is, beyond preschool or kindergarten) varies significantly among countries. Each block on the graph represents an age cohort, with age six indicated in the shaded column. The varying block heights indicate years of preprimary, compulsory, and noncompulsory education. Years of compulsory schooling, the tallest block, range from 0 in Singapore, to 7 in Argentina and the Philippines, to 12 in Belgium, with most countries—22 of the 38—requiring children to attend school for either 9 or 10 years. England, Israel, the Netherlands, and Scotland require 11 years of compulsory schooling. Argentina, Bulgaria, Denmark, Iran, Italy, Kuwait, the Philippines, Romania, the Russian Federation, Slovakia, and Slovenia require 8 years or fewer.

The shading within the blocks indicates the approximate percent of the age cohort actually enrolled in school; disparities between requirements and enrollment are shown in this way. Colombia, Argentina, and South Africa do not achieve 91 percent enrollment for any of their compulsory grades, and Iran only achieves 91 percent in five of its eight compulsory grades. Greece, Hungary, and Romania experience declining enrollment in the final years of compulsory schooling. Singapore, the only country that does not specify compulsory attendance for any grade or age, is also the only country to have 100 percent enrollment in all grades.

The countries are clustered according to the years of compulsory schooling and ordered within the clusters according to years of preprimary education and enrollment within those years. The Netherlands, for example, offers one year of noncompulsory, preprimary schooling, which is attended by over 91 percent of the cohort. Below it on the graph is England, which has two years of preprimary attended by 31 to 40 percent of the cohort. Scotland also offers two years of preprimary, but the first is attended by less than 30 percent of the cohort and the second by 31 to 40 percent.

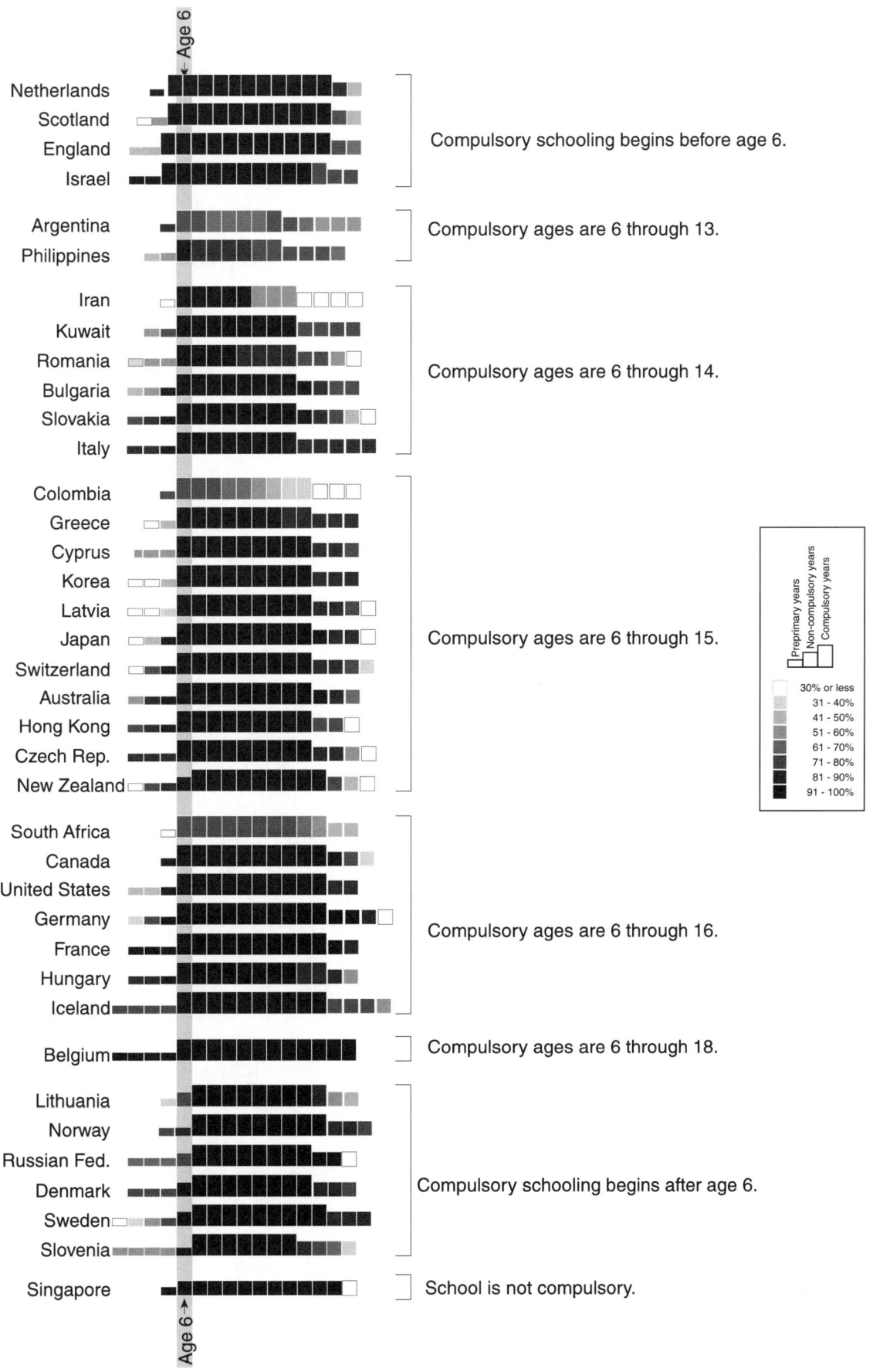

Figure 1: Years of Compulsory Schooling

The School Year

Two areas in which there seem to be considerably more uniformity across countries than there was twenty or more years ago when IEA's second mathematics study was conducted are the length of the school year and the structure of the school week. Almost 80 percent of the school years reported were between 190 and 209 days long, with a mean of 194. Some countries reported more than one school year when, for example, the length of the year in elementary schools was different from that of secondary schools. Some countries also reported a range of days; for example, the school year is between 180 and 200 days in Bulgaria.

Almost all countries have either moved or are in the process of moving to a five-day school week. In the past, in some European countries and in Japan, there was a six-day school week; however, the popularity of that option has declined. Only the Russian Federation reported that some children still went to school six days a week.

PROFILES OF TEACHERS

During the design phase of TIMSS there was a great deal of interest shown in finding out how teachers in the participating countries compared to one another on a wide variety of factors. These ranged from simple comparisons for a variety of demographic variables to questions of status and the kind of professional and academic preparation that was required in order to obtain professional certification.

Gender

Data on the gender of teachers at the elementary level were provided by 33 countries. In all but one case, women constituted the majority—and usually the vast majority—of elementary teachers. The exception was Greece, where it was reported that the split between male and female teachers at the elementary level was almost equal. The most extreme case was reported by Israel, where fewer than 5 percent of elementary teachers are male. The average across the 33 countries was between 70 and 75 percent female teachers at the elementary level.

There was a much closer balance between male and female teachers of mathematics and science at the sec-ondary level. While female teachers were reported to be in the majority in 11 countries, male teachers were in the majority in 15. Sweden, Japan, and the Netherlands indicated that more than two-thirds of mathematics and science teachers were male at the secondary level; while in Israel, Latvia, Lithuania, and the Philippines more than two-thirds were female. The countries with the highest proportions of male teachers of mathematics and science in secondary schools were Japan (71 percent), the Netherlands (68 percent), and Sweden (77 percent). Those with the highest proportions of female teachers were Latvia (90 percent), Lithuania (86 percent), and the Philippines (80 percent).

Age

Only 18 of the countries included information about teachers' ages. In those, the average age of teachers ranged from about 37 to 46 years, with an overall average of about 42. Higher proportions of younger teachers were reported in Hong Kong and Kuwait, each with a mean age less than 40. On the other hand, Denmark, Norway, and Slovakia reported average ages of 45 or more.

Assuming that, in most cases, teachers enter the profession when they are between 20 and 25 years old, these results indicate that the typical teacher in many countries is someone with more than 15 years of experience. Several countries reported aging teaching forces with few positions available for new teachers.

Certification Requirements

Prospective elementary school teachers in most of the TIMSS countries are required to complete between two and four years of postsecondary education as a prerequisite to professional certification. In some cases teacher preparation programs are offered in teachers colleges or normal schools; in others, they are offered at universities. A number of countries require prospective elementary teachers to complete a bachelor's degree—typically in arts or science—followed by a one- or two-year teacher education program.

Among the countries requiring the least number of years of postsecondary education for prospective elementary teachers were Argentina, Bulgaria, Hong Kong, and Japan. Each of them required two or three years of postsecondary education. The countries with the most stringent academic requirements for elementary teachers were Canada, France, Germany, Slovenia, and Sweden.

In most countries, teaching secondary school mathematics or science requires specialization in two or more subject areas, as well as a university degree. In some cases, professional education is completed in the same number of years as for elementary teachers, whereas in others, an additional year of postsecondary education is required.

Canada, Korea, Singapore, and Sweden had the highest requirements, each with either a five- or a six-year program. Most other countries had a four- or five-year requirement, including a university degree in education or a degree in a field of specialization, combined with teacher education.

Once certified, teachers in most countries are expected, but not required, to keep up with changes in their fields through in-service training. In Germany, for example, teachers are provided with a wide range of opportunities to attend in-service sessions in subjects they teach as well as in broader fields such as psychology and sociology in education and in didactic methods. These sessions are organized regionally, and time off from school to attend is provided. In the Netherlands, similar opportunities are provided. However, funds for this purpose, which have been allocated directly to schools since 1994, are not well utilized. A recent study showed that most schools in the Netherlands do not have a clear in-service policy, and many teachers are unresponsive to in-service training.

Economic Status of Teachers

A wide range of comments was made about the economic status of teachers. Among those countries where teachers were reported to enjoy a relatively high social status were Australia, Cyprus, Denmark, New Zealand, and Switzerland. Among those which reported much lower status levels for teachers were Bulgaria, Columbia, Greece, Hungary, Latvia, Lithuania, Philippines, Russia, and Slovakia.

It is clear that the overall status of teachers and their salary levels are closely associated: low status was associated with low salaries for teachers. Exceptions to this trend were reported in Japan and Slovenia, where the status of teachers is reportedly high while salaries are low, at least relative to those of other professionals.

The status of teachers in most of the eastern European countries included here was reported to be low, both in terms of status and salary relative to other government workers. It seems to be the case that teaching is accorded the lowest status in those countries where the proportion of female teachers is highest, particularly at the secondary level.

CURRENT TRENDS AND ISSUES IN MATHEMATICS EDUCATION

A major objective for this volume has been to obtain current, accurate, and detailed information about what is happening in mathematics and science education around the world. The format provides a means of getting beyond the kind of data typically provided by questionnaire responses, and allows authors to give more of the detail needed to flesh out a picture of current trends and issues in their countries. Among the topics the authors were asked to address in their chapters were the goals of the mathematics curriculum, changes which had occurred through shifts in content and approach, current issues in the field, pedagogy, and textbooks. Some of these topics were also addressed in the Expert Questionnaires, completed for mathematics and science earlier in the study, and authors were urged to refer to the responses for their countries when writing this section.

Goals for the Mathematics Curriculum

In most countries, goals for the mathematics curriculum are published in some form of curriculum guide prepared by the national or regional ministry of education. There are only a few instances of goals being developed at the local or school levels.

Not all of the chapters include explicit statements of goals for mathematics education; however, the majority do. Among the goals most frequently mentioned were developing students' mental abilities (for example, critical thinking), developing students' interest in mathematics, increasing the relevance of mathematics through a focus on real-world applications, and teaching children the skills and processes needed for successful problem solving.

Changes in the Mathematics Curriculum

Most of the chapters include a description of significant changes in the teaching and learning of mathematics that have occurred during recent years. Some involve a change in content or focus, others dealt with new applications in the area of technology, and a few dealt with the relevance and accessibility of the mathematics curriculum for all students.

Several countries described recent or ongoing revisions to their mathematics curriculum resulting in major shifts in direction or focus; other revisions were limited to changes in content. Among those reporting that a major revision of their mathematics curriculum was either currently under way or recently introduced were England, Germany, Israel, Japan, Kuwait, the Netherlands, and New Zealand. Several other countries also described curriculum revision initiatives, but on a more modest scale. In those cases, the content of the curriculum remained relatively stable, but there was increased emphasis on processes and applications. In a few cases the level of decision making was altered significantly, with teachers and schools being given increased flexibility and discretion. Some detail of the kinds of changes that are under way are described in the following paragraphs.

Utilization of Technology

Eighteen countries reported an increase in the use of calculators and computers in mathematics classrooms, although the use of computers for teaching and learning is still not widespread. In a number of countries calculators have only recently been permitted at the secondary-school level and their use is still not permitted in elementary schools. In most countries, both elementary and secondary students are permitted at least some use of calculators in mathematics, and secondary students in all of these country have access to calculators.

The use of technology was frequently mentioned as a pedagogical issue with significant implications for teaching and learning. In cases where calculators and computers are being used for teaching purposes, there have been important changes in strategies used to develop concepts, the nature of exercises and assignments, and assessment practices. In Denmark, for example, both of these tools are introduced as early as Grade 1,

resulting in new ways of dealing with calculations and graphing. In contrast, teachers in Argentina are reluctant to make use of them at either the elementary or secondary level. In spite of numerous projects designed to encourage their use, there has been little impact on teaching practice in that country. Extensive use of these aids, on the other hand, is made in England, where every primary and secondary school has had at least one microcomputer since the mid-1980s, with recent figures showing an increase to an average of 10 computers in primary schools and 85 in secondary schools. However, the increased use of calculators and computers has resulted in some concern about a possible detrimental effect on competency in mental arithmetic.

Computer use is considerably less widespread than that of calculators. Six countries reported widespread application of computers in schools, while another four indicated they were used little, if at all. Among the limitations to computer use reported by these authors are cost and the availability of good software. The need for software becomes even more apparent in countries where it must be written in different languages.

Enrollment in Mathematics

Quite a few countries reported initiatives aimed at increasing the numbers of students studying mathematics. Among the strategies being employed were making mathematics compulsory at all grade levels, encouraging girls to continue with the subject at higher grade levels, emphasizing the links between mathematics and culture and society, and providing a more interactive learning environment. In Australia, for example, there is an increased emphasis on the relationship of mathematics to culture and society, on applications, and on relevance for everyday needs.

Shifts in Content or Focus

Specific changes in the mathematics content being taught at various levels are reported in many of the chapters. In Bulgaria, for example, mathematics has been changing from a heavily structured curriculum consistent with the rest of Eastern Europe, to one more closely approaching that in Western European countries. Among the most notable of those changes are the following:

- The topic that was most frequently mentioned as having been added to the mathematics curriculum was data analysis. This change was reported for Colombia, the Czech Republic, France, Greece, Italy, Latvia, and Slovenia.

- In Canada, work with decimals is introduced earlier in the elementary school mathematics curriculum than it was previously.
- In Israel, there has been a reduction in the amount of geometry in the curriculum and an earlier introduction of calculus topics.
- In Italy, reasoning, spatial geometry, and concepts from statistics are being taught at the primary level; coordinate geometry and geometric transformations at junior secondary; and probability, statistics, and computer science at senior secondary.
- Several countries identified as a major trend a shift from a formal axiomatic approach to an applied approach in the teaching of mathematics. This involved a change, not only in content and focus, but also in beliefs, philosophy, and methodology. In a number of cases, this shift has apparently not been accompanied by the necessary resource materials and the kinds of professional development programs needed to implement it. For example, it was reported that teachers frequently continued to use a lecture-style approach in situations that were most effectively taught in an interactive learning environment.
- Assessment and evaluation issues were prominent in many countries. In some cases, there were calls for clearly defined standards or benchmarks to ensure consistency in achievement and reporting within and across jurisdictions. This was particularly the case in countries that had moved toward a more decentralized curriculum, where only general goals, or else core curriculum, were determined at the national or state level. Needs in this area were also identified by countries undergoing change in reporting procedures, by using scales corresponding to standards.
- Some countries identified the integration of the curriculum as an area of concern. In some cases this related to integrating across areas of the mathematics curriculum—algebra, geometry, discrete mathematics; in others, it concerned integration across disciplines: that is, more integration of mathematics with the natural and social sciences, and with other subjects in the curriculum.
- Seven countries reported either a restructuring of their curricula or the establishment of a new focus. The Philippines have adopted a spiral approach to

the mathematics curriculum at the secondary level. Sweden has moved toward a more goal-oriented curriculum. In Argentina and Australia there is now a greater emphasis on problem solving, while in Norway both problem solving and the integration of curriculum are being given increased emphasis.

A number of other issues were raised in some countries. Among them were the need for mathematics to contribute to culture, the role of teacher as a facilitator of learning, the back-to-basics movement, the accessibility of mathematics, transition from one level of the system to another, student attitudes toward mathematics, needs of gifted students, streaming by ability, participation rates of females, and organization of curriculum using a thematic as opposed to a modular approach.

Changes in Mathematics Teaching

Countries identified several major issues of interest regarding pedagogy. They included applications of mathematics, preparation of teachers, the inductive approach, rote learning, and communication.

Applications of Mathematics

In several countries there has been a clear shift in direction from teaching mathematics with an emphasis on a theoretical approach, using a more or less formal, deductive method, to one stressing applications of mathematical concepts in real-world settings. Among the areas of application being emphasized are finance, applied engineering, and consumer-related topics. Significant shifts in this direction were reported in Australia, Belgium, the Czech Republic, Denmark, Germany, Latvia, Korea, and Romania. In the Czech Republic, for example, this shift has been a significant departure from a closely prescribed approach to teaching mathematics, resulting in greater freedom for schools and teachers in deciding what will be taught and how much time to devote to topics.

Such changes were not reported explicitly in a number of other countries; however, it seemed clear from other information they provided that they were moving in similar directions. A trend to a more applications-oriented approach to mathematics is apparent in many countries, and this shift in emphasis or direction is frequently accompanied by an increased emphasis on constructivist approaches to teaching and learning mathematics.

Preparation of Mathematics Teachers

A shortage of adequately prepared mathematics specialists was identified as a major concern. It was highlighted as a particularly urgent issue in Iceland and the Philippines, where it has been a concern for a number of years. In Iceland, for example, only one in five mathematics teachers at the lower secondary level majored in the subject during their teacher education program, and at the upper secondary level the shortage of trained mathematics teachers is considered severe. A similar but perhaps more severe shortage of trained mathematics teachers exists at all levels in the Philippines. The lack of specialists in this area gives rise to concerns that innovative teaching methods are not utilized and that teachers hold a perception of mathematics as a difficult subject, resulting in similar opinions among their students.

There is also a need for more professional development programs for teachers to address changing goals of the curriculum. Teachers need to become familiar with topics such as alternative strategies for teaching and evaluation, and the use of computers in the mathematics classroom.

Inductive Approach

The use of teaching strategies associated with an inductive approach was among the trends most frequently reported. They reported significant increases in the use of projects and investigations, the use of manipulative materials, and hands-on activities in small group settings. Thirteen countries identified a definite shift toward use of such strategies in mathematics classrooms. Several others indicated that their use was limited to the elementary level, and that a deductive approach still prevailed at the secondary level.

Rote Learning

A number of countries reported that rote learning, the use of repetitive exercises, and direct instruction were still popular, especially at the secondary level. Among those reporting that such an approach was prevalent were Cyprus, Greece, Hong Kong, Hungary, Japan, Norway, the Philippines, and Slovenia.

Oral and Written Communication

Oral and written communication were frequently mentioned as a means of teaching and learning mathematics. Activities around this process included the use of presentations and reports, the organization and reporting of data using different formats, and applications involving a variety of media. This area was considered particularly important in England, Iceland, Italy, Bulgaria, and New Zealand.

Several other pedagogical approaches or issues were identified by countries. Among them were teaching mathematics through problem solving, cooperative learning, real-life problems and applications, integration of the curriculum, ability groups, an holistic approach to learning, active learning, and peer tutoring.

Nature and Role of Mathematics Textbooks

Almost all countries reported that mathematics textbooks were used extensively by students and teachers, although there was considerable variation in the ways in which they were used. In some countries they serve almost solely as sources of exercises and problems for students to solve. In other countries they are used by teachers to prepare lessons and as a source of exercises; and, in still others, they serve as the de facto curriculum guide or syllabus. A few countries reported a recent shift away from using a single textbook as the primary learning resource. Australia and Sweden, for example, reported that at the elementary level textbooks are only one of a number of different print and media resources used by teachers and students.

In Denmark, newer textbooks can be characterized by a number of features, including their focus on everyday themes, practical mathematics activities, cross-curricular work, pupil-teacher cooperation, and differentiated teaching. Some textbooks deal with special topics and include thematic approaches. In spite of significant changes to recent textbooks, however, the replacement of old, outdated books is slow. Many textbooks in Germany, on the other hand, are considered to be lacking in number and types of projects, historical highlights, real applications, and information about mathematics in careers

In most cases, textbooks are provided to elementary school students free of charge. Secondary students in quite a few countries, on the other hand, have to purchase or rent them. Textbooks are published by private companies in most countries, while in some others they are developed, published, and distributed by the Ministry of Education.

Several countries reported a considerable degree of decentralization in the selection of textbooks, with schools being able to choose from a list of approved books. Although the selection of textbooks by schools has been permitted for some time in many countries, it is a recent innovation in the Czech Republic, Hungary, Latvia, and the Russian Federation.

Several countries noted that efforts had been made to make textbooks more attractive and appealing to students. At the elementary level most countries indicated that recent books contained pictures, graphs, and activities designed to interest students. Color was used frequently and narrative text kept to a minimum. Similar features are included in many of the textbooks at the secondary level. Some include references and activities related to applications of mathematical concepts in the real world and to careers. Only a small number of countries reported that mathematics textbooks still lacked color, contained a great deal of text, and tended to be dry and uninteresting.

CURRENT TRENDS AND ISSUES IN SCIENCE EDUCATION

A similar set of questions was investigated with respect to science education. They included goals for the science curriculum, changes in science curricula, changes in the teaching of science, and the nature and role of science textbooks.

Goals for the Science Curriculum

Goals for the science curriculum are developed in the same manner as those in mathematics. Typically, either the Ministry of Education at the national or regional level is responsible for developing new or revising existing curricula, including a statement of philosophy or rationale, goals and objectives, and implementation strategies. Among the goals frequently mentioned by countries were development of a scientific point of view, protection of the environment, the role of science and technology, and the development of related skills and processes. Goals tended to be fairly general at the elementary level and more specific at the secondary level.

At the elementary level in most countries, science is taught either as a single, discrete subject or else integrated with a number of different subject areas. For example, in some countries science is taught along with civics, health, or social studies.

At the secondary level in most countries, science is organized into several identifiable subjects, each corresponding to a different discipline. Most countries offer science courses in chemistry, physics, and biology. Other choices include geology, zoology, earth science, and geography. A few countries continue with an integrated science course until Grade 9 or 10.

Changes in Science Curricula

Eleven countries reported that a revision in their science curricula had either recently taken place or was currently underway. Major emphasis in these revisions was being given to development of science concepts, process skills, and problem solving; environmental issues; the role of investigations in science learning; and science-technology-society issues.

Responsibility to Society

Ten countries identified a new curricular emphasis on the responsibility of science and scientists to society. Among the trends in this area was an increased amount of attention to developing an awareness of the need for solutions to environmental issues such as pollution and the management of natural resources. As an example, the United States has stressed scientific literacy for all citizens since 1983. This focus has resulted in increased emphasis on earlier and more science instruction in elementary school, focusing on social and environmental issues. Other changes dealt with understanding the relationships of science to natural phenomena and applications of science to everyday life.

Course Focus and Content

Several changes were reported in course content, focus, and organization. Seven countries noted that curriculum change had resulted in less required content, thereby giving teachers greater freedom of choice in the selection of topics to be studied. Three reported that the focus of the science curriculum had changed to better serve the needs of a wider range of student aptitudes and abilities.

Use of Technological Aids

Several countries reported increasing use of technological aids such as video, audio, and computer software in the teaching of science. This trend seems likely to continue as computers become less expensive, and as more appropriate software becomes available.

A concern shared by several countries about the wide range of expertise and readiness of teachers to utilize these aids was expressed by Iceland. It reported that a considerable gap exists between teachers who have made progress in this area and those who do not know how to apply technology in their teaching. It was surmised that progress in this field was hindered by a lack of clear policy on expectations for the application of these aids in education.

Shift from Theory to Utility

In a number of countries there has been a shift away from a highly structured, academically oriented science curriculum toward one giving greater emphasis to applications and the relevance of science. This raises a number of issues having to do with teachers' beliefs, philosophy, and instructional approaches. On the one hand, it has generated concerns about the end result of placing less emphasis on structure and theory. On the other hand, it has given increased priority to societal needs and concerns about the environment.

Applications of Technology

Although there was widespread recognition of the importance of effectively utilizing technology, corresponding needs were identified for better software, educating teachers about the applications of technology, and development of compatible curriculum materials. In England, for example, computers and other technology are widely available in schools and teachers are encouraged to use information technology in the teaching of science. However, the use of information technology is less widespread than might be expected. A recent report, for example, found that in reality there were relatively few schools where it is used well to support learning in science.

Integration Issues

A more effective means by which to integrate science was listed as an issue by six countries. They identified a number of areas in which integration should occur, including the following: between science and technology, among the disciplines of science, between science and other subjects, and between science and the environment.

Several other curricular initiatives were reported. They included more emphasis on laboratory work and experimentation, increased efforts to encourage more girls to continue to study science at higher grade levels, improving students' attitudes toward science and the study of science, a spiral approach to the curriculum, courses addressing science-technology-society issues, and a focus on process skills such as observing, classifying, inferring, and communicating.

Trends in Science Teaching

Among the most widespread trends in science teaching reported by the chapter authors were an increasing focus on inductive approaches at the elementary level, the use of investigative techniques, and the role of the teacher as a facilitator of learning.

Inductive and Deductive Approaches

Most countries reported frequent use of an inductive approach based on concrete examples and experimental demonstrations in the teaching of science at the elementary level. Students frequently work in small groups and use examples from everyday life. In France, the official curriculum guide requires that science learning be based on observations and activities, rather than lectures. In Italy, science lessons are based on students' identification of problems or direct observation.

Although some countries reported that there was an effort to balance inductive and deductive approaches at the secondary level, it was not a common practice. Most reported that a deductive approach was primarily used at the secondary level.

Group Work and Investigative Techniques

Several countries reported that investigative activities involving problem solving and creative thinking were commonly used approaches to science teaching. Two referred specifically to the use of cooperative learning as one technique used in this area. Kuwait, for example, reported that cooperative learning is stressed, as are communication skills.

The Teacher as a Facilitator of Learning

Eight countries made reference to the changing role of the teacher, shifting from that of the major disseminator of information to that of facilitator. In this role the teacher directs students to sources of information and provides them with direction for analysis, interpretation, and reporting of findings. Students are expected to assume greater responsibility for learning.

Several other pedagogical approaches were identified as important by some countries. They included integration of content, focus on oral and written communication, use of more practical activities, and a student-oriented approach that takes into account students' interests, aptitudes, and abilities. Many of these approaches are being used in attempts to become more effective in the teaching of science through greater relevance and social awareness. Being able to compete in a highly competitive and technological market is also an important objective. New Zealand, for example, stresses the importance of science knowledge for career purposes, as a necessary grounding for effective living and citizenship, and to meet needs in the world of trade and commerce.

Nature and Role of Science Textbooks

Procedures for the development, selection, and distribution of science textbooks were similar to those for mathematics. Textbooks in most countries are produced by publishers who follow a prescribed curriculum outline. When selections are made at the school level, they are chosen from an approved list.

Science textbooks, similar to those in mathematics, are now very likely to include pictures, diagrams, and activities, and widespread use of color. Few textbooks are used in science at the elementary level, likely due to the integrated approach used in many countries.

The use of textbooks at the elementary level differs widely among countries. In Norway, for example, elementary teachers rely heavily on textbooks since many have limited knowledge of science, particularly physics and chemistry. This has resulted in fewer student activities in this area. In contrast, textbooks are usually not used at the elementary level in England and Scotland. They rely more on photocopied resources, work cards, or worksheets, which are integrated into planned thematic work. Through this approach, students are expected to play a greater part in their own learning by following written instructions and accessing a variety of resources.

EVALUATION: POLICIES AND PRACTICES

In all countries, student assessment and evaluation is carried out most frequently by teachers in their own classrooms, using both formative and summative techniques. Formative evaluation is commonly used for tracking students' learning on a more or less continuous basis and to assist in planning for instruction or helping teachers decide what methodology to use. Summative techniques are used primarily to determine students' standing in a course or program. Among the evaluation methods being used in science and mathematics are written and oral tests, portfolios, performance assessment, self-assessment, and teacher observation.

Most countries require students to write examinations at the end of each major level of the system to determine which students have qualified for entry to the next level. Such examinations are most commonly set and marked at the school level. In a number of countries, universities have their own entrance examinations.

The various types of examination programs vary widely among countries, ranging, for example, from a series of examinations at national and regional levels in the Russian Federation to a single national examination in Greece. The program in the Russian Federation, for example, includes a compulsory national examination in mathematics and Russian, as well as a compulsory regional examination in a subject selected locally for students who receive a basic school-leaving certificate. Those who earn a secondary school-leaving certificate must pass five compulsory examinations: national examinations in mathematics and Russian language and literature, a regional examination, and two examinations in subjects selected by the student.

About 10 countries reported that program assessments were administered on a regular basis, using either a census or a sample of students at particular grade levels throughout the system. Results of these assessments are used to monitor the system, to address issues of accountability and change over time, to identify areas of strength and weakness for purposes of program planning, and to provide direction for curriculum revision.

Associated with some of these assessments is the establishment of standards or expected levels of achievement. For example, considerable dialogue has occurred around these expectations in Belgium and Hong Kong. The introduction of minimum standards and goals is considered a major innovation in Belgium's education system. They are intended to improve consistency in the teaching of curriculum outcomes and to provide a clear definition of what should be focused on and what evidence is available for monitoring the system. In Hong Kong, a target-related assessment scheme, using a criterion referenced focus in which students were assessed against targets at progressive levels, was introduced in 1990. It has since been changed to a target-oriented curriculum, stressing task-based learning.

MAJOR AREAS OF SIMILARITY

School systems in different countries differ from one another in many ways, and this underscores the absolute necessity of being aware of the significance of context in comparing national systems or in making decisions about new directions in curriculum and instructional practice. The fact that an innovation takes hold and is successful in Country A is no guarantee of the success of that same innovation in Country B. A major goal of this book has been to show the wide range of policies and practices that exist around the world with respect to the teaching and learning of mathematics and science.

However, one cannot read the 38 national chapters without being struck by the fact that there are many important similarities in these countries, both as concerns the intended curriculum in mathematics and science and the kinds of initiatives that are currently under way. The strong impression that is conveyed by these authors is that, in many ways, countries are borrowing from and learning from one another about new directions in mathematics and science education. The following are a few examples of this overall trend.

- Very few national systems are moving toward centralization. The trend is in the other direction, toward decentralization. The countries of eastern Europe provide the most dramatic evidence of the strength of this trend, and there are no examples of a move in the opposite direction. This trend toward more decentralization has been accompanied by increased attention to the question of accountability. As educational decision making becomes decentralized, the central authority or government feels a need for increased and better sources of information about what is happening in schools.

- Many countries commented on the fact that they had aging teaching populations and that many teachers of mathematics and science lacked the relevant academic qualifications for their work. In the not-too-distant future there could well be an international crisis of teacher supply in these areas unless appropriate remedial action is taken fairly soon.

- There is a widespread shift in teaching approach in mathematics toward a greater emphasis on applications and away from more axiomatic approaches. The importance of using manipulative materials for teaching, cooperative learning, and earlier introduction of calculators all provide evidence of this trend.

- In the case of science, there is greatly increased interest in and attention paid to issues related to the needs of society and the environment. More time is being devoted to topics such as pollution and resource management.

Argentina

Carlos A. Mansilla, Universidad del Chaco

1 COUNTRY PROFILE

Argentina is located at the southern end of South America. The country is bordered by Bolivia, Paraguay, and Brazil to the north, Uruguay and the Atlantic Ocean to the east, Chile and the Atlantic Ocean to the south, and Chile to the west. It occupies an area of 2,767,00 km^2. An additional 969,464 km^2 of territory is claimed by Argentina, including the Malvinas, South Orcadas, Sandwich and Georgia Islands, and portions of Antarctica.

Argentina is divided into 24 political jurisdictions: 23 provinces and a federal capital district. The city of Buenos Aires is the country's capital, and most of the country's political and economic activity is concentrated there, including commercial, financial, industrial, educational, and administrative activities. More than 85 percent of Argentina's population lives in large urban centers. Three million people live in the capital and an additional seven million reside in its suburbs. In 1993, Argentina's total population was 33.8 million, with a density of 12 persons/km^2. The population showed an annual incremental growth rate of 15 per thousand. Life expectancy is 71 years.

Argentina's population is principally of Spanish and Italian origin. A small percent of its population includes peoples of aboriginal descent and emigrants from other European countries, the Middle East, and Asia. Spanish is the official language. The country is politically organized as a representative federal republic, governed by a constitution established in 1953 and reformed in 1994.

Argentina's gross national product was approximately US$244,036,000 in 1993, with a per capita GNP of US$7220 in the same year. In 1995, almost 4 percent of the GNP was devoted to education, an increase of 1 percent over 1980. The adult literacy rate is 95 percent.

Argentina is a member of many international organizations, including the United Nations and the Organization of American States. Since 1990, the country has been a member of the Free Trade Agreement, MERCOSUR, which links Argentina, Brazil, Paraguay, and Uruguay.

2 THE EDUCATION SYSTEM

Governance and Decision Making

Elementary and secondary education are provincial responsibilities, with each province controlling all aspects of its own system. Provincial ministers of education have established the Consejo Federal de Cultura y Educación, the Federal Council of Culture and Education, which provides a forum for discussion and coordination of educational policies and issues such as national projects in curriculum and assessment. However, the Council does not have administrative authority; its decisions are simply recommendations that may or may not be followed by provincial governments.

The intended curriculum is defined by the Federal Council of Culture and Education. Some provinces implement curriculum guides based on the Council's recommendations and containing educational policies, rationale, objectives, contents and methodological suggestions. These guides are developed by provincial curriculum departments, often with the participation of teachers. Final decisions regarding classroom planning are made at the school level. Textbooks are selected by schools or individual teachers, although by law schools cannot require students to purchase them. Nevertheless, teaching content and methods are strongly influenced by textbooks.

These are times of transition for the Argentine education system. Profound changes, recommended by the Council in 1993, were slated for implementation within a year. They are now gradually being implemented, according to each province's capabilities.

Structure of the System and Participation Rates

Present Structure

Currently, the education system is divided into three levels: *inicial* (preprimary), *primario* (primary), and *secundario* (secondary). These levels, as well as the approximate enrollment rates, are shown in Figure 1. The preprimary level is of one year's duration for pupils aged five. Primary school comprises three cycles: first cycle, Grades 1, 2, and 3 for ages 6, 7, and 8; second cycle, Grades 4 and 5 for ages 9 and 10; and third cycle Grades 6 and 7 for ages 11 and 12. Secondary school comprises two cycles: *ciclo básico*, Grades 8, 9, and 10 for ages 13, 14, and 15; and *ciclo superior*, Grades 11 and 12 for ages 16 and 17.

At this time, school is compulsory from Grades 1 to 7. There is no minimum leaving age; students are required to complete Grade 7 before leaving school.

New Developments

A new organization of school levels was introduced in 1995: *educación general básica* (general basic education)

and *educación polimodal* (polymodal education). The first level includes Grades 1 to 9 for students 6 to 14 years old, and is compulsory. The second includes Grades 10 to 12 for students 15 to 17 years old, and offers both academic and general programs. When the new structure is fully implemented, the minimum leaving grade will be Grade 9.

Public and Private Schools

Approximately 70 percent of students are enrolled in public schools funded by the government. The remaining 30 percent are enrolled in Catholic schools, most of which also receive government assistance. In the recent past, there has been increased interest in private schools; this tendency appears to be slowing down as parents become increasingly aware of the costs of private schooling and as its purported benefits come into question.

Mathematics and Science Programs

Students at all levels in all schools study a compulsory program of mathematics and science courses. When the new curriculum is implemented, optional courses will be offered at the polymodal level.

Schools in the System

The School Year

The school year in Argentina ranges from 170 to 180 school days at the elementary school level, and from 160 to 170 school days at the secondary school level. The year begins in mid-March and continues to the first week of December, with a two-week winter break in July. There are several holidays during the year, usually on Monday. Students attend 40-minute classes for four hours per day from Monday to Friday. Most schools operate from 7:30 until noon and from 1:30 to 6:00 p.m., with students attending either the morning or the afternoon session. Physical education is offered outside this time frame. Approximately 17 percent to 20 percent of class time is devoted to mathematics, and the same percent to science. Class sizes range between 20 and 40 students in both elementary and secondary schools.

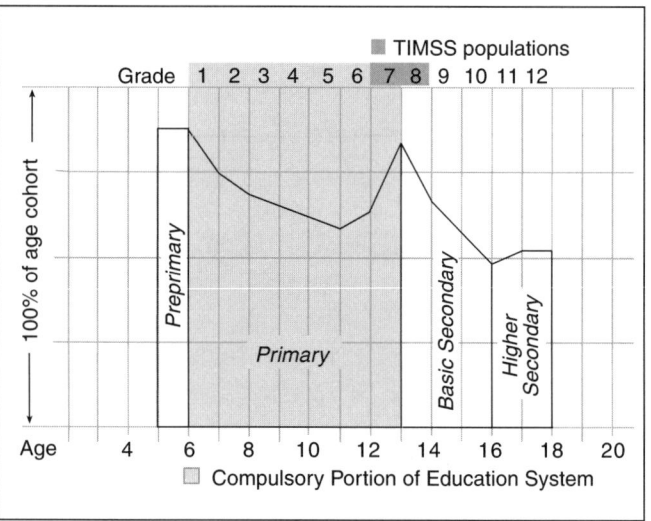

Figure 1: Structure of the Education System

Certification of Mathematics and Science Teachers

Elementary school teachers take either two and a half or three years of teacher education at the college level. Elementary school teachers are prepared to teach all subjects; there are no specialists for mathematics or science. Secondary school teachers prepare at college or university for three or four years, depending upon the institution. Secondary school teachers must be certified as specialists in one or more subjects, and they follow a program with emphasis on mathematics, physics, chemistry, biology, geography, or other subjects.

Teacher Profile

Elementary school teachers earn a salary of between US$3500 and US$6500 per year, comparable to salaries of administrative government employees in urban areas. In rural regions with few amenities and poor living conditions, teachers receive extra payment. At the secondary school level, teachers are paid by the hour. For 30 weekly hours, the maximum allowed, salaries range between US$7,000 and U.S.$15,000 per year.

3 TIMSS POPULATIONS

Argentina, IEA, and TIMSS

Argentina has not participated in any previous IEA studies; however, tests from the Second International Science Study and the Second International Mathematics Study were piloted in the provinces of Chaco and Corrientes.

Argentina is participating in TIMSS at the Population 2 level only. Ten provinces are participating in the study: Buenos Aires, Santa Fe, Entre Rios, Misiones, Catamarca, Corrientes, Santa Cruz, Jujuy, Chaco, and Formosa.

Argentine TIMSS Populations

According the definition established by IEA, Population 2 in Argentina includes Grades 7 and 8. The lower grade is part of the elementary school system, and the upper, of the secondary system.

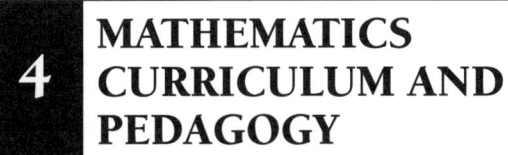

4 MATHEMATICS CURRICULUM AND PEDAGOGY

Goals for the Mathematics Curriculum

The Federal Council of Culture and Education influences curriculum development throughout Argentina, and as a result the intended mathematics curricula show many similarities across regions. The same topics are covered in all schools, but provinces are free to produce their own curriculum guides. Mathematics textbooks, which are published in Buenos Aires and used in all jurisdictions, are a second factor in the development of consistent curriculum throughout the country.

The *Contenidos Básicos Comunes para la Educación General Básica,* Basic General Contents for Basic General Education, prepared by the Federal Council of Culture and Education, has established a series of goals for students in mathematics education:

- Gain conceptual comprehension and a belief in the utility and value of mathematics.
- Formulate and solve problems about everyday situations.
- Perceive connections between mathematics and the other sciences.
- Use technologies such as calculators, computers, and multimedia in the mathematics classroom.
- Comprehend the internal structure of mathematics and understand the value of mathematics in culture and society.

Following recent suggestions made by the Council, the provinces are developing goals and curriculum guides that reflect regional needs and issues.

Major Changes in the Mathematics Curriculum

Significant changes in mathematics education have been introduced since the Federal Council of Culture and Education's 1993 recommendations, and these are in the process of adoption by provincial jurisdictions. The mathematics curriculum will change from an emphasis on an extremely formal approach to an approach consistent with current trends in mathematics education. Special emphasis will be given to problem solving, the significance of mathematics in culture and society, and the increased use of technology such as calculators, computers, and multimedia.

The mathematics content for General Basic Education has been organized into eight strands: numbers; operations; graphic and algebraic language; geometry; measurement; statistics and probability; mathematical procedures such as reasoning, communication, and problem solving; and attitudes toward mathematics.

Mathematics teachers are reluctant to use technologies such as calculators, computers, and overhead projectors at both elementary and secondary school levels, and indeed these technologies barely exist at the elementary level. Some secondary school teachers allow students to use calculators, but no teaching program is based on the use of calculators and computers. Numerous projects encouraging the use of computers and calculators in the mathematics classroom have been developed at an experimental level, but as yet these have had little impact on teaching practice. The Federal Council of Culture and Education recently recommended that special emphasis on new technology be incorporated with mathematics in the Basic General Education program in order to both simplify calculus and facilitate the learning that the use of new technology allows.

Current Issues in the Mathematics Curriculum

The Federal Council of Culture and Education's publication *Contenidos Básicos Comunes para la Educación General Básica* gives special attention to current issues in the mathematics curriculum. Its first recommendation is that school mathematics emphasize conceptual, procedural, and attitudinal aspects of mathematics. Furthermore, it is expected that mathematics education will contribute to the transmission of culture, as well as providing necessary mathematical tools and an understanding of modern technology. It is recommended that mathematics instruction be integrated with other disciplines such as the natural sciences, social sciences, technology, arts, and physical education. The teacher should act as a learning facilitator, and students should be encouraged to search for their own sources of information.

Mathematics Textbooks

Textbooks are not widely used by students in Argentine schools, and teachers use them almost exclusively as sources for lesson planning and pedagogical ideas. This may change in the future, however, since significant changes have been made to mathematics textbooks in the last five years. Both content and methodology now follow guidelines suggested by the *Contenidos Básicos Comunes para la Educación General Básica*. More pictures, diagrams, and color are used in textbooks, and topic presentation is enhanced through real-world examples and applications.

The new textbooks accord less importance to formal mathematics and the routine application of operations without context. There are suggestions for projects and practical activities for the use of calculators and computers in class. Textbooks now include topics such as numbers, operations, units of measurement, geometry, transformations, data representation, statistics and probability, equations, functions, and trigonometry.

Specialist teams of qualified teachers and mathematicians write mathematics textbooks based on the curriculum, which are then edited and published in Buenos Aires. They are selected by teachers or school mathematics departments, but because of economic constraints their use is not compulsory for students in public schools. When they are required, as they are in some private schools, textbooks are paid for by the students.

5 SCIENCE CURRICULUM AND PEDAGOGY

Goals for the Science Curriculum

The *Contenidos Básicos Comunes para la Educación General Básica* divides the science curriculum into two portions: natural sciences and social sciences. The goals listed in the document reflect a desire to provide students with the scientific knowledge necessary for an understanding of life and the world. The study of science should encourage the formulation of questions and explanations, educational research, and the links to life in the community. Emphasis on memorization of facts and content, on repetition and subject division has given way to a tendency toward integrated areas of knowledge and more creative project work.

Currently, the science curriculum includes practical activities for the study of the environment, and some of these are performed outside the classroom. Topics emphasizing social content are studied, including ecology and current social, economic, and political issues. Special emphasis is given to environmental issues. The curriculum aims for the integration of topics among the sciences and gives less emphasis to the memorization of facts.

Technology is not widely used in science classrooms. Calculators are allowed for secondary school science students, but computers are not considered a teaching tool for science classes. The National Ministry of Culture and Education has developed science education videos, and these are used extensively in schools.

Major Changes in the Science Curriculum

Curricular changes promoted by the *Contenidos Básicos Comunes para la Educación General Básica* are still in the implementation process, and include a shift to an interactive learning environment, with extensive use of investigative projects, cooperative work, data collection, and oral and written communication. The changes are not easy for all teachers, but most are predisposed to accept them.

Current Issues in the Science Curriculum

According the *Contenidos Básicos Comunes para la Educación General Básica,* today's world relies heavily on science and technology, and it is therefore necessary for citizens to have scientific and technological capabilities. Citizens of the twenty-first century will need to understand the structure of science and technology, as well as the internal coherence of the natural sciences. Furthermore, it is believed that students should have knowledge and understanding of social concepts such as democracy and systems of government, as well as a strongly formed sense of the responsibilities of citizenship.

Science Textbooks

As with mathematics textbooks, science textbooks are not widely used by students in Argentine schools. At the secondary school level, there is no single science textbook. The sciences are divided into biology, physics, chemistry, geography, civic instruction, and health education. The following topics are included:
- life and its properties;
- the physical world;
- matter, structure, and changes;
- the Earth and its changes;
- research into the natural world;
- attitudes toward the world and natural sciences;
- society and geographic spaces;
- society through the time, change, continuity, and cultural diversity;
- human activity and social organization;
- the understanding and explanation of the social reality.

Pedagogy

The *Contenidos Básicos Comunes para la Educación General Básica* states that scientific literacy should include a conception of the structure and processes of the universe; the capabilities of research, exploration and experimentation; problem solving and communication capabilities; and the use of specific symbolic representations. It recommends the use of various approaches for classroom processes, including investigative projects and the use of both oral and written communication. These guidelines have been distributed throughout the education system, and the corresponding development courses for teachers and supervisors have been established. Actual classroom processes are gradually improving although there is still a feeling of reluctance among some older teachers.

6 EVALUATION POLICIES AND PRACTICES

At the classroom level, evaluation is performed to determine student knowledge, understanding, and skills in the topics developed in class. Qualitative reports are made to parents based on the results of these evaluations. In general, evaluation consists of written tests and oral expositions, but other aspects such as cooperative learning, student behavior in class, and class participation are taken into consideration. The main purpose of evaluation at the classroom level is to grade students, and neither the curriculum nor the program is evaluated. There are no widespread practices of evaluation at the provincial level, although some jurisdictions administer entrance examinations in mathematics and language to students entering secondary school.

A significant problem in Argentina is that the educational projects are not evaluated. Teachers are evaluated by a subjective annual report, but this is not performed systematically.

In 1993, the *Dirección Nacional de Evaluación de la Calidad de la Educación* (National Direction for the Assessment of the Quality of Education) was established to perform evaluations at the national level. Two studies of student achievement have so far been conducted, one in mathematics in 1993 and one in Spanish in 1994. The results are being processed at an international center, and will be used only to assess student performance and as another parameter for comparative purposes.

7   REFERENCES AND SOURCES FOR FURTHER READING

COMMON Sources

1   Husén T, N T Postlethwaite 1994 *International Encyclopedia of Education*. Pergamon Press, Oxford

2   Organisation for Economic Co-operation and Development 1995 *Education at a Glance: OECD Indicators*. Organisation for Economic Co-operation and Development, Paris

3   United Nations Educational, Scientific and Cultural Organization 1995 *Statistical Yearbook*. United Nations Educational, Scientific and Cultural Organization, Paris

4   World Bank 1995 *World Development Report 1995*. Oxford University Press, New York

Other Sources

5   República Argentina *Ley Federal de Educación* N: 24.195

6   Consejo Federal de Educación de la República Argentina *Recomendación N: 26/92*, noviembre de 1992

7   Consejo Federal de Cultura y Educación de la República Argentina 1994 *Contenidos Básicos Comunes para la Educacion General Basica*

8   Ministerio de Cultura y Educación 1993 *Sistema Nacional de Evaluación, Resultados Nacionales, Primer Informe*

9   Ministerio de Cultura y Educación 1993 *Dirección Nacional de Planificación Educativa y Programación*.

10 Ministerio de Cultura y Educación 1993 *Dirección Nacional de Estadística*

11 Ministerio de Cultura y Educación 1993 *Dirección Nacional de Curriculum*

12 Ministerio de Cultura y Educación 1993 *Secretaría de Programación Económica*

Statistical References

Section	Statistic	Reference	Page	Table	Year of Statistic
Country Profile	2,767,000 km2	4	163	1	1993
Country Profile	33.8 million	4	163	1	1993
Country Profile	15/1000; 71	12			
Country Profile	US$244,036,000	4	163	1	1993
Country Profile	US$7220	4	163	1	1993
	4%	4	163	1	1993
Country Profile	95 %	4	163	1	1993

All other data is taken from unpublished Ministry sources.

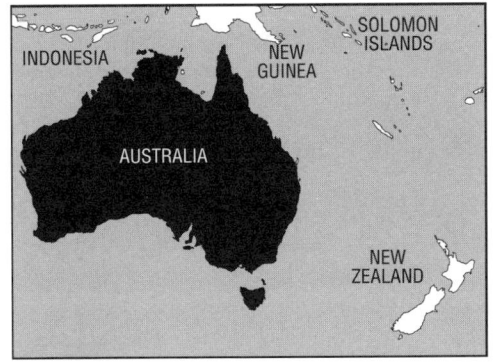

Australia

John G. Ainley, Australian Council for Educational Research

1 COUNTRY PROFILE

Australia is an island continent with an area of approximately 7.7 million km², located close to Southeast Asia and the South Pacific. In 1993 the population of Australia was 18 million, having grown from 15 million in 1980. Growth is projected by the World Bank to continue at a slightly reduced rate, reaching 23 million by the year 2025. Although the overall population density is low, 2 persons/km², it is a highly urbanized society with two-thirds of the population living in cities of more than 100,000 people. The two largest cities, Sydney and Melbourne, contain 40 percent of the population. Outside the cities, the country is sparsely populated with small communities separated by vast distances.

About 2.5 percent of Australian school children, some of whom live in isolated communities, are of Aboriginal descent. The Australian population is principally of European background, although recent immigration has produced greater ethnic and cultural diversity. Largely as a consequence of immigration, the population has more than doubled since 1950. One-third of Australians are either first or second-generation immigrants. In a departure from previous patterns, fewer than half of the post-1945 immigrants came from Britain or Ireland. During the 1950s and 1960s many immigrants arrived from southern European countries, and more recently the Middle East and Southeast Asia have been important sources of immigrants. English remains the language of most activities in education.

The Australian system of government is a parliamentary democracy that developed out of its colonial background and is based on the British system. Prior to 1901 there were six colonies administered from Britain. In 1901 the colonies federated as states of the Commonwealth of Australia. Under the constitution, the states retained responsibility for education, although the educational financing and policy roles of the federal government have increased over the past 40 years. There are also two territories with educational powers similar to the states. Political responsibility, federally and within each state, is exercised by a minister who is accountable to the relevant Parliament. Federal and state ministers of education and training meet regularly to discuss major policy issues. Although there is no single Australian education system, there are few major policy differences between the states. The policies of the major political parties, a social democratic and a conservative party, differ in some respects, and this is sometimes reflected in state practices.

The World Bank classifies Australia as a high-income country. In the 1992–93 financial year, per capita GDP was just under US$17,250. Within this generally affluent position, the past few years have seen unemployment and overseas debt remain high. This situation has resulted in some attention being focused on education. The Australian economy was once dominated by agriculture and mining, but these industries now employ less than 10 percent of the labor force. About 20 percent are employed in manufacturing, utilities, and construction, while the remaining 70 percent work in service industries such as transportation, communications, sales, finance, education, and health. Curriculum policies are placing an increasing emphasis on higher order skills. In 1993, just over one-half of the labor force held a postsecondary educational qualification. Literacy among the adult population is nearly universal.

For the financial year 1990–91 total public and private expenditure on education was US$16 billion, which represented less than 6 percent of the GDP, a decline from a peak of just over 6 percent in the mid-1970s. The decline can be attributed largely to relatively static school enrollments and a drop in education costs. State and federal governments supply more than 90 percent of the funds for education. The states provide about 60 percent of all public expenditure on education, or 90 percent of the expenditure on school education, and the federal government provides 40 percent, or most of the expenditure on universities.

2 THE EDUCATION SYSTEM

Governance and Decision Making

School education is the responsibility of the individual states and territories, although the influence of the federal government has grown in recent times. State education departments recruit and appoint the teachers in government schools; supply buildings, equipment and materials; and provide limited discretionary funding for use by schools. In most states, some responsibility for administration, staffing, and curriculum has devolved to regional education offices and schools. Devolution of responsibilities to schools is likely to become more extensive as most states are now moving toward self-managing schools in the government school sector.

There is no common school curriculum across the country, although almost all students are exposed to a curriculum that provides coverage of English language arts, mathematics, science, social studies, humanities, the creative and performing arts, physical education and, less frequently, a foreign language. Within states, the general pattern is that central authorities specify broad curriculum guidelines, and schools have considerable autonomy in deciding curriculum detail, textbooks, and teaching methodology. This situation applies particularly at the primary and junior secondary levels. In Grades 11 and 12, the curriculum is more likely to be specified in detail by a state authority responsible for examining and certifying student achievement. At this level students generally have more scope to specialize, and a range of elective studies is provided.

Learning materials and tests are prepared by a variety of agents including the curriculum sections of state education departments, academics, commercial publishers, and teachers' subject associations. In a signifi-

cant development in 1990, the state and federal education ministers established the Curriculum Corporation, a semi-autonomous body with a charter to develop curriculum materials on a commercial basis.

Structure of the System and Participation Rates

Structure of Education

Education is compulsory from ages 6 to 15 in most states and 6 to 16 in Tasmania, and between these ages there is virtually 100 percent attendance at school. Figure 1 shows the structure of education and approximate enrollment rates. Most children start primary school at 5 years of age and a majority of 4-year-olds attend kindergarten on a part-time basis. Primary education lasts for either six or seven years, depending on the state, and students complete that stage of schooling at the age of 11 or 12. Almost all government primary schools are coeducational. Attendance at primary schools is almost universal, and in 1993 there were 1.8 million primary students.

Secondary education is provided for either five or six years depending on the length of primary education in the state. Students normally commence secondary school at about 12 years of age. Most secondary schools are comprehensive; the technical and vocational secondary schools that previously existed in some states have been phased out. There is a small number of selective-entry secondary schools in some states. These schools follow the same general program but with op-

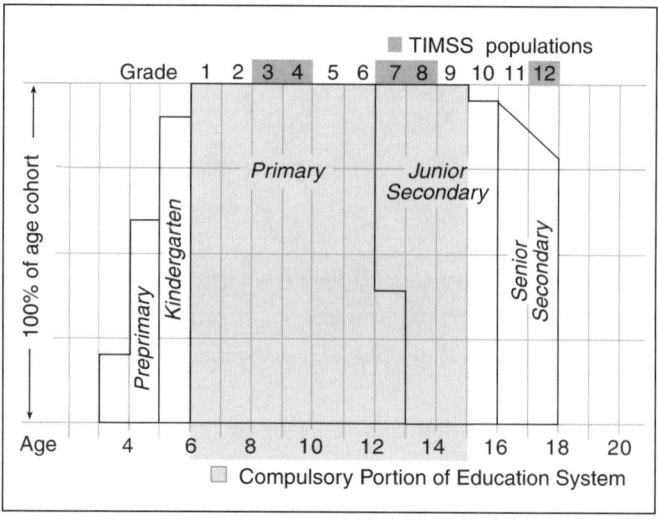

Figure 1: Structure of the Education System

portunities for more advanced study or accelerated promotion through grades. Almost all government secondary schools are coeducational, but the majority of private secondary schools are single sex.

Participation Rates

One of the most marked changes during the 1980s was an increase in the percentage of students who remained beyond the minimum school-leaving age to complete secondary school. The percentage of commencing secondary students remaining to the final year of school rose from 35 percent in 1980 to 77 percent in 1993. Since 1976 the secondary school completion rate has been higher for girls than boys and, in 1993, 81 percent of girls, compared to 72 percent of boys, remained to the final year of school.

Special education programs are provided for students with physical, intellectual, and emotional disabilities. A national study estimated that in 1992 about 2 percent of students in Australian schools had some form of disability. At that time about 30 percent of students with disabilities were enrolled in special schools, 25 percent were enrolled in special classes or units attached to other schools, and 45 percent were enrolled in regular school classes.

Public and Private Schools

Private schools are an important feature of the education system and in 1993 enrolled 28 percent of students, including 25 percent of primary and 32 percent of secondary school students. Almost all private schools have some religious affiliation. The most common affiliation is with the Catholic church, and 69 percent of private school students are enrolled in Catholic schools. Other private schools vary from long-established, prestigious schools to relatively new, fundamentalist and alternative education schools.

Private schools are supported by a range of funding sources, including government grants. In 1992, 37 percent of private school income was derived from fees and charges, 7 percent from private donations and income, 19 percent from state government grants, and 37 percent from federal government grants. Government grants were much more important for Catholic schools (72 percent of income) than other nongovernment schools (33 percent of income). The fees levied by private schools range from about US$300 a year in some parish primary schools to about US$6000

in some secondary schools. State and federal taxation revenues provide almost all the financial resources for the operation of government schools. Although parents are not officially required to pay fees for students to attend government schools, many schools seek voluntary contributions from parents and raise funds from other local sources. In some secondary schools, this can amount to 5 percent of the operating costs of a school.

Mathematics and Science Programs

Almost all students study some mathematics and science up to Grade 10. In Grades 11 and 12, a large percentage of students are still enrolled in a mathematics class: 97 percent of males and 95 percent of females in Grade 11, and 90 and 82 percent respectively in Grade 12. Seventy percent of males and 72 percent of females are enrolled in a Grade 11 science class, as are 70 percent of males and 68 percent of females in Grade 12. Mathematics and science are designated as key learning areas in most states and all students are expected to undertake some studies in those areas up to Grade 10. In primary schools these subjects are taught by a general classroom teacher and it is difficult to establish how much time is allocated to their teaching. A recent study in one large state suggested that in primary schools 20 percent of instructional time was spent on mathematics and 5 percent on science. From Grades 7 to Grade 10, science is typically taught as an integrated subject by one science teacher. The time allocation is typically 200 minutes per week, or 13 percent of total class time. In the senior secondary years, students choose five or six subjects, depending on the state, from a wide range of electives.

Table 1: Enrollment in Science Subjects

	Grade 11		Grade 12	
Science Subject Enrollments in Grades 11 and 12: 1994 (Percent of Grade Cohort)				
	Male	Female	Male	Female
Biology	23	38	22	40
Chemistry	27	24	26	20
Physics	32	13	31	11

According to a national study of subject choice conducted in 1994, 96 percent of Grade 11 students study mathematics and 73 percent study a science course. At Grade 12, 86 percent study mathematics and 69 percent are enrolled in one or more science courses.

Table 1 indicates the percent of Grade 11 and Grade 12 students enrolled in each of the three main science subjects. In Grade 12, biology is studied by 32 percent of students, chemistry by 23 percent of students and physics by 20 percent of students. The figures for Grade 11 students are similar.

Schools in the System

The School Year

For both primary and secondary schools, the length of the school year is approximately 200 days, with small variations between states. The school year typically begins at the end of January in most states, and mid-February in Tasmania, and ends the third week of December. In most states the school year is divided into four terms of 9 to 11 weeks' duration with a 2-week vacation between terms. The summer vacation is 5 or 6 weeks. Students in primary and secondary schools attend school five days per week from Monday to Friday. In primary schools there are about five hours' tuition per day, and in secondary schools there are about five and a half hours per day. There are about one and a half hours in addition to this for lunch, recess breaks, and other activities.

Class Size

In 1994, the average size of primary schools was 171 students, since a scattered rural population necessitates a large number of small primary schools. For secondary schools the average was 688 students. The ratio of students to teachers in primary schools was 18 to 1 and in secondary schools was 12 to 1. These figures include principals, deputy principals, librarians, and senior teachers involved administrative duties, along with some guidance counseling and careers advisers. Official information on the average size of classes is not available on a national basis. For public schools in the two largest systems in 1993, the average sizes of classes in primary schools were 27 students in New South Wales and 25 in Victoria; and in secondary schools the average sizes of English classes were 23 students in New South Wales and 22 in Victoria.

Streaming and Tracking

There is no common national policy on ability grouping, streaming, tracking, or setting of students in Australia. Streaming is a school-based decision and is not officially promoted in any state. Some schools choose to stream students according to ability and some have special enrichment and remedial programs for groups of students. In primary schools, there is almost no systematic grouping and most classes are heterogeneous in their composition.

Up to Grade 10 in secondary school, grouping practices vary between schools and between states. In two states, and in some schools in other states, a unit curriculum system operates under which students choose semester-length units within certain requirements, including specified studies in mathematics and science. This practice can result in *de facto* ability grouping through the combinations of units chosen, since some students choose more science or mathematics than the minimum required. Other schools provide enrichment studies in mathematics and physical sciences for those students likely to pursue studies in the sciences, particularly in Grade 10. In at least one state, setting in mathematics is common.

In Grades 11 and 12 students choose subjects from a range of possible studies. It is common for mathematics to be provided at different levels and for students to choose the level appropriate to their future plans.

Certification of Mathematics and Science Teachers

Teacher education occurs in universities but states are responsible for determining acceptable teacher qualifications. Four years of higher education is now the normal length of initial training for secondary teachers. Generally this comprises a three-year degree in a major discipline, followed by a one-year diploma in education. Primary teachers normally complete a three-year diploma in teaching. Many primary teachers later upgrade their qualifications by undertaking specialist one-year diplomas or a general fourth year resulting in the award of a degree.

A national review of teacher education in mathematics and science conducted in 1989 concluded that 44 percent of entrants to primary teacher education programs had completed one mathematics subject to Grade 12 level, while 8 percent had completed two. Even so, staff at the training institutions regarded the mathematical background of teacher education students as superficial. It also reported that most students in primary teacher education lacked confidence in science knowledge and had mastered few laboratory skills. Although 60 percent had studied a science subject in Grade 12, only 14 percent of students had studied a physical science.

Government school teachers usually serve a one- or two-year probationary period before becoming eligible for permanent employment. The states vary in the extent to which teachers are required to undergo formal appraisal after this time. Private school teachers are normally employed directly by the school concerned, although, like government school teachers, their salaries and conditions are determined by industrial tribunals.

Teacher Profile

A little over 200,000 (full-time equivalent) teachers were employed in schools in 1993: 97,000 teachers in primary and 103,000 in secondary schools. Teaching is an occupation with relatively high prestige, being in the third-highest of 16 categories in the best established occupational prestige scale, as rated by a sample of the adult population. Although salaries are determined at the state level, there are in practice only small differences between states. The annual starting salary is typically US$20,250 and for a teacher with 10 years experience is US$30,400. In Australia starting salaries for teachers are 1.2 times the per capita GDP and midcareer salaries are nearly twice the per capita GDP.

In 1993 the average age of teachers was 40 years. Over recent years the average age of the teaching profession has increased. Overall 63 percent of teachers were female, including 74 percent in primary and 51 percent in secondary schools. Among teachers of mathematics and science in Grade 12, the percentage of females is rather lower. According to a 1990 national survey among Grade 12 teachers, 35 percent of those teaching mathematics and 41 percent of those teaching science were female.

3 TIMSS POPULATIONS

Australia, IEA, and TIMSS

Australia has had a long involvement with IEA. It participated in the First International Mathematics Study in 1964 and has since participated in the First International Science Study in 1970 and the Second International Science Study in 1982. Australia also collaborated in the Second International Mathematics Study.

Australian TIMSS Populations

Population 1 includes students in Grades 3 and 4 in four states: New South Wales, Victoria, Tasmania, and the Australian Capital Territory; and Grades 4 and 5 in the remaining states: Queensland, South Australia, Western Australia, and the Northern Territory. This selection includes 96 percent of the nine-year-old cohort.

Population 2 contains students in Grades 7 and 8 in four states: New South Wales, Victoria, Tasmania and the Australian Capital Territory; and Grades 8 and 9 in Queensland, South Australia, Western Australia, and the Northern Territory. These grades contain 98 percent of the 13-year-old cohort. Fewer than 1 percent of 9-year-olds or 13-year-olds were excluded because they were in special schools.

Population 3 corresponds to Grade 12 in all Australian states. The modal age for students in Grade 12 is 17. Mathematics and physics specialists were defined as subsets of the general Population 3. Students enrolled in mathematics or physics courses that prepared them for post-secondary study in those areas were identified as specialists. This definition identified 20 percent of Population 3 as physics specialists. Mathematics specialists were harder to define because of differences between states in the structure and nomenclature of mathematics courses. On the basis of enrollment in designated courses, approximately 18 percent of Population 3 were identified as mathematics specialists.

4 MATHEMATICS CURRICULUM AND PEDAGOGY

Goals for the Mathematics Curriculum

There is no national curriculum for mathematics in Australia because curricula are defined at the state level. In most states the curricula for kindergarten to Grade 10 are determined in broad terms by the State Education Department, while a statutory body determines the curricula for Grades 11 and 12. During the early 1990s, a national statement and profile on mathematics was developed cooperatively through the National Council of Ministers of Education. Even though it was eventually not adopted on a national basis by the council, many aspects were incorporated by the states in their own curricula statements. Because they were developed cooperatively, the national statement provides an indication of general thinking about mathematics curricula in Australia. The national statement on mathematics indicates that as a result of learning mathematics, all students should:

- realize that mathematics is relevant to them personally and to their community;
- gain pleasure from mathematics and appreciate its fascination and power;
- realize that mathematics is an activity requiring the observation, replication, and application of patterns;
- acquire the mathematical knowledge, ways of thinking, and confidence to use mathematics to conduct everyday affairs such as money, measurement, and planning; make individual and collaborative decisions at the personal, civic, and vocational levels; and engage in mathematical study for further education and employment;
- develop skill in presenting and interpreting mathematical arguments;
- possess sufficient command of mathematical expressions, representations and technology to interpret information in which mathematics is used, continue to learn mathematics and communicate mathematically to a range of audiences;
- appreciate that mathematics is a dynamic field with its roots in many cultures.

Major Changes in the Mathematics Curriculum

There have been major changes in mathematics curricula in all the states of Australia over the past 10 years. Almost all the states have introduced new curriculum guidelines or syllabi and many plan further changes over the next few years. The most important change has been a new approach to mathematical learning and teaching. This approach emphasizes the relationship of mathematics to our culture, student motivation, and active learning. The relevance of mathematics and its application to everyday life have been given greater emphasis in order to make mathematics learning more meaningful to students. Problem solving, modeling, and investigative approaches are now being used instead of the earlier emphasis on formal algorithms.

Modern technology has had a strong impact on mathematics teaching and learning. The use of calculators as teaching and learning tools to support the development of concepts is encouraged in both primary and secondary schools. This has resulted in an increased emphasis on estimation and automatic response and has enabled teachers to make use of real rather than contrived data. Programmable and graphics calculators enable more investigative approaches but cost is an impediment to widespread use. Computers enable students to investigate and apply mathematical ideas in a way not easily achieved by other means, and their use is being encouraged in all states. In secondary school mathematics, the use of spreadsheets, graphing and calculus packages, and data management and analysis is becoming more widespread, although limited by the availability of resources in many schools. These forms of modern technology support the common goal shared by the states of making mathematics more meaningful to students and more related to everyday life.

Information on the use of these technologies is hard to establish, but one state has estimated that in 1994 there was one computer for every 8 students at the secondary level and one computer for every 18 students at the primary level. Approximately 80 percent of secondary schools reported using computers in the teaching of mathematics. In science, around 60 percent of schools reported using computers in teaching in Grades 11 and 12, with the corresponding figures for Grades 7 through 10 being approximately 40 percent.

Organizationally, a recent change has been the development of more formal curriculum frameworks at the state level, around which schools are expected to organize their teaching programs. The frameworks are not as specific as a syllabus. They are typically divided into strands that reflect the major elements of learning in each area, and bands corresponding to different stages of schooling. Many reflect the national statement in which the strands are attitudes and appreciation, mathematical inquiry, choosing and using mathematics, space, number, measurement, chance and data, and algebra. Corresponding profiles describe outcomes and show typical progression in the achievement of outcomes. By describing what students are expected to achieve at each level, the profiles provide a common language for reporting and for the development of teaching and learning.

Current Issues in the Mathematics Curriculum

Many of the changes in mathematics curricula that have been introduced are as much concerned with approaches to teaching and assessment as with content. One current issue focuses on the adoption of new approaches and the implementation of statewide curriculum frameworks. Another concerns the rapid growth in the use of modern technology in teaching at all levels of schooling. This involves not just the use of technology for computation but also the incorporation of a range of multimedia resources in a way that relates to the real world. In relation to the implementation of new curricula, the adoption of new approaches and the utilization of new resources, the provision of appropriate professional development for teachers of mathematics and other teachers has emerged as an important issue. In addition, there is concern about ensuring that mathematics curricula are inclusive of the full range of groups in the community. Despite a number of changes, females still have lower participation rates in advanced mathematics courses in Grades 11 and 12 than males.

Mathematics Textbooks

Over the past 10 years there has been a reduction in the emphasis placed on the use of textbooks. This is particularly true of primary schools, where textbooks serve mainly as resource material for teachers, if they are used at all. Consequently, many primary teachers use a number of textbooks rather than a single book, some with additional resources such as black line masters. Occasionally, class sets may be used but generally primary school students are not required to purchase textbooks. In secondary schools, although the degree of emphasis on textbooks has decreased somewhat, the teaching of mathematics is still commonly centered around the use of a textbook, or a range of textbooks.

The type of material found in textbooks has changed, with a stronger emphasis on problem solving, mathematical modeling, real-world applications, and the use of calculators rather than routine exercises. There is evidence that the presentation of material has changed, with more space devoted to graphs, illustrations, and tables. Textbooks are generally used under teacher guidance as a source of activities for the reinforcement of concepts and the practice of skills. Students are not generally expected to learn on their own from textbooks. Textbooks are not developed or prescribed by state departments of education. Decisions about textbooks, and the extent of their use by students, are made at school level.

Pedagogy

As noted above, many of the recent changes in mathematics curricula have been concerned primarily with approaches to teaching and assessment. The national statement on mathematics for Australian schools emphasized that learning is best thought of as an "active and productive process on the part of the learner." This was seen as recognizing that learners construct their own meanings from the ideas, objects, and events that they experience, that learning takes place when existing conceptions are challenged, and that learning requires action and reflection on the part of the learner. In terms of teaching, it is recognized that although there is no definitive approach to the teaching of mathematics, teaching should be informed by an understanding of how learning occurs. It indicated that mathematics teaching should build upon and respect students' experiences, involve activities that the learner regards as purposeful and interesting, be enhanced by feedback, use and develop an appropriate language, and provide challenge within a supportive framework. The national statement's comments on pedagogy can be seen as representing what is desirable rather than what necessarily occurs in practice. Pedagogical practice varies between schools and between teachers within schools. Changes in mathematics teaching from an emphasis on arithmetic skills and formal algorithms to a more constructivist approach have evolved slowly over many years. There is some evidence that approaches to teaching that support active learning are more prevalent in primary schools than in secondary schools.

5 SCIENCE CURRICULUM AND PEDAGOGY

Goals for the Science Curriculum

As was noted for mathematics, there is no national curriculum for science in Australia. A national statement and corresponding profile on science were developed during the early 1990s through the National Council of Ministers of Education. Many aspects of the statement were incorporated by the states in their own curricular frameworks even though the document was not adopted on a national basis. Because the statement was developed cooperatively, it provides a good indication of science curricula in Australia. In terms of goals, it states that science education should "develop an understanding of the role of science and technology in society, together with scientific and technological skills." More specifically the goals indicated that through a science education:

- students should develop the confidence, optimism, skills, and abilities to satisfy their curiosity about the workings of the physical, biological, and technological world and devise solutions to problems arising from their own experiences in daily life;
- all students, as members of society, should be increasingly able to take part confidently in public debate and in decision making about science and public science policy;
- all students, in preparing for post-secondary options, should gain enough understanding of the work of science to allow them to make decisions about further education and careers, and gain an appreciation of scientific knowledge, processes, and attitudes to enable them to contribute to building a more productive, ecologically sustainable, and prosperous economy.

A number of unifying ideas are contained in the national statement on science in schools. They are that, in order to understand something or solve a problem, it is helpful to analyze its parts; change occurs because of interaction; interaction and change are the result of energy transfer; science is conducted partly to create meaning and partly to improve the world; working scientifically is an effective way of generating understanding and solving problems.

Major Changes in the Science Curriculum

For kindergarten to Grade 10, the science curriculum is determined in broad terms by the State Education Departments. The curricula for these grades are generally guidelines for school-based curriculum development. Those guidelines are mandatory in most states at the secondary level and in half the states at primary level. In primary schools, much of the teaching of science takes place as a multidisciplinary activity in the context of projects. Up to Grade 10, science is taught as an integrated subject. The curricula for separate science subjects in Grades 11 and 12 are specified by a statutory body in each state except for some school-based courses.

One of the most important changes in science curricula in all the states of Australia over the past 10 years has been a greater emphasis on the relevance of science and the applications of science in everyday life. Schools have been encouraged to incorporate local resources, interests, and needs into the curriculum. The human aspects of science and the impact of science on the environment and on society as a whole form an integral part of new syllabi. Issues such as global warming, greenhouse effect, ozone depletion, recycling, energy alternatives, and pollution are being addressed in the science curricula in many states.

Current Issues in the Science Curriculum

An important current issue concerns the influence of the national statement on science for Australian schools on curriculum frameworks in each state and in turn on school teaching programs. The national statement sets out the goals for the science curriculum at each of eight levels and specifies what students at that level should be able to do and should know. The science profile, primarily intended for use in reporting student achievement and progress, maps science outcomes in five strands: earth and beyond, energy and change, life and living, natural and processed materials, and working scientifically.

The initial and continuing education of science teachers remains another important issue for the implementation of science curricula. A national inquiry reported in 1989 that there was pressing need to revise and upgrade both initial and continuing education of teachers in science, especially for those teaching in primary schools. The more recent national statement on science for Australian schools noted that science in primary schools did not show a balance between the biological and the physical sciences and that there was an imbalance among primary schools in the amount of science taught.

The national statement on science for Australian schools also identified a number of equity issues in science education. It noted that although usefulness had been a powerful element in the rationale for science in the school curriculum, many students had little appreciation of its role in society or its relevance to their daily lives. It identified skewed patterns of participation in science as an issue, as evident in the underparticipation of females in the physical sciences, the underparticipation by Aboriginal students in school science, and the lack of resources for geographically isolated students and students living in poverty.

Science Textbooks

The degree of emphasis given to science textbooks in Grades 1 to 10 has decreased, particularly in primary school where textbooks serve mainly as resource material for teachers. In some states new science curricula encourage teachers to design their own courses based on a set framework. Textbooks are still used by students in the early secondary grades in most states but often as a class set rather than as an item purchased individually by students. In addition, a range of textbooks is used rather than a single book. In Grades 11 and 12, a main textbook is used for each science course but these textbooks are not prescribed by the state; the choice is left to each school.

At the primary and lower secondary level textbooks have changed from an emphasis on content to an ac-

tivity-based problem-solving approach with an emphasis on applications based on real-life examples. In Grades 11 and 12 the books are generally based on the material in the syllabus. Although there has been less change in the textbooks at this level, most have incorporated issues concerned with science and society. Textbooks that have been written for the new syllabi relate knowledge to wider contexts and emphasize the applications of science and technology to a greater extent than those produced some years ago. At all levels there has been an increase in the amount of graphic and pictorial material that is used in text books. Science textbooks are not produced by state education departments, and the ways in which they are used are determined at school level.

Pedagogy

Over the past 10 years there have been changes in the methodology of science teaching and learning. Based on the principle that students learn best when they are actively engaged in the learning process, new teaching strategies include cooperative hands-on and problem-solving activities. Theory is related to practice and the introduction of concepts is linked to students' experiences and real-world contexts. A previous strong emphasis on science content is now balanced by an emphasis on scientific processes. The Australian Science Education Project produced a range of curriculum and teacher-support materials in the late 1970s that provided a foundation for the more recent changes in pedagogy in junior secondary school science. Similar projects influenced the teaching of science in primary schools, and other materials from that era, such as an Australian adaptation of BSCS Biology, influenced teaching in the specialist science disciplines for Grades 11 and 12. Almost all states increasingly use new technologies for science teaching. Programmable and graphics calculators, computers, and interface devices are used for the measurement, analysis, and presentation of data. The use of modern technological equipment has increased the scale and variety of activities that can be carried out in classrooms.

6 EVALUATION POLICIES AND PRACTICES

Evaluation policies and practices in Australia can be best considered at the state (system) or the local (school and classroom) level. At the national level, national studies of literacy and numeracy were conducted in 1975 and 1980 but they have not been repeated since. There is almost no formal assessment in mathematics or science at the school-district or regional level.

Statewide Assessment

In recent years, most Australian states have introduced, revived, or continued statewide assessment and monitoring programs for primary schools. These have involved mathematics but have rarely included science. In New South Wales, Victoria, and South Australia there is population testing of all students in government schools in English and mathematics in Grades 3 and 5. In Queensland, there are statewide tests in English and mathematics for all Grade 6 students in government schools. In Tasmania, students who are 10 years old are tested in these areas. In Western Australia, there is a program of testing samples of students in Grades 3 and 7 in all eight key learning areas, including mathematics and science. Population assessment in science of Grade 3 and Grade 5 students in Victoria was undertaken in 1996. Victoria conducted sample surveys of achievement in science and social education as well as English and mathematics, prior to introducing population testing.

The purposes of these testing programs are to both report information about student achievement based on a common test to parents, teachers, and schools and to monitor the overall performance of the education system. There has been an increased emphasis on evaluation for accountability purposes and these testing programs reflect that emphasis. Most of the assessment instruments used in these programs of testing incorporate a range of items, including multiple choice and open-ended response items, and make use of modern measurement techniques and approaches to scaling, which have allowed more informative forms of criteria-referenced reporting to be implemented.

Formal statewide assessment programs are less common in the lower secondary grades than in primary school. In New South Wales, all students in Grade 10 are tested in English, mathematics, and science through reference tests. These are used to moderate school assessments in the School Certificate program. In the Northern Territory, all students in Grade 10 in government schools, and in Tasmania all 14-year-olds, are tested in English and mathematics as part of monitoring programs. In Western Australia, sample surveys of achievement in key learning areas, including mathematics and science, are conducted at Grade 10 as part of a monitoring standards program.

At the end of Grade 12, all states conduct formal assessments of performance in students' subjects. The purpose of these assessments is to certify student achievement at the end of school. They also provide the basis for selection to courses in higher education. In most states the assessments are based on a combination of curriculum-specific formal examinations conducted by a state authority and school-based assessments of student performance on specified tasks or assignments. In Queensland and the Australian Capital Territory, there are no external examinations at all: internal school assessments are adjusted against students' scores on an aptitude test to achieve comparability across schools.

Local Assessment

School-based assessment is the most common mode of assessment at the primary and lower secondary levels of schooling. In primary schools, assessment is mainly informal in nature. making use of checklists, observations, projects, and portfolios. In the lower secondary grades, teacher-made tests are used, including multiple-choice items that may be from item banks, short response, and extended answer formats. Projects, laboratory assignments, and seminar presentations also form part of the assessment process. Over the past ten years there has been greater use of a wider range of assessment instruments rather than just traditional written tests, and continuous assessment rather than end of term tests. At the local level, assessment is used for a variety of purposes: evaluating student progress, reporting to students and parents, evaluating programs, and at the lower secondary level providing guidance about further courses of study.

7 REFERENCES AND SOURCES FOR FURTHER READING

Common Sources

1   Husén T, N T Postlethwaite 1994 *International Encyclopedia of Education*. Pergamon Press, Oxford
2   Organisation for Economic Co-operation and Development 1995 *Education at a Glance: OECD Indicators*. Organisation for Economic Co-operation and Development, Paris
3   United Nations Educational, Scientific and Cultural Organization 1995 *Statistical Yearbook*. United Nations Educational, Scientific and Cultural Organization, Paris
4   World Bank 1995 *World Development Report 1995*. Oxford University Press, New York

Other Sources

5   Ainley J, L Robinson, A Harvey-Beavis, G Elsworth, M Fleming 1994 *Subject Choice in Years 11 and 12*. Australian Council for Educational Research, Melbourne
6   Australian Bureau of Statistics 1993 *Schools, Australia 1993*. Australian Bureau of Statistics, Canberra
7   Australian Bureau of Statistics 1993 *Year Book, Australia 1993*. Australian Bureau of Statistics, Canberra
8   Australian Education 1993 *National Report on Schooling in Australia 1992*. Curriculum Corporation, Melbourne
9   Curriculum Corporation 1994 *A National Statement on Mathematics for Australian Schools*. Curriculum Corporation, Melbourne
10 Curriculum Corporation 1994 *A National Statement on Science for Australian Schools*. Curriculum Corporation, Melbourne
11 de Lemos M 1994 *Schooling for Students with Disabilities*. Melbourne, Australian Council for Educational Research
12 Fensham P J (ed.) 1995 *Science and Technology Education in the Post-Compulsory Years*. Australian Council for Educational Research, Melbourne
13 Logan L 1990 *Teachers in Australian Schools. A 1989 Profile*. Australian College of Education, Canberra
14 McKenzie P 1994 Education in Australia. In T Husén & N T Postlethwaite (eds.), *International Encyclopedia of Education, Research and Studies* (pp. 415–23) Pergamon, Oxford
15 Ministerial Council on Education, Employment, Training and Youth Affairs 1994 *National Report on Schooling in Australia 1993*. Curriculum Corporation, Melbourne
16 Shears L 1995 *Computers and Schools*. Melbourne, Australian Council for Educational Research
17 Speedy G W, C Annice, P J Fensham 1989 *Discipline Review of Teacher Education in Mathematics and Science*. Australian Government Publishing Service, Canberra
18 Willis S (ed.) 1990 *Being Numerate: What Counts*. Australian Council for Educational Research, Melbourne

Statistical References

Section	Subject	Source
Country Profile	Population and work force statistics	7
	Income and expenditure	14, 2, 4
The Education System	Enrollment and enrollment distributions and expenditure by school sector	6, 8, 15
	Students with disabilities	11
	School science & mathematics subject enrollments by grade sector	5
	Schools and student grouping	14, 15
	Certification of teachers	17
	Teacher profile	13, 14
Mathematics Curriculum and Pedagogy		16, 18

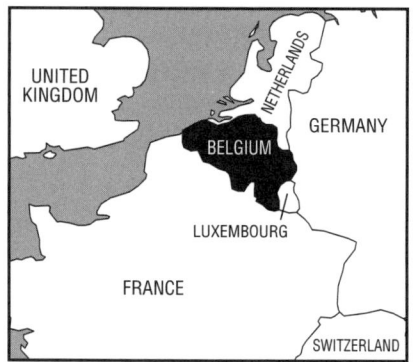

Belgium

Christian Monseur, University of Liège

Christiane Brusselmans-Dehairs, University of Ghent

1 COUNTRY PROFILE

Belgium lies in western Europe and shares frontiers with France, Germany, Luxembourg, and the Netherlands. It has a short coastline on the North Sea, where the port of Ostend is located. Brussels is the capital of both Belgium and the European Union.

Belgium covers an area of 30,528 km^2. With a population of 10 million people in 1992, it has the second highest density in Europe at 328/km^2. The density varies, however, from 429/km^2 in the north to 194 in the south. Since 1970, the annual growth rate of the population has been less than 1 percent. Today, approximately 9 percent of the population is made up of non-native Belgians. This figure is expected to increase to 18 percent by 2040, with 60 percent of new citizens coming from European Union countries. From the beginning of this century until 1970, the number of immigrants was relatively stable. Indeed, for a long time Belgium sought workers from outside its borders, particularly for the traditional industries in the southern part of the country. A wave of immigration between 1970 and 1981 increased the total number of foreigners by more than 25 percent and brought many Turks and Moroccans to Belgium. Their children now account for half of the non-native Belgian population under the age of 15.

Three languages are spoken in Belgium: Dutch, French, and German. These three linguistic communities occupy different territories, and only in Brussels are both French and Dutch spoken.

Belgium is a constitutional monarchy with a representative, parliamentary system of government. Two types of federal entities were established in the 1970s: linguistic communities and economic regions. In 1993, Belgium became a federal state uniting communities and regions. The three communities, Flemish, French, and German, are based on the official language and distinct culture of the people. There are also three economic regions: Brussels, Flanders, and Wallonia. Each of these federal entities has a government with executive power and a council with legislative power. In Flanders, the community and the region coincide politically, and there is one Flemish Parliament and one Flemish government. Since January 1989, each community has had, except for a few restrictions, full autonomy in the area of education. Each community operates its own educational system, with the Flemish system being responsible for 58 percent of school-age children in Belgium, the French system for 42 percent, and the German system for less than 1 percent. The economic regions are responsible for housing, employment, the environment, and economic development. The national government has retained responsibility for the treasury, law, social security, and foreign policy, as well as several other areas.

Belgium's gross national product is ranked slightly below the average of high-income economies of OECD countries. The average annual growth rate was 2 percent between 1980 and 1993, and the average annual inflation rate during this period was 4 percent. The most important economic activities are services, manufacturing industries, trade, banking, and insurance. Expenditure on education in 1991 was 15 percent of total public expenditures and more than 5 percent of the GDP.

2 THE EDUCATION SYSTEM

Governance and Decision Making

Education, a national responsibility until 1989, is now the province of the linguistic communities. Only three matters pertaining to education remain with the national government: setting the beginning and end of compulsory schooling, the minimum conditions for awarding diplomas, and pensions.

One of the most important principles governing the system is freedom of education, meaning that educational institutions may not be submitted to restrictive measures and parents may choose the school or type of education they wish for their children. A consequence of this freedom of education is the diversity of educational networks that exist. An educational network may come under the authority of the community, province, municipality, or public official, as well as under private persons or associations. Traditionally, there have been three networks:

- community education, which comprises schools that were originally set up by the state but are now the responsibility of the community. In 1989, the Flemish community delegated its organizing authority to an autonomous community education council, known as *Argo*.

- subsidized official education, which is organized by the provincial authorities, and municipal education, which is set up by the municipal authorities. The schools of this network may be denominational or nondenominational.

- subsidized private education, which operates on the initiative of a private person or organization. It consists of denominational (mainly Catholic), nondenominational, and pluralistic education.

Education organized within the first two networks is called official education; education provided by the third network is called private education. The networks are largely autonomous: free to choose their teaching method, timetables, and curricula, subject to government approval.

Important work has recently been accomplished in quality control in education. To avoid curricular fragmentation, a set of common objectives has been developed for primary schools and for the first two grades of secondary education. These attainment targets set the minimum standard for what must be learned at school. Before their actual application, the objectives were the subject of broad-based general debate. The objectives provide teachers with a clear definition of what they should focus on and provide authorities with tangible criteria for quality control.

The introduction of minimum standards and goals in Belgium's education system is the most important innovation that the authors have witnessed. The new policy has recast the relationship between schools and

Table 1: Enrollment in the French and Flemish Communities

Networks	Community		Subsidized Official		Subsidized Private	
School levels	French	Flemish	French	Flemish	French	Flemish
Elementary						
Nursery	9	12	51	19	40	69
Primary	11	13	44	23	45	64
Secondary	27	17	18	8	55	75
Higher						
Nonuniversity	19	20	31	16	50	64
University	28	30	–	–	72	69

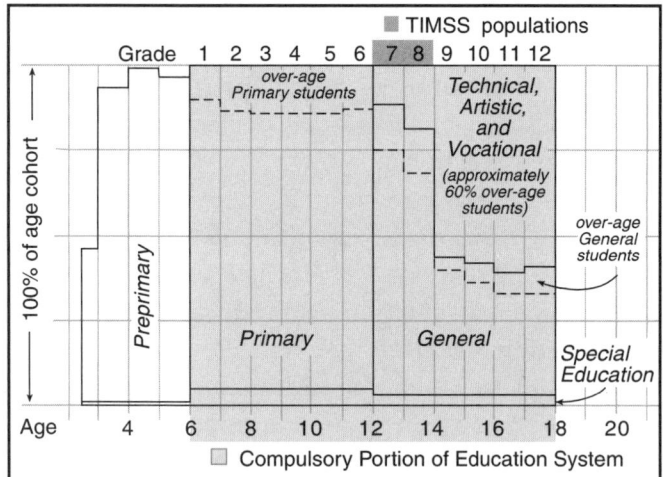

Figure 1: Structure of the Education System, Belgium-Flemish

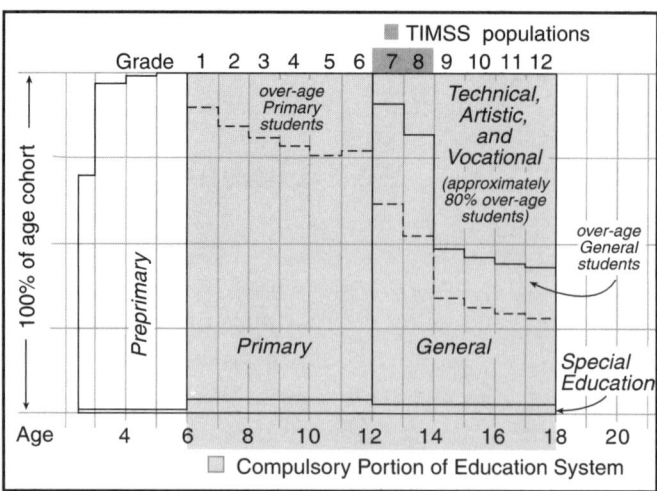

Figure 2: Structure of the Education System, Belgium-French

society, and this is probably the reason for some of the strong negative reactions during the public debate. Nevertheless, the debates should continue until it is determined whether the policy has a negative impact on the quality of education. A defense of the status quo in education for its own sake reminds the authors of Peter Lawrence's saying, "Bureaucracy defends the status quo, long after the moment that the quo has lost its status."

Structure of the System and Participation Rates

Compulsory Education

In Belgium, compulsory school attendance covers a period of 12 years, starting the year the child reaches 6 and ending the year the child reaches 18. It is full-time up to age 16. Between ages 16 and 18, at least part-time education is compulsory. Education is free of charge for children from nursery school to Grade 12. In 1992, 20 percent of the population 25 to 64 years of age had obtained a tertiary education certification, and 25 percent of this population had an upper secondary education certification. At that time, over 40 percent of the 23-year-old age group completed a formal tertiary education program. Table 1 shows student enrollment in the French and Flemish communities according to education level and network.

In spite of major efforts by the government, there are still unacceptably high numbers of students lagging behind. Moreover, there is an unequal participa-

tion of children from socioeconomically weak backgrounds. These problems have yet to be successfully addressed.

Structure of Education

There are three levels of education: elementary, secondary, and higher education. Figure 1 shows the structure of the education system in the Flemish part of Belgium, and Figure 2 shows the French. In addition to the regular program of education, there is a special education program for children and adolescents who are not well served in regular classrooms. Approximately 4 percent of the total school-age population attend special education. Part-time education is also available.

Elementary education consists of nursery education, which is attended by most children aged two-and-a-half to six years, and primary education which comprises Grades 1 to 6 is attended by children of the six to twelve age group.

Secondary education is provided for students aged twelve to eighteen and consists of Grades 7 to 12. These six years are divided into three parts, each lasting two years. There are four branches:

- general secondary education, which emphasizes broad theoretical training and provides a strong base for higher education.
- technical secondary education, which emphasizes general and technical-theoretical subjects. Students finishing this level may enter the job market or attend higher education.
- artistic secondary education, which consists of a

general and broad education and is linked with active art practice. Students finishing art education may enter the job market or make the transition to higher education.

- vocational secondary education which is a practical education in which students acquire specific skills and simultaneously receive general education. Students may proceed to higher education when certain conditions are met.

At the secondary level, the choice of which branch to attend is made as late as possible, allowing students contact with many subjects. Great weight is attached to basic education, and part of the curriculum is identical for all pupils of the same year. In addition to the basic curriculum, students may select several optional subjects.

Higher education consists of university education and nonuniversity education for three years of study (short-term) or four to five years of study (long-term). Only university-level engineering science requires an entrance examination. Net enrollment in public and private tertiary education for the 18 to 21 age group is 31 percent: 15 percent at nonuniversity institutes and 17 percent at universities.

In our rapidly changing society, regular retraining and in-service education have become a necessity in all sectors. The education department, therefore, organizes part-time adult education, part-time artistic education, distance learning, basic education, and second-chance education to cater to this need. Many students within the regular school system study programs in subjects they are not suited for. In later life, these groups are confronted with reduced opportunities in the labor market. Adult education is aimed at remedying this problem.

Funding the System

Since 1959, the basic principle has been that the state, and now the community, must provide all funding, in the form of a block grant, for the schools within its jurisdiction. The grants issued to the subsidized networks are intended to cover teachers' salaries, and the running, maintenance, and replacement costs of equipment and buildings. The subsidy received by a given school is directly related to the number of pupils, weighted according to the type of education provided.

For several reasons, education in Belgium is not cheap. One of the main reasons is probably freedom of education, which results in very dense school and discipline coverage, particularly in secondary and higher education. Not only do the different networks offer the same discipline, but within the networks the offerings are often very similar. Controlling the education budget will remain a priority in the coming years, and a broader application of budget financing will be envisaged. This will likely lead to further decentralization in the future, with individual schools and groups of schools assuming more tasks and greater responsibility. Finally, increased economies of scale, by rationalizing both the number of schools and the number of subjects, are obviously needed.

Enrollment in Mathematics and Science

Mathematics and science are compulsory for all students aged 6 to 18. In primary education, the sciences are integrated in a general course. In secondary education, the type of mathematics and science curriculum offered varies with the educational branches and the study options organized within the branches.

Schools in the System

The School Year

Primary and secondary schools must be open 182 days each school year, from September 1 to June 30. The year is divided into three terms, with a two-week vacation between each term in the form of Christmas and Easter holidays. There are also week-long spring and autumn holidays, and several national, community, and religious holidays. Summer vacations take place during July and August.

Instructional time is organized into periods of 50 minutes. The weekly timetable varies according to grades, networks, tracks, and schools. There are, however, minimum requirements to which all schools must adhere. Grades 1 to 4 are obliged to teach exactly 28 instructional periods per week. In Grades 5 and 6, two optional periods are available for second-language learning. The timetable for Grades 7 to 9 may vary from 28 to 32 instructional periods, and for Grades 10 to 12 from 28 to 36. The school week is from Monday to Friday. Each day is approximately six to eight periods long, with the exception of Wednesday, which is four or five periods long.

Table 2: Routes to Teacher Certification

Qualixfication titles	Location of training	Course structure
Preschool education Nursery school (2.5 to 6 years of age)	Nonuniversity education (short term)	Three years of concurrent academic study, teacher education, teaching observation, and practice increasing over the three years to about 50 percent in final year.
Primary education (Grades 1 to 6)	as above	as above
Lower secondary education (Grades 7 to 9)	as above	Three years of academic and education courses; specified subject study and further optional practice increasing over three years to about 50 percent in the final year.
Upper secondary education (Grades 7 to 12) and higher nonuniversity education	University	Four to five years academic training plus teacher education either during the final one or two years of university study, parallel to the degree course, or as a two-year part-time course after completion of a degree.
University	University	Academic training plus a doctoral degree.

Class Size

The ratio of students to teaching staff is about 14 to 1 for primary education and 8 to 1 for secondary education. Although it is possible to compute the average class size for the primary level, it becomes more hazardous to do so for the secondary, due to modifications of the class groups depending on the different courses.

Streaming and Tracking

As syllabi are nearly identical for students in primary education, and as there are rarely more then two classes per grade, there is no grouping or tracking policy. In secondary education, students are oriented into different educational branches depending on their achievement level or ability at the end of Grade 8 and again at the end of Grade 10. Officially, there is no grouping policy.

Certification of Mathematics and Science Teachers

An upper secondary school certificate is required for entrance to all teacher education programs. Table 2 pro-

vides details on the qualifications for teaching at the various levels.

In the Belgian education system, there are no official in-service training requirements. Teachers have the right to pursue in-service training but are not obliged to do so. For some functions, such as inspector or principal, training and examinations are required at some levels for some networks. Before 1989, the available in-service courses were divided among the different school networks. Since 1989, the government has instituted a legislative structure allowing teachers to follow official recognized courses or workshops during instructional time for up to 10 days per year. Outside this framework, teachers may still take courses during vacations or weekends. Objectives of these official sessions are determined by the government together with teacher associations.

Teacher Profile

As in most high-income countries, the educational profession has been attracting more and more women for several decades. Today, more than 98 percent of nursery school teachers and 62 percent of primary teachers are women. An equal distribution has nearly been reached for secondary and higher nonuniversity education. Twenty-two percent of educational staff at the preprimary level and 36 percent at the secondary level work half-time. This percentage increases to 45 for higher nonuniversity education. The approximate average age of teachers for all levels is just over 40. At all levels, the male averages are somewhat higher.

Teacher salaries depend on their certification and on their length of service. In short, nonuniversity certified teachers begin their professional lives with an after-tax and social security monthly income of about US$1400 and end with US$2050. There are some differences between the incomes of preprimary, primary, and lower secondary teachers.

Another challenge is the reevaluation of the teaching profession. The future of education indeed depends to a large extent on the quality of teachers, their professional and social commitment, and their ability to participate in continuing education. It is important for society to fully recognize the value of the teaching profession, both at a social, financial, and material level. Not only initial teacher education but also continuing education should be improved, and career prospects and working conditions should also be reviewed.

3 TIMSS POPULATIONS

Belgium, IEA, and TIMSS

Belgium has been a member of IEA since 1965 and has participated in most IEA research projects. The legal seat of IEA is located in Belgium.

Belgian TIMSS Populations

In Belgium, the test was administered only at the Population 2 level, which includes the adjacent Grades 7 and 8. These two grades are equivalent to the first two years of the secondary school system and contain more than 90 percent of 13-year-old students. Students enrolled in special education and students enrolled in 1B (Grade 7 students who did not succeed in primary education) were excluded.

4 MATHEMATICS CURRICULUM AND PEDAGOGY[1]

Goals for the Mathematics Curriculum

In Belgium, mathematics is taught as an integrated course. The intended curricula are developed by the education networks by grade, and are then approved by the Ministry of Education before implementation. All syllabi for mathematics share the same approximate structure, including:
- a description of the initial pedagogical situation, focusing on the mathematics objectives acquired in the previous grade;
- general objectives for mathematics;
- mathematical content;
- basic objectives for mathematics that must be acquired;
- differential objectives that are to be acquired if possible;
- methodological issues;
- bibliographic references.

The implementation of these curriculum guides in the classes is directly controlled by the inspectors of mathematics.

As in most European countries, standards for mathematics are set centrally. Within each linguistic community, new objectives for mathematics have been developed for the elementary level and for the first two grades of secondary school. The emerging trend is not only to examine the objectives common to all educational networks, but also to formulate goals for the performance standards to be achieved. In order to promote vertical cohesion and continuity in education, the objectives of consecutive levels of education are geared to the preceding level. Finally, grade structure has been replaced by a cyclical structure. Networks are still responsible for their programs but they must respect this conceptual framework.

In the future, the mathematics curriculum will be approached from four different angles:
- mathematics as a tool for problem solving;

---

[1] Information in the next two sections is based on the intended curriculum only.

- mathematics as a means of communication;
- mathematics as a link with science;
- mathematics as a source of reasoning and abstraction.

The most important specific competencies for elementary education are computation, measuring, and space-structuring abilities. Lower secondary education essentially focuses on algebra and geometry. Other areas are progressively introduced at the upper secondary grades. Mathematics is not only perceived as a link with science but also as a key to science. Apart from its own educational value, mathematics should support the sciences.

Major Changes in the Mathematics Curriculum

Recent changes in the mathematics curriculum concern general aims, methodological approaches, and content, while mathematical knowledge is no longer considered an outcome in itself. Nevertheless, this important shift in focus has yet to be implemented. In previous programs, there were clashes between this goal and some methodological approaches or content. Today, the different elements of the programs are more consistent. Some formal knowledge, for example, has been de-emphasized, and a real-life problem-solving approach adopted in the guidelines. Methodology, however, remains the weak point in the guides and should become a priority in the near future.

Current Issues in the Mathematics Curriculum

In primary schools, several modern mathematical concepts such as sets and correspondence have been introduced, more or less at the cost of computation. In secondary school, however, a purely modern approach has been replaced by a compromise between modern and traditional mathematics, with more stress on algorithms and geometry and less attention on set theory, groups, and vector spaces. In geometry, the initial focus on the study of structures and invariants for transformations has now shifted to the study of geometric figures. In the final years of secondary education, an emphasis on statistics and probability is gradually increasing.

Technology has had no obvious impact on the curriculum. Calculators are allowed almost everywhere from the beginning of secondary school. Other tools such as computers are occasionally used.

Mathematics Textbooks

In the Flemish community, all students and between 90 and 95 percent of teachers use textbooks. In the French-speaking community, although the use of textbooks is strongly recommended, many teachers use the textbook to set exercises, leaving their pupils to copy the theory in their notebooks.

For each educational type and each grade level, textbooks are developed in accordance with the intended mathematics curricula. Usually, the textbooks contain more information than the official syllabi. Compared to some twenty years ago, the layout of textbooks has been improved. Most of the time, pupils do exercises from the textbooks. Students usually lease their textbooks from the school library, paying about one-fifth of the value of the book annually. Generally textbooks are written by groups of mathematics teachers, with inspectors of mathematics as coauthors. Teachers decide which textbook will be used by the class.

Pedagogy

Current issues in mathematics pedagogy have been expressed as general objectives as well as guidelines for the outcomes. These objectives reflect the knowledge that students should acquire by the end of secondary education. Objectives are organized in terms of content, thinking processes, skills, and attitudes, and include:

- the development of adequate mathematical concepts, as mathematics is considered a language;
- the acquisition of computational techniques as important mathematical skills;
- the use of cooperative learning, introduced through class discussions on mathematical problems;
- the concept of mathematics as an integrated science;
- the acquisition of the necessary skills in calculator and computer use for solving certain kinds of mathematical problems;
- the use of both deductive and inductive approaches (inductive approaches are considered more important, particularly at younger age levels);

- the introduction of manipulative materials at the preprimary level only;
- the development of thinking processes necessary for conducting investigations and open-ended projects;
- the acquisition of written communication skills through homework and written exercises in class (oral communication is acquired through teacher-pupil interaction in class and through group discussion).

5 SCIENCE CURRICULUM AND PEDAGOGY

Goals for the Science Curriculum

At the primary level, biology is the science most frequently taught, accounting for approximately 7 percent of total instructional time. At the secondary level, a number of specialized science courses are taught, including biology, natural sciences, physics, practical scientific work, and chemistry. The timing of the introduction of these science courses, as well as the number provided, can vary according to educational network. Each educational network develops its own grade-by-grade curriculum guides, which must be approved by the minister of education. The general structure of the guides is similar to those described for the mathematics curriculum.

The implementation of the syllabi in the classes is directly controlled by the inspectors of biology, chemistry, and physics. In Belgium, the inspectorate is responsible for all educational levels, from nursery school to short-term higher education. The inspection corps is composed of official education and subsidized or private education, with equal representation. The inspectorate is not subject-related and does not check on individual teachers; it aims at examining the whole school. The tasks of the inspectorate include:

- checking whether minimum timetables and approved curricula are complied with;
- checking whether the final aims are achieved, ensuring the application of laws on the use of languages, ensuring the hygiene of classrooms, and supervising the care of didactic materials and school equipment;

- issuing recommendations on the financing or eligibility for financing of institutions and divisions;
- issuing policy recommendations on education.

Major Changes in the Science Curriculum

As with mathematics, objectives are being developed for each of the science courses. The objectives are regarded as standards to be attained by all pupils, regardless of educational network. Objectives have been developed for primary education and the two first grades of secondary education; the remaining grades are under development.

Current Issues in the Science Curriculum

Selected issues include the following:
- A less encyclopedic approach to the teaching of physics will be stressed, and more attention will be given to basic knowledge.
- In biology courses, a more experimental and functional approach will predominate over a systematic and descriptive approach.
- Biology will provide an introduction to an ecological view of the living world.
- The minimum equipment facilities for science classrooms will be stipulated.
- In all areas of science more attention will be given to the learning of concepts. Graphical solutions will be stressed, and more audiovisual means will be introduced.
- Although the use of computer applications in the diverse fields of science is still limited, attention will be given to technological developments in this area.

Science Textbooks

In the French-speaking community, as a rule, students usually work from hand-outs prepared by the teacher. Nevertheless, science textbooks are more often used in the private network than in the community one. Inside this network, this learning medium is more developed for the upper grades, particularly in chemistry and biology. In the Flemish community, students and teachers use textbooks and worksheets at all levels.

Science textbooks are developed in accordance with the intended curriculum. Compared to some twenty years ago, a greater variety of textbooks has been published. These new books usually contain more information, and layouts now include more graphical elements, photographs, and diagrams, as well as many exercises for students to practice. In the physical sciences, a recent trend is the use of manuals as guidelines for teaching practice.

Students have their own textbooks, which are either bought by the students themselves or leased from the school library. Textbooks are generally written by a group of specialized science teachers, with science inspectors as coauthors. The class teacher decides which textbooks will be used.

Pedagogy

General objectives in science pedagogy were developed at the same time as the learning objectives for students. As the sciences are taught as separate subjects, separate lists of objectives were developed. Current trends in science pedagogy include the following:

- Renewed attention is given to the development of concepts in all sciences.
- Cooperative learning is stressed through class discussion in all sciences.
- An effort is presently being made to integrate topics across the curricula so that an interdisciplinary approach is stimulated. Also, within certain science disciplines like biology, several subtopics like anatomy, physiology, and morphology are now taught as a unit.
- Laboratory work was recently made compulsory for chemistry and physics. In the teaching of physics there is a tradition of demonstration of experiments, but in chemistry the use of demonstrations is only now becoming important. Recently, in all the sciences, more emphasis has been placed on student-led investigations.
- In Belgium, students are accustomed to doing homework. In the sciences, this usually consists of problems.
- Science-technology-society is perceived as a very important area of study. In the chemistry and physics curricula teachers are expected to include technological applications. The chemistry curriculum states that the emphasis should be on the danger

and the safe use of chemicals and on waste problems. Ecology is an mandatory topic in biology.

6 EVALUATION POLICIES AND PRACTICES

In 95 percent of schools, examinations and the transfer from one grade to the next are school-based. Awarding certificates and qualifications also lies solely with the schools themselves, even qualifications for university entrance. There are no compulsory external examinations during either primary or secondary education. Each school organizes an autonomous examination for each subject in the curriculum, which is taken by all students every trimester or semester. During the year, results of formative tests that occur on an ongoing basis are used to determine the final grade. The instruments used in testing are constructed by the teacher and are nearly always open-ended.

Two optional external examinations are organized at the end of primary education: the district examination and the interdiocese examination. Some local boards also organize examinations for their students. At the end of secondary education there is no external examination. About 90 to 95 percent of students pass the final year of secondary education, and their diplomas are officially endorsed by government committees. The role of these committees is to verify that all topics included in the official curriculum were taught.

This system of examinations yields a greater variance in student achievement among schools than exists in the other European countries. Nevertheless, this variance cannot be related only to some instructional effects. Indeed, student characteristics, such as socioeconomic status, differ from one school to another.

At this time, the Flemish and French-speaking authorities are preparing lists of objectives to provide a clear picture of what pupils are to assimilate at school. The lists of objectives will make the aims of education clearer for parents and provide opportunities for general discussion as the options are outlined. They will provide tangible criteria for quality control and inspection by the authorities, and will provide teachers with a clear definition of what they should focus on in their teaching.

CONCLUDING REMARKS

On a recurrent basis, IEA surveys have revealed wide variations in pupil performance in Belgian schools in both mathematics and science. This is an important issue for policymakers, particularly when assessing evaluation policies and practices. It is expected the introduction of new standards in mathematics and science education, expressed as final objectives, will affect the actual interschool variance considerably in the future. The final objectives are statements of attainment targets that will have to be reached by the majority of students in specific levels and disciplines. All curricula will include the final objectives.

Although Belgium compared rather well internationally on mathematics in the past (SIMS), a careful analysis of the TIMSS data is necessary. Indeed, situated in the heart of Europe, Belgium cannot ignore international comparisons that might help avoid excessive disparities in regard to its partners. As the OECD examiners rightly expressed in their review of Belgian national policies on education, "Increasing labor mobility in Europe will require improved knowledge of the working of the educational systems"(OECD 1993).

Belgium did not take part in SISS. The Third International Mathematics and Science Study will undoubtedly provide interesting international perspectives for science education in Belgium. Indeed, we already know that for at least one structural feature of our education system, important differences can be observed in relation to the indicator "teaching time per subject" (OECD 1995). This indicator refers to the percent of the total available teaching time per subject category according to the intended curriculum (school year of reference: 1991–92) for data covering the secondary Grades 7 to 11. In comparison to other OECD countries, science in Belgium receives the smallest percentage of time, 6 percent, the OECD mean being 11 percent.

7 REFERENCES AND SOURCES FOR FURTHER READING

Common Sources

1   Husén T, N T Postlethwaite 1994 *International Encyclopedia of Education*. Pergamon Press, Oxford

2   Organisation for Economic Co-operation and Development 1995 *Education at a Glance: OECD Indicators*. Organisation for Economic Co-operation and Development, Paris

3   United Nations Educational, Scientific and Cultural Organization 1995 *Statistical Yearbook*. United Nations Educational, Scientific and Cultural Organization, Paris

4   World Bank 1995 *World Development Report 1995*. Oxford University Press, New York

Other Sources

5   Commission Européene 1995 *Les chiffres clés de l'éducation dans l'Union Européenne*. Luxembourg, E.U.

6   De Groof J, T Van Haver and others 1993 *De school of rapport: het Vlaams onderwijs in internationale context*. Pelckmans, Kapellen

7   Department for Educational Development 1994 *Objectives and Activities of the Department for Educational Development*. Department for Educational Development, Brussels

8   Dienst Voor Onderwijsontwikkeling 1993 *Voorstel eindtermen secundair onderwijs - eerste graad*. Ministerie van de Vlaamse Gemeenschap, Departement Onderwijs, Brussels

9   European Commission 1994 *Preschool education in the European Union—Studies no.6*. Office for Official Publications of the Euroform Communities, Luxembourg

10 European Commission 1995 *Education Training Youth: Key Data on Education in the Euroform Union 94*. Office for Official Publications of the European Communities, Luxembourg

11 Eurydice European Union 1993 *Administrative and Financial Responsabilities for Education and Training in the European Community*. Eurydice European Unit, Brussels

12 Eurydice European Union 1993 *Requirements for Entry to Higher Education in the European Community*. Eurydice European Unit, Brussels

13 Eurydice European Commission 1994 *Measures to Combat Failure at School: A Challenge for the Construction of Europe*. Office for Official Publications of the European Communities, Luxembourg

14 Koen M 1994 *Memento statistique de la Belgique*. Eds Labor

15 Ministère de l'education, de la Recherche et de la

Formation 1993 *Age du personnel 1991–1992*. Service des Statistiques, Bruxelles

16 Ministère de l'education, de la Recherche et de la Formation 1994 *Annuaire statistique 1992–1993*. Service des Statistiques, Bruxelles

17 Ministerie Van De Vlaamse Gemeenschap, Departement Onderwijs 1992 *Education in the Dutch-speaking Part of Belgium*. Centrum voor informatie en documentatie, Brussels

18 Ministerie Van De Vlaamse Gemeenschap, Departement Onderwijs, Bestuur Statistiek 1995 *Statinfo B.S.T. Schooljaar 1994–95*. Ministerie van de Vlaamse Gemeenschap, Brussel

19 Ministerie Van De Vlaamse Gemeenschap, Departement Onderwijs 1995 *Basisonderwijs: ontwikkelingsdoelen en eindtermen. Decretale tekst en uitgangspunten*. Ministerie van de Vlaamse Gemeenschap, Brussel

20 Nationaal Instituut Voor de Statistiek 1993 *Statistisch zakjaarboek*, Nationaal Instituut voor de Statistiek, Brussels

21 Organisation for Economic Co-operation and Development 1993 *Reviews of National Politics for Education—Belgium*. Organization for Economic Cooperation and Development, Paris

22 The Flemish Community 1994 *Educational Developments in Belgium 1992–1994. The Flemish Community, Report for UNESCO*. Centrum voor Informatie en Documentatie, Brussels

23 Van Damme J, A Van Beveren and others 1995 *Het Vlaams onderwijs in de kijker: een internationaal perspectief*. Ceuterick Leuven

Statistical References

Section	Statistic	Reference	Page	Table	Year of Statisitc
Country Profile	30.528	20	7	VI	1993
Country Profile	10,045	20	1	1	1992
Country Profile	$328/km^2$, $429/km^2$, $194/km^2$	20	7	VI	1993
Country Profile	0.16% (<1%)	4289	T26	–	1993
Country Profile	9%	20	19	IV	1991
Country Profile	18%	14	15	3.2.1.T.	1994
Country Profile	60%	20	19	IV	1991
Country Profile	25%	21	59	–	1992
Country Profile	15	21	60	–	1992
Country Profile	58%, 42%, 0.5% (<1%)	x7	–	–	1994
Country Profile	2%, 4%	4239	1	–	1980–1991
Country Profile	15 %	4259	11	–	1980
Country Profile	5,4 %	21	66	P1	1991
Structure of the System	20 %, 25 %	220	C01	–	1992
Structure of the System	40 %	10	35	E1	1990–1991
Structure of the System	BFr Table 1	16	165/451	–	1994

Continued

Statistical References continued

Section	Statistic	Reference	Page	Table	Year of Statisitc
Structure of the System	Bfl Table 1	18	14/47	1/2	1994–1995
Structure of the System	4%	13	–	–	1994
Schools in the System	31%, 15%, 17%	2	153	P06	1992
Schools in the System	182, 50, 28, 28–32, 28–36	10	96-98	1/2/3	1995
Schools in the System	14, 8	2	179	P32	1992
Teacher Profile	98%	2	194	P36	1992
Teacher Profile	62%	2	194	P36	1992
Teacher Profile	22%, 36%, 45%	16	567	–	1994
Teacher Profile	40	14	14/Sq	–	1994
Teacher Profile	US$1400, US$2050, US$1650, US$2600	(x)		–	

(x) La tribune (CGSP) mensuel novembre 1994.

ROMANIA
YUGOSLAVIA
BULGARIA
F.Y.R MACEDONIA
GREECE
TURKEY

Bulgaria

Kiril Bankov, Foundation for Research, Communication, Education, and Informatics (INCOBRA)

1 COUNTRY PROFILE

Bulgaria occupies about 111,000 km² of the Balkan peninsula in southeastern Europe. The 1994 census showed a total population of 8,427,418 people with a density of 76 persons/km². Two-thirds live in urban areas and the process of urbanization continues. The capital, Sofia, is home to 1.2 million people, and 48 towns of 20,000 or more mean an additional 52 percent of the population is concetrated in urban areas.

A decreasing birth rate and an increasing mortality rate have characterized Bulgaria's demographic situation for the past five years, resulting in negative annual population growth. Substantial emigration is a second factor in the lower population figures of recent years. In 1992, some 476,000 people left Bulgaria to live in other countries.

Native Bulgarians make up 86 percent of the country's population. The Turkish minority, which currently makes up 9 percent of the population, is decreasing in number principally because of emigration. Ethnic Romany make up an additional 4 percent of the population, a figure that has remained constant in recent years. Several other ethnic groups live in Bulgaria, but their number remains less than 1 percent of Bulgaria's total population. The country's official language is Bulgarian, which uses the Cyrillic alphabet.

Bulgaria is a parliamentary republic governed by a president and a prime minister. Both parliament and the president are elected by the population every four years.

According to World Bank data, the economy of Bulgaria is ranked as middle-income. During the 1990s it had a low per capita gross national product, US$1140, and gross domestic product, US$10,369 in 1993. Citizens with a university education make up 11 percent of Bulgaria's population, those with a college education, 4 percent, and those with secondary education, 38 percent. The adult literacy rate is 98 percent. The Ministry of Education, Science, and Technology spends approximately 3 percent of the state budget, or about US$105,414,000 annually.

2 THE EDUCATION SYSTEM

Governance and Decision Making

In Bulgaria, education is a state responsibility, and the Ministry of Education, Science, and Technology controls virtually all aspects of the system from elementary to the end of secondary school. Since 1990, policymakers at the Ministry have worked toward the decentralization of the system, but the process has been a slow one. Policies relating to curriculum, textbooks, and accreditation of schools are still set by the Ministry. Regional education councils, made up of civil servants appointed by the Ministry, are responsible for the administration of schools, the hiring of teachers and support staff, and for some of the financial support for local schools.

The intended curricula in mathematics and science are determined by the Ministry and are implemented in all schools across the nation. The Ministry publishes a curriculum guide that describes the content to be taught by grade level and course, accompanied by a statement of course goals and the intended learning outcomes. The development of textbooks is overseen by the Ministry, which establishes a competition in which both government and private corporations may participate. Of all the mathematics and science textbooks developed, two or three are usually approved by the Ministry. Some years ago, only one series of textbooks for mathematics and science existed. In recent years, however, several series have been produced and schools may choose the series they prefer. These new textbook series cover the same content by grade, but differ in the presentation of material.

Individual teachers and schools make decisions on instructional techniques and classroom processes used. Teaching methods are taught during teacher education courses offered by faculties of education within universities.

Structure of the System and Participation Rates

Structure of Education

There are three levels of education in Bulgaria, following a 4-4-4 model. Figure 1 shows the structure of education and the approximate enrollment rates. Primary school, Grades 1 to 4, is offered for students aged 6 to 10 years; junior secondary school, Grades 5 to 8, for students aged 11 to 14; and senior secondary school, Grades 9 to 12, for students aged 15 to 18. Children under 6 may attend a noncompulsory preschool, but these are not part of the state system. Most children enter Grade 1 in September if they are 6 years old at the end of the previous June. Some parents keep their children at home for an additional year if their seventh birthday falls in the following calendar year, and Grade 1 classes thus contain both 6- and 7-year-olds.

Some schools offer vocational education beginning in Grade 7 or 8. Some secondary schools, mainly technical and vocational ones, allow students to complete secondary education one year earlier than normal by offering short versions of courses. The shorter route is sometimes chosen by students with lower ability levels who nevertheless wish to receive a diploma for secondary education.

The minimum age for leaving school is 16, although this age does not correspond to a particular education level. Those who are not able or do not wish to continue their education may obtain vocational or technical qualifications during their compulsory period at school, at night school, or by distance education after leaving school. As of 1990, 96 percent of the age cohort was enrolled in primary school and 74 percent in secondary school.

Public and Private Systems

Almost all schools in Bulgaria are funded by the state. There are 31 registered, fee-charging private schools that receive no government funding and that are run by individuals or organizations. Private schools have been in existence only since 1990, and their numbers appear to be increasing.

Mathematics and Science Programs

Mathematics is a compulsory subject from Grades 1 to 12. Integrated science is compulsory until the end of Grade 5; from Grades 6 to 9, courses in physics, chemistry, and biology constitute the compulsory science curriculum. From Grades 10 to 12, when the sciences are no longer compulsory, enrollment drops from 100 percent to approximately 20 percent for physics, 15 percent for chemistry, and 10 percent for biology. The enrollment of male and female students is roughly equal during these years.

Schools in the System

The School Year

The school year in Bulgaria begins September 15 at all levels and ends May 30 for primary school, June 15 for junior secondary school, and June 30 for senior secondary school. The year ranges from 180 to 200 days and is divided into two terms: September to the end of January, and February to year-end. Schools are closed twice during the school year: for 14 days at the end of December and beginning of January, and for 10 days at the beginning of April.

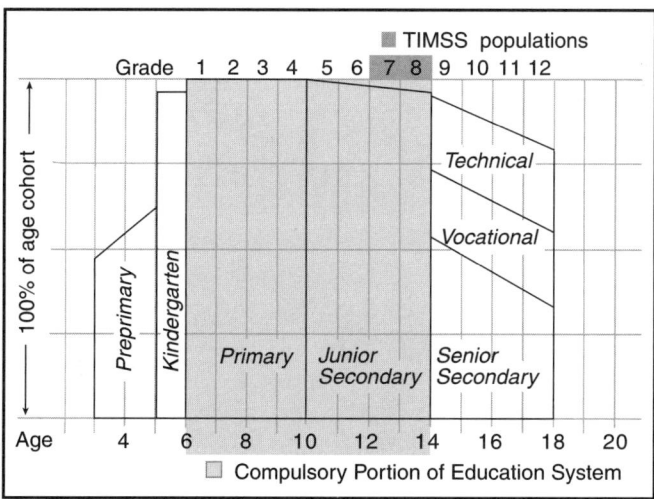

Figure 1: Structure of the Education System

Instructional Times

Students attend school five days a week, from Monday to Friday. The school day ranges from four hours of instructional time in primary school to six hours in senior secondary school. The amount of mathematics instruction ranges between 17 and 19 percent of the week's class time at all grade levels, while science instruction ranges from 3 percent to 22 percent per week.

The average class size in primary and junior secondary school ranges between 25 and 30 students, while in senior secondary school it ranges between 20 and 25.

Grouping in Mathematics and Science

There is little variation in educational opportunity among schools up to Grade 7. Mathematics is compulsory up to the end of secondary school. Science as a single, integrated subject is compulsory from Grades 2 to 5. Physics, chemistry, and biology are taught as separate, compulsory subjects from Grades 6 to 9 for students of all ability levels.

Schools offer elective courses in mathematics from Grades 4 to 12, and in science from Grades 6 to 12. These courses are chosen by students having a particular interest in the subject. After Grade 7, some students enter schools offering advanced programs in specific subjects such as mathematics, science, foreign languages, arts, and sports. About 30 senior secondary schools across the country specialize in mathematics and science, and these are highly sought after by students with both interest and ability in these courses. Candidates must write an entrance examination and compete with each other for places in these schools. Students entering the program choose to specialize in mathematics, physics, chemistry, or biology.

Certification of Teachers

Most primary school teachers are graduates of a teacher training institution or university, where they have completed two to three years of formal education and pedagogy. Teachers may teach all subjects or focus on two or three specialties.

Secondary school teachers must complete three years' study of a particular subject at a university, followed by two years of teacher education at the same faculty. This program is equivalent to a bachelor's degree and qualifies the candidate to teach any grade.

Primary school teachers are required to take two or three years of postsecondary education. Half this time is spent on the study of general pedagogy and the other half on the chosen specializations. Secondary school teachers complete three years of university education in subject matter, followed by two years of full-time study at the same faculty. During the final two years, preservice teachers focus on practice teaching, subject-specific courses, and methodology and pedagogy as applied to their chosen subject.

Upgrading and special training programs are offered by university departments of education and by teacher education institutions. Courses and workshops include seminars on specific subjects, study of the history and curriculum of the subject, computer training, and evaluation methods, and lead to qualifications equivalent to a master's degree. Few teachers receive a Ph.D.

Teacher Profile

The socioeconomic status of teachers in Bulgaria is relatively low. Their income is about 80 percent of the average salary of government employees, while employees in private sector companies earn two to three times the salary of teachers.

Most teachers at all levels of education are women. Although there has recently been a slight increase in the proportion of men teaching mathematics and science in secondary school, the majority of teachers of these subjects are female.

3   TIMSS POPULATIONS

Bulgaria, IEA, and TIMSS

Bulgaria has been a member of IEA since 1991, at which time it participated in the second stage of the Computer Education Study. Bulgaria has been participating in TIMSS since it began in 1991.

Bulgarian TIMSS Populations

Bulgaria is participating in TIMSS at the Population 2 level. The international definition for Population 2 includes Grades 7 and 8, which contain the majority of 13-year-olds at the time of testing. In Bulgaria, these grades are the last two grades of junior secondary school.

4 MATHEMATICS CURRICULUM AND PEDAGOGY

Goals for the Mathematics Curriculum

The intended curriculum in mathematics for Bulgaria is determined by the Ministry of Education, Science, and Technology. The Ministry appoints a group of experts from universities, schools, and education institutes to update the curriculum regularly and publish the revised document. Goals and content for the mathematics curriculum are described by grade in the guide. Grade-specific aims are included along with some general goals:

- Learn and apply mathematical facts.
- Train workers to be mathematically literate.
- Develop logical thinking and analysis skills.
- Become confident in the ability to present practical problems and to solve them mathematically.

Mathematics is taught as a single subject up to Grade 6, and from Grades 7 to 10 is divided into algebra and geometry. Analysis is taught in Grade 11 and a general mathematics course in Grade 12. Some additional mathematics may be studied as extracurricular topics, a practice that is widespread in some secondary schools. In the special schools for mathematics, students cover some additional topics, study the content at a more profound level, and spend a great deal of time problem solving.

Major Changes in the Mathematics Curriculum

For several decades the mathematics curriculum in Bulgaria was heavily influenced by those of the former socialist countries. The entire educational system, in fact, was closely modeled on and under the influence of the rest of eastern Europe. Since 1990, the mathematics curriculum has been slowly changing, coming closer to western European standards, while preserving within itself the unique cultural and social issues that have been its intellectual underpinnings for centuries.

The shift toward western European standards has influenced virtually every grade in Bulgarian schools. Since 1990, Grade 12 has been added to the secondary system, and the curriculum now introduces a number of topics in this new final year. The new curriculum has been implemented in schools since the 1992–93 school year, and the first-, second-, and third-graders of that year were the first students to follow it. Those in higher grades are finishing their mathematics education following the old curriculum. It is worth noting that at the time of the TIMSS main data collection, Population 2 students were still following the old curriculum, which was described in the Curriculum Analysis.

Use of calculators is encouraged and has increased somewhat, although class usually proceeds without their use. With the exception of a few schools, computer use is limited. In order to become better acquainted with computers, students have for the last decade been introduced to a new subject called "informatics." Unfortunately, because of a lack of equipment, only theory, supplemented by teacher demonstrations, is taught.

Current Issues in the Mathematics Curriculum

The trends described in the previous section are expected to continue for the next several years. More attention will be paid to the relationships between mathematics, other school subjects, and practical problems. Attempts will be made to make mathematics more interesting and more accessible to students.

A problematic issue is the search for modern assessment methods appropriate for mathematics education in Bulgaria. This has an impact on the accreditation of the students graduating from secondary school, particularly those in advanced mathematics programs. At this time, the Bulgarian secondary school diploma is not recognized in other countries, and it is thought that the use of different assessment methods might rectify that problem.

Other issues in the mathematics curriculum include:

- A new movement toward creating a "learning by doing" environment in the classroom, allowing students to do more problem solving.

- Maintaining traditional aspects of the curriculum, particularly as they relate to Bulgarian cultural and social issues. Euclidean geometry, for example, is emphasized in the curriculum at the expense of coordinate and transformational geometry.

Mathematics Textbooks

For many years textbooks have been the most important tool in the mathematics classroom, and they are used daily in both primary and secondary schools. Textbooks help teachers in preparing lessons, as well as providing teaching examples, and exercises, problems, and applications for students. Textbooks are, therefore, the vehicle of the intended curriculum.

Mathematics textbooks focus on different topics for each grade, as determined by the Ministry. They are usually written by teams of university professors and teachers, and are produced by government or private publishing companies. Up to Grade 8, each student has a mathematics textbook paid for by the government.

A new emphasis on color, pictures, and photographs makes primary school mathematics textbooks more attractive to the students. At the secondary level, textbooks are more formally laid out, and the writing style is often quite academic and the mathematical reasoning very advanced. Textbooks written for junior secondary school, in particular, are not geared to the students' ability level. Although these textbooks are professional from a mathematical point of view, students find them unreadable. Usually they do not like them and do not read them, using them only for exercises and problem solving.

When the new mathematics curriculum was instituted in 1992, new textbooks were needed. The shortcomings of the old textbooks were analyzed, and the ones most recently published for Grades 4, 5, and 6 are more attractive. The mathematical reasoning contained in them is better suited to the students, making the books more appealing to read.

During recent years two or three sets of textbooks per grade have been published. All textbooks receive the approval of the Ministry, and therefore conform to the curriculum guide in coverage of material. The manner of presenting the topics may differ from textbook to textbook. The mathematics textbooks to be used are chosen by the teacher, although many teachers use more than one textbook for reference purposes.

Pedagogy

The Ministry of Education, Science, and Technology has determined that up to Grade 6, mathematics instruction should be more empirical than theoretical. In these grades, classroom teaching is usually inductive and manipulative materials are used frequently. Computational techniques are strongly encouraged and play an important role in classroom processes. Although calculators are allowed, students must acquire computational skills without them.

Beginning in Grade 7, mathematics teaching is more theoretical and systematic, and deductive approaches form an increasing component of learning. Some algebraic computational skills must be acquired, but it is expected that the focus will be on developing mathematical concepts rather than on learning procedures and rules. Manipulative materials are not normally used in these grades.

Oral and written communication activities are expected at all levels. Problem solving is an important component of curriculum delivery, in contrast to open-ended projects and cooperative learning groups, which are seldom used. Much effort has been made to integrate topics within the mathematics curriculum, but little attention has been paid to integration with other school subjects.

For the last decade computers have been used in schools in a limited way. It is expected, however, that soon they will significantly influence pedagogy. Other educational technologies such as videodisk players, CD-ROM players, and computer networks are not yet used in schools.

5 SCIENCE CURRICULUM AND PEDAGOGY

Goals for the Science Curriculum

The Ministry of Education, Science, and Technology determines the intended curriculum in science for the whole country. The Ministry appoints a group of experts from universities, schools, and educational institutes to regularly update the curriculum and publish the revised document. Goals and content for the science curriculum are described by grade in the guide.

Science is taught as a single subject in Grades 3, 4, and 5, and the goal is to help students to develop knowledge of nature. Beginning in Grade 6, science is divided into three different subjects, physics, chemistry, and biology, which are compulsory for all students up to Grade 9 and elective in the upper grades.

A new framework for the science curriculum in Bulgaria, under development for the past two years and recently approved, elaborates two levels of science education at the secondary level. The first level includes compulsory courses in physics, chemistry, and biology. The goals are:

- to develop knowledge and a coherent understanding of the unity of nature;
- to create interest in and positive attitudes towards science as a basis for promoting careers in science;
- to develop students' understanding of the links between man and nature;
- to create the skills to study and use science.

The second level includes different elective science courses that offer an opportunity for profound content study. The goals of the second level are more concrete and often depend on the school profile.

Major Changes in the Science Curriculum

The new framework for the intended science curriculum places more emphasis on:

- increasing the proportion of common knowledge at the expense of theory;
- relating environmental issues to science;
- promoting careers in the sciences;
- developing the skills to use tools and do research while conducting investigations unaided by the teacher;
- providing the opportunity for each student to choose an elective course in science during the second level of science education, in contrast to the study of the three formerly mandatory subjects. This leads to better differentiation of students' interests and emphasizes individual capabilities;
- using teaching methods to develop scientific habits in thinking.

The use of audio and video materials has increased during recent years although these aids are available almost exclusively in urban areas. Computers have even less influence on the science curriculum because the

appropriate software and hardware are not available in schools.

Current Issues in the Science Curriculum

The new science curriculum requires a substantive change in teaching methods, and a new program is under development. The importance of the relationships between the different branches of science, physics, chemistry, and biology, as well as between science, technology, and society is now recognized.

It is expected that the use of computers will increase as software becomes more sophisticated and accessible. For this purpose, multifunctional software that allows learning and assessment at different levels is now being developed to complement each science course.

As discussed in Section 4, newer assessment methods are being developed for the mathematics curriculum. The same is being done for the science curriculum, and a particular area of concern is the accreditation of students taking advanced programs in science subjects.

Science Textbooks

Science textbooks focus on different topics for each grade, as determined by the Ministry. They are usually written by teams of university professors and teachers, and are produced by government or private publishing companies. The same team writes the corresponding teacher's guide. Up to Grade 8, each student has a science textbook paid for by the government.

As with mathematics textbooks, new science textbooks have appeared in recent years, but many of the old problems remain. Although the use of illustrations has increased, these visual aids are not sufficient. Dry, textual information still predominates, and students are often bored by the material. Furthermore, there are not enough problems and experiments covered in the books.

Two or three new sets of textbooks per grade have recently been published. These textbooks all conform with the curriculum guide in covering the same material by grade and differ only in the presentation of topics. Science textbooks are chosen by the teacher, although many teachers use more than one textbook for reference purposes.

Pedagogy

Science in Grades 3, 4, and 5 focuses on acquainting students with the laws of nature and is taught inductively, based on concrete examples and experiments demonstrated by the teacher. A greater emphasis is placed on formal scientific knowledge beginning in the upper grades. From Grade 6 onward, when physics, chemistry, and biology are taught as separate subjects, classroom teaching is more deductive than inductive. Laboratory work forms an important component of the curriculum in these grades.

Students are expected to communicate both orally and in writing in all grades, although written activities have increased in recent years because of the use of laboratory-based activities. Working in small groups is not widely done. More attention must be paid to the development of scientific skills and thinking, even though teachers are advised to use approaches that promote scientific habits of mind.

6 EVALUATION POLICIES AND PRACTICES

National Examinations

There are two major school leaving examinations in Bulgaria. One is administered at the end of Grade 8, the final year of compulsory education, and tests mathematics and Bulgarian. Results from this examination may be used to qualify applicants to senior secondary schools with entrance requirements, such as schools with a large number of applicants. The second national examination is administered at the end of Grade 12. All students write the compulsory sections on mathematics and literature, and they choose two other subjects as well. Results from the second examination are used to rank students for entry into universities and other institutions of higher education.

The mathematics component of both examinations is written. Students are given four mathematical problems and four hours in which to solve them. They take the form of extended free-response and open-ended problems, and there is no short-response or multiple-choice component to the examination. The examinations are written at the end of the school year, on the same day across the country. The Ministry of Educa-

tion, Science, and Technology prepares these examinations as well as the curriculum and there is, consequently, a high level of correlation between the two.

Schools offering advanced programs in specific subjects administer entrance examinations in mathematics and Bulgarian. Students taking these tests are not exempt from the national examinations. The format of the mathematics entrance examination is similar to that of the nationals: students are given four hours to solve four curriculum-based problems.

Classroom-Level Evaluation

At the classroom level, schools administer both formative and summative assessment. The goal of formative assessment is to collect information about the learning process on a continuing basis: how students understand concepts, what skills they have acquired, and what the common difficulties are. Interpreting this information enables teachers to keep abreast of the needs of the class. The goal of summative assessment is to evaluate knowledge and skills at specific stages of the learning process. This information is used by teachers to form final grades at the end of the school year.

7 REFERENCES AND SOURCES FOR FURTHER READING

Common Sources

1 Husén T, N T Postlethwaite 1994 *International Encyclopedia of Education*. Pergamon Press, Oxford

2 Organisation for Economic Co-operation and Development 1995 *Education at a Glance: OECD Indicators*. Organisation for Economic Co-operation and Development, Paris

3 United Nations Educational, Scientific and Cultural Organization 1995 *Statistical Yearbook*. United Nations Educational, Scientific and Cultural Organization, Paris

4 World Bank 1995 *World Development Report 1995*. Oxford University Press, New York

Other Sources

5 National Statistical Institute of Bulgaria 1995 *Statistical Reference Book*. Statistical Press, Sofia (in Bulgarian)

6 Bulgarian Parliament 1995 *An Official Gazette*. No. 46, 19 May 1995 Sofia (in Bulgarian)

7 National Statistical Institute of Bulgaria 1994 *Results of the Census of the Population and Housing*. Vol. 1 Demographic characteristics. Sofia (in Bulgarian)

8 Ministry of Education and Science 1992 *White Book for the Bulgarian Education and Science*. University Press "St. Kliment Ohridski," Sofia (in Bulgarian)

9 Ministry of Education, Science, and Technology 1994 *Mathematics Curriculum*. Sofia (in Bulgarian)

10 Ministry of Education, Science, and Technology 1994 *Physics Curriculum*. Sofia (in Bulgarian)

11 Ministry of Education, Science and Technology 1994 Chemistry Curriculum. Sofia (in Bulgarian)

12 Ministry of Education, Science, and Technology 1994 *Biology Curriculum*. Sofia (in Bulgarian)

13 Ministry of Education, Science, and Technology 1994 *Geography Curriculum*. Sofia (in Bulgarian)

Statistical References

Section	Statistic	Reference	Page	Table	Year of Statistic
Country Profile	111,000 km^2	2	239	1	1991
Country Profile	8,427,418	5	27 XIII	1	1994
Country Profile	2/3, 52%	7	–	–	1992
Country Profile	476,000	7	XII	–	1992
Country Profile	86% 4%	7	XXVI	1	1992
Country Profile	9%	7	XXVI	11	1992
Country Profile	US$1140	2	162	1	1993
Country Profile	US$10,369	2	166	3	1993
Country Profile	11%, 4%, 38%, 98%	7	XXI	8	1992
Country Profile	3% US$105,414 (original in Bulgarian currency)	6	4	–	1995
Structure of the System	96%, 74%	2	295	29	1990
Structure of the System	31	5	49	11	1994
Schools in the System	17%-19% 3%-22%	8	21	–	1992

Canada

Alan R. Taylor, University of British Columbia

1 COUNTRY PROFILE

Canada occupies more than half of the North American continent, including a large proportion of the land north of the forty-ninth parallel. On the east and west it is bounded by the Atlantic and Pacific oceans respectively. It is the second largest country in the world, covering an area of almost 1,000,000 km². However, Canada is not densely populated. In 1991, there were fewer than 3 persons/km², the majority of whom live in urban centers located within 200 km of the border of the United States.

Many different ethnic groups make up Canada's population. Aboriginal people, the first inhabitants, have contributed significantly to the exploration and culture of the country.[1] The first Europeans to settle in Canada left a dual heritage of French and English languages, both of which have official status. In addition to its aboriginal population and the descendants of the original European immigrants, Canada has experienced a steady flow of immigration from other countries. In the 1986 census, 25 percent of Canadians reported ethnic origins other than aboriginal, British, or French. This is a significant factor in education, particularly in urban areas, which tend to absorb the majority of new immigrants.

Canada is a confederation of 10 provinces and two territories. It is a parliamentary democracy with its citizens represented in an elected House of Commons. The government is headed by a Prime Minister who is the leader of the political party in power. Areas of government jurisdiction are divided between the federal government and the provinces. Federal powers include jurisdiction over areas such as defense, external affairs, monetary and fiscal matters, fisheries, and energy. Provinces and territories are responsible for matters such as education, social services, health, forestry, and highways.

A founding member of OECD, Canada also is a member of the G7 Economic Group of countries. The World Bank rates the Canadian economy as high income; its per capita gross national product for 1993 was US$19,970, with an average annual growth rate of just over 1 percent between 1980 and 1993. The average annual rate of inflation during this period was under 4 percent.

Canadians place a high priority on education; spending on education is the second largest public expenditure, at 20 percent of all public expenditures in 1992–93. Spending on education relative to its economy is higher than other G7 countries. According to OECD, Canada's spending on education as a percentage of GDP was over 7 percent in 1992, somewhat above the mean of other OECD countries at 5 percent. In 1992–93, expenditures on education reached US$30 billion, having doubled in the previous decade. The adult literacy rate in 1992 was greater than 95 percent.

Canada is a country of great diversity in geography, political organization, and cultural makeup. All of these factors have had an impact on the structure and nature of its education system. It is within this context that the main features of the education system are discussed next.

---

[1] In the early 1990s, the number of aboriginal people in Canada was almost one million. The breakdown, as calculated by the First Nations peoples themselves, includes the following: 315,000 status Indians living on reserves, 200,000 status Indians not living on reserves, 150,000 Métis, 250,000 nonstatus Indians, and 35,000 Inuit. There are no verifiable estimates of how many aboriginal people lived in Canada at the time of the arrival of the first Europeans.

2 THE EDUCATION SYSTEM

Governance and Decision Making

In Canada, education is a provincial or territorial responsibility. Each province controls all aspects of the education system up to the end of secondary school.[2] Policies relating to curriculum, teacher certification, accreditation of schools, and the reporting of student progress are set at the provincial or territorial level. School boards set local policy within parameters defined by the Ministry; decisions about the operation of schools, implementation of curriculum, and the hiring of teachers and support staff are made at this level. The intended curriculum in mathematics and science is defined at the provincial or territorial level. The Ministry of Education produces curriculum guides outlining intended learning outcomes by grade level and course. Most guides include a philosophy and rationale for the teaching of mathematics and science, a description of the content to be taught, intended learning outcomes, and time allocations for each subject. These guides are consistent with educational policy in each jurisdiction and are commonly developed by teams of teachers under the direction of the Ministry of Education.

Textbooks and other curriculum support materials are reviewed first at the provincial or territorial level; a list of recommended materials is then supplied to the districts. Currently, there is a trend away from a single textbook for each subject toward a variety of materials from which to choose. Decisions about which materials to use in classrooms are made at the district or school level.

Decisions about instructional techniques and other classroom processes are left to individual teachers and schools. Teaching methods can be influenced at the school district and school levels through the provision of in-service training and the development of corre-

sponding support materials.

There has been a recent trend within most jurisdictions to decentralize decision making in the area of curriculum resourcing. This has resulted in a greater array of support materials becoming available to teachers and schools. Teachers now use of a wide variety of print, video, software, and manipulative materials in mathematics and science instruction.

Structure of the System and Participation Rates

Structure of Education

The public education system is divided into two levels: elementary and secondary. The two stages of education are in turn subdivided into two components: primary and intermediate at the elementary level, and junior and senior high school at the secondary level. Compulsory education in Canada begins at age 6; most provinces offer kindergarten beginning at age 5 as an option, but preschool programs involving younger students are typically not part of the public school system. The minimum age for leaving school is either 15 or 16 depending on the jurisdiction. Figure 1 shows the structure of the education system, as well as aproximate enrollment rates.

The primary level includes kindergarten and Grades 1 to 3. Typical age ranges for this level are 5 to 9. Ages of students at the intermediate level, Grades 4 to 6 or 7, typically range from 9 to 12 or 13. In 1992, 98 percent of all children of elementary school age attended school.

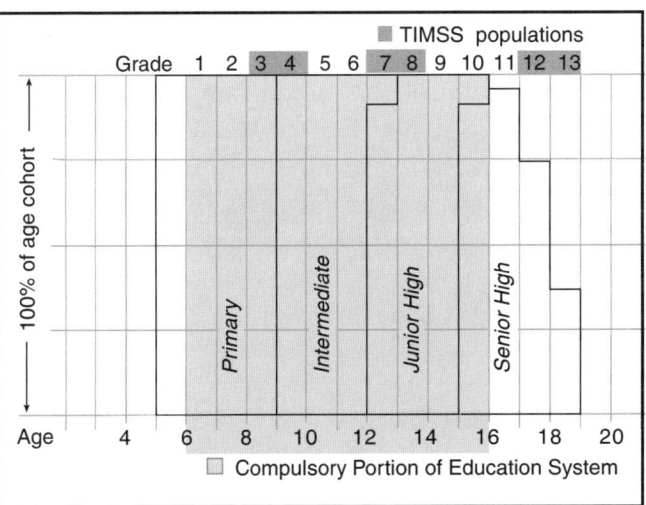

Figure 1: Structure of the Education System

---

[2] The ministers of education from each province, however, have established a national secretariat, the Council of Ministers of Education of Canada, to ensure communication on issues such as funding, programs, and student assessment. From time to time the Council undertakes national projects in curriculum and assessment.

Junior high school includes Grades 7 to 9 in all jurisdictions except British Columbia, where it covers Grades 8 to 10. Senior high school includes Grades 10 to 12 in most jurisdictions. Exceptions are Quebec, where it ends at Grade 11, and British Columbia, where it includes only Grades 11 and 12. As of 1992, 90 percent of 12- to 17-year-olds were enrolled in school.

Educational opportunities in the public school system go beyond elementary and secondary school. Students can complete courses or programs in several ways, including night school, correspondence education, and distance learning through the use of telecommunications.

School Types

Canada offers a comprehensive school system in which students can attend schools running both academic and general programs. In most cases, students at the secondary level planning to proceed to postsecondary studies or planning to find employment attend the same school.

Most students are enrolled in public schools, which are funded through taxation revenues. These resources are intended to cover all capital and operating costs. There has been increasing interest in private schools in recent years, generated by some parent groups concerned about traditional values and a perceived decline in standards in the public system. Private or independent schools in most provinces receive only partial funding from the government.

Schools in the System

The School Year

The school year in Canada ranges from 180 to 200 days, beginning the day after the first Monday in September and continuing for 10 months, ending in the third or fourth week of June. Schools in most jurisdictions close for approximately two weeks at the end of December and one week in March. In addition to these holidays, there are several national or local holidays on which schools are closed between September and June.

Instructional Times

Students attend school from Monday to Friday. Elementary schools are typically in session for 30 hours per week. Of this, 23 to 24 hours are spent on instruction,

and the rest divided between daily opening exercises, lunch, and recess. At the secondary level, schools are typically in session for between 30 and 35 hours each week. A typical school day for an elementary school student would be from 9:00 in the morning until 3:00 in the afternoon. Secondary school students have a longer day, often from 8:30 in the morning until 3:30 in the afternoon.

Class Size

There is no typical size for a Canadian classroom. Class size may range from under 10 students to over 40. In Quebec, the average elementary class was 27 and the average secondary class was 30 for the year 1993–94. In British Columbia in 1992–93, the average elementary class size was 23 students, while the average secondary class size was 25 students. In both provinces, these sizes have varied little over the past 10 years.

Streaming in Mathematics and Science

Mathematics and science are compulsory up to Grade 9 or 10; in some provinces this requirement includes Grade 11. Students at the elementary level study a compulsory program of mathematics and science. Elective courses usually begin in junior high school for mathematics and senior high school for science. Students choose which courses they wish to take, often with the assistance of a school guidance counselor.

Two mathematics courses are most frequently offered in junior high school, one with an academic focus and the other of a more general nature. In Grades 11 and 12, additional courses are offered; these typically focus on precalculus topics for academic students and business, consumer, and trade applications for others. General science courses are usually offered up to Grade 10. In the senior years they are differentiated among disciplines such as biology, chemistry, physics, and earth science.

The concept of streaming is a debatable issue, tied closely to the goals and purposes of the public school system. It has both proponents and detractors across the country, both of whom base their claims on philosophical and pragmatic perspectives. On the one hand, supporters claim streaming can better meet the needs of students through ability groupings, where bright students proceed at a faster pace and slower students more successfully meet expectations geared to their ability. Those opposed argue that ability groupings will result in lower expectations and achievement in groups of slower students. In addition, the opportunity for all students to proceed in areas of further study will be diminished.

Certification of Teachers

Teachers in Canada follow one of two certification programs: a four-year university course, usually completing a bachelor of arts or science degree, plus one year of teacher education, or a bachelor's degree followed by two years of teacher education. Teacher education is conducted by faculties of education within universities, and the Ministry or Department of Education grants certificates to successful candidates.

In order to qualify as an elementary school teacher, a candidate usually must complete between four and six years of postsecondary training. Most of the time spent on training in pedagogy focuses on general applications across curricular areas. Although some universities require at least one course in mathematics methodology, others do not.

At the secondary level, requirements for teacher certification include between 5 and 6 years of postsecondary training. The number of required pedagogy courses specific to mathematics or science normally range from none to one or two. However, teachers of secondary mathematics or science are expected to have an academic major in the area, which includes completion of at least three or four subject-matter courses.

In-service development and upgrading involve additional training after teachers are practicing in the field. These are usually done through courses offered by departments of education, school boards, or faculties of education. Courses and workshops are also offered by teachers' associations and other professional organizations of educational and administrative officials.

Although additional training is not mandatory, it is essential in many cases due to curriculum changes and reassignment of teaching subjects. Areas such as second language instruction, computer training, and multicultural orientation are common subjects of study. Other areas of focus include alternative teaching strategies such as cooperative learning and the expansion of student evaluation methods such as observational techniques, criterion-referenced evaluation, student-parent-teacher conferences, and portfolio assessment.

Incentives for teachers to remain current in their specialty areas or to prepare for changes in assignments include more than just professional expectation or obligation. They frequently involve higher certification levels with corresponding salary increases and potential promotion to department head or other school administrative positions.

Teacher Profile

Teachers are well paid in comparison with other government and business employees but their income lags behind that of self-employed professionals. The average after-tax income of teachers and professors in Canada in 1991 was US$26,129, whereas federal, provincial and municipal government employees received between US$20,182 and US$20,850. Self-employed professionals such as doctors, dentists, lawyers, and accountants earned considerably more than teachers; teachers earned only 41, 46, 51, and 66 percent of the after-tax income respectively of each of these professional groups. The average age of teachers is increasing. In 1972–73, the average teacher was 34; in 1981–82, 38; and in 1992–93, 42. The proportion of teachers 45 years of age or older increased from 20 percent in 1972–73 to 23 percent in 1981–82 to 41 percent in 1992–93.

Given a continuation of this trend, there will be a serious shortage of teachers in the future, as many currently teaching reach retirement age. It is projected that a large number of teachers will be retiring or nearing retirement in 2001–2010; 45 percent of the 1992–93 national teacher workforce could be in this category. This issue is compounded further by a stagnant employment market for teachers in many regions of the country. As a result, many teachers new to the profession are unable to get permanent positions and others are either leaving or not attempting to enter the profession.

Women constitute the majority of full-time teachers in Canada, at 61 percent. However, women are less likely than men to teach mathematics and science. In Ontario in 1994–95, 39 percent of secondary school mathematics teachers and 34 percent of secondary school science teachers were women. In British Columbia, only 23 percent of secondary school mathematics teachers were women in 1992–93. In the same year, only 22 percent of secondary school science teachers were women. While these figures are low, they show a marked increase from 1986-87, when 15 and 13 percent respectively were women.

Given the need to encourage more girls to enroll in mathematics and science courses at the senior level, it is important that the trend toward more female teachers in this area increase. It is expected that more girls would be encouraged to study in these areas with female teachers as role models. Perhaps this can be accomplished through some provision of incentive for women to teach mathematics and science.

3   TIMSS POPULATIONS

Canada, IEA, and TIMSS

British Columbia, Ontario, and Quebec (inactive) have been members of IEA since 1981, at which time the first two participated in the Second International Mathematics Study. Canada is participating in TIMSS at all population levels. In addition, five provinces (British Columbia, Alberta, Ontario, New Brunswick, and Newfoundland) are participating independently in one or more populations.

Canadian TIMSS Populations

In Canada, the international definition for Population 1 includes Grades 3 and 4, where the latter is the target grade in the two-grade design. Population 2 includes Grades 7 and 8. These grades are both part of the secondary school system in all provinces and territories except British Columbia, where Grade 7 is part of the elementary program.

In all provinces except Ontario and Quebec, Population 3 corresponds to Grade 12. In Ontario, some students complete secondary schooling at the end of Grade 12, whereas others continue for an extra year to complete the Ontario Academic Credits necessary for admission to university. Students in Quebec continue from Grade 11 to either a one- or a two-year training program prior to entry into tertiary education or the workplace.

Mathematics and physics specialists were defined as subsets of the general population for each province and territory. Students enrolled in mathematics or phys-

ics courses that prepared them for postsecondary study were identified as specialists. This definition identified approximately 20 percent of Population 3 students as mathematics specialists and 10 to 15 percent as physics specialists.

4   MATHEMATICS CURRICULUM AND PEDAGOGY

Goals for the Mathematics Curriculum

Mathematics curricula across Canada are similar in many respects. Common purposes include topics covered by grade and course level, and a new focus on active learning and mathematical processes. These similarities are partly the result of efforts of the Council of Ministers of Education. A second factor in the development of consistency across curricula is textbook publishers; these often supply several jurisdictions and for reasons of economy publish materials that satisfy more than one province.

Goals for mathematics are contained in curriculum guides published by each province. They reflect the rationale behind the structure and content of the curriculum. Goals in mathematics common to most provinces and territories are similar to those contained in *Standards for School Mathematics*, published by the National Council of Teachers of Mathematics. These include, at the student level, learning to value mathematics, becoming confident in the ability to do mathematics, becoming mathematical problem solvers, communicating mathematically, and reasoning mathematically. Goals at the system level, on the other hand, include producing mathematically literate workers, promoting lifelong learning, and providing opportunity for all. The British Columbia curriculum guide, for example, lists the following goals: the provision of mathematics to develop skills in logical analysis and to present problem solutions in a clear and precise manner, and to provide the mathematics necessary to function in society, engage in lifelong learning, and pursue further formal study in mathematically related areas.

Major Changes in the Mathematics Curriculum

Major shifts in the approach to the teaching of mathematics have occurred in most provinces and territories during the past 10 years. These have included a movement toward a more interactive learning environment in which students actively engage in open-ended tasks involving hands-on learning. Direct instruction still plays an important role in the delivery of the mathematics curriculum, as does the use of open investigations. Greater attention is also paid to problem solving and the use of manipulative materials.

Curriculum changes are resulting in some mathematical material being taught at an earlier age. Decimal fractions, for example, are now introduced at earlier grade levels with a corresponding de-emphasis on common fractions. A more noticeable shift has been in the area of data analysis, which has become a major strand across all levels of mathematics curricula, beginning in elementary school. Other areas of emphasis at the elementary level include concept formation, estimation skills, and transformational geometry. At the secondary level, introductory calculus and geometric proof have become more important in recent years.

Electronic aids such as calculators and computers have increased in use among mathematics classrooms during the past 10 years. At the elementary level, the use of calculators has increased significantly. Areas of greatest application at this level are checking answers, exploring number patterns, and problem solving. The use of scientific calculators at the secondary level has resulted in little or no need for mathematical tables such as powers of numbers, nth roots, logarithms, and trigonometric functions.

Calculator use at both levels of the system has changed the nature of mathematical applications. In problem solving, for example, the emphasis has been on real-world applications in which numbers do not usually fall into neat patterns that factor or calculate easily. In addition, the increased use of graphing calculators has the potential to revolutionize the teaching and learning of graphing concepts and applications.

Computers have, as yet, limited usage in the classroom. However, as computer software improves, teachers are making greater use of demonstrations with overhead viewers. This is particularly true in examining patterns of numbers or objects and in graphing. The use of computers for remedial instruction is also gaining in popularity as more sophisticated programs enter the marketplace.

During the past 10 years, researchers have shown a great deal of interest in active learning, where the student is directly involved in the process or activity under investigation. Consequently, greater attention has been paid to how students approach problems, what procedures they use, and how they communicate results. As a result, less reliance is placed on traditional paper and pencil tests in the design of many research projects, with greater attention paid to observational techniques in collecting information.

Classroom processes have also received a great deal of attention. There is greater interest, for example, in how teachers organize their classrooms, how they group students, and what methodologies they use, than in their backgrounds and training.

These shifts in attention should result in the identification of variables accounting for significant amounts of variance in achievement. Given findings from such studies, further direction could be gained in providing more effective learning environments.

Current Issues in the Mathematics Curriculum

Several recent trends in content, focus, and methodology will continue or gain momentum over the next five years. Among these are continued emphasis on problem solving and real-world applications in mathematics.

It is expected that the curriculum will become more integrated as real-world applications gain in prevalence and interrelationships among the disciplines become more apparent. This focus on relevance will change the role of the teacher from that of an information source to one of a facilitator, as students are directed to alternative sources of information.

As a more pronounced focus on real-world applications in mathematics is implemented, students should view the subject with greater relevance and appeal. An improvement in attitude and in the acquisition of research skills will likely result.

Mathematics Textbooks

As a result of the shift away from reliance on a single

Table 1: Major Topics in Typical Mathematics Textbooks

Population 1	Population 2	Population 3
Operations: whole numbers	Common/Decimal Fractions	Patterns, Relations, Functions
Meaning: whole numbers	Percentages	Equations and Inequalities
Common Fractions	Integers	Vectors
Measurement Units	Perimeter, Area, Volume	Change
Perimeter and Area	2-D Geometry	1-, 2-D Geometry
3-D Geometry	Transformations	3-D Geometry
Data Representation	Equations & Inequalities	Transformations
Basic Geometry	Data Representation	Infinite Processes

textbook in mathematics, to the use of a variety of curriculum materials, teachers now select from print and video resources as well as textbooks for use in their classrooms. Many schools also utilize technology in curriculum support through applications of new computer software and access to international databases.

Mathematics textbooks have changed significantly in format and approach during the past 10 years. They now contain more pictures and diagrams, and color is used more frequently. Noticeable attempts are also made to enhance the relevance of topics through examples of real-world applications. In developing concepts, greater attention is paid to phenomena and relationships among them; formerly, textbooks relied on the routine application of algorithms or long and involved factual explanations. There is also greater focus on careers in mathematics, "hands-on" activities, problem solving, and applications. Many textbooks contain themes or applications of special interest such as relationships between tasks associated with jobs or consumer activities, and specific concepts in mathematics such as graphing or probability models.

Textbooks for mathematics focus on different topics at each population level. The major topics included in typical mathematics textbooks are shown in Table 1.

Pedagogy

Practices defined in curriculum guides mandate that mathematics be organized in separate courses at each level of the system, in particular in Populations 2 and 3. It is expected, however, that within each subject, topics will be integrated. The focus is expected to be on the development of concepts rather than on proce-

dure or rules taught by rote. Extensive calculator use is required at all levels, particularly at Population 3. The use of computers, while encouraged, is not expected. Manipulatives are to be used extensively in Population 1, in contrast to practice at Population 2, where little use is expected. Methods of curriculum delivery such as open-ended investigations, cooperative learning groups, and communication activities are expected at all levels.

A direct consequence of interactive learning has been a shift in pedagogical approach. Although instruction remains an important means of delivery, teachers are making greater use of hands-on activities, manipulatives, and small group instruction. Cooperative learning, for example, is now employed in many mathematics classrooms. In addition, projects and assignments make greater use of community involvement, where students apply mathematical knowledge and procedures to issues around them.

5 SCIENCE CURRICULUM AND PEDAGOGY

Goals for the Science Curriculum

Common goals for science are depicted in the publication *Science for Every Student: Educating Canadians for Tomorrow's World*, published by the Science Council of Canada. In this report, the Council states four broad aims for all students:

- to encourage full participation in a technological society;
- to enable further study in science and technology;

- to facilitate entry into the world of work;
- to promote intellectual and moral development of individuals.

These broad aims or goals are reflected in recent science curricula through increased emphasis on understanding scientific processes and principles, conceptual understandings approached from a phenomenological basis, integration within the sciences, a focus on the impact of technology on society with respect to environmental and resource issues, and the promotion of science-related careers.

Major Changes in the Science Curriculum

There has been a continued emphasis on both science processes and on knowledge and skills during the past 10 years. The focus on processes such as observing, classifying, inferring, and communicating has begun at lower grade levels, moving towards a greater emphasis on knowledge and skills in the upper grades.

A shift has occurred, however, in the content and method of delivery of the science curriculum, grounded in changes in philosophy and the changing needs of society. They have included the following:

- understanding scientific processes and principles;
- solving problems rather than memorizing facts and performing traditional laboratories;
- conceptual understanding approached from a phenomenological rather than a mathematical basis;
- integration within the sciences;
- integration of sciences within other subject areas;
- issues surrounding the impact of science and technology on society; particularly with respect to environmental and resource issues;
- teaching from constructivist principles;
- issues of gender equity and cultural bias.

There has been general acceptance of the use of calculators and computers by students and teachers of science at all levels of the system. Use of these aids has facilitated better understanding of principles, as students do not need to spend inordinate amounts of time on tedious calculations.

The availability and use of audio, video, and software materials have increased significantly. Many of these aids are offered as integrated packages and focus on emerging areas such as environmental issues and resource management.

Technology has also had an impact on the nature of courses offered in science. For example, new Science and Technology courses have been developed in several provinces for use at the secondary level.

Current Issues in the Science Curriculum

The use of electronic aids will grow further as software becomes more sophisticated and the power of computers increases to accommodate it. There will also be greater use of a variety of information sources as information processing skills become essential to manage rapid growth in technical and scientific information.

Through the use of technology such as interactive video, students can instantly visit locations of particular interest and interact with individuals located offsite. These benefits will do much to bring the real world into the science classroom and facilitate interactive learning.

Science Textbooks

Many of the recent changes in format, use of graphics, and applications in the real world reported for mathematics textbooks are similar to those found in science curriculum support materials. In addition, there is an emphasis on how science and technology interact with society, addressing goals related to their impact on the environment and the utilization of resources. As in mathematics, greater attention is paid to phenomena and relationships among them with a focus on careers in science.

Science topics contained in textbooks also differ by TIMSS population. Major topics included in typical textbooks are shown in Table 2.

Pedagogy

Approaches to the teaching of science are similar to those for mathematics in most areas. Among the differences is a greater focus on the integrated use of computers, particularly in Populations 2 and 3. In addition, the use of hands-on materials is significantly greater in Populations 2 and 3 science than in math-

Table 2: Major Topics in Typical Science Textbooks

Population 1	Population 2	Population 3
Plant Fungi Types	Rocks and Soil	Subatomic Particles
Animal Types	Animal Types	Energy Types
Organs, Tissues	Classifying Matter	Wave Phenomena
Physical Properties	Energy Types	Light
Life Cycles	Heat and Temperature	Electricity
Sound & Vibration	Physical Change	Magnetism
	Biomes, Ecosystems	Time, Space, Motion
	Science & Technology in Society	Dynamics of Motion
	Scientific Method	

ematics. This difference is due to the use of laboratory activities in science at these levels.

Interactive learning, involving process-focused student activities, has gained in popularity in recent years. This trend has been due largely to findings from research on effective learning environments and to advances in technology resulting in an increased awareness of the importance of information processing skills and the use of simulations and modeling techniques for problem solving.

A major shift is also under way in the teacher's role in the teaching-learning process. Student access to information on external databases, such as the Internet, and the availability of more effective software programs for computer-assisted instruction have turned the teacher into more of a facilitator of the learning environment instead of a disseminator of information. Given this change, there is need to redefine roles and responsibilities of teachers and students. The student, for example, will need to take greater responsibility for learning, while the teacher must develop more effective means of monitoring and evaluating student performance.

6 EVALUATION POLICIES AND PRACTICES

Purpose and Goals of Assessment

At the classroom level there are two main purposes for assessment and evaluation: formative and summative. Formative assessment involves the systematic collection of information during the learning process and identifies which concepts students understand and which ones they are having difficulty with. Based on the interpretation of this information, teachers can adjust their methodology to meet the needs of individual students or the class. At the summative stage, teachers evaluate how well students understand and can demonstrate their knowledge and skills at the completion of an educational sequence. Information at this stage provides students with an opportunity to demonstrate their knowledge and understanding, and forms the basis for reporting to students and parents.

Assessment Practices

Student assessment practices at the classroom level are similar across all jurisdictions in Canada. Teachers make extensive use of paper and pencil tests, class projects, and individual assignments in the assessment and evaluation of student achievement. Examples in textbooks and exemplars in curriculum guides have considerable effect on the type and nature of questions asked on teacher-made quizzes and examinations.

Due to the shift toward more active learning in many classrooms, teachers are beginning to expand upon traditional assessment procedures. For example, the use of observational techniques is becoming more important as students become directly engaged in the learning process. Using this approach, teachers learn which specific student behaviors are important in both individual and group learning situations, how to document this information, and how to interpret it. Other important techniques gaining in popularity include portfolio assessment, self-assessment, and peer assessment.

Assessment and Testing Programs

Involvement in external assessments differs among the provinces and territories. At the program level, most if not all jurisdictions participate in selected international and national assessments such as TIMSS and the School Achievement Indicators Project, sponsored by the Council of Ministers of Education of Canada.

Several provinces supplement national and international assessments with provincial program reviews. Five provinces administer program level assessments in mathematics and science every three or four years. These reviews collect information on the achievement, background, and attitudes of students, as well as on the perceptions, backgrounds, and classroom practices of teachers. They involve all students at Grades 3, 6, and 9 in four of the provinces, and in Grades 4, 7, and 10 in the fifth. Saskatchewan performs provincial learning assessments in mathematics in Grades 5, 8, and 11 every two years.

System-wide examinations at the end of secondary school are administered in core subjects in four of the provinces. Results from these examinations are combined with assigned marks to determine the final standing of students. In British Columbia, required examinations in academic subjects account for 40 percent of a student's final mark. Saskatchewan, on the other hand, requires an examination in academic subjects only if the respective teacher is not accredited in the subject.

Anticipated Changes

The trend towards performance assessment, or assessment of students in active learning situations, is gaining momentum and will likely continue to do so. However, the use of such broadly based measures will continue to be challenged by demands for system-wide accountability by the public and politicians. Central testing programs are the obvious answer to this; existing programs will likely be maintained and grow. These will include continued involvement in both national and international assessment programs.

CONCLUDING REMARKS

The diverse nature of Canada's physical geography and the many cultures of its people have provided a unique challenge to its education system. It has succeeded in providing relatively consistent and high-quality education programs, in spite of servicing large regions of great expanse and dealing with issues of governance associated with a number of different jurisdictions.

These accomplishments are evidenced through the similar content, focus, and pedagogical approaches employed in mathematics and science education across the country. In fact, it could be contended that mathematics and science programs are more alike than different, across provinces and territories in Canada.

Some credit for this consistency could be claimed by the effects of restricted numbers of textbooks, due to the small market available to publishers. Other factors affecting this result could be attributed to the curriculum and assessment initiatives undertaken by the Council of Ministers of Education.

As a result of changes in focus and in pedagogy, it is likely that students will benefit though gaining a more comprehensive understanding of concepts and their applications. Conceptual understanding should be enhanced through increased use of manipulative materials and hands-on activities, improvements in attitudes and interest result from more relevant applications in everyday life, and problem-solving and research skills enhanced through more effective use of technological aids.

Canada intends to benefit in a number of ways from its participation in TIMSS. It is expected that direction for curriculum planning and development, and in classroom organization and procedures, will be gained through an examination of practices in other jurisdictions. Further, it is expected that the opportunity to examine relationships between these practices and student achievement and attitudes will provide further insight into the teaching and learning process.

Given many common issues faced in mathematics and science education across country jurisdictions, it will be helpful to examine those most relevant to Canada within an international context. Among the issues of common interest are those involving funding; the structure and governance of educational institutions; decision making for policy, curriculum, and instruction; and standards and accountability.

Another opportunity of particular interest in Canada is the ability to examine within-country differences in curriculum, textbooks, and student achievement and attitudes. Since several provinces participated at a country level, the sample for Canada involved oversampling in five jurisdictions. Given the nature of the sample, Canada will be able to examine similarities and differences both across countries, and across several provincial jurisdictions within Canada as a whole.

7 REFERENCES AND SOURCES FOR FURTHER READING

Common Sources

1 Husén T, N T Postlethwaite 1994 *International Encyclopedia of Education*. Pergamon Press, Oxford

2 Organisation for Economic Co-operation and Development 1995 *Education at a Glance: OECD Indicators*. Organisation for Economic Co-operation and Development, Paris

3 United Nations Educational, Scientific and Cultural Organization 1995 *Statistical Yearbook*. United Nations Educational, Scientific and Cultural Organization, Paris

4 World Bank 1995 *World Development Report 1995*. Oxford University Press, New York

Other Sources

5 Bateson D 1993 *Science Curriculum Expert Questionnaire for Canada*. Third International Mathematics and Science Study, University of British Columbia, Vancouver.

6 Burgess W D April, 1990 New Mathematics is an essential part of new technology. *Ontario Mathematics Gazette*. 28(3): 6–7

7 Canadian Teachers' Federation 1994 *Economic Service Bulletin*.

8 Carrodus J F 1993 Co-operative learning: Transforming the math classroom. *Vector*. 35(1): 48–51

9 Gayfer M 1991 *An Overview of Canadian Education* Canadian Education Association, Toronto.

10 Hunter D September, 1989 Where are we and where do we go from here? *Ontario Mathematics Gazette*. 28 (1): 6–10

11 Liedtke W Winter, 1990 Mathematics curriculum changes—implications for diagnosis and remediation. *Vector*. 31 (1): 12–16

12 Maloney J April, 1993 Specialization years committee comments. *Ontario Mathematics Gazette*. 31(3): 10–11

13 Ministers of Education 1991 *Secondary Education in Canada. A Student Transfer Guide, 6th Edition*. Toronto.

14 Putz B Winter/Summer, 1991 Elementary mathematics in the 90s. *Saskatchewan Mathematics Teachers' Society Journal*. 27(3): 37–39

15 Sherrill J 1991 *Participation Questionnaire Parts I and II*. Third International Mathematics and Science Study, University of British Columbia, Vancouver

16 Statistics Canada 1993 *Canada Yearbook 1994*, Ottawa

17 Statistics Canada 1992 *Education in Canada: A Statistical Review for 1991–92*, Ottawa

18 Statistics Canada 1992 *Advance Statistics of Education*, Ottawa

19 Statistics Canada 1995 *A Statistical Portrait of Elementary and Secondary Education in Canada*. Ottawa

20 Taylor A 1993 *Mathematics Curriculum Expert Questionnaire for Canada*. Third International Mathematics and Science Study, University of British Columbia, Vancouver

Statistical References

Section	Statistic	Reference	Page	Table	Year of Statistic
Country Profile	1,000,000 km^2, 1%, US$19,970, 4%	4	163	1	1993
Country Profile	7%, 5%	2	73	F01	1992
Country Profile	US$30 billion	3	4.8	4.1	1992
Country Profile	95%	4	162	1	1992
Structure of the System	98%, 90%	3	3-35	3.2	1992
Teacher Profile	US$26,129, US$20,182, US$20,850, 41%, 46%, 51%, 66%	7	15	4	1991
Teacher Profile	61%	19	28	4.2	1993
Teacher Profile	34 38 42 45 20% 23% 41% 45%	19	28	4.2	1972–73 1981–82 1992–93 1972–73 1972–73 1981–82 1992–93 1992–93

Colombia

Carlos J. Díaz, Universidad del Valle

Efrain Solarte, Universidad del Valle

Jorge Arce, Universidad del Valle

1 COUNTRY PROFILE

Colombia lies in the northwest corner of South America, bordered by Panama, Venezuela, Brazil, Peru, Ecuador, the Atlantic Ocean to the north and the Pacific Ocean to the west. It occupies an area of 1,139,000 km² and is the fourth largest country in South America. Although Colombia straddles the equator, it has a varied climate ranging from very hot at sea level to very cold at the high altitudes of the Andes.

The 1993 census showed a population of nearly 33 million, with a growth rate that had decreased to less than 2 percent per annum. Seventy-three percent of the population lives in urban areas, while 27 percent lives in rural areas. The average population density is about 29 persons/km², with most of the population concentrated in the Andean mountain range and on the Caribbean coast. The literacy rate is about 86 percent.

When the northwest part of South America was conquered by the Spanish in the sixteenth century, small tribes of indigenous peoples were scattered throughout the country. Today less than 3 percent of the population is descended from aborigines, and most live in small rural communities. Most of the population is a mixture of Spaniard and aborigine, locally known as *mestizo*. The official language, Spanish, is spoken by 99 percent of the population, although there are over 180 languages and dialects of Indian origin that are official languages in their own territories. In the Caribbean islands of San Andreas and Providence, about 60,000 people speak an English dialect. There are small communities of German, French, and Anglo-Saxon people; but, in general, immigration to Colombia has been low. Emigration has been high for the past 25 years, with most emigrants choosing to live in North America or Europe.

Colombia is a decentralized democratic republic with several autonomous regions and a constitution that guarantees basic human rights for all. The government is a constitutional democracy led by an elected president and a two-chamber congress. Since 1992 the governors of the provinces have also been elected, as have city and town mayors since 1989.

The World Bank considers Colombia a middle- to low-income country. The per capita GNP was US$1900 in 1994, from a total estimated GNP of US$64 billion and an expected annual growth rate of 5 percent in 1995. The inflation rate has remained in the low twenties for the past decade, with an expected annual rate of inflation of 19 percent for 1995. Twenty-five percent of the economy is based on agriculture, with coffee and bananas as the main export crops. Oil exports have increased significantly in the past few years and are expected to grow in the future. Exports of manufactured goods have increased, with textiles and leather as the leading products.

Government spending on education has increased slightly at the national level. As a percent of the GNP, spending on education has grown from 2.85 percent in the 1970s to 2.99 percent in the 1980s, 3.07 percent in 1993, and 3.65 percent in 1994. It was 4.04 percent for 1995 and is projected to be 4.88 percent by 1998. These spending levels are lower than in other Latin American countries at a similar stage of development. The distribution of spending within the levels of education in 1994 shows 31 percent of the total education budget allocated for primary education, 28 percent for secondary education, and 19 percent for higher or university education. The other 22 percent is spent on other forms of education and in administration of the system. The private sector contributes significantly to expenditure on education, with some sources estimating that as much as an additional 30 to 35 percent of the budget is contributed by the private sector.

2 THE EDUCATION SYSTEM

Governance and Decision Making

The education system in Colombia is still being developed to the level and coverage desirable by a modern society that demands quality and relevant education for all. Colombia's central government includes a Ministry of Education that controls all aspects of the education system from primary to tertiary. Policies relating to planning, development, curriculum, evaluation, the ranking system of teachers, school accreditation, quality control, and government expenditure are set by the Ministry of Education. The system is in transition from a highly centralized to a decentralized one giving autonomy to the regions and to the schools in defining the educational needs and goals of their people. The education secretaries of cities and towns now carry out many administrative duties and have increased decision-making powers. In addition, provincial secretaries of education now carry out some functions relating to the supervision and control of the system.

The education system is composed of the Ministry of Education, its specialized branches, and all public education institutions. There are four main subsystems in the education system: education, culture, recreation and sports, and science and technology. Each has a complex administrative structure with specialized functions, and all operate under the control of the Ministry of Education.

The intended curriculum in mathematics and science is determined at the national level. The Ministry of Education provides curriculum guides that outline the expected outcomes by grade and subject matter. Most guides include the philosophy and rationale behind the development of the curriculum, general and specific objectives, content, indicators of evaluation, and methodological advice.

The 1993 General Law of Education proposed a new organization of the education system as regards decision making and evaluation. The new law changes the role of the Ministry of Education in determining the curriculum: the Ministry now provides an overview of curriculum goals for further development at the school level. The law also emphasizes the importance of the Institutional Education Plan, which comprises diagnostics, planning, programming (including curriculum design, contents, instructional methods, activities, assessment, and resources), organizational structure, execution, control and supervision, and evaluation. These ideas are in the process of implementation and have had little impact as of yet.

Textbooks and other curriculum support materials are selected by individual schools, either by the school council, the principal, or the teachers' committee. Because there are no official textbooks, publishing companies offer a wide variety of books for schools to choose from. Teachers determine the pedagogical method and form of instruction, occasionally guided by a school coordinator or department head. There is little or no advice or orientation from professional or teachers' associations.

Structure of the System and Participation Rates

Structure of Education

The structure of the school system in Colombia spans 11 years, following a 5-4-2 pattern. Children begin Grade 1 at six years of age and attend five years of primary school, four of lower secondary, and two of upper secondary. Noncompulsory preschool education exists in the form of a one-year kindergarten program for five-year-olds.

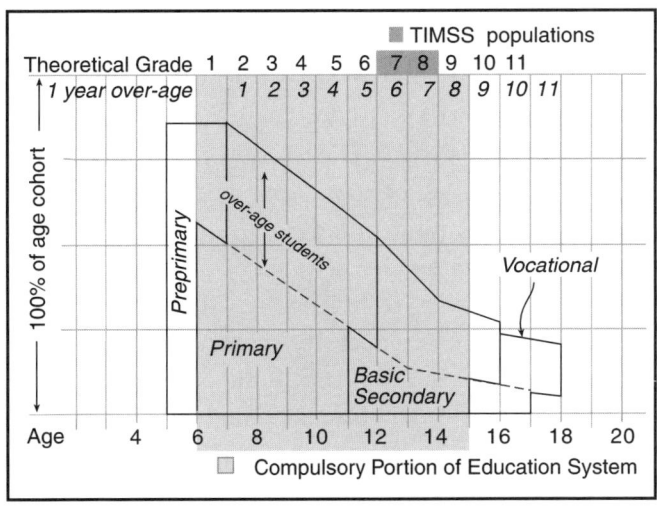

Figure 1: Structure of the Education System

Participation Rates

Coverage of the education system is still low, and the illiteracy rate is still high, at 14 percent. Although the constitution requires nine years of compulsory schooling, the system does not have the capacity in either human or material resources to meet the current demand. In 1995 the coverage for primary education was 86 percent, with a high rate of drop-outs. As of 1993, the percent of over-age students in schools was also high: 26 percent of children are seven years old when they enter Grade 1 and 7 percent are over eight.

Figure 1 shows the education pyramid in Colombia as the number of children registered in school in 1994. The vertical axis also shows the theoretical age of the children in each grade. For every 100 students beginning primary school, only 59 complete it, and only 30 complete the compulsory Grade 9. The goal of the development plan is to increase coverage to 78 percent and 55 percent, respectively, by 1998.

Public and Private Systems

In Colombia, private education is an important part of the system. In 1992, 18 percent of primary school children and 37 percent of secondary school students attended private schools. Funding for public schools comes from the central government's allocation budget, contributions from provincial, city, or town budgets, and from registration and monthly fees paid by parents. Private education is financed by parent contributions, tuition, and fees. Although the constitution affirms compulsory education financed by the state up to Grade 9, it also suggests that parents and the private sector contribute as well.

Mathematics and Science Programs

In Colombia most schools are coeducational, and male and female students attend the same mathematics and science classes. Mathematics and science courses are diversified beginning in Grades 10 and 11, and students in these grades choose to study the science, mathematics, humanities, technical, vocational or pedagogical modality. In each modality, science and mathematics vary in content, level, and hours of instruction per week. More males than females register in technical education, while more females than males register in business education.

School Types

Elementary schools offer kindergarten to Grade 5, while secondary schools provide Grades 6 to 11. Some schools provide the entire range from kindergarten to Grade 11, but most of these are private schools. Secondary schools may be academic, technical, or vocational in nature. Up to Grade 5, students typically attend school in self-contained classrooms and are taught by one teacher, although specialists may teach physical education, music, art, or science.

Schools in the System

The School Year

For historic and climatic reasons, the education system in Colombia has two operating calendars. School calendar A corresponds to the school year of the southern hemisphere, running from the first week of February to the third week of November and divided into two semesters of 20 weeks. School calendar B corresponds to the school year of the northern hemisphere and runs from the first week of September to the third week of June with two semesters of 20 weeks. Eighty-five percent of schools are on Calendar A and 15 percent are on Calendar B. All schools operate five days a week, Monday to Friday, for a total of 200 days a year, of which about 10 days are religious or national holidays. Teachers and administrators spend an additional two weeks planning at the beginning of the school year and two weeks evaluating at the end.

The school day varies from six to eight classroom periods, with each period lasting 45 minutes. In primary schools, students spend about 30 hours a week in school, while in secondary schools students may attend up to 40 hours of classes per week. To increase efficient use of facilities, some schools operate in shifts, with a morning period from 7:00 a.m. to 12:45 p.m., an afternoon period from 1:00 p.m. to 6:30 p.m., and an evening period from 6:30 p.m. to 9:00 p.m. Many teachers choose to teach more than one session per day as a means of supplementing their low salaries.

Class Size

Class sizes vary widely. Primary school classes accommodate 30 to 35 students, while secondary school classes accommodate 40 to 50. Private schools usually have somewhat smaller class sizes.

Streaming and Tracking

There is no policy for within-school tracking, and schools are not tracked. In Grades 1 to 9 all students take mathematics and general science courses. Students in Grades 10 and 11 choose from a range of options, some of which emphasize mathematics or science. There are no restrictions regarding the selection of options.

Certification of Teachers

In Colombia, teacher education is quite varied, as are the requirements to enter the official teacher's ranking of the Ministry of Education. Teachers are ranked at Levels 1 through 14 according to academic background and teaching experience. Secondary school graduates enter the ranking at Level 1, while college or university graduates in education enter at Level 7. In order to rise to Level 14, teachers must have at least 15 years' teaching experience and have written a textbook, or have a graduate degree in education. In 1995, the annual starting salary was US$2800 for Level 1, progressing to US$9800 for Level 14. Some regions of the country have difficulty hiring qualified teachers because of climatic or health reasons, violence in the area, or perceived isolation from the rest of the country. The new Education Law stipulates that, beginning in 1996, all primary and secondary school teachers must have a License in Education.

Before the passing of the 1993 General Law of Education, there were no requirements for becoming a kindergarten teacher, although there are institutions that provide one to two years of special training after high school for kindergarten teachers. Primary school teachers may complete their preparation in secondary school. After completing the nine years of basic schooling, they may follow either a three-year program at a teacher education school, or a two-year program with pedagogical emphasis at a vocational high school. These candidates enter the teachers' ranking at Level 1 and teach are all subjects. Primary school teachers receive instruction in methodology and pedagogy as well as subjects such as science and mathematics.

Over the past 10 years the system has changed. Many schools of education now offer a four-year program after high school to qualify primary school teachers in two main areas: mathematics and natural sciences, and language and social sciences. Upon completion they receive the License in Education degree, that is, a certificate with an emphasis on primary education, enabling them to enter the ranking at Level 7. These teachers are well prepared not only in science and mathematics, but in theoretical courses such as sociology, psychology, cognitive sciences, curriculum development, and pedagogical practice.

Secondary school teachers must follow a four-year preparation program at schools of education within a university to obtain a License in Education. This qualification involves a specialization in mathematics and physics, or biology and chemistry. They enter the teacher's ranking at Level 7.

Other secondary school teachers obtain a college degree in a scientific or technical area, such as mathematics, physics, or engineering. Following this, they must take one year of preparation in science or mathematics teaching, a master's degree in education, or pedagogical training to be hired as high school teachers. In general, secondary school teachers have substantial training in science and mathematics education, special education, psychology, sociology and curriculum development courses. All secondary school teachers must practice teaching for one year under the supervision of an experienced teacher.

Continuing education and in-service training are required to rise in the ranking system. Courses may include areas other than those in mathematics and science; in recent years new teaching and learning techniques have been introduced, as well as new programs in the use of computers in education.

Teacher Profile

Compared to other professions in Colombia, the socioeconomic status of teachers is not high and their salaries are lower than other government, business, or industry employees. Salaries are assigned according to the public education teacher's ranking system, which has 14 levels. Teachers are positioned in the ranking by academic background, years of experience, in-service training courses, and productivity. In 1995, the annual salary scale started at US$2800 and progressed to a maximum of US$9800 per year. The mean salary for 1995 was estimated to be approximately US$5600 a year. Private school teachers are ranked in the same manner, but most are paid at a rate that is approximately 80 percent of the public scale.

Most kindergarten and primary school teachers are female. In secondary schools the ratio of male to female teachers is greater than 1 to 1. In science and mathematics education, however, the male to female ratio is greater than 3 to 1.

Plans for the expansion of education at the primary and secondary levels may lead to an increased demand for teachers within the next few years. Rural schools currently experience difficulties in recruiting teachers in spite of economic incentives to work outside cities.

3 TIMSS POPULATIONS

Colombia, IEA, and TIMSS

Colombia is not a member of IEA. Its participation in TIMSS stems from the interest of other Latin American countries, including Mexico and Argentina. Funding for the study in Colombia comes from the Ministry of Education, and the work has been carried out at the Universidad del Valle in Cali.

Colombian TIMSS Populations

While participating in TIMSS, Colombia encountered a number of difficulties not usually found in developed countries. Some of them derive from the lack of current, accurate statistics on the education system; others derive from the over-age phenomenon of children in our schools. A national census conducted in 1993 still had not released official results late in 1995, and was in any case not accurate enough to give approximate figures for levels of coverage in education. Ministry of Education statistics are not complete in the number, location, and size of all schools in the country, creating particular difficulties in defining the target populations for TIMSS.

Colombia is participating in TIMSS at the Population 2 level only, which includes students in Grades 7 and 8. The latter grade is the target and includes children in the 12- to 14-year-old age group. Both grades belong to lower secondary school. In the target grades for Population 2, 31 percent of students are one year older than the age 13 target, 23 percent are two years older, and 22 percent are three or more years older.

Only 24 percent of students in the target grades were 13 years old at the time of testing, with the high repetition rate being the principal reason for this situation.

4 MATHEMATICS CURRICULUM AND PEDAGOGY

Goals for the Mathematics Curriculum

Goals for mathematics are set out in curriculum guides and include a theoretical frame of reference for the development of mathematics curriculum, the structure and organization of the content, and the methodology and the specific development of each unit for each grade.

The curriculum guide states that the main objectives of mathematics education are that the students:

- develop the capacity for logic and critical thinking;
- develop the skills needed to make generalizations;
- develop skills in using algebraic and geometric procedures;
- develop the ability to communicate in the language and symbolic representation of mathematics;
- become familiar with the basic concepts of mathematics;
- be able to interpret reality using mathematical models;
- develop a variety of approaches for solving problems in mathematics, or involving mathematics in science, technology, and everyday life;
- recognize the contribution of mathematics to the development of science and technology, and the improvement of life.

Major Changes in the Mathematics Curriculum

In the early 1980s, a revised curriculum promoting changes in the organization and development of the mathematics curriculum was initiated. Mathematics concepts are now developed by system, meaning that the curriculum for all grades includes numeric, geometric, metric, data, and logic systems. Numeric systems, for example, include the study of integers as a set of numbers, the operations {+ , *}, and the relationships: {<, ≤ , >, ≥}.

Emphasis has been placed on a shift in teaching methods from a traditional "chalk and talk" method to a hands-on approach with more active participation on the part of students. Concept formation, drill and practice, geometric manipulatives, problem solving, and projects are practices recommended by the curriculum guides.

Current Issues in the Mathematics Curriculum

A radical change is occurring in the process of curriculum development, as outlined in the new education law, Law 115. The law states that every school in the country must have an institutional education plan that includes curriculum design, content, instructional methods, activities, assessment, resources, and evaluation guidelines. The planning process is achieved with the participation of school administrators, teachers, parents, and the community. Expectations about the effects of these changes are high, particularly in regard to the quality and relevance of education for children.

For mathematics, the Ministry of Education has published a booklet giving a general orientation for curriculum development, methodology, and evaluation, but no core curriculum has yet been defined. It will take some years before schools are in a position to formulate their mathematics curricula, and during that time the Ministry of Education's old curriculum guides will be used.

Mathematics Textbooks

The curriculum redesign that changed the approach to developing the mathematics curriculum effected a change in textbook organization and presentation. Colombia does not have official textbooks, but authors must follow the curriculum guides in organization and content. Textbooks are produced by publishing companies, with about 10 companies dominating the market. Textbooks are neither recommended nor officially approved by the Ministry of Education, but must compete for adoption at the school level based on their quality and the promotional skill of the publisher.

Elementary school mathematics textbooks have recently undergone some changes. Many now contain more graphics and some now contain exercises to be solved in the books themselves. Some textbooks reproduce Ministry of Education objectives and evaluation guidelines as an orientation for teachers, and this is often the only information about the official curriculum that teachers receive. Textbooks, more than the curriculum guides, are the tool used by teachers to implement the curriculum in the classroom. Although textbooks all follow the content stipulated by the Ministry guides, there are differences in the presentation, depth, and teaching methodology.

The Ministry of Education provides a limited quantity of free mathematics textbooks to public elementary school students in Grades 1 to 5. Priority is given to children in schools located in rural areas, regions experiencing violence, and populous suburban regions.

Pedagogy

The curriculum guides suggest a methodology for the teaching and learning of mathematics, supported by teaching and learning practices centered around an active learning approach. The methodology stresses the importance of concept formation and the application of learned concepts in different contexts. To this end, hands-on experiences, working in groups, problem solving, and manipulative materials (for geometry) are suggested as new methods in teaching mathematics.

The new education law stresses the importance of developing a school curriculum that integrates the school into the community; that is, that takes into account school context and environmental, economic, and social variables.

The curriculum guides suggest the use of both inductive and deductive methods to develop logical thinking and reasoning. Students are expected to achieve higher thinking and reasoning, and to develop independence in studying, learning, and problem solving.

Calculators are widely used in schools, particularly in upper secondary school where calculations with logarithms, exponentials, nth roots, and trigonometric functions are commonly performed with calculators. Computer use is more limited; some private schools have introduced computers into the classroom, but very few public schools own computer equipment. A Ministry of Education program exists to develop information technology for education, but this material has been little used as of yet.

5 SCIENCE CURRICULUM AND PEDAGOGY

Goals for the Science Curriculum

Goals for natural sciences and health are outlined in the curriculum guides, which include a theoretical framework for the development of the science and health curriculum, the structure and organization of the content, the methodology, and the specific development of the units for each grade. The curriculum guide states that the main objectives of the science and health curricula are that the students:

- develop a scientific view of the world around them, enabling them to investigate and interact with nature, make observations, perform experiments, draw conclusions, and explain answers;
- be aware of protecting the environment and the preservation and proper use of natural resources;
- recognize that humans as living things are part of a system involving themselves and the environment, preserving a biological and social equilibrium for well-being and good health;
- identify the influence of the interactions between humans as social beings and nature that contribute to the transformation or preservation of the environment;
- value the health of the population as an indicator of the quality of life;
- use scientific knowledge and methods for solving problems;
- be aware of the evolution of scientific knowledge;
- understand the role of science and technology in the economic and social growth of the country;
- develop logic and skills in critical thinking.

Major Changes in the Curriculum

In 1980, the official curriculum guides promoted a change in the development of the science curriculum, known as the curriculum renewal. The science curriculum is organized into two main, unifying concepts, the structure of the universe and interactions, which encompass the contents of the Natural Sciences and Health courses. The structure of universe includes matter, energy, space, time, and all systems composing this structure. Interactions include change, conservation, and the evolution of the systems.

Emphasis has been placed on a shift in teaching methods, from a traditional "chalk and talk" method to a hands-on approach with more active participation on the part of students. As a result of the new method, students are able to develop skills in observation, the use of materials, and the conducting of small-scale investigations. Curriculum guides contain many suggestions for such practical activities.

Current Issues in the Science Curriculum

The 1993 education law, Law 115, radically changed the development process of the science curriculum. The law states that every school in the country must develop an institutional education plan, which must include curricular design, content, instructional methods, activities, assessment, resources, and evaluation guidelines. The planning process is achieved with the participation of school administrators, teachers, parents, and the community.

The Ministry of Education has published a booklet giving a general orientation for science curriculum development, methodology, and evaluation, but a core curriculum has yet to be defined. A new emphasis has been placed on ecology and environmental education. It will likely take some years before schools are able to formulate their own science curricula, and until then the old Ministry of Education guides will be used.

Science Textbooks

As with mathematics textbooks, science textbooks are neither recommended nor officially approved by the Ministry of Education, but are adopted at the school level. The curriculum renovation that changed the approach to curriculum development also effected a change in textbook organization and presentation. A major change has been the inclusion of practical activities using materials, equipment, and measuring devices. Elementary school textbooks are more pictorial, and some contain out-of-school activities for exploring the natural environment and for solving problems.

Pedagogy

The curriculum guides suggest a methodology for the teaching and learning of science, supported by practices centered on student activity. Methodology must stress the importance of concept formation and the development of investigative skills in order to conduct small investigations and to apply the acquired concepts in a different context. Hands-on experiences, group work, problem solving, and the handling of materials and equipment are suggested as the new trends in teaching science.

Calculator use is widespread in upper secondary schools, and is common at other levels. Computer use is more limited; while some private schools are introducing them into the classroom, public schools in general do not have the equipment. Formal laboratory equipment is used in many schools, although there is considerable variation in the quality of this equipment.

6 EVALUATION POLICIES AND PRACTICES

The evaluation policies issued by the Ministry of Education (Resolution 17486, 1984) establish the parameters and forms of evaluation for both the learning process and the promotion of students to the next grade. This norm, however, does not include curriculum evaluation or the evaluation of school activities as a whole, nor does it give a particular orientation to evaluation in science and mathematics.

From kindergarten to Grade 3, children are automatically promoted to the following grade, except in exceptional circumstances. For Grades 4 to 11, academic achievement is charted during four periods of the school year, and this information is used to promote students to the following grade. Evaluation instruments are defined by individual teachers, and typical tests in both mathematics and science include both short-answer questions and problem solving.

Secondary schools administer entrance examinations to Grade 6 students as a mechanism for selecting appropriate candidates for admission. A lack of space and limited resources preclude the admission of all students who apply.

There is a national or state examination at the end of secondary school that is used by some universities

as an admission criterion. The examination is also intended to be an indicator of the academic quality of the schools. This examination covers mathematics, science, history, geography, humanities, civics, and an optional foreign language.

The new education law promotes new forms of evaluation. The Institutional Education Plan must include evaluation of all school activities, including curriculum development. The Ministry of Education is developing quality indicators for science and mathematics education, and plans to establish specific attainment targets for certain grades.

CONCLUDING REMARKS

Colombia's participation in TIMSS is an important step in the process of detecting the deficiencies of the education system. Accurate and relevant data on our education system are vital if decisions that will solve Colombia's education problems are to be made. The international comparisons may not be of great relevance in this context. What we hope to learn from the study is the extent of the difference between the intended curriculum set out by the Ministry of Education and the attained curriculum learned by the student, identifying as many factors in the delivery of the implemented curriculum as possible. TIMSS will provide us with a rich source of data to investigate the factors that are affecting the quality and efficiency of science and mathematics teaching in Colombia.

7 REFERENCES AND SOURCES FOR FURTHER READING

Common Sources

1   Husén T, N T Postlethwaite 1994 *International Encyclopedia of Education*. Pergamon Press, Oxford

2   Organisation for Economic Co-operation and Development 1995 *Education at a Glance: OECD Indicators*. Organisation for Economic Co-operation and Development, Paris

3   United Nations Educational, Scientific and Cultural Organization 1995 *Statistical Yearbook*. United Nations Educational, Scientific and Cultural Organization, Paris

4 World Bank 1995 *World Development Report 1995.* Oxford University Press, New York

Other Sources

5 Departamento Nacional de Estadística, DNP 1995 *1993 Census Tables,* Santafé de Bogotá

6 Ministerio de Educación Nacional 1995 *El Salto Educativo. (Education, the Country's Development Axis.)* Colombia

7 Ministerio de Educación Nacional de Colombia 1994 *Sistemas Educativos Nacionales—Colombia.* Organización de Estados Iberoamericanos, para la Educación, la Ciencia y la Cultura (OEI)

8 Ministry of Education 1994 *Resolution 8700.* Colombia

9 Ministry of Education Secretaría Técnica 1992 *Dirección de Servicios Técnicos, National Statistics, Teachers by Docent Ranking and Sex.*

10 Ministerio de Educación Nacional 1986–1991 *General Frameworks and Curriculum Programs, Grades 1 to 9, Mathematics and Natural Sciences.*

11 Ministerio de Educación Nacional de Colombia 1993 *General Frameworks: Natural Sciences and Environmental Education.*

12 Ministry of Education 1984 Resolution 17486 Colombia

13 Diario Oficial 1993 *General Law of Education, Law 115.* Santafé de Bogotá

Statistical References

Section	Statistic	Reference	Page	Table	Year of Statistic
Country Profile	33 million, 73%, 29/km^2, 86%	5	diskette		1993
Country Profile	2.85%, 2.99%, 3.07%, 3.65%, 4.04%, 4.88% 31%, 28%, 19%, 22%	6	41	6	1995
The Education System	structure of system	13			1993
Structure of the System	26%, 7%	13	12	3	1995
Structure of the System	59%, 30%, 78%, 55%	13	12	3	1995
Teacher Profile	1:1, 3:1	9			1992

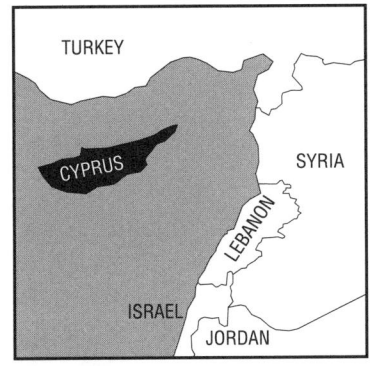

Cyprus

Constantinos Papanastasiou, University of Cyprus

1 COUNTRY PROFILE

Cyprus is the third largest island in the Mediterranean, covering an area of 9251 km², of which 1733 km² are forested. The maximum length of the island is 226 km, and it is 98 km wide at the widest point. It is situated 380 km north of Egypt, 105 km west of Syria, 75 km south of Turkey, and 380 km east of the nearest Greek island, Rhodes.

In 1994, the population of Cyprus was 729,800, compared with 722,800 in the previous year, an increase of 1 percent. The ethnic composition of the population is 84 percent Greek Cypriots, 13 percent Turkish Cypriots, 3 percent foreign residents, and small numbers of Maronites, Armenians, and Latins. The total population figures do not include illegal settlers and military troops from Turkey, who are estimated at about 85,000 and 40,000, respectively. The overall population density is 79 persons/km², with the majority of the population living in urban centers. The proportion of the population living in urban areas is 68 percent. According to the census of 1992, during the period 1982 to 1992, the portion of the population in rural areas dropped by more than 4 percent.

Prior to the Turkish invasion in 1974, in which Turkey occupied 37 percent of the island, the Greek-Cypriot majority and the Turkish-Cypriot minority lived together in a ratio of about four Greeks to one Turk, in all six administrative districts. After the invasion, the total population experienced negative growth up to the middle of 1977. In the ensuing years, population growth resumed, but with pronounced differences between the population in the government-controlled area and the Turkish-Cypriot community in the Turkish-occupied area. During the period from 1990 to 1993, the population of the free area of Cyprus increased at a rate exceeding 2 percent, while in the Turkish-occupied area, the Turkish-Cypriot community decreased due to emigration. According to Turkish-Cypriot sources, more than one-third of Turkish-Cypriots emigrated between 1974 and 1992.

The constitution of Cyprus recognizes Greek and Turkish as the official languages of the Republic of Cyprus. The Greek spoken and written is Hellenic Demotiki, which is used in Greece and is the language of education and daily communication. The Greek-Cypriot dialect is closely related to standard Greek. The Turkish minority uses Turkish in communication, education, and religion. The official language of the Maronite minority is Greek, while the Armenian population is bilingual in Armenian and Greek.

Cyprus is an independent republic with a presidential system of government. Under the constitution, executive power is exercised by the president, who is elected by universal suffrage for a five-year term of office. The president exercises executive power through a council of appointed ministers. Legislative authority is exercised by a house of representatives in which members are elected by the population.

Cyprus is classified as a middle-income country, with a per capita income of about US$12,000. It is considered to have a record of economic management that bodes well for the future. In 1992, the economy achieved a 10 percent growth rate in its gross domestic product, the highest rate in the last 10 years.

During 1993, expenditure on all levels of education, both public and private, accounted for 13 percent of the country's budget and 4 percent of the GNP. According to the census of 1992, the number of persons aged 15 and over who had never attended school or had not completed primary education was 15 percent, while 26 percent had completed primary education, 12 percent completed the gymnasium, 31 percent completed secondary education, and 16 percent were college or university graduates. The overall literacy rate among males was greater than 97 percent, while that of females was 90 percent. As age increases, the rate of literacy decreases, dropping to 73 percent in the over-65 age group.

2 | THE EDUCATION SYSTEM

Governance and Decision Making

The public education system in Cyprus is highly centralized, with a Ministry of Education and Culture responsible for the enforcement of educational laws and the preparation of new legislation. Public schools are financed from public funds, while private schools raise their funds primarily from tuition fees. At the secondary level, private schools receive a small state subsidy. In a few cases, funds are generated from foreign sources or through religious organizations. Private schools are administered by private individuals or bodies, but are liable to supervision by the Ministry of Education. Teachers in public schools are appointed, transferred, and promoted by the Educational Service Commission, which is an independent body appointed by the president. Educational policies are formulated by the Ministry of Education on the advice of the Educational Council and approved by the Council of Ministers.

The intended curriculum for all subjects is formulated by the Ministry of Education. Syllabi, curricula, and textbooks are prescribed to a large extent by governmental agencies. Schools at all levels are visited by inspectorates that offer in-service training, advice, and supervision. School evaluation also lies with the inspectorate. Teachers determine what instructional techniques and classroom processes to use in their teaching but they are guided in these choices by an inspector.

Structure of the System and Participation Rates

Structure of Education

In Cyprus, education is provided in preprimary, primary, general secondary, technical and vocational secondary schools, and special schools. Figure 1 shows the structure of education and approximate enrollment rates for each level. Preprimary education occurs in public, community, and private kindergartens. Approximately 54 percent of children aged three to five are enrolled. Thirty-four percent of these are enrolled in public, 20 percent in community, and 46 percent in private kindergartens.

Children begin free, compulsory, primary education during their sixth year, and leave when they have completed the prescribed six-year course. During the 1993–94 school year, 67 percent of primary school students were enrolled in urban schools and 33 percent in rural schools.

Secondary education takes place principally at public schools but there are a few private ones. Public general secondary education is divided into two cycles, which provide a six-year course for students in the 11 to 17 and above age group. Secondary education is compulsory to the end of Grade 9.

The gymnasium comprises Grades 7 to 9 of secondary education, during which all pupils follow a uniform course of general education. At the lyceum, which comprises Grades 10 to 12, there are three categories of subjects: compulsory core subjects, specialized subjects, and elective supplementary subjects. Instead of the lyceum, students may enroll in a technical and vocational school for a program of three years' duration; these programs often have a particular emphasis on the sciences and mathematics. Private secondary schools are oriented towards commercial and vocational education and last for six years.

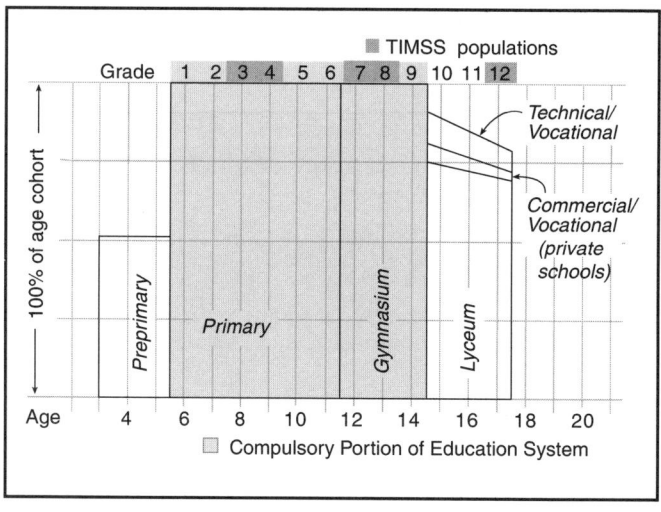

Figure 1: Structure of the Education System

Public and Private Schools

During the 1993–94 school year, 11 percent of students were enrolled in private schools, 82 percent in public general secondary education, and 7 percent in technical and vocational schools. Less than 1 percent of students attended special schools for students having visual, hearing, or mental disabilities that precluded their needs being met in a regular classroom. In the 1992–93 school year, 90 percent of the students who enrolled in secondary school went on to complete compulsory education (Grade 9), and 77 percent graduated from secondary education at the end of Grade 12.

Mathematics and Science Classes

Mathematics is a compulsory subject up to the end of Grade 12. Science as a single, integrated subject is compulsory until the end of Grade 7. After Grade 7, science is split into several subjects, all of which are compulsory at different levels. Anthropology is taken for one year in Grade 8, biology for two years in Grades 9 and 10, and physics and chemistry for all five years. Biology is offered in Grades 11 and 12, but only a small proportion of students study it.

Schools in the System

The School Year

Primary schools must open on the second Monday in September and close on the last Friday in June. Secondary schools open September 10 and close on June 5. The school year is divided into three terms of about equal duration. At the end of each term, secondary students are given an evaluation. School and national examinations end June 20. Both primary and secondary schools are open for between 175 and 180 instructional days per year.

Instructional Times

The school week is from Monday to Friday and each school day is six hours long. The school day begins at 7:30 in the morning and ends at 1:30 in the afternoon. Schools have holidays at Christmas and Easter of approximately 10 working days. In addition to these, there are 10 national and religious holidays during the school year on which schools are closed.

Class Size

The average class size in primary school was 25 students during the 1993–94 school year, while the average secondary class size was 26 students. Class size may range, however, from under 10 students to a maximum of 35 students in primary and secondary schools.

Grouping in Mathematics and Science

Mathematics and science are compulsory for all students in both elementary and secondary schools. Students at the primary and gymnasium levels study a compulsory program of mathematics and science. Primary and secondary classes, up to Grade 9, generally include students of varying abilities. There is no official policy on streaming or tracking.

Elective cycles of subjects begin in Grade 10 in the lyceum, the upper secondary school. Students may study one of the six optional cycles: classics, science and mathematics, economics, commerce, foreign languages, and technical and vocational. All students in a given cycle take the same courses, and only in Grades 11 and 12 are optional courses like biology and technical design offered.

Certification of Teachers

Certification

Until 1992, elementary teachers in Cyprus were required to take three years of formal education and pedagogy at a teacher education institution, leading to a certificate for teaching children up to the age of 12. Teachers could teach any subject in any grade at the primary level. After 1992, elementary school teachers began following four-year programs at universities, leading to certification for primary teaching. These graduates are normally generalist teachers. In order to qualify as a secondary school teacher of mathematics or science, a candidate must have a four-year university degree specializing either in mathematics or science.

In-service and Upgrading Requirements

Additional education for elementary school teachers is not mandatory. For secondary school teachers the only mandatory course is a one-year, in-service course taken two days per week. Schools, an in-service training institution, and the Ministry of Education offer a variety of professional development programs for teachers. These usually consist of ongoing guidance from inspectors, principals, and head teachers, and courses taken on a voluntary basis during afternoons. Teachers are not required to upgrade, except in order to become an assistant principal. Teachers wishing to upgrade a diploma or degree may undertake study through the University of Cyprus or abroad.

Teacher Profile

Teachers in Cyprus are well paid and enjoy a relatively high socioeconomic status. According to 1993 labor statistics and the 1968 occupation code ISCO, among the subgroup professional, technical, and related workers, secondary teachers are ranked in the second and primary teachers in the fourth of six categories of occupation. The average income of secondary school teachers in Cyprus in 1993 was US$28,000, and of primary school teachers US$21,000. By contrast, the average salary for computing professionals is US$22,600 and the average salary for architects, engineers, and related professionals is US$27,000.

In 1993, 64 percent of primary school teachers were women and 51 percent of secondary school teachers were men. Among mathematics teachers, 62 percent were men. Among science teachers, 64 percent were men.

3   TIMSS POPULATIONS

Cyprus, IEA, and TIMSS

Cyprus has been a member of IEA since 1990. Cyprus participated in the Reading Literacy Study and is participating in the Language Education Study, the Civic Education Study, and TIMSS at all population levels.

Cypriot TIMSS Populations

In Cyprus, the definition for Population 1 included Grades 3 and 4, with Grade 4 being the target grade in the two-grade design. These grades contained the largest portion of students in the nine-year-old cohort at the time of testing.

Population 2 included Grades 7 and 8, which were both part of the gymnasium, or junior secondary school. At the time of testing, 83 percent of nine-year-old students were in Grades 3 or 4, and 85 percent of thirteen-year-old students were in Grades 7 and 8.

Population 3 included all students in their final year of secondary education, with the sample being drawn from Grade 12. For mathematics and science literacy, the sample was drawn from 82 percent of students. All mathematics and science specialist students, 18 percent of the sample, were included in the study.

4   MATHEMATICS CURRICULUM AND PEDAGOGY

Goals for the Mathematics Curriculum

The intended curriculum in mathematics is determined by the Ministry of Education. A committee of the curriculum development unit, consisting of teachers and curriculum experts from the Ministry, is responsible for the creation of the national curriculum in mathematics. All schools, whether primary or secondary, are required to implement it as given. Supervisors from the Ministry of Education visit schools, attend lessons, assess teacher performance, and ensure that the curriculum is implemented properly.

According to the Ministry of Education, the goals of mathematics teaching in primary schools are to:
*   develop thought processes;
*   teach students to solve mathematics problems in everyday life ;
*   help pupils appreciate the usefulness of mathematics and enjoy the disciplined thought and harmony inherent in it;
*   develop positive attitudes toward mathematics;
*   help pupils to find mathematical relations in their environment;
*   use mathematical terminology effectively;
*   encourage and create interest in mathematics.
Mathematics teaching in secondary schools aims to:
*   develop skills in logical thinking, analysis, abstraction, and generalization, and develop an acquaintance with mathematical proof;
*   develop the ability to grasp the concepts,

magnitudes, and properties and the relations that are necessary for the solution of practical problems;

- teach students to use the language of mathematics and to appreciate the contributions of mathematics to the development of civilization;
- develop mental abilities, enabling students to see links and relations within the various disciplines of mathematics.

Major Changes in the Mathematics Curriculum

A number of innovations have been introduced into the curriculum during the past 10 years, including a specific mathematics curriculum for the 9 years of compulsory education. The most substantial point to note is the reallocation of subject matter to the primary and lower secondary curricula; this was done to maintain balance and to provide satisfactory mathematics education to those who do not attend upper secondary school. These changes in the curriculum reflect a desire to make mathematics accessible and interesting to all students.

Among other changes in the curriculum are the following:

- particular attention was given to the development of problem-solving skills;
- special attention was given to constructions with the use of geometric instruments;
- the teaching of computational techniques is to be completed by a certain age to ensure a basis for further learning.

Overall, the mathematics curriculum has been arranged in such a way as to take into account the needs of other subjects.

The use of computers as a teaching aid was recently launched, leading to in-service education of teachers in this area. Calculators and computers are encouraged but their usage in the classroom is limited.

In the curricula for gymnasium and lyceum, new units were introduced and some were excluded. At the lyceum, the units that have been introduced are differential equations, matrices and transformation, and three-dimensional analytic geometry.

Current Issues in the Mathematics Curriculum

Most changes in the intended curriculum have more to do with content than approaches to teaching and learning mathematics. Although inspectors have made efforts to promote instruction that focuses on student enjoyment and conceptual understanding, only small changes in actual teaching have taken place. The fact is that the accreditation requires students to show competence in all subtopic areas, and this is what teachers tend to emphasize in their teaching.

Mathematics Textbooks

The intended curriculum recommends the textbooks to be used, and these are provided free to all students. To a large extent, textbooks determine what and how students learn, since they reflect the intended curriculum. In primary and secondary public schools, the principal resource used is a standard textbook for each grade. Textbooks are prepared by special committees of the curriculum development unit, consisting of experienced teachers and curriculum officials from the Ministry of Education. Many teachers, however, produce their own supplementary materials. The presentation of information in textbooks has changed, and books now incorporate more pictures, graphs, and other illustrations. Real-life applications of mathematics are included. Exercises are classified according to degree of difficulty, and more difficult exercises are included.

Pedagogy

The curriculum guides mandate that mathematics be organized by grade in both primary and secondary schools. The guides state that teachers must focus on the development of concepts and skills, so that in primary and junior secondary schools students learn to work with applications and solve real-life problems. The use of real-life problems is now considered a basic element in the development of mathematical thinking.

Lecturing or whole-class instruction is the most important means of transmitting mathematical knowledge. A great deal of teaching fosters rote learning and repetitive practice exercises. In general, the inductive method is used in primary schools and the deductive method is used in secondary schools.

5 SCIENCE CURRICULUM AND PEDAGOGY

Goals for the Science Curriculum

The major goals of the intended science curriculum are described in the national science curriculum. Schools are required to implement the curriculum as given, although this is emphasized more strongly in Grades 7 to 12 than in Grades 1 to 6.

The primary science curriculum states a belief in the need for experimental science, regular and frequent science instruction, more emphasis on thinking, and less rote learning. The implementation of the science curriculum depends to a large extent on the proficiency of the classroom teacher.

For secondary schools, the general goals are:
- to help students understand the basic concepts of science;
- to enable students to use scientific knowledge and methods to solve everyday problems;
- to recognize the contribution of science in the improvement of living conditions;
- to prepare students to face various problems that arise from the applications of science and technology;
- to enrich students' scientific knowledge so they can communicate in a scientific and technological environment;
- to prepare high-ability students for further academic study in the sciences and for science-related careers.

Major Changes in the Science Curriculum

The new curriculum apportions the content of the science program over nine years. The material is presented in spiral form, covering the same content at increasing levels of complexity each year, based on the abilities of each age group. In teaching methods, more emphasis is given to scientific procedures; that is, the method science applies in the discovery of knowledge, rather than to the mere dispensing of existing knowledge. The new science curriculum caters to the abilities and needs of pupils and future citizens and uses modern pedagogic principles.

Current Issues in the Science Curriculum

Science, which is compulsory for all students, is seen as an essential subject in the national curriculum. Curriculum changes are in the process of being introduced. More emphasis is given to science applications with reference to technological achievements. Interactive media and videodisks are used for enrichment of science teaching.

Science Textbooks

The intended curriculum recommends the textbooks to be used, and these are provided free to all students. To a large extent, textbooks determine what and how students learn, since they reflect the intended curriculum. In primary and secondary public schools, the principal resource used is a standard textbook for each grade level. A curriculum development group made up of science teachers is working toward developing supplementary material to be used in science teaching. This material will be used in addition to the textbooks, which are mandatory for all teachers. New textbooks use examples of science applications, referring more and more to technological achievements. The human biology textbook is based on a series of educational videodisks, and other videodisks are available to enrich science teaching.

Pedagogy

The curriculum of the natural sciences suggests that instructors do the following during their teaching:
- use experiences and existing knowledge and examples from everyday life in developing new concepts;
- separate the class into groups using criteria appropriate to the specific characteristics of each class and topic;
- study certain topics with an approach based on the subject, particularly in the gymnasium cycle;
- use computers for instruction whenever the appropriate software and hardware are available;
- use both inductive and deductive teaching methods;
- assign small research projects that provide students with the opportunity to study topics of interest.

6 EVALUATION POLICIES AND PRACTICES

In Cyprus, the basic purpose of assessment is for student accreditation. In primary schools, no systematic testing is performed, but diagnostic tests are used during the school year. Essay questions and fill-in-the-blank questions are the most common format of these tests. At the end of the school year, all students are promoted to the next grade. The school keeps records of each student's progress.

In secondary schools, the school year is organized into three terms and students are tested at least once during each term. Portfolios of students work and oral contributions during class are also considered in assessment. At the end of each term grades are given to students. At the end of the school year, final examinations are organized at the school level.

Final examinations in mathematics and science for gymnasium students consist of 15 and 16 questions respectively, with most questions being essay-type questions. The test is school based, and is constructed by a team of two teachers. Mathematics final examinations at the lyceum consist of two parts, of 15 and 6 questions each. Both physics and chemistry examinations at the lyceum consist of three parts of 12, 8, and 3 questions each. These lyceum examinations are national examinations accounting for 25 percent of the yearly grade, and are prepared by the Ministry of Education. The grade for the year is the mean of four marks in that particular subject: three term marks and the external final examination. Candidates who fail must write the examination again during the first week of the next school year. If they fail again, they do not receive a leaving certificate.

7 REFERENCES AND SOURCES FOR FURTHER READING

Common Sources

1   Husén T, N T Postlethwaite 1994 *International Encyclopedia of Education*. Pergamon Press, Oxford
2   Organisation for Economic Co-operation and Development 1995 *Education at a Glance: OECD Indicators*. Organisation for Economic Co-operation and Development, Paris
3   United Nations Educational, Scientific and Cultural Organization 1995 *Statistical Yearbook*. United Nations Educational, Scientific and Cultural Organization, Paris
4   World Bank 1995 *World Development Report 1995*. Oxford University Press, New York

Other Sources

5   Department of Statistics and Research 1994 *Census of Population 1992*, Vol I, Series I, Report No 8 Ministry of Finance, Cyprus
6   Department of Statistics and Research 1995 *Demographic Report 1994*, Series II, Report No 32 Ministry of Finance, Cyprus
7   Department of Statistics and Research 1994 *Economic Report 1992*, Series I, Report No 37 Ministry of Finance, Cyprus
8   Department of Statistics and Research 1994 *Labor Statistics 1993*, Series II, Report No 12 Ministry of Finance, Cyprus
9   Papanastasiou C 1991 Current issues—The Minister of Education's views. *Education Newsletter/Faits Nouveaux, 3/91*. Council of Europe, Strasbourg
10 Papanastasiou C 1991 Developments in primary, secondary and technical education. *Education Newsletter/Faits Nouveaux, 4/91*. Council of Europe, Strasbourg
11 Papanastasiou C 1994 Curriculum of the nine-year school. *Education Newsletter/Faits Nouveaux, 2/94*. Council of Europe, Strasbourg
12 Papanastasiou C 1994 Cyprus: System of Education. In T Husén and N T Postlethwaite (Eds). *The International Encyclopedia of Education* (2nd ed.) Pergamon Press, Oxford
13 Ministry of Education 1994 *Primary School Curriculum*, Ministry of Education, Cyprus
14 Department of Statistics and Research Statistical Abstract 1992 Series I, Report No 38, Ministry of Finance, Cyprus
15 Department of Statistics and Research 1994 *Statistics of Education 1993–94*, Series I, Report No 26, Ministry of Finance, Cyprus
16 Ministry of Education *Annual Report 1994*, Ministry of Education, Cyprus

Statistical References

Section	Statistic	Reference	Page	Table	Year of Statistic
Country Profile	9251	6	13	–	1994
Country Profile	98km	13	1367	–	1994
Country Profile	226km	13	1367	–	1994
Country Profile	729,800	7	13	–	1995
Country Profile	722,800	7	13	–	1995
Country Profile	37%	6	21	–	1994
Country Profile	13%	6	13	–	1994
Country Profile	32%	5	18	2.1	1994
Country Profile	68%	5	18	2.1	1994
Country Profile	US$12,000	8	79	2	1994
Country Profile	13%	15	94	10	1994
Country Profile	4%	15	94	10	1994
Country Profile	15%	5	25	–	1992
Country Profile	31%	5	25	–	1992
Country Profile	16%	5	25	–	1992
Country Profile	97%	5	26	–	1992
Country Profile	90%	5	26	–	1992
Country Profile	73%	5	26	–	1992
Structure of the System	34%	16	107	1	1994
Structure of the System	20%	16	107	1	1994
Structure of the System	46%	16	107	1	1994
Structure of the System	67%	16	126	2	1994
Structure of the System	33%	16	126	2	1994
Structure of the System	11	16	43	1	1994
Structure of the System	7	16	43	1	1994
Structure of the System	82	16	43	1	1994
Schools in the System	25	16	151	1	1994
Schools in the System	26	16	151	1	1994
Teacher Profile	US$28,000	9	139	XI	1994
Teacher Profile	US$21,000	9	139	XI	1994

Czech Republic

Jana Svecová, Charles University

Jana Straková, Research Institute of Education, Ministry of Education

1 COUNTRY PROFILE

The Czech Republic is located in central Europe, and shares borders with Germany, Austria, Slovakia, and Poland. It covers an area of 78,864 km² and in 1994 had a population of more than 10 million people, with a population density of 131 persons/km². The capital city, Prague, has a population of 1.2 million. The population has declined steadily though not dramatically over the past five years due to a decreasing birth rate. Between 1989 and 1994, the number of live births per woman dropped from 1.9 to 1.4. During the same period, the number of deaths exceeded the number of births.

Ninety-five percent of the inhabitants of the Czech Republic are Czechs, that is, they identify themselves as Czech, Moravian, or Silesian. Slightly over 3 percent are Slovaks. There are also small numbers of Poles, Germans, and Romanians.

The Czech Republic is resuming its prewar tradition of advanced parliamentary democracy. After the revolution of 1989, legislative, executive, and judicial power were separated. Legislative power is now exercised by Parliament, which consists of two chambers, the Chamber of Deputies and the Senate. The president acts as head of state and is elected by parliament for a five-year term. The president and the cabinet share executive power. The president appoints the prime minister and, on the latter's suggestion, the other members of the government. Responsibility for education rests with the Ministry of Education, Youth, and Sport.

The Czech Republic is gradually transforming its economy to a modern market economy through deregulation, privatization, and tax reform. Careful monetary and budgetary policies have made it possible to maintain macroeconomic stability, an inflation rate of 10 percent, an unemployment rate of 3 percent, and a balanced budget in 1994. A low exchange rate, 29 Czech crowns for US$1 in 1994, has provided protection from difficult external conditions and has stimulated exports.

In 1994, the Czech economy began to overcome the decline that had occurred as a result of economic changes since 1990. The gross domestic product has grown by 3 percent and further increases are anticipated. Economic change has brought about shifts in the structure of employment, increasing the proportion of employment in the service sector while reducing employment in agriculture and industrial production

Gross expenditure on education has risen from US$815 million in 1989 to US$2,333 million in 1994, or from 4 percent in 1989 to 6 percent in 1994. Over 98 percent of expenditure on education is in the form of pubic funds. Education is free in all state schools and universities, and students are entitled to reduced tram, bus, and train fares, as well as free medical care. The adult literacy rate is virtually 100 percent.

2 THE EDUCATION SYSTEM

Governance and Decision Making

Until 1989, education was administered through regional and national committees, controlled centrally by the Ministry of the Interior and the Communist Party. Schools and local authorities had limited autonomy. After 1989, a new system was introduced that allowed the transfer of all education matters to the Ministry of Education and other ministries responsible for specific sectors of education. The 1990 act concerning state administration and self-government in education gave greater independence to individual schools through their directors, the municipalities, and the regional school offices.

Other participants in governance are the Ministry of Education, the school inspectorate, and other central bodies of the state administration. Among these are the Ministries of the Economy and of Agriculture for the training of apprentices and the Ministry of Health for the administration of health-service schools. The 1990 act significantly strengthened the authority of the school director, who can fundamentally affect the school in terms of personnel and equipment. The school director is responsible for the implementation of curricula, for the standard of education within the school, and for the efficient use of funds.

Activities within schools and other educational institutions are managed by regional school offices, which are directly supervised by the Ministry of Education. They allocate and monitor the use of funds for the salaries of teachers and other educational personnel, for textbooks, teaching materials, and any other costs covered by the state. The costs of building maintenance and equipment are covered by the community budget.

Structure of the System and Participation Rates

Public and Private Systems

Education is provided primarily by state schools. Just over 1 percent of elementary and 12 percent of secondary schools are private or ecclesiastical. Both ecclesiastical and private schools are outside the state system.

Structure of Education

Figure 1 shows the structure of the education system in the Czech Republic and the approximate enrollment rates.

In the Czech Republic education is provided for preschool children from a very young age. Children up to the age of three are educated in crèches administered by the health-care sector. Almost 87 percent of children between the ages of three and six are enrolled in kindergarten.

Elementary schools provide a compulsory nine-year education, divided into two levels. In Grades 1 to 4, pupils study the Czech language, mathematics, the natural sciences, and elementary civics. Art and physical education are part of this curriculum, as is work education, where pupils acquire basic work skills and habits. A single teacher teaches all classes.

The second level includes Grades 5 to 8 (9) and is attended by 10- to 14- (15)-year-olds. At this level, there is a specialist teacher for every subject. Talented students may attend elementary schools with extended course in language studies, mathematics, sports, and other areas. Students who have successfully completed the first or second level of elementary school may apply to attend the secondary school of their choice. Ninety-six percent of all children in the elementary school age cohort are enrolled in school. The minimum age for leaving school is 15.

The division between the two levels of elementary school is currently being changed. According to a recent parliamentary decision, beginning in 1996 the first level will consist of Grades 1 to 5, and the second level of Grades 6 to 9. Until this decision was made, Grade 9 was an optional grade, attended by 14 percent of the age cohort in 1993–94. Under the new system, Grade 9 will be compulsory.

There are three types of secondary schools. The gymnasium is a four-, six-, or eight-year general secondary school attended by 16 percent of all secondary students. Secondary technical schools, which combine general and vocational education at different levels, enroll 37 percent of students. Secondary vocational schools, which are practically oriented and provide vocational training as well as a foundation for general education, enroll 47 percent of students. Ninety-five percent of the age group are enrolled in various types

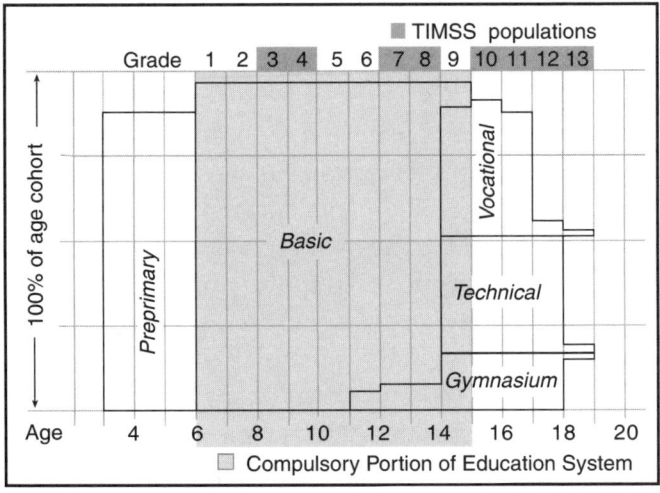

Figure 1: Structure of the Education System

of secondary education. A newly developed type of secondary school is the integrated technical and vocational school. Students successfully completing gymnasium or secondary technical school and passing the final examination, or *maturita*, are entitled to apply to institutions of higher education. Eighteen percent of all 18-year-olds are enrolled in higher education.

In 1994–95, there were 4216 elementary schools, 349 gymnasia, 731 secondary vocational schools, and 965 secondary technical schools, of which 195 were integrated technical and vocational schools.

Enrollment in Mathematics and Science

At elementary schools and gymnasia, mathematics and science are compulsory for all students. At most other secondary schools, mathematics is compulsory in at least the lower grades.

Schools in the System

The School Year

In the Czech Republic, the school year runs for approximately 200 days, from September 1 to June 30. School vacations include 2 days in the fall, 10 days at Christmas, 1 midterm day, 5 days in the spring, and a long summer vacation of two months.

The school week runs from Monday to Friday. Students attend elementary school for four to seven hours per day, with approximately 23 lessons per week in the lower grades and 30 lessons in the higher ones. Fifteen to 20 percent of the total instructional time is devoted to mathematics and approximately the same amount to science.

In the 1994–1995 school year, the average class size was 23 in kindergarten, 21 in the first level of elementary school, 24 in the second level, 29 in the gymnasium, 26 in secondary technical school, and 24 in secondary vocational school.

Mathematics and Science Classes

All students in Grades 1 through 8 receive a virtually identical mathematics and science education. They are taught elementary science in Grades 3 and 4, biology and geography in Grades 5 to 8, physics in Grades 6 to 8, and chemistry in Grades 7 to 8. Some freedom is given to school directors to decide how many lessons will be allocated to different subjects, but there are no substantial differences between schools. The only exception is among students in special mathematics classes in Grades 5 to 8, constituting less then 1 percent of the population.

At the secondary school level, gymnasia have the largest proportion of mathematics and science lessons. Time allocations, however, are not the same in all gymnasia, especially for science subjects. Secondary technical schools and secondary vocational schools differ substantially in their curricula depending on the vocation taught. Some offer few mathematics and practically no science courses.

Certification of Teachers

Teachers at elementary and secondary schools must be university graduates. They must have studied for four or five years either in a university faculty of education, or five years in faculties of philosophy, science, mathematics and physics, physical education, or theology. Technical, economic, art, and agricultural colleges educate teachers of vocational subjects for secondary schools. Teacher education courses end with a final examination and thesis. Graduates are awarded the degree *Magister*.

Teachers in the first level of elementary school teach all elementary school subjects. Teachers in the second level of elementary school and for secondary school specialize, usually in two subjects.

Both mathematics and science teachers study in faculties of mathematics, physics, or natural sciences. Besides general pedagogy and psychology, the focus is on the specialty subject areas. Preservice teachers study an average of 6 to 10 subject matter courses in each of two chosen fields. In all science specializations, a great emphasis is put on laboratory courses. All students are also required to spend 15 weeks assisting at a school during their course of study, first observing in a classroom and progressing to practical teaching.

Teachers in the Czech Republic may take advantage of continuing education courses and lectures offered by departments and faculties of education. However, additional study is not mandatory.

Teacher Profile

Seventy-two percent of Czech teachers are female. The proportion of women teachers differs by school type and level: 83 percent in elementary school, 67 percent in gymnasia, 57 percent in secondary technical schools, and 56 percent in secondary vocational schools. The average monthly teacher's salary is US$349, or 95 percent of the average salary in the Czech Republic.

The average age of Czech teachers is 42. Only 15 percent of all teachers are younger than 29 years of age.

3   TIMSS POPULATIONS

The Czech Republic, IEA, and TIMSS

The Czech Republic has been a member of IEA since 1991, and TIMSS is the first IEA study in which it has participated.

Czech TIMSS Populations

In the Czech Republic, Population 1 includes Grades 3 and 4, which contain the largest proportion of nine-year-olds. Population 2 includes Grades 7 and 8, which contain the largest proportion of thirteen-year-olds.

Population 3 includes all students in their final year of secondary school. All types of secondary vocational schools and all secondary technical schools and gymnasia were included in the testing. Students enrolled in vocational secondary schools complete their education at different grades, depending on the type of vocation. In almost all secondary technical schools and gymnasia, students complete their education at the end of Grade 12; a few complete their studies at the end of Grade 13.

The general definition of Population 3 includes all students from secondary vocational schools and secondary technical schools. All students at gymnasia were defined as mathematics and physics specialists, meaning that approximately 18 percent of all students were included in Population 3 specialists.

4   MATHEMATICS CURRICULUM AND PEDAGOGY

Goals for the Mathematics Curriculum

In the Czech Republic, curricula are prepared by committees consisting of curriculum experts, teachers from schools and universities, and representatives from professional groups, as established by the Ministry of Education. At present, the curricula published by the Ministry of Education in 1991 are compulsory for all state schools, and the school inspectorate ensures that they are followed.

A new compulsory core curriculum was published in 1995. Because it is a core curriculum only, there is opportunity for modification at the school level. The curriculum document outlines the general aims of mathematics education at the elementary level:
- Provide students with basic knowledge and skills in arithmetic, geometry, and algebra.
- Provide a foundation for the understanding of functional and other relationships among quantitative phenomena.
- Help students to understand quantitative relations in natural and social processes.
- Teach students to apply mathematical skills in everyday life.
- Develop in students the habit of verifying the reasonableness of results.
- Provide a foundation for using new technologies.
- Use problem solving for the development of independent logical thinking and spatial perception.

Major Changes in the Mathematics Curriculum

The new curriculum includes a minimum of required content and gives great freedom to schools and teachers in deciding what will be taught and how much time will be devoted to it. The mathematics curriculum no longer includes a set approach to teaching mathematics, which was a dominant trait in past mathematics curricula. An effort has been made to make mathematics accessible to every student, necessitating a departure from a formal academic approach. More emphasis has been put on real-world applications. New topics, such as statistics, probability, and financial mathematics, are being included in the curriculum. Less emphasis is put on specific topic content and more on the attainment of methods and skills.

According to the compulsory curriculum of 1986, calculators are to be used from Grade 5 onwards at all schools. At this level the students have attained sufficient computational skills, and may concentrate on problems from the real world, where results cannot easily be calculated by hand.

Although computers are widely available in schools, their use in mathematics instruction is limited and depends entirely on individual schools and teachers.

Current Issues in the Mathematics Curriculum

A system of educational standards, of which the core curriculum is a part, for all subjects at the elementary and secondary levels is currently being developed. On the basis of these standards, educational programs will be designed by professional groups, curriculum experts, and schools. After approval by the Ministry of Education, each school will be able to choose which program to follow. Mathematics curriculum experts will be involved in both the development of the core curriculum and the proposals for educational programs.

A second issue is the recent parliamentary decision to increase the number of elementary school grades from eight to nine. The redistribution of subject matter must be addressed before this occurs.

Mathematics Textbooks

A variety of textbooks are now available for particular grades and subjects. Until 1990, only one set of textbooks existed, but this changed along with the decentralization of some aspects of education. The old textbooks closely followed the prescribed curriculum and were accompanied by a set of teaching guides. Since 1990, some freedom has been given to schools in choosing textbooks.

Teachers and students at all levels use textbooks daily. Each student has his or her set of textbooks to read, study exercises and solutions, and do practice exercises from. The new textbooks are more attractive

for students, containing more pictures and supplements with paper models. Worksheets and working textbooks are also being used extensively.

Mathematics textbooks are written by university teachers and curriculum experts. Recently, many school teachers have begun taking part in the process. Textbooks are published by private enterprise publishers, but must correspond to the national curriculum and be approved by the Ministry of Education in order to be used in schools. Elementary schools buy textbooks for their students from their own budgets, while secondary school students buy their own textbooks.

Pedagogy

As mentioned previously, the main goal of mathematics education is to make mathematics more accessible to every student, particularly at the elementary level; to avoid formal and theoretical approaches, if possible; and to include more real-life problems and problem-solving approaches. Items connected with ecology and a healthy life-style have also been introduced. Integration of topics within mathematics and between mathematics and other subjects is strongly recommended. The focus is placed on an inductive approach, while a deductive approach is used in a limited way only in the final years of elementary education. Manipulative materials are also now used a great deal, particularly in the lower grades of elementary school.

5 SCIENCE CURRICULUM AND PEDAGOGY

Goals for the Science Curriculum

In the Czech Republic, an integrated science course is taught only in Grades 3 and 4. At the higher grades, separate curricula are published for biology, geography and earth science, physics, and chemistry.

The goals stated in the official curriculum document for all science subjects are as follows:

- Help students understand the phenomena and processes encountered in nature, everyday life, and technology.
- Proceed from objective to abstract knowledge.

- Help students acquire complex skills by using different scientific methods such as observation, description, comparison, analysis, systematization, experimentation, and measurement.
- Investigate causes and relationships in the sciences.
- Solve practical problems.
- Ensure that students become aware the value of natural resources, human knowledge, and the environment.

Major Changes in the Science Curriculum

As in other subjects, the most important change in the science curriculum is a reduction in content and an increase in the choices given to schools and teachers in deciding what will be taught and how time will be allocated.

A greater emphasis is now placed on environmental issues, and on developing a sense of responsibility toward society and the local environment. The traditional emphasis placed on theoretical knowledge and the memorization of facts and formulae has, to some extent, been replaced by problem solving and real-life applications. Sex education, and information on drug and alcohol misuse and AIDS problems have been introduced into the curriculum

Computers are widely available in schools. Whether they are used in the science classroom depends entirely on individual schools and teachers. Their usefulness in science education is limited by the quality and quantity of software available. Computers are principally used for gaining practice in contemporary topics.

Current Issues in the Science Curriculum

As stated previously, the most important issue in elementary and secondary education is the development of standards and the prolongation of elementary school. Following a core curriculum rather than a specific compulsory curriculum will substantially change science education in the Czech Republic. Educational programs will differ in the number of instructional hours in science, as well as in the content taught and approaches used.

Science Textbooks

The way science textbooks are developed and the rules for their use are identical to those discussed for mathematics textbooks. The most important change over the past few years is the fact that several textbooks are now available for each grade and subject.

The newer textbooks are more appropriate for the learning needs of students. They are colorful, readable, and contain many pictures and humorous applications. The content itself, however, has not changed very much in the past 10 years. The nature of exercises has changed somewhat, with more emphasis on creativity, problem solving, and real-life applications, as opposed to the traditional system of memorizing facts. Independent work has become an important component of science teaching, and many schools are using worksheets supplied by textbook publishers for this.

Pedagogy

Science pedagogy has not changed much in recent years. Currently, there is a greater focus on the integration of topics from different science subjects. Independent and creative work is stressed more. The laboratories in sciences are more investigative in nature, moving away from a "follow the cookbook" style. More emphasis is placed on written and oral communication. Reasoning as opposed to mechanical memorization of facts is stressed. There is an effort to balance deductive and inductive approaches in the curriculum.

6 EVALUATION POLICIES AND PRACTICES

The Czech Republic has no system of national examinations, but students in all school types are regularly assessed for instructional purposes. Achievement in elementary and secondary schools is determined by teacher-developed oral and written examinations, to which teachers give marks ranging from one to five. In mathematics and the sciences, tests with open-ended items are widely used. Students receive achievement reports at the end of the first term and at the end of the school year. Primary school teachers are also allowed to use verbal assessment.

Students are admitted to secondary school on the

basis of their achievement in both elementary school and their entrance examination. Approximately 95 percent of elementary students, most of them aged 14, continue to secondary school. The content and process of the examination is the responsibility of the secondary school director. In most schools the students write tests consisting of open-ended items in mathematics and the Czech language.

Education at gymnasia and technical schools ends with the examination known as the *maturita*. In most cases it consists of a four- or five-subject oral examination; a written portion may also be included in some subjects. Passing this examination is a precondition to admission to higher education. At secondary vocational schools, the training course concludes with an apprenticeship examination.

About 18 percent of students are admitted to an institution of higher education. Most of these students are 18 years of age. Admission procedures to higher education vary from one institution to another; in general, admission is based on passing the entrance examination successfully and on overall achievement in secondary school, including the final examination or *maturita*.

7 REFERENCES AND SOURCES FOR FURTHER READING

Common Sources

1   Husén T, N T Postlethwaite 1994 *International Encyclopedia of Education*. Pergamon Press, Oxford
2   Organisation for Economic Co-operation and Development 1995 *Education at a Glance: OECD Indicators*. Organisation for Economic Co-operation and Development, Paris
3   United Nations Educational, Scientific and Cultural Organization 1995 *Statistical Yearbook*. United Nations Educational, Scientific and Cultural Organization, Paris
4   World Bank 1995 *World Development Report 1995*. Oxford University Press, New York

Other Sources

5   *Statistical Yearbook of the Czech Republic*. 1994 Czech Writen, Prague

6   *Statistics of the Educational System of the Czech Republic*. 1994 Institute for Information in Education, Prague
7   Svecová J 1994 Czechoslovakia. In: Karsten S and D Major (eds.) 1994 *Education in East Central Europe: Educational Changes after the Fall of Communism*. Waxmaann, Münster/New York
8   Kotásek J, J Svecová, 1995 The Czech Republic—System of education. In: *The International Encyclopaedia of National Systems of Education*, Second Edition. Elsevier Science Ltd, Oxford
9   Standards of elementary education 1995. In: *Ninth Bulletin of the Ministry of Education, Youth, and Sport*. Ministry of Education, Youth, and Sport, Prague (in Czech)
10 Transforming Education 1995 *Background Report to the OECD Review of Education in the Czech Republic*. Education Policy Center, Prague
11 Czech Statistical Office 1991 *Population, Houses, Apartments, and Households*. Czech Statistical Office, Prague (in Czech)
12 Fritenská H, A Sulitka 1994 *Guide to the Rights of National Minorities in the Czech Republic*. DAS, Prague (in Czech)
13 Ministry of Education, Youth, and Sport 1996 *Annual Report on the System and Development School Year 1994–95*. Ministry of Education, Youth, and Sport, Prague (in Czech)
14 Ministry of Education, Youth, and Sport 1994 *Quantitative, Structural, and Economic Analysis of the Development of Czech Education from 1989 to 1993*. Ministry of Education, Youth, and Sport, Prague (in Czech)
15 Czech Statistical Office *Statistical Information—Employment and Salaries of Employees*. Czech Statistical Office, Prague (in Czech)
16 Ministry of Education, Youth, and Sport 1991 *Curriculum Guide for Elementary School*. Ministry of Education, Youth, and Sport, Prague (in Czech)

Statistical References

Section	Statistic	Reference	Page	Table	Year of Statistic
Country Profile	78,864 km^2	5	46	1	1993
Country Profile	10 million	5	46	1	1993
Country Profile	131/km^2	5	46	1	1993
Country Profile	1.2 million	5	46	1	1993
Country Profile	5	5	83	–	1993
Country Profile	1.9 to 1.4	5	83	9	1994
Country Profile	95%	12	13	–	1991
Country Profile	3%	12	13	–	1991
Country Profile	10%	10	4	1.2	1994
Country Profile	3%	10	4	–	1994
Country Profile	29 Czech crowns = US$1	10	4	–	1994
Country Profile	3 %	10	4	1.2	1994
Country Profile	US$815 million	10	99	8.3	1994
Country Profile	US$2,333 million	10	99	8.1	1994
Country Profile	4 to 6%	10	99	8.1	1994
Country Profile	98%	10	99	–	1994
Country Profile	100%	11	25	–	1991
Structure of the System	1.3%	13	51	–	1994
Structure of the System	12%	13	53	17	1994
Structure of the System	87%	10	34	–	1994
Structure of the System	14%	14	15	–	1993
Structure of the System	96%	14	14	–	1993
Structure of the System	16%	10	42	–	1994
Structure of the System	37%	10	43	–	1994
Structure of the System	47%	10	43	–	1994
Structure of the System	95%	14	18	–	1993
Structure of the System	18%	10	131	11.1	1994
Structure of the System	4216	6	27	–	1994
Structure of the System	349	6	9	–	1994

Continued

Statistical References continued

Section	Statistic	Reference	Page	Table	Year of Statistic
Structure of the System	731	10	57	4.3	1994
Structure of the System	965	10	57	4.3	1994
Structure of the System	195	6	7		1994
Schools in the System	23 lessons 30 lessons 15 to 20%	16	–	–	1991
Schools in the System	23, 21, 24, 29, 26, 24	10	113	9.4	1994
Teacher Profile	72%, 83%, 67%, 57%, 56%	10	108	9.1	1994
Teacher Profile	9431; US$349	13	–	–	1995
Teacher Profile	8171; US$303	15	35	–	1995
Teacher Profile	42	10	109	9.3	1994
Teacher Profile	15%	10	109	9.3	1994

Denmark

Peter Weng, Danish National Institute for Educational Research

INTRODUCTION

In Denmark, the TIMSS results will attract much attention. This is in part due to current general education debate, and in part to the IEA Reading Literacy Study, in which Denmark also participated. The results from this study caused much debate about Danish pupils' reading comprehension.

A corresponding debate about attitudes to and achievements in mathematics and science can be of importance both to teaching in basic school as well as to teacher education. Information from TIMSS will be part of an evaluation of ideas about education and about the various subjects contained in the new act concerning primary and lower secondary education, which came into force in 1994. The TIMSS results will be used in the revisions of the *Folkeskole* Act, in new developments in youth education, and in the upcoming revision of teacher education.

1 COUNTRY PROFILE

Denmark, one of the Scandinavian countries, consists of a large peninsula—Jutland—two large islands—Funen and Zealand—and 404 smaller islands. The capital city, Copenhagen, is situated on Zealand and, with 1.7 million inhabitants, is the country's only large city. Apart from a German minority group in southern Jutland and a limited number of immigrants from various cultures, the Danish people are a homogeneous group, with their own language, Danish.

Denmark's population in 1994 was 5.2 million inhabitants distributed over roughly 43,000 km^2, with a density of approximately 121 persons/km^2. From 1981 to 1992 the population increased by less than 1 percent, and during the middle 1980s there was a period of negative growth. In 1993, almost 45,000 persons immigrated and almost 35,000 emigrated. Greenland and the Faroe Islands, with populations of approximately 56,000 and 47,000 respectively, are two self-governing areas belonging to the Danish monarchy.

Denmark is a constitutional monarchy. Formally, the monarch has executive power, but in practice the government exercises legislative power. Parliament and the government are established through general elections held every four years. One hundred seventy-nine members are elected to parliament, of whom two are elected from Greenland and two from the Faroe Islands. The courts of law wield judicial power. Both ministers and judges are appointed by the crown.

Denmark is a welfare state where, in the words of Danish clergyman, poet and school reformer, N.F.S. Grundtvig (1783–1872), "few have too much and fewer too little." The Danish democracy and educational system are still marked by Grundtvig's ideas about the equality of human beings and the importance of education to the development of society.

The strict financial policies of recent years are yielding results and after some years with an adverse trade balance, the economic situation has become favorable. The tax load on the Danish wage earner is heavy, but social security benefits for unemployment, illness, and old age are relatively high.

Every year 2.3 million Danes, of whom 1.3 million are adults, enroll in the education system. Adult education covers a very large field and includes both general and further education. Further education can be either general or vocational in orientation. Adult education takes place in many different educational institutions, including evening schools, folk high schools, adult-education centers, labor market centers, and universities. In 1991 about US$1 billion was spent on education, which is more than 10 percent of public expenditure, or US$1950 per person.

2 THE EDUCATION SYSTEM

Governance and Decision Making

Through legislation, the Danish parliament or *Folketinget*, sets the objectives and frameworks for the various types of education. In 1994, a new act concerning primary and lower secondary education came into effect. The *Folkeskole* Act empowers the Ministry of Education to lay down learning objectives and frameworks for primary and lower secondary schools. The Ministry issues guidelines for curricula and teaching for all subjects and compulsory topics mentioned in the *Folkeskole* Act. On the basis of this description local education authorities and schools decide how to arrange their teaching.

Each school designs a curriculum based on the official curriculum guides and the *Folkeskole* Act. The draft curriculum, describing the content to be taught, the number of lessons, and the placement of the various subjects within the timetable, is sent to the local school board for approval. The school board then submits the proposal to the local education authority for final approval before implementation. In practice, very few schools prepare local curricula for mathematics and science. The curriculum established by the Ministry of Education is used when no local curriculum has been proposed. In such cases, teachers are obliged to teach to Ministry curriculum. Public and private schools may teach special programs and subjects of their own design, provided they cover the material outlined in the *Folkeskole* Act.

Individual teachers are bound by the contents of both the Act and the local curriculum, but teachers select their own teaching method. Textbook choices are made at the school level, by either the teacher or a group of subject teachers. This choice must be formally approved by the school board.

Upper secondary education is divided into two fields, general and vocational. General upper secondary school includes the gymnasium and the higher preparatory examinations, or HF-courses, both of which provide a general education as well as preparation for higher education. Vocational training courses qualify students for both direct entry into the labor market

and for higher education. The higher commercial examination, the HHX, and the higher technical examination, the HTX, both qualify students for tertiary education.

The gymnasium and HF-courses are administered at the county level, except in the cases of a few private and one state institutions. The Ministry of Education administers the higher functions of gymnasia and vocational schools, describing the curriculum framework and issuing subject guidelines. The school principal manages teaching and administration and is responsible to both the county council and the Ministry. Budgets and class size are decided by each school board. Vocational schools have wide scope in deciding content to be taught, and representatives from the labor market often act as consultants in the preparation of teaching guidelines.

Structure of the System and Participation Rates

Structure of Education

Denmark had approximately 3000 schools and other educational institutions during the 1992–93 school year. Figure 1 shows the structure of education in Denmark and the approximate participation rates at all levels. At the *Folkeskole* or primary and lower secondary level there were 1688 public schools, 403 private schools, and 222 continuation schools. Continuation schools are private, residential schools for students aged 14 to 18. These schools offer the same curriculum as

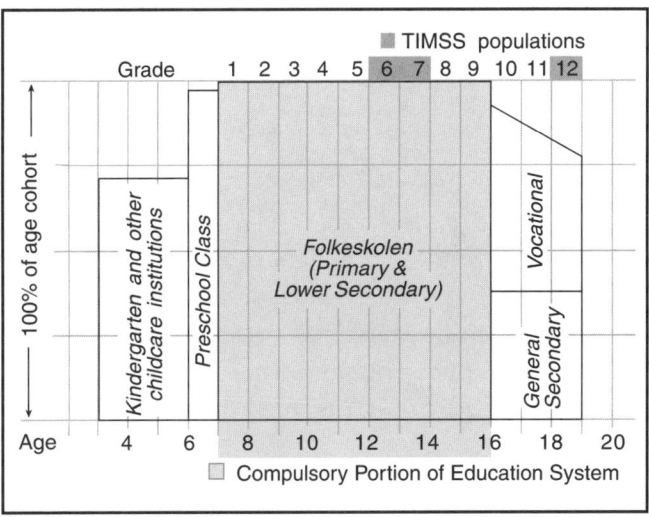

Figure 1: Structure of the Education System

the *Folkeskole*, with an emphasis on social and recreational activities. At the upper secondary level there were 152 gymnasia and 202 vocational schools. In addition, there are 225 institutions for further education.

Primary and lower secondary schools, including Grades 1 to 9, are generally housed in the same building. This level is known as *Grundskole*, or basic school. Normally, most children begin school in a preschool class in the year of their sixth birthday. Children begin Grade 1 in the year of their seventh birthday. Education is compulsory for all children aged seven to sixteen. Children may attend a public or private school, or parents may choose to educate their children at home. In 1993, 88 percent of seven-year-olds were enrolled in public schools and 12 percent in private schools.

Compulsory education stops when the student has had nine years of education, usually at the end of Grade 9. Beyond that it is possible to continue one more year, Grade 10, at the *Folkeskole* level before leaving school or proceeding to upper secondary school or vocational training. Upper secondary education consists of Grades 10 to 12. Of the 60,405 pupils completing *Folkeskole* in 1993, approximately 5 percent left the educational system, over 37 percent continued in general upper secondary school, and 58 percent in vocational education. About 34 percent of these students are expected to go through further or higher education.

Enrollment in Mathematics and Science

In Grades 1 to 9, mathematics and science are compulsory. In the gymnasium, various mathematics and science courses are required at different levels during Grades 10 and 11. Enrollments among Grade 12 students at the gymnasium in 1995–96 are shown in Table 1.

Public and Private Systems

Educational expenses are covered by local taxation. Parent groups or other religious or political groups can establish private schools. Such schools are obliged to maintain teaching standards corresponding to those within the public school system. The state supports private schools by covering up to 85 percent of teaching expenses.

Table 1: Enrollment in Mathematics and Science Subjects, Grade 12 Students

Subject	Percent of males	Percent of females
Mathematics	65	35
Physics	20	4
Chemistry	15	10
Biology	15	21
Geography	5	5

Schools in the System

Folkeskole

In *Folkeskole*, the school year begins on the first of August and lasts for approximately 200 school days. Each year the Ministry of Education issues guidelines concerning holidays, summer vacations, and days off. A school week has five days, from Monday to Friday.

There is no streaming in *Folkeskole*. Students are taught compulsory subjects together throughout the 9 or 10 years of schooling. Teachers may, however, periodically split a class if it is expedient for pedagogical reasons. Class size in *Folkeskole* must not exceed 28, and in the 1993–94 school year the average was 18 pupils. From preschool to the end of Grade 10, between 14 and 20 percent of class time is spent on mathematics, with the lower number representing the final grades of *Folkeskole*. Science classes occupy a wider range, from 5 percent in preschool to Grade 2 to 24 percent in Grades 7 and 8. Lessons are 45 minutes long.

Gymnasium

At the gymnasium level, the school year has 199 days and 31 to 32 lessons per week. During their first year in the mathematics program, students spend 16 percent of class time on mathematics and 28 percent on science. In the languages program, students spend 19 percent of class time on the subject of natural science, which is an integration of mathematics, physics, and chemistry. In their second year students in the mathematics program spend at least 16 percent of class timemathematics and 19 percent on science, whereas those in the languages program spend at least 22 percent of class time on mathematics and science.

By their second year students may choose among different levels of subjects and, therefore, may have more lessons in the subjects than the minimum noted above. A student who chooses to study mathematics and physics at a high level, for example, will study these subjects for 31 percent of the weekly timetable.

Certification of Mathematics and Science Teachers

Teachers in *Folkeskole* and those teaching in upper secondary school are educated in very different ways. *Folkeskole* teachers must pass a certification examination at 1 of the 18 teacher training colleges in Denmark. The program lasts four years, with two years of general subject-matter studies, including mathematics and science for all students. The final two years involve the study of two specialty subjects. Between 12 and 14 percent of preservice teachers specialize in mathematics. In addition to the academic requirements, twenty weeks of practice teaching are required for certification. Throughout the four years, pedagogy and psychology play an important role.

Teachers at the upper secondary level must have a degree from one of the five universities in Denmark. Degree programs usually involve five to six years of study in a major and a minor subject. The final stage for certification as an upper secondary school teacher requires completion of a five-month program with both theoretical and practical components.

In-service training of *Folkeskole* teachers takes place at the Royal Danish School of Educational Studies, which has divisions all over the country. This institution is a university for teachers, where educational research as well as in-service programs are carried out. Many teacher education colleges also offer short and long in-service courses. The time required for participation in in-service courses is regarded as part of the total working time of the teacher.

There is no institution corresponding to the Royal Danish School of Educational Studies for in-service training of upper secondary school teachers. The Ministry of Education, however, offers a series of courses which can be one day or longer. The courses were designed by subject-matter advisory committees consisting of representatives from the Ministry and several teachers' associations.

Teacher Profile

According to the socioeconomic rankings performed by the Danish Institute for Social Research, *Folkeskole* teachers are ranked in the second category and upper secondary school teachers in the first category of five categories of occupations. Teachers' salaries are not large compared to those of other occupational groups. In 1994, the annual initial salary of a *Folkeskole* teacher was about US$33,000, and the highest salary was US$42,500. The corresponding figures for upper secondary school teachers were US$34,000 and US$50,000. Engineers, chartered accountants, and correspondence clerks, for example, earned US$61,000, US$53,000, and US$44,000 respectively in 1994.

In the 1994–95 school year, the average age of *Folkeskole* teachers was 46 years. Sixty-three percent of teachers were women and 37 percent men, according to the Danish teachers' union. In 1993, 50 percent of the general upper secondary school teachers were 40 to 50 years of age. Forty percent were women and 60 percent men. Among mathematics teachers, 23 percent were women and 77 percent men. The corresponding figures for the sciences were: biology, 38 percent female and 62 percent male; physics, 14 percent female and 86 percent male; chemistry, 27 percent female and 73 percent male; and for geography, 24 percent female and 76 percent male.

3 TIMSS POPULATIONS

Denmark, IEA, and TIMSS

Denmark has not participated in previous IEA studies of mathematics and science. Denmark became a member of IEA in 1986 and participated in the IEA Study of Reading Literacy. The Danish National Institute for Educational Research represents Denmark in IEA.

Danish TIMSS Populations

Denmark is not participating in TIMSS at the Population 1 level. Population 2 includes students from Grades 6 and 7. These grades contained the largest number of 13-year-olds at the time of testing. Population 3 includes students in their final year of upper secondary education, generally Grade 12. Both generalists and specialists in mathematics and physics have been tested. The specialists are defined as that part of the general Population 3 which is enrolled in mathematics, physics, or both at a high level.

4 MATHEMATICS CURRICULUM AND PEDAGOGY

Goals for the Mathematics Curriculum

In the spring of 1995, when the TIMSS data collection took place, the new *Folkeskole* Act had been in place for a little less than a year. The *Folkeskole* Act was, therefore, not of much significance in relation to students participating in TIMSS.

The important fields of knowledge and achievement in mathematics are determined by the Ministry of Education and are binding for all teachers. Mathematics content should be connected to everyday life and surroundings, so that students can obtain mathematical knowledge and skills by using numerals, calculation and geometry in investigation, systematization, and reasoning.

Goals for mathematics education require that students:

- be able to understand and use mathematics in connection with daily life, society, and nature;
- experience mathematics both as a means of problem-solving and a creative subject;
- experience and acknowledge the role of mathematics in a cultural and social context.

In addition to objectives and content frameworks, the Ministry of Education issues curriculum guidelines. The guidelines are divided into elementary (Grades 1 to 3), middle (Grades 3 to 7) and final stages (Grades 7 to 9 and Grade 10). A full description of subject-matter content and teaching guidelines is given for each stage. The teaching guide describes the mathematics teacher's role and provides examples of teaching procedures.

The Ministry's description of goals, content, curricula, and teaching methods is drafted with the assistance of a curriculum committee supported by a secretariat. Mathematics and education experts, as well as representatives from teacher and parent organizations, participate in the curriculum committee.

In the gymnasium, the goals of mathematics are described in a 1987 executive order from the Ministry of Education. The learning objectives state that students are to acquire a mathematical foundation, thereby increasing their understanding of the surrounding world and their participation in the development of society. Content is broken down into main topics and related aspects, including numbers, plane and solid geometry, functions, differential calculus, statistics and probability, vectors in the plane and space, integral calculus, and differential equations. The teaching of these main topics must include historical aspects, modeling, and the interior structure of mathematics.

Higher commercial examination mathematics has moved from being very much like the program taught at the gymnasium to a program more closely related to trade and society. To some extent this is also true for higher technical examination mathematics courses, where teaching is project oriented and includes practical aspects wherever possible.

Major Changes in the Mathematics Curriculum

The *Folkeskole* Act requires that teaching be based on the individual pupil's requirements and take the form of a dialogue between teacher and pupil. These general requirements have resulted in changes in the content and planning of mathematics teaching. Mathematics classes now emphasize that numerals and algebra are useful in verbal as well as written communication, and are an important part of our culture. This new focus on a more informal treatment of algebra has been under development for a long time.

Calculators and computers are important elements in the mathematics classroom, beginning in Grade 1, and have resulted in new ways of dealing with calculations and graphing. Electronic data processing is to be integrated in all subjects, in order that all *Folkeskole* students may become familiar with information technology and computer use.

Mathematics, besides being a tool subject, is now also viewed as part of a general education. Mathematics teaching includes a history component, linking mathematics to cultural development and providing it with a humanistic aspect distinctly different from the formal approach of earlier times.

Mathematics teaching in upper secondary education is designed to relate mathematics to problems that students might encounter in everyday life. Mathematics, however, must also contain a preparatory aspect from which students will be able to continue their education within fields where mathematical knowledge and skills are required. Mathematical concepts and reasoning, therefore, still play an important part in teaching.

Current Issues in the Mathematics Curriculum

During recent years the role of mathematics at all levels has been debated. Secondary schools claim that the effort of making mathematics a subject for everybody in *Folkeskole* has involved general costs in mathematical skills and thinking. The current debate now focuses on the education and in-service training of mathematics teachers, the teacher being a critical factor in how individual students benefit from teaching, with little regard for other potentially important issues.

Mathematics Textbooks

In both *Folkeskole* and gymnasium, teachers choose the textbooks and other teaching materials most suited to the kind of teaching they intend to carry out. Many textbooks are published for both these levels. They are published either as textbook series with one or more books to be used at the various grade levels, or as topic or theme books to be used at one grade level only. Several new textbook systems offer possibilities of integrating electronic data processing into the curriculum as early as Grade 1.

Newer textbooks can be characterized by their focus on everyday themes, practical mathematical activities, cross-curricular work, pupil-teacher cooperation, and differentiated teaching. Recently, a number of mathematics textbooks dealing with special topics have been published. A great many theme textbooks have been published, for example, that aim at upper secondary students who work with noncompulsory topics at a high level.

Despite the many newly published textbook systems and updated editions of old systems, many textbooks still in use are outdated or have left an indelible mark on teaching practices. Even when a group of teachers within a school has decided to purchase a new textbook system, the replacement process takes several years because of strict financial control of the educational system.

Pedagogy

The approach to teaching is based on the experience individual students possess as they enter school; so too is the process of mathematical conceptualization. Through the student's intuitive understanding of mathematics, mathematical concepts are developed by means of dialogue, play, games, experiments with concrete materials, and investigations. Students are offered the possibility of developing their own methods of calculation instead of being presented with definite algorithms to follow.

Calculators and computers are used in teaching from the earliest years of schooling. This offers the possibility of working more deeply with the four basic arithmetical operations, because drilling in basic skills can be reduced. In addition, working with models and practical problems with large figures is possible at an earlier stage.

Applicability forms the basis of activities designed to develop concepts, reasoning on different levels of abstraction, and communication using mathematical concepts. Based on the curricula, *Folkeskole* is setting the stage for exemplary teaching, in which inductive teaching is heavily weighted in relation to deductive; not until the gymnasium is the deductive approach used.

Important aspects of planning for compulsory mathematics at the upper secondary level include the following:

- Student prerequisites and student input should be considered when planning teaching.
- The work may be organized as procedures where the intuitive aspect forms the basis, whereas others set the stage for emphasizing the deductive character of the subject.
- Adequate areas of mathematics should be selected, offering opportunities for working with the craftsman aspects of the subject and with reasoning and proofs.
- Different methods of work should be used. The

teacher might, for instance, apply a method of working that directly involves the girls of the class, if he or she considers it expedient.

- The class should work with mathematical language, making class dialogue on the subject matter essential.
- Written work might, apart from homework, also consist of communications tasks in the shape of articles or contributions to a debate.
- Electronic data processing and graphic calculators should be integrated whenever they offer support to the concepts and methods taught.
- Integration across other subjects is considered important to demonstrate the applicability of mathematics.

5   SCIENCE CURRICULUM AND PEDAGOGY

Goals for the Science Curriculum

Science as a single subject is not taught in Danish schools. The compulsory science subjects in *Folkeskole* are nature and technology in Grades 1 to 6, geography and biology in Grades 7 and 8, and physics and chemistry in Grades 7 to 9. Optional subjects in upper secondary school include biology, geography, physics, chemistry, technical studies, and natural science. Cultural geography is included in nature and technology and geography courses at both levels.

The goals of these subjects are set in the same way as for mathematics (Section 4). Thus, the goals for science teaching encompass a number of subjects and only some of the common features can be described:

- The new *Folkeskole* Act contains "a strain of green," which promotes students' understanding of the interaction of human beings and nature.
- Science is relevant to all subjects, and should form a component of every subject taught.
- In order to establish a basic understanding of science subject matter, local, national, and global questions concerning nature, technological development, conditions of life, and the environment form part of the curriculum from the earliest grades. Through experience within these areas students de-

velop thinking, language, and concepts enabling them to raise questions about society's interaction with nature. Questions about resources, the environment, and gene technology contain both ethical and technical aspects, and students must have some basic knowledge of these questions in order to form opinions.

Major Changes in the Science Curriculum

Resource and environmental problems have become part of everyday life; consequently nature and technology are now introduced in Grade 1. It is intended that knowledge of society's interaction with and utilization of nature should be based on students' experience with local surroundings.

At all levels, teaching should be arranged so that the significance of science and technological development in our everyday life are explored. The importance of technology in society is to be included whenever possible. It is expected that the number of lessons used for observation and investigation of nature or for factory visits will increase at all grade levels in the near future. The history of science and the importance of science to our world now have a special emphasis.

As calculators, electronic data processing, and videotapes are now used at all grades, the application of these tools will increase in importance for the analysis of observations from investigations and experiments.

Current Issues in the Science Curriculum

The science curriculum sets the stage for deeper study of the subject outside school. Experimental aspects of science should therefore be given a higher priority in the future. Equipment for investigations, experiments, and electronic data processing material should be supplemented in order to meeting the curricular guidelines.

As both *Folkeskole* and the gymnasium attach great importance to general education, student influence on and participation in teaching should be real. The problem of girls' early rejection of some science subjects, for example, might be reduced by paying more attention to student input in course development.

Science Textbooks

Both in *Folkeskole* and the gymnasium, importance is attached to the fact that, in the science subjects, students should learn through experiments covering specific subject matter as well as more general phenomena. Textbooks play a strong role in this movement. For some years there has been a trend in *Folkeskole* away from blackboard physics toward teaching based on investigation and experiment.

New textbooks include many suggestions for student activities supplemented by text, drawings, and photographs. There are textbook series in which each book deals with a different topic, and the use of these is becoming more common, beginning in Grade 1. This happens even though Nature and Technology, which is the first science subject taught, is not supposed to be a book subject. Familiar materials from the students' everyday lives, not expensive laboratory equipment, should be sufficient teaching material.

Textbooks are often published with a teacher's guide and supplementary material in the form of copy folders, work sheets, and boxes of material. The guidelines for chemistry teaching at the gymnasium emphasize that students have the opportunity to use sources of information other than textbooks. Collections of statistical tables, works of reference, periodicals, other subject-related literature, newspaper articles, and literature in foreign languages should be used in this regard.

Pedagogy

A broad constructivist view of education forms the basis of science curricula. The role of the teacher is developing from a transmitter of knowledge into a facilitator of situations furthering the urge to learn, construct knowledge, and accumulate understanding. Such situations are brought about through the teacher's knowledge of the individual student's prerequisites and motivation, that is, the principle of differentiated teaching. This demands a closer cooperation between pupil and teacher than was previously the case.

Students should experience the importance of cooperation in the development of concepts, language, and thinking. This can be obtained by introducing open ways of experimenting at the expense of the more traditional standard experiments. The development of classroom cooperation should further students' participation in learning and, as a consequence, their feelings of responsibility for their own education.

As mentioned earlier, science teaching now includes the use of newspaper articles, television, video, and electronic data processing. Students thus have the opportunity of discussing and debating topics and themes, so that their everyday language includes technical terms and concepts used in communication within and about the subjects.

6 EVALUATION POLICIES AND PRACTICES

In 1989, the Ministry of Education appointed a committee to describe and evaluate the various forms of tests and examinations used in the Danish education system. The following is a summary of some of the conclusions of the committee:

- Tests and examinations are an important part of the concept of quality in education and teaching.
- Tests and examinations do not in themselves constitute evidence of qualification for further education or occupation.
- The development of personal qualifications, which is considered important throughout the Danish education system, should not be given a lower priority because such qualifications cannot be measured by tests and examinations.
- An education system without tests and examinations is not desirable.
- Examinations should not use large blocks of time or money at the expense of the teaching process.

Students receive no marks during the first seven years of their schooling in *Folkeskole*. The school must, however, brief students and their parents on student progress. Students are automatically promoted to the next grade throughout *Folkeskole*.

In *Folkeskole*, students, in consultation with their parents and teachers, decide whether they will take one or more leaving examinations. The leaving certificate for primary and lower secondary school covers specific subjects and is not a compulsory examination. Consequently, there is no average mark required for passing. Students may write the leaving examinations in Danish, mathematics, English, German, and physics and chemistry after Grades 9 and 10.

Previously, the leaving examination in mathematics consisted of two examinations written at the end of Grade 9. According to new regulations, this has been changed to one oral and one written examination. The physics and chemistry examinations will also be changed, eliminating the written component and leaving only a practical, oral laboratory test performed individually or by a group of pupils.

In order to pass the upper secondary school leaving examination, or studentereksamen, students must write an examination in Danish and the other subjects studied at the senior level. At the end of the second year of the gymnasium, students write a compulsory examination in either English or mathematics, depending on whether the student followed the languages or mathematics stream. There are oral evaluations in all subjects, and students must take five or six examinations during their three years at the gymnasium. By the end of the gymnasium, students have been examined in 10 subjects. Furthermore, during their final year, the students are given one week to complete a major paper in Danish, history, or another high-level subject. Marks for this task are noted on the graduation certificate.

The *Folkeskole* Act calls for an evaluation of how individual students benefit from teaching, and formative evaluation is constantly being developed in order to address this need. Formative evaluation forms the basis of differentiated teaching. Through the teacher's dialogue with pupils, colleagues, and others about goals and expectations, evaluation should involve new individual goals for the education of each pupil. Formative evaluation is meant to make teaching relevant to all students and to take into consideration the educational needs of all students.

CONCLUDING REMARKS

The speed with which the world changes makes it necessary that education continue as a lifelong activity. The objectives of education are not solely to give knowledge and skill to students, but also to promote cooperation, creativity, and a desire to learn. The desire to learn is the key to the development of the student, both as an individual and as a member of a democratic society.

Teaching in individual subjects, therefore, cannot take place isolated from the world within and surrounding the school. In their teaching, teachers must incorporate other school subjects and aspects from everyday life. The single subjects are not to be considered unchangeable—as for instance is or has been the belief about mathematics—but something living, created by human beings, and therefore changeable and to be developed together with social awareness. The teacher must maintain the integrity of the subject within the teaching, but must also involve both pupils and their parents in educational planning. Parental influence is more direct in basic school and less direct in youth education. Thus, the content and organization of the teaching are becoming more varied.

For mathematics and science, this has meant a gradual decrease in theoretical teaching and an increase in aspects of applications and practical use of the subject. In science, this is most apparent in the new subject Nature and Technology, which is taught from Grades 1 to 6. The subject promotes the importance of getting in touch with nature and technology in a way that allows practical and exploratory methods of working. The hope is that this development will contribute to a general strengthening of interest in mathematics and science in such a way that these subjects will no longer be regarded as difficult, but as subjects contributing to a better life. The more pupils obtain knowledge and skill in these subjects, the greater the chance of reducing the distance between "the two cultures," to the benefit of the individual and the development of society.

7 REFERENCES AND SOURCES FOR FURTHER READING

Common Sources

1 Husén T, N T Postlethwaite 1994 *International Encyclopedia of Education*. Pergamon Press, Oxford

2 Organisation for Economic Co-operation and Development 1995 *Education at a Glance: OECD Indicators*. Organisation for Economic Co-operation and Development, Paris

3 United Nations Educational, Scientific and Cultural Organization 1995 *Statistical Yearbook*. United Nations Educational, Scientific and Cultural Organization, Paris

4 World Bank 1996 *World Development Report*. Oxford University Press, New York

Other Sources

5 Danmarks Statistik 1995 *Statistical Yearbook [Statistisk Årbog]*. Danmarks Statistik, Copenhagen

6 Datakontoret 1993 *Education Crisscross [Uddannelse på kryds og tvers)*. Undervisningsministeriet, Copenhagen

7 Dehn-Nielsen H 1994 *Who Earns How Much 94/95? [Hvem tjener hvor meget 94/95]*. Holkefeldt, Copenhagen

8 Undervisningsministeriet 1995 *Figures Speaking. Education Key Figures [Tal, der taler, Uddannelsesnøgeltal]*. Undervisningsministeriet, Copenhagen

9 Folkeskoleafdelingen 1995 *Primary Education in Figures, Administration Figures [Folkeskolen i tal, planlægningstal]*. Undervisningsministeriet, Copenhagen

10 *Newsletter from the Ministry of Education [Undervisningsministeriets Nyhedsbrev]*. 1995, 9(18)

11 Undervisnings-og Forskningsministeriet 1990 *Mathematics Quality in Education [Matematik Kvalitet i Uddannelse og Undervisning]*. Undervisnings-og Forskningsministeriet, Copenhagen

12 Gymnasieafdelingen Key Figures 1993/94. *Upper Secondary Education, Two-Year Higher Preparatory Examination, and Adult Upper Secondary Level Course. [Nøgletal 1993/94, Gymnasier, 2-årige hf-kurser og studenterkurser]*. Undervisningsministeriet, Copenhagen

13 Undervisningsministeriet 1994. *Goal & Central Knowledge and Achievement Fields. The Subjects of Folkeskole. [Formål & centrale kundskabs- & færdighedsområder]*. Undervisningsministeriet

14 Undervisningsministeriet 1995 Curricula. *The Subjects of Folkeskole [Læseplaner Folkeskolens Fag]*. Undervisningsministeriet, Copenhagen

15 Gymnasieafdelingen 1993 Mathematics. *Teaching Guides for Upper Secondary Education No. 22 [Matematik Undervisningsvejledninger for Gymnasiet No. 22]*. Undervisningsministeriet, Copenhagen

16 Undervisningsministeriet 1991 Test and Examination. *Quality in Education and Teaching [Prøver og eksamen. Kvalitet i uddannelse og undervisning]*. Undervisningsministeriet, Copenhagen

17 Direktoratet for Gymnasieskolerne og HF 1988Chemistry. *Executive Order and Guidelines [Kemi. Bekendtgørelse og vejledende retningslinjer]*. Undervisningsministeriet, Copenhagen

Statistical References

Section	Statistic	Reference	Page	Table	Year of Statistic
Country Profile	1.7 million	5	14	14	1995
Country Profile	5.2 million	5	14	14	1995
Country Profile	43,000 km2	5	14	14	1995
Country Profile	121/km2	5	14	14	1995
Country Profile	less than 1%	5	42	42	1992
Country Profile	56,000 47,000	5	14	14	1995
Country Profile	2.3 million 1.3 million	6	10	–	1993
Country Profile	US$1 billion, US$1950	8	10	–	1993
Structure of the System	3000, 1688, 403, 222, 152, 202, 225	8	36	–	1993

Continued

Statistical References Continued

Section	Statistic	Reference	Page	Table	Year of Statistic
Structure of the System	88%, 12%	9	12	–	1995
Structure of the System	60,405, 5% 37%, 58% 34%	10	3	–	1995
Schools in the System	18	9	1	–	1995
Teacher Profile	US$33,000 US$42,500 US$34,000 US$50,000 US$61,000 US$53,000 US$44,000	7	87 & 109	–	1994
Teacher Profile	50% 40% 60% 23% 77% 38%/62% 14%/86% 27%/73% 24%/76%	12	119–200	–	1994

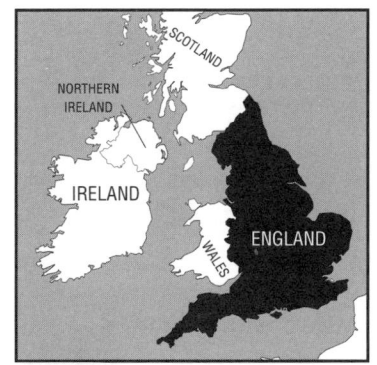

England

Claudia J. Davis, National Foundation for Educational Research

1 COUNTRY PROFILE

England, part of the United Kingdom along with Wales, Scotland, and Northern Ireland, is situated less than 40 km off the northwest coast of mainland Europe. The United Kingdom's closest neighbors are Belgium, France, and the Netherlands to the east, and the Republic of Ireland to the west.

Although England is quite small in terms of land area, 130,423 km^2, in 1993 it had the highest population density in Europe, with an average of 372 persons/km^2. The highest concentration of people is found in the southeast of England, and the lowest in East Anglia. England had a total population of 48,208,000 in 1991, of which about 20 percent were under the age of 16.

English is the official language spoken in England. The majority (94 percent) of the population in England is Caucasian, with a number of ethnic minority groups mainly from India, Africa, and the Caribbean.

England is part of a constitutional monarchy in which the sovereign is head of state. The government consists of the Parliament, the executive or cabinet, and the judiciary. Parliament, the law-making body, comprises the sovereign, the House of Lords, and the House of Commons. Most of the work of Parliament is conducted in the House of Commons which is composed of 650 elected members, known as members of Parliament or MPs.

England's economy is ranked as high-income by UNESCO; the per capita gross domestic product for the United Kingdom was US$14,058 in 1993. The per capita gross national product in 1993 was US$14,923 with an average annual growth rate of over 3 percent between 1980 and 1991, a figure that was one of the highest among the high-income group of countries as ranked by the OECD. Education expenditure in 1993–94 was 16 percent of net government expenditure and 5 percent of the United Kingdom GDP in 1991. England's literacy rate is about 99 percent.

2 THE EDUCATION SYSTEM

Governance and Decision Making

The Education Act of 1944 and its subsequent amendments provide the basis for the management of all schools and colleges in England. The Act states that the statutory system of education must be a continuous progression divided into three distinct stages: primary, secondary, and further or higher education. The Education Reform Act (1988) established a national curriculum and national curriculum assessment in England.

The central government holds the authority and responsibility for the complete provision of education services, for determining national policies, and for planning the direction of the system as a whole. The Secretary of State for Education and Employment is appointed by the Prime Minister and is accountable to Parliament for giving direction to and controlling the public education system and employment. A series of education acts and statutory instruments or orders made by the Secretary of State for Education and approved by Parliament forms the basis of England's education legislation. At the local level, policies are implemented by local education authorities and schools' governing bodies, together with further and higher education institutions.

Following the introduction of local management of schools, a provision of the Education Reform Act of

1988, most school administration and management functions are now carried out by the local institution. All state schools in England must have a school governing body consisting of representatives from the local education authority, the community, the parents, and the teaching staff of the school. It is the responsibility of the school's governing body to allocate the budget and to determine the general direction of the school and its curriculum, subject to the requirements of the national curriculum. In practice many of these powers are delegated to the head teacher, who is responsible for the school's internal organization and management. The head teacher usually delegates specific aspects of curriculum organization, teaching methods, and pastoral care to senior members of the teaching staff.

A four-year cycle of school inspections was established in the Education (Schools) Act of 1992. A nonstatutory body, the School Curriculum Assessment Authority, is responsible for overseeing the implementation and assessment of the national curriculum. The Office for Standards in Education is responsible for the administration of the school inspection program.

Structure of the System and Participation Rates

Structure of Education

Figure 1 shows the structure of the education system and approximate enrollment rates. Compulsory education in England takes place between the ages of 5 and 16. Children generally start school at the beginning of the term following their fifth birthday. However, some schools admit four-year-olds for the term during which they will become five.

Most schools in England are part of the state system. These schools, which in January 1993 were attended by about 93 percent of the school-age population, are funded by the central government, but administered locally. Parents have the right to express a preference as to which state school they wish their child to attend. There are, however, a number of registered private schools maintained by public funds. These schools exist at all levels of education and are often known as independent schools. Some long-established secondary residential private schools are known as public schools.

In England, nursery education for pupils under five is voluntary. The Education Act of 1944 defined primary education as the education of children aged 5 to 11 years. The Education Reform Act of 1988 further divided this phase into two key stages. Key Stage 1 caters to pupils aged 5 to 7 years; Key Stage 2 caters to those aged 7 to 11. Secondary education covers schooling for students aged 11 to 18 years, with Key Stage 3 catering to 11- to 14-year-olds and Key Stage 4 to 14- to 16-year-olds. Postcompulsory education is also provided in institutions for further education. The first five years of secondary education fall within the period of compulsory education, and thereafter schooling is voluntary.

Enrollment in Mathematics and Science

Mathematics and science are compulsory courses until the end of Grade 11. In the final two years of high school, enrollment in mathematics drops to approximately nine percent for males and four percent for females, while science enrollment drops to approximately 23 percent for males and 16 percent for females.

Schools in the System

The School Year

In 1981, it became compulsory for all state-maintained schools in England to be open for a minimum of 380 half-days per year, according to Education (Schools and Further Education) Regulations. The local education authority decides the dates of the school terms for county schools. The school governing body, in con-

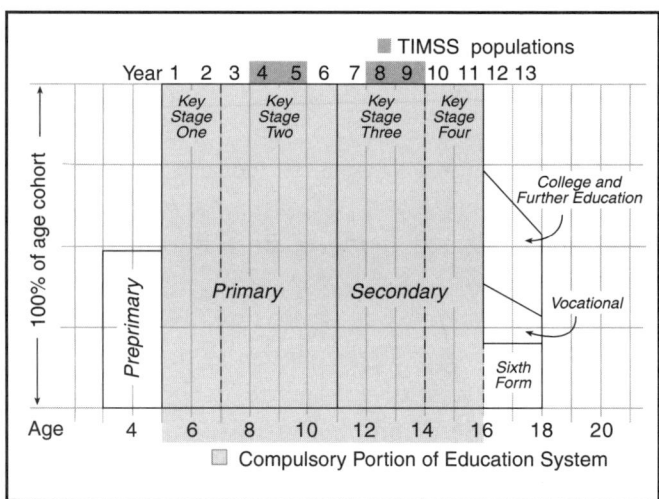

Figure 1: Structure of the Education System

sultation with the local education authority, determines school term dates for voluntary-aided schools and special agreement schools. In the case of grant-maintained schools, the school's governing body makes the decision.

The school year is divided into three terms, with a long summer break of about six weeks in July and August and shorter breaks of two and three weeks at Christmas and at Easter. The school week is from Monday to Friday, and each teaching day is approximately six hours long.

Class Size

The average class size in primary schools in January 1995 was 27 students. In secondary schools, the average class size was around 22 students, although classes are typically smaller at senior levels.

Streaming and Tracking

There is no official policy on within-school streaming, and most schools are not streamed. Primary school classes generally include students of varying abilities. Mathematics and science form part of the curriculum of all students of compulsory school age. Streaming is quite common for secondary mathematics classes, but less common for science.

Certification of Teachers of Mathematics and Science

Teachers in England may be qualified by two routes: by completing a four-year bachelor of education degree, or by earning a postgraduate certificate in education through one year of study after a three-year degree in another discipline. The former course includes curriculum, pedagogical and educational studies, practical teaching, university-level study of the main subject, and teaching methodology for the main subject(s) in primary or secondary schools, as appropriate. The postgraduate certificate in education focuses on curriculum, pedagogical and educational studies, practical teaching skills, and the application of the students' degree subject(s) to school teaching. Bachelor of education courses for future primary teachers include at least 25 weeks of practical teaching experience. The same program for future secondary teachers will include at least 32 weeks of practical teaching experience. Postgraduate certificate courses for future pri-

mary teachers will include at least 15 weeks of practical teaching experience, and for future secondary teachers at least 32 weeks is required.

A scheme of formal appraisal to take place in a two-year cycle is currently being introduced for all teachers. Experienced teachers may apply for posts as head of department, head of year, and eventually as deputy head teacher or head teacher. A teacher's qualifications, experience, and performance are taken into consideration whenever he or she applies for a new post. Relevant professional development and participation in in-service education and training of teachers may therefore assist a teacher in securing promotion.

Teacher Profile

Over 80 percent of primary school teachers are female, whereas in secondary schools, the split between the genders is about equal. The average age of full-time teachers in England in 1992 was between 40 and 49. The pay scale currently runs from US$17,357 to US$46,985 per annum. In practice, the majority of teachers are paid between US$28,971 and US$38,129 per annum.

3 TIMSS POPULATIONS

England, IEA, and TIMSS

England and Wales have together been a member of IEA since its founding in 1959. England alone, or England and Wales together, have taken part in almost all previous IEA studies, including the First and Second International Mathematics (FIMS, SIMS) and Science (FISS, SISS) Studies.

English TIMSS Populations

Population 1 includes all students enrolled in years four and five in primary schools. These are the school years which contained over 95 percent of students in the nine-year-old age cohort at the time of testing. Population 2 includes all students enrolled in years eight and nine in secondary schools. These school years contained over 95 percent of thirteen-year-olds at the time of testing.

4 MATHEMATICS CURRICULUM AND PEDAGOGY

Goals for the Mathematics Curriculum

In 1988 a national curriculum was introduced in England for the first time. It applies to all pupils of compulsory school age, 5 to 16, in state-maintained schools. Independent schools are not required to follow the national curriculum, but most choose to do so.

There have been two revisions of the original 1988 national curriculum; the most recent one took effect in September 1995. There are no further plans for changes until 2000.

The national curriculum is organized on the basis of four key stages: Key Stages 1 and 2 are carried out at primary school, and Key Stages 3 and 4 at secondary school. The curriculum consists of three core subjects, English, mathematics, and science, supplemented by various foundation subjects. The table below indicates, by shaded area, at which key stage subjects are studied.

For each subject and for each key stage, programs of study set out what pupils should be taught. Attainment targets set out the standards for student performance.

For each key stage, the programs of study are set out in sections. Table 2 shows the five sections in mathematics, two of which apply to Key Stages 3 and 4 only.

Each section within the program of study describes what pupils should be taught. The program of study for Key Stages 2 and 3, as outlined in the national curriculum, is shown in Table 3.

Major Changes in the Mathematics Curriculum

Significant changes have been made to the mathematics curriculum the last eight years. England is now experiencing a period of stability after two revisions to

Table 1: Schooling, Key Stages, and Pupils' Ages

Subject	Primary		Secondary	
	Key Stage 1 Ages 5 to 7	Key Stage 2 Ages 7 to 11	Key Stage 3 Ages 11 to 14	Key Stage 4 Ages 14 to 16
English	▓	▓	▓	▓
Mathematics	▓	▓	▓	▓
Science	▓	▓	▓	▓
Physical Education	▓	▓	▓	▓
Design and technology	▓	▓	▓	▓
Information technology	▓	▓	▓	▓
Modern foreign language			▓	▓
History	▓	▓	▓	
Geography	▓	▓	▓	
Music	▓	▓	▓	
Art	▓	▓	▓	

Table 2: Mathematics Program by Key Stage

Program of Study	Relevant Key Stages
Using and Applying Mathematics	All key stages
Numbers	All key stages
Algebra	Key Stages 3 and 4 (Secondary)
Shape, Space and Measures	All key stages
Handling Data	Key Stages 3 and 4 (Secondary)

the original, national curriculum for mathematics. No further changes to the national curriculum are planned for five years.

In 1988 a new General Certificate of Secondary Education was introduced as a replacement for two separate examinations: the General Certificate of Education Ordinary level, for above average pupils; and the Certificate of Secondary Education, for average and below average pupils. The General Certificate of Secondary Education was therefore built on the foundation of these earlier examinations. The certificate makes extra demands on students during their courses. They must be more organized both in the production and timing of their course-work assignments and in the maintenance of the necessary files than was previously necessary. The new examination also had a profound effect on teachers, since new techniques for the planning and monitoring of courses were needed. In addition, teachers must assist their pupils in making suitable choices of differentiated papers. The idea of dif-

ferentiated papers for students with differing abilities was largely boosted by the publication of the 1982 Cockroft report *Mathematics Counts,* which proposed differentiated courses for secondary school mathematics.

Electronic aids such as calculators and computers have increased in use within mathematics classrooms in the past 10 years. This has affected the nature of the subject, to the extent that the national curriculum recommends teaching the appropriate use of calculators as well as the use of spreadsheets.

Current Issues in the Mathematics Curriculum

Current concerns in the mathematics curriculum are diverse:

*   The emergence of a large number of low attainers, early in their secondary school careers, who show little progression in mathematics. These students often hold negative attitudes towards mathematics and therefore do not go on to study the subject at higher levels. Only a small percentage of students continue to study mathematics past the age of sixteen, and an even smaller percentage read the subject at university.
*   The relatively low proportion of girls opting to study mathematics at higher levels.
*   The possibly detrimental effect of widespread calculator use on pupils' levels of mental arithmetic competency. There is a need to adapt the training of teachers to take account of new technology.

Table 3: Mathematics Program in Key Stages 2 and 3

Program of Study	Pupils should be taught
Using and Applying Mathematics	(1) making and monitoring decisions to solve problems (2) communicating mathematically (3) developing skills in mathematical reasoning
Numbers	(1) understanding place value and extending the number system (2) understanding and using relationships between numbers and developing methods of computation (3) solving numerical problems
Algebra	(1) understanding and using functional relationships (2) understanding and using equations and formulae

Mathematics Textbooks

Textbooks are normally commercially produced. With the introduction of the national curriculum and subsequent revisions, many textbooks have become out of date. Since no changes in the national curriculum are planned for five years, however, publishers are now rewriting their books to fit in with the new curriculum.

A vast choice of textbooks exists in England at both the primary and secondary levels. Within primary schools, textbooks are often supplemented by work cards for the pupils to use. Primary school teachers usually decide collectively which scheme the school will adopt.

One of the most widely used of the secondary textbooks is the *School Mathematics Project* series, developed in the 1960s. The *School Mathematics Project* scheme is the predominant one in England, and is used in about half of schools. This series of textbooks relies heavily on the input of practicing teachers. The *School Mathematics Project* series for students aged 11 to 16 was designed in two parts. The first two years, for students aged 11 to 13, incorporate short booklets that support independent learning on the part of the students. In the final three years, for students aged 13 to 16, the course is presented in textbook form, but at levels to suited to different levels of attainment. The emphasis, however, remains on demonstrating topics through tasks. These are sequences of simple problems that gradually take the students through the topic, all the time incorporating recent research findings on student misconceptions and common errors. In theory, these texts offer ample opportunity for discussion. However, such discussion only occurs if the students come together to take advantage of the opportunity.

Pedagogy

The General Certificate of Secondary Education was introduced in 1988, but investigations and project work did not become compulsory until 1991. As a result of the introduction of compulsory project work for the General Certificate, project work was introduced in the lower grades. In lower grades pupils are encouraged to work in groups until, in later years, they are encouraged to work more independently.

As a result of the introduction of the General Cer-

tificate of Secondary Education, pupils are taught at an early age to write explanatory accounts in mathematics classes and not to simply produce figures. They are being taught, for example, to write descriptions of methods they have used to achieve their result or answer.

There is normally some discussion each time a new topic is introduced into the classroom. Discussion is, however, usually confined to the beginning of new topics and does not occur throughout the teaching of the unit.

Computers are widely available in schools and students are becoming familiar with their use and applications. As a result of the government's "Micros in Schools" and "Micros in Primaries" schemes, which were completed in 1984, every primary and secondary school in England had at least 1 microcomputer by the mid-1980s, and recent figures show this has increased to an average of 10 in primary schools and 85 in secondary schools. Calculators are widely used in mathematics classes.

5 SCIENCE CURRICULUM AND PEDAGOGY

Goals for the Science Curriculum

The intended curriculum in science for schools in England is described in *The National Curriculum* (1995). This publication is produced by the Department of Education and advises teachers on aims, content, teaching approaches, and methods of assessment. The science curriculum is mandatory for all public-sector schools, and the majority of independent schools also teach it.

As with the national curriculum for mathematics, the document is not prescriptive as far as teaching methods are concerned. Teachers are invited to select from the suggestions offered, to bring together material from a wide range of resources, and to accommodate the learning needs of particular classes or individuals.

The organization of the school curriculum within the statutory framework is the responsibility of the head teacher. As with mathematics, schools decide how

Table 4: Science Program by Key Stage

Program of Study	Pupils should be taught:
Experimental and Investigative Science	(1) planning experimental work (2) obtaining evidence (3) considering evidence
Life Processes and Living Things	(1) life processes (2) humans as organisms (3) green plants as organisms (4) variation and classification (5) living things in their environment
Materials and their Properties	(1) grouping and classifying materials (2) changing materials (3) separating mixtures of materials
Physical Processes	(1) electricity (2) forces and motion (3) light and sound (4) the earth and beyond

much time should be spent on teaching the program of study for science, how science is timetabled, and what the most appropriate teaching methods and materials are.

As with the national curriculum for mathematics, programs of study are set out in sections. The four sections, which are all applicable at all key stages, are shown in Table 4.

In addition to the aspects of science shown above, five requirements apply to all four key stages within the programs of study, at levels suited to the key stage. These are as follows: systematic inquiry, science in everyday life, the nature of scientific ideas, communication, and health and safety. These requirements apply in the context of the other content that is taught at the appropriate key stages.

Major Changes in the Science Curriculum

In 1989 the national curriculum for science was introduced into two primary years and one secondary year and was phased into different years in successive years (in order to ease the burden on schools, as it would be several years before all the students would be affected by the changes). In practice, however, some schools found it easier to introduce all the changes at once.

In the original national science curriculum (1989), there were 17 attainment targets for the secondary level and 14 for the primary level. The national curriculum for science was revised in 1991, reducing the number of attainment targets to 4 and more clearly defining the division between physics, chemistry, and biology. Following the Dearing Review in 1993, a revised science order was implemented, which, among other changes, reduced the prescribed content at all key stages and moved some areas of the curriculum from the primary to the secondary level. The programs of study for experimental and investigative science at all key stages were substantially revised to reflect a broader range of experimental and investigative work, and to give increased prominence to qualitative work. In terms of formal assessment at age 16, there are three different options as far as science courses are concerned: the single award course, which tends to be taken by less able pupils; the double award course, which is taken by a wide cross-section of pupils; and the separate science course, which tends to be taken by the more able pupils. Following the Dearing Review, the material in the program of study for single science was reduced.

Current Issues in the Science Curriculum

Current concerns are as follows:

- In most primary schools in England the teacher teaches all subjects to the class, but is assisted in particular subjects by a subject coordinator, whose job it is to give colleagues guidance on specialized subjects. The need for more specialist science teaching in primary schools is currently being debated nationally.

- Traditionally, science teaching at the primary level has been approached within the context of topic or thematic work, in which several areas of the curriculum are covered using a theme such as water, transport, or materials. Recently, there have been suggestions that curriculum content could be taught more effectively as specific subjects rather than within the framework of an all-embracing topic. The arguments for this move are most forceful at the upper primary level, Grades 6 and 7.

- The need for a balance between the amount of time spent developing the skills of scientific investigation (attainment target 1) and on the knowledge and understanding of science (attainment targets 2 to 4). Currently, 50 percent of the time is spent on attainment target 1 at primary school, and 25 percent at secondary school.

- An increasing occurrence in both primary and secondary science classes is the relating of science lessons to everyday situations, with environmental, resourcing, and conservation issues being particularly topical.

Science Textbooks

At the primary level, science textbooks are not usually used; it is more common for materials such as photocopied resources, work cards, or worksheets to be integrated into planned thematic work. These are supplemented with books which rely on colorful material as a stimulus to the student. In secondary schools, however, there is a more widespread use of textbooks, with a small number dominating a large market. Attempts are increasingly being made to relate contexts of textbooks at all levels to everyday-life situations.

In the primary grades, it is common practice for the teachers to plan their lessons to fit in with other work that is being done by the class, often selecting resources from several different schemes. Secondary school science lessons in contrast, have a stronger reliance on what appears in the textbook, with the students tending to stay with a particular scheme throughout a key stage. In Key Stage 4, for example, the chosen scheme would be one that has a close relation to course requirements as students approach the General Certificate of Secondary Education assessment. Textbooks and schemes produced for secondary school pupils increasingly provide for a spread in the ability of students, with supporting materials for those students of lower ability and extending topics for more able students. It is left to the school's discretion which textbook is used, although they are offered guidance and recommendations by advisers.

The main professional association for science teachers, the Association for Science Education, produces a range of support materials for teachers. These include resources which involve practical investigations and emphasize the relevance of science and technology to everyday life.

Pedagogy

A strong emphasis is placed on practical activities across all age groups in both primary and secondary schools. Practical science lessons in primary schools use everyday material and equipment with which the pupils are familiar. In contrast, all secondary schools are equipped with a science laboratory, where pupils use Bunsen burners, acids and alkalis, and other specific science equipment.

Teachers at primary and secondary levels have increased their efforts to cater for the different abilities present in their classes, providing extension work for those who require it and supporting those who are less able. These accommodations are necessary in science lessons, since students are not ability-grouped as they often are in mathematics.

Within primary schools, cross-curricular topics embracing science are taught for extended periods of time, usually a half or whole term. In secondary schools, science departments frequently divide the curriculum for each year into "modules" focusing on a particular area. These usually last four to six weeks and are covered by groups of students in rotation. Different teachers are commonly used for the different modules, with chemistry teachers teaching chemistry-applicable modules, and likewise for the physics and biology modules.

Primary school pupils are sometimes asked to produce an oral rather than a written report of an activity they have carried out. In secondary schools, there is less opportunity for oral work, and written reports are usually prepared.

Teachers are expected to include the use of information technology in the context of science, and one of the attainment targets in the 1989 science curriculum concerns the use of microcomputers in science. Computers are widely available in schools and students are becoming familiar with their use and applications. The ways in which computers are used for the teaching of science vary from class to class and from school to school. Although teachers are encouraged to include the use of information technology in the teaching of science, in reality it is less widespread than it might be. A recent Office for Standards in Education report has found that, in reality, there are relatively few schools where information technology is well used to support pupils' learning in science. The reasons for this are unclear, but may include shortages of hardware or software for use in science classes.

6 EVALUATION POLICIES AND PRACTICES

A number of laws passed since 1980 in England oblige school governing bodies and local education authorities to make information available to parents and others on the performance of individual institutions and their students. This information is based on the results of national curriculum assessments and public examinations.

The assessment arrangements for the new curriculum were proposed in 1987 by the government-appointed task group on assessment and testing. The changes were introduced and supervised from 1988 to 1993 by the school examinations and assessment council, which was superseded in 1993 by the school curriculum and assessment authority. The aim was to provide an assessment system that was formative, summative, evaluative, and informative, and that enhanced teachers' professional development. To meet these objectives, the task group on assessment and test-

ing suggested an innovative assessment system. Attainment is measured in a continuous scale of 10 levels, which covers the entire age range from 5 to 16. Pupils acquire a particular level by demonstrating the performance set out in the criteria, and are not ranked in order in comparison to others of their age, as would be traditional.

The assessment system consists of two complementary and separate strands: continuous assessment by teachers and assessment by externally devised assessment tasks or tests. There are four ages at which the legislation requires external assessment according to national curriculum criteria: 7, 11, 14, and 16. Each corresponds to the end of a key stage of education lasting for two, three, or four school years.

Mathematics and science are two of the three core subjects in the curriculum, English being the third. At Key Stages 1, 2, and 3, the level of performance in the core subjects is assessed in two ways: tests set by the government agency responsible for curriculum and assessment, the school curriculum and assessment authority, and separate assessments made by each pupils' teacher. At the end of Key Stage 4, pupils may sit for a number of subjects at the General Certificate of Secondary Education examinations set by one of five independent examination boards. The nature and content of these examinations is overseen by the school curriculum and assessment authority. New General Certificate of Secondary Education syllabi that reflect the revised national curriculum will be introduced beginning in 1996.

The National Council for Vocational Qualifications is not an examining body in its own right, but was established in 1986 to monitor existing awarding bodies and to ensure that the qualifications they awarded met the standards drawn up by the Council. Those examinations that it approves are designated National Vocational Qualifications, or General National Vocational Qualifications. General qualifications are mainly for students aged 16 to 18 in full-time education, and, as their name suggests, are directed at broad vocational areas rather than at particular occupations, and are uniform throughout the country. They were introduced in September 1993, and may in the future be combined with national vocational qualifications, General Certificate of Education A and AS Levels, and General Certificate of Secondary Education courses and qualifications.

There are a number of issues in evaluation standards and practices today:

- The standards of General Certificate of Secondary Education and General Certificate of Education A and AS Levels have fallen during the last 10 or 20 years. In the autumn of 1995, the government set up an inquiry into comparative standards over time, on which they are to report at the end of 1995.
- Assessment is taking up too much time: assessment of pupils by their teachers is ongoing for all subjects, and is supplemented by formal assessment at the end of key stages. Pupils' levels of attainment are reported to parents in formal written reports.
- The curriculum is excessively influenced by the requirements for ongoing and formal assessment. Some would argue that there is a danger of teachers becoming too preoccupied with their pupils achieving high levels of attainment in tests at the expense of experiencing a wide and balanced curriculum.
- Although there are currently no plans to implement league tables at Key Stages 1 and 3, they will be brought in at Key Stage 2 when the system has settled down sufficiently.

7 REFERENCES AND SOURCES FOR FURTHER READING

Common Sources

1   Husén T, N T Postlethwaite 1994 *International Encyclopedia of Education*. Pergamon Press, Oxford
2   Organisation for Economic Co-operation and Development 1995 *Education at a Glance: OECD Indicators*. Organisation for Economic Co-operation and Development, Paris
3   United Nations Educational, Scientific and Cultural Organization 1995 *Statistical Yearbook*. United Nations Educational, Scientific and Cultural Organization, Paris
4   World Bank 1995 *World Development Report 1995*. Oxford University Press, New York

Other Sources

5   Central Office of Information 1995 *Britain 1995: An Official Handbook*. London, HMSO
6   Chartered Institute of Public Finance and Accountancy 1995 *Education Statistics 1993–94 Estimates*. London, CIPFA
7   Education 1995 Class size. *Education*. 186 (15) 13 October
8   Ekninsmyth C, J Brynner 1994 *Basic Skills of Young Adults*. London, ALBSU
9   Epic Europe 1995 *National Dossier of the Education System of England, Wales and Northern Ireland 1994*. Slough, NFER, Epic Europe
10 Great Britain, Department for Education 1995 *Survey of IT in Schools*. (Statistical Bulletin 3/95) London, DFE
11 Great Britain Central Statistical Office 1992 *Annual Abstract of Statistics 1992*. London, HMSO
12 Great Britain Central Statistical Office 1994 *Annual Abstract of Statistics 1994*. London, HMSO
13 Great Britain Central Statistical Office 1994 *Regional Trends 29*. London, HMSO
14 Great Britain Central Statistical Office 1995 *Annual Abstract of Statistics 1995*. London, HMSO
15 Great Britain Central Statistical Office 1995 *Capital Expenditure: Provisional Results—1st Quarter 1995*. (News Release 102) London, HMSO
16 Great Britain Central Statistical Office 1995 *Economic Trends*. London, HMSO
17 Great Britain Central Statistical Office 1995 *Social Trends 25* London, HMSO
18 Great Britain Office of Population Censuses and Surveys 1993 *1991 Census. Ethnic Group and Country of Birth Great Britain*. London, HMSO
19 Great Britain Parliament, House of Commons 1995 *The Department for Education and Office for Standards in Education Departmental Report: The Government's Expenditure Plans 1995–96 to 1997–98*. (Cm. 2810) London, HMSO
20 Great Britain Parliament, House of Commons 1995 *School Teachers' Review Body: Fourth Report 1995*. (Cm. 2765) London, HMSO
21 Great Britain Treasury 1995 *Public Expenditure: the Statistical Supplement to the Financial Statement Budget Report (1995–1996)*. London, HM Treasury
22 Hunter B (ed) 1995 *The Statesman's Yearbook 1995–96*. London, Macmillan
23 Office for Standards in Education 1993 *Mathematics Key Stages 1, 2, 3 and 4. Fourth Year 1992–93 (The Implementation of the Curricular Requirements of the Education Reform Act)*. London, HMSO
24 Office for Standards in Education 1993 *Science Key*

Stages 1, 2, 3 and 4. Fourth Year 1992–93 (The Implementation of the Curricular Requirements of the Education Reform Act). London, HMSO

25 Payne J 1995 *Routes Beyond Compulsory Schooling*

(Youth Cohort Report No. 31). Sheffield, ED

26 School Curriculum and Assessment Authority 1995 *1994 A and AS Examination Results.* London, SCAA

Statistical References

Section	Statistic	Reference	Page	Table	Year of Statistic
Country Profile	130,423 km^2	14	11	2.5	1995
Country Profile	372/km^2	5	6	–	1993
Country Profile	48,208,000	14	16	1.2	1991
Country Profile	20%	9	2	–	1991
Country Profile	94%	22	1306	–	1995
Country Profile	$14,058. $14,923.	16	T6	2.4	1995
Country Profile	3%	Derived from 11, 12	237, 240	14.1	1992, 1994
Country Profile	16%	21	85	7.7	1995
Country Profile	5%	17	59	–	1995
Country Profile	99%	8	20	2.2	1994
Structure of the System	9, 4, 23, 16,	Derived from 14, 25, 26	–	–	1995
Schools in the System	380	9	41	–	1995
Schools in the System	27, 22	7	17	–	1995
Teacher Profile	80, half and half, 40, 49	20	61	3	1995
Teacher Profile	$17,357. $46,985. $28,971. $38,129.	20	56	Appendix D	1995
Mathematics Pedagogy	10	10	–	Table 2	1995
Mathematics Pedagogy	85	10	–	Table 2	1995

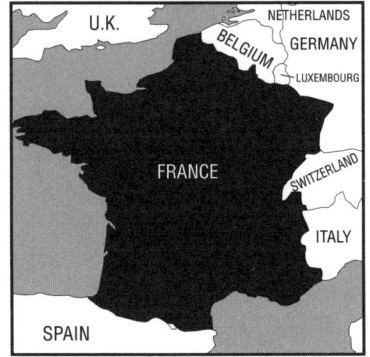

France

Anne Servant, Ministry of Education[1]

---

1 COUNTRY PROFILE

France is a country of 549,000 km² located at the western end of Europe, bordering Spain, Italy, Switzerland, Germany, Luxembourg, and Belgium. In addition to continental France there are four overseas *départements*, Guiana, Guadeloupe, Martinique, and Réunion, and two territorial collectives, Mayotte and Saint-Pierre-Miquelon. Overseas territories include French Polynesia, New Caledonia, Wallis and Futuna Islands, and the French Southern and Antarctic territories.

France's population in 1995 was 58 million, with a density of 105 persons/km². The total annual growth rate of the population has been less than half a percent for several years. The birth rate is relatively stable and the death rate is decreasing marginally. Immigrants, most of whom are from North Africa, comprise about 8 percent of the population. The rate of immigration has decreased in recent years.

About 75 percent of the population lives in urban areas. In addition to Paris there are 10 cities with more than 200,000 inhabitants. Once a predominately rural country, France has undergone an industrial transformation and agriculture now employs less than 7 percent of the working population. France is nevertheless the leading agricultural power in the European Union.

France's employed population totals 22 million. Services such as administration, commerce, banks, transportation, and the army employ about 60 percent of the population. Six percent of the active population work in the education system. The unemployment rate is high, comprising three million persons.

France is a republic headed by a president who is elected for seven years by universal suffrage. The president names the prime minister, and following the prime minister's recommendations, the members of the government. Parliament consists of the *Assemblée nationale* and the *Sénat*. The government is highly centralized, but the country has undergone a strong decentralization since 1982. France is divided administratively into 21 *régions,* and these are subdivided into 100 *départements*. The smallest unit is the *commune*, of which there are a large number.

France is a member of the OECD and is ranked as a high-income country. Spending on education corresponded to 7 percent of the gross domestic product in 1994, an increase of 2 percent over 1993. Eighty-eight percent of expenditures on education come directly from public funds, including state, territorial, community, and other administrations. This percentage, the highest among comparable countries, is due in part to the fact that tertiary education is largely covered by the state. The Ministry of Education financed 57 percent of spending on education in 1994, and its expenditures represent 21 percent of the state's total budget.

2 THE EDUCATION SYSTEM

Governance and Decision Making

In France, all schools, public and private, must conform to national legislation on education with decrees and rules established by the Ministry of Education. The only exceptions are some agricultural schools, which are controlled and administered by the Ministry of Agriculture.

The Ministry of Education is a body consisting of a minister, the *Bureau du cabinet*, the General Inspector-

---

[1]This article was written with the assistance of members of the *Inspection générale de l'Education nationale* and of the *Direction des Lycées et des Collèges.*

ate, high-level executives from the Department of Defense, 13 Directorates, and several other subsidiary departments. In addition to these, there are several institutions under Ministry supervision, including the National Institute of Educational Research, the International Center for Pedagogical Studies, the National Center for Pedagogical Documentation, the National Office of Information on Training and Professions, the National Center for Distance Education, and others.

At the regional level, education in France is divided into 28 administrative units called *Académies*. Since the institutional reform of 1982, a process of decentralization has been implemented and expanded, giving more autonomy at the local level. Nevertheless, it is the Ministry that defines educational goals, programs, levels of diplomas, and appoints teachers.

Programs are elaborated at the national level by special commissions with members of the General Inspectorate, the Pedagogical Directorates, together with the National Council of Programs, and university professors. Programs are published in the Ministry's official bulletin. An important new aspect is that programs are submitted to a process of general consultation with all teachers before final implementation.

Teachers may choose their pedagogical approach, provided they follow the national curriculum. Textbooks are published by private corporations, but they follow most of the course of study outlined in the national curriculum. In primary school, books for students' use are usually purchased by the *commune*, but not replaced every year. At the lower secondary level, *collège*, the state pays for all textbooks and they are returned to the school at the end of the year. Teachers decide collectively which textbooks will be used, and their decision must be approved by the school administration council. Until recently, the state did not provide any funds for the purchase of textbooks at the upper secondary level, and as a consequence some students did not have books. Since that time, a fund has been created to rectify these situations.

Structure of the System and Participation Rates

One of the main goals of the education system is to provide every young person with a minimum level of education. This program has met with considerable success. During the last 15 years, the number of stu-

dents who are qualified to pursue a *baccalauréat* has grown from 34 to 70 percent of the cohort. Regional disparities in access to education are less pronounced than 20 years ago.

France has a long tradition of preprimary schooling, and today most preprimary schools are public. Between 1960 and 1995, the number of children two to five years old increased from 50 percent to 85 percent. Nearly all three-year-olds are now in school.

Figure 1 shows the structure of education in France and the approximate enrollment rates for each level. The compulsory entrance age for school is six years. Children must turn six by December 31 of the year they enter school, and some children thus enter Grade 1 at the age of five. By January 1, 1995, 91 percent of Grade 1 pupils were six years old, 7 percent were seven years old, about half a percent were eight or more, and 1.5 percent were five years old. School is compulsory until the age of 16.

Eighty-six percent of primary schools are public. The five years of primary education are organized into *cycles* that are linked to preprimary education, with aspects of the curriculum being learned over several years. Repeating a year at primary school, which used to be common in the 1960s, is rare today. Virtually 100 percent of the age cohort is enrolled in primary school.

Thirty-three percent of secondary schools are private. In general, private schools enroll fewer students than public schools; private schools had a mean size of 325 students in 1994–95, while the corresponding figure for public schools was 619. Thus, nearly 80 percent of secondary school students are enrolled in public schools.

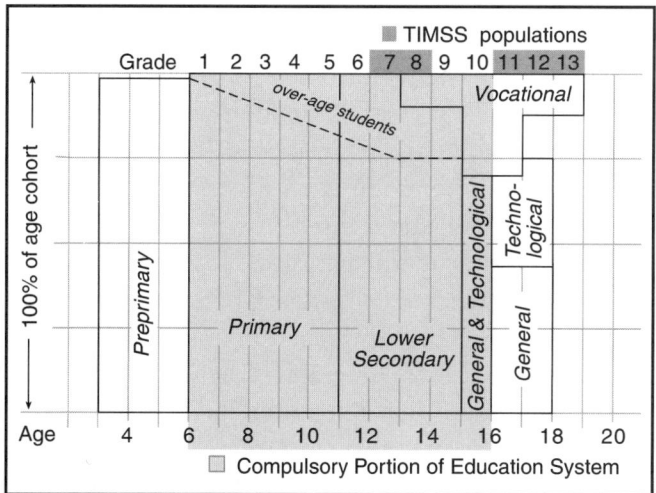

Figure 1: Structure of the Education System

There are three different types of secondary schools: *collèges* or lower secondary school for Grades 6 to 9, *lycées* or upper secondary school for Grades 10 to 12, and *lycées professionnels* or vocational upper secondary school. Programs in vocational upper secondary may end at Grade 11 or 13. In the public sector, *collèges* represent 66 percent of schools, a higher percentage than in private sector, which invests more in upper secondary schools, including vocational schools. The average number of pupils in a private lower secondary school is about one-third that of a public one.

The number of lower secondary schools has increased enormously since 1960. The size of secondary schools varies from fewer than 300 students to more than 1500. The division between the size of the public and private sectors, and also between types of schools is extreme: about 57 percent of public *lycées* have more than 900 students. The size of vocational schools varies according to the nature of the training provided.

Most private schools are under contract to the state, which is responsible for paying teachers. The state sets the same pedagogical requirements for private as for public schools and provides the same financing. Private schools are able to draw on their own funds.

In Grades 6 and 7, all students follow the same curriculum. Vocational training is available after Grade 7, but enrollment at this level has decreased dramatically in recent years. Most students in Grades 8 and 9 attend general studies, but there are also technological classes leading mainly to higher-level vocational studies. About 10 percent of students attend these classes, and about two-thirds of these are boys. All pupils at this level study mathematics and science. Classes in physics and chemistry begin in Grade 8. At the lower secondary level, the repetition rate is about 10 percent. This figure is slightly higher after Grade 7.

In Grade 10, a clear distinction emerges between vocational studies and general or technical education. There are both common areas of study and optional courses in general and technical Grade 10. All students at this level study mathematics and science courses. In 1994–95, 68 percent of Grade 10 students were enrolled in general and technical programs, and the remainder in vocational programs. Calculated in another way, 92 percent of that generation has access to either Grade 10 or *Brevet d'études professionnelles*.

In Grade 11 the different tracks are strongly differentiated, leading to corresponding types of *baccalauréats*. The *baccalauréat général* has three main tracks, and the *baccalauréat technologique* four.

Vocational Grade 10 is the first year of a program leading to the *Brevet d'études professionnelles* or to the *Certificat d'aptitude professionnelle*. Most pupils achieve a *Brevet d'études professionnelles*, which is granted after Grade 11. About 50 percent of students achieving this diploma decide to continue their studies, either by joining the technological track through a *classe d'adaptation* or by continuing in vocational upper secondary for an additional two years to achieve the *baccalauréat professionnel*. Their choice depends mainly on the area of their studies and employment prospects with a *Brevet d'études professionnelles*. The *baccalauréat* leads directly to university studies.

Young people between 16 and 25 who are preparing for a vocational diploma within a work contract are known as apprentices. Training within an enterprise is usually associated with courses at regional apprenticeship centers, which are under the aegis of the Ministry of Education. Apprentices come from diverse backgrounds. Fewer than one-third enter the apprenticeship program from Grade 9.

Schools in the System

The School Year

The law states the school year must be 36 weeks long, spread over five terms of comparable length, divided by four periods of school holidays. The longest holiday is during the summer, beginning at the end of June. The 1995–96 school year began on September 5. Not all regions have holidays on the same dates, but the length of the school year is identical.

Primary school pupils have 26 hours of class per week and lower secondary students, an average of 27. The organization of these hours through the week varies. At the primary level, for example, more and more schools are adopting a four-day week, excluding Wednesday and Saturday, and having shorter holiday periods. At the secondary level, schools organize their time in various ways and the length of school day can also vary significantly during the week.

Class Size

The average number of pupils in preprimary classes is 27. In public primary schools, the average class size is

23, in private-sector schools, 24. The average number of students in a lower secondary school class is 25. Classes in the public *lycée* have on average 30 students, while in private *lycées* the figure is 26. Vocational classes tend to be smaller: 22 in public and 21 in private upper secondary schools.

Certification of Mathematics and Science Teachers

As a result of a drive to reassert the value of the teaching profession, recent years have been marked by increased qualification levels, a simplification and harmonization of the general statutory situation, and more homogeneity in salaries. Since 1991, requirements and conditions for teacher qualification have been standardized within and between the primary and secondary levels. The basic requirement for primary and secondary teachers is a three-year university degree, called a *licence*, or the equivalent. After obtaining the *licence,* preservice teachers must pass competitive examinations. Some prepare for the examinations individually, while others attend a teacher education institute associated with a university to prepare for the exam. These examinations lead to different types of certification, depending on what subject the candidate intends to teach. The primary school examination takes into account the many subjects taught at this level by the teacher, while the secondary school examinations focus on specific disciplinary fields. After passing the competitive examinations, preservice teachers must attend a teacher education institute associated with a university for one year. This period of time, which may be the second year at such an institute for some candidates, involves additional professional education and leads to final certification. Final placement in a teaching position is very competitive: In 1994, 27 percent of mathematics and science candidates were placed through external competition, and 41 percent through internal competition. Internal placement refers to placement of those who are already teaching, but who do not have all the new qualifications. There are opportunities for them to upgrade, but it is a difficult process. Some teachers prepare for a more difficult competitive examination called *Agrégation,* for which an additional year of university study is needed.

Teacher Profile

In recent years, there have been concerted efforts to reassert the value of teaching as a profession, increasing qualification levels and equalizing salaries. Primary school teachers holding the new certification now earn the same salary as secondary school teachers, or about US$32,000 in 1994. The salary of primary school teachers with older qualifications is somewhat lower; in 1990, the average salary at this level was US$22,700.

Teachers in secondary school are only slightly older than those in primary schools, with average ages of 42 and 41 respectively. In primary schools, 76 percent of teachers are women. In secondary schools, 56 percent of teachers are women. Fifty percent of mathematics teachers are women, as well as 65 percent of biology and geology teachers and 44 percent of physics and chemistry teachers.

The work week for teachers varies a great deal. Teaching time in primary school is 26 hours per week; the figure ranges between 15 and 22 hours at the secondary level, depending on the discipline and the status of the teacher.

3 TIMSS POPULATIONS

France, IEA, and TIMSS

Since 1996, the *Direction de l'Evaluation et de la Prospective* at the Ministry of Education has represented France at IEA. France has participated in several IEA studies, including the Second International Science Study and the Reading Literacy Study.

French TIMSS Populations

France is participating in TIMSS at the Populations 2 and 3 levels. Population 2 includes students enrolled in Grades 7 and 8 in both general and technological classes. Population 3 generalists includes all students in the final year of preparation for the *baccalauréat.* This involves students in Grade 12 preparing for the *baccalauréat général ou technologique*, and in Grade 13 for the *baccalauréat professionnel* (vocational). In addition, Population 3 includes students who are in their final year of preparation for the *Brevet d'études professionnelles* or the *Certificat d'aptitude professionnelle* at *lycées professionnels* who will not continue toward a *baccalauréat*. Population 3 specialists includes students in the scientific track of the *baccalauréat général*. These students represent 47 percent of the *baccalauréat général*, 28 percent of all *baccalauréats*, and about 22 percent of Population 3.

Goals for the Mathematics Curriculum

Official national goal statements are given with every curriculum and syllabus. They include:

- mastering basic mathematical knowledge and skills in order to use them in various fields;
- developing logical thinking and reasoning abilities;
- grasping problem-solving methods;
- learning scientific processes and attitudes.

As in other fields, the program of study is elaborated by national groups. In mathematics, the *Association des professeurs de mathématiques de l'enseignement public* sometimes offers suggestions and proposals to the Ministry of Education. Official statements are prepared at the Ministry level.

From Grade 8 on, and most particularly in upper secondary school, when the tracking system defines diverse routes, goals are adapted to the different roles that mathematics will play in the lives and further studies of students. In addition, goals are articulated according to the different intellectual or practical implications of the program of study. This corresponds to current discussions about the place of mathematics in the education system in general.

Major Changes in the Mathematics Curriculum

The years from 1970 to 1980 were known as the era of "modern mathematics," with an emphasis on abstraction and mathematical structures. The general trend now is to teach application-oriented knowledge and reduce the amount of theoretical knowledge. Recently, there have been some attempts to take into account research on new teaching methods, particularly those efforts completed within the framework of teacher education institutes.

Calculators and computers have been integrated into everyday teaching. In primary schools, the use of four-function calculators has slightly decreased the importance of written calculations and has enabled teachers to pose problems with an emphasis on methods rather than operations. At the lower secondary level, equal importance is given to calculators, hand calculation, and mental calculation. In upper secondary schools, the use of programmable and graphics calculators has brought about changes in the calculus curriculum and in the study of functions.

Computers are also used, without being part of the intended curriculum except in a few specialized classes. Nevertheless, data processing is becoming increasingly important. Statistics and probability have been added to the mathematics curriculum.

Current Issues in the Mathematics Curriculum

The variety of tracks, particularly after lower secondary school, is important in the context of defining curricular issues in mathematics. Important reforms have taken place, particularly in upper secondary schools in recent years, that allow the elaboration of new programs.

Current teaching programs plan the study of concepts over several years. Proportionality, for instance, is systematically studied in all the years of lower secondary school, as are orthogonal and central symmetries. Arithmetic is not taught anymore. At the upper secondary level, the official program states that it is important to choose a progression that will allow the development of new concepts.

At the upper secondary level, new programs were designed to be easily adapted to different situations and needs. For instance, in the *terminale économique et sociale* track in Grade 12, the mathematics curriculum stresses applications within economics and social sciences. It states that mathematics should promote the coherent training of pupils. For the scientific tracks, the official policy states: "To meet the national goal of providing more engineers, researchers, teachers, and technicians with a solid scientific grounding, we want to pursue the policy of opening scientific tracks, while continuing to offer quality mathematics training."

Much attention is given to the development of scientific activities, methods, and communication. As in other fields, modules have been created for Grades 10 and 11 as optional materials for teachers to use according to the specific needs of the students.

Mathematics Textbooks

There are no official mathematics textbooks. Textbooks are written by individuals, usually following the curriculum closely, and published by private companies. There have been no significant changes in the number of topics and pages. The way of presenting material, however, has been modified and greater importance is now given to activities rather than lectures. Some historical highlights can also be found, and especially at the lower secondary level, more pictures and illustrations.

Teachers use textbooks to plan lessons and particularly to set exercises and problems for students. Textbooks are designed for students to read and learn from on their own, but in fact students tend to rely on teachers for instruction and very often use textbooks only for exercises.

Pedagogy

Mathematics education cannot be limited to formal knowledge. A three-step program of teaching is sometimes used, stressing approach activities, synthesis, and finally activities concerning applications and deeper knowledge. At the lower secondary level, the curriculum states: "These conditions are essential if we want to bring pupils into the intuitive comprehension of concepts and their appropriate use in simple situations, while allowing them to deepen and enrich their mathematical education."

Among other suggested pedagogical approaches are the following, all of which should be adapted to the individual needs of students:

- cooperative learning should be of moderate importance;
- deductive approaches should be used in mathematics instruction;
- scientific ways of thinking should be developed, along with an emphasis on students' individual work, investigations, and problem solving;
- new knowledge and skills should be introduced in the context of the students' frame of reference, building on concepts studied both within mathematics classes and within other disciplines.

The need for skills in reading and understanding mathematics texts is strongly emphasized at all levels. Students must be able to express themselves in a precise and clear way. Communication capacities are important in the new programs at the upper secondary level, where students encounter various situations involving listening and individual expression, both written and oral.

5 SCIENCE CURRICULUM AND PEDAGOGY

Goals for the Science Curriculum

The sciences are differentiated between biology and geology, also called life and earth sciences, and physics and chemistry, also called physical sciences. Official national goal statements are issued with the curricula, with texts specific to the different scientific fields.

The first basic concepts of science are taught in primary school. Biology and geology are taught beginning in Grade 6, and since 1992, physics and chemistry beginning in Grade 8. At the lower secondary school level, biology and geology "must enable pupils to acquire the scientific knowledge essential to understanding the contemporary world and its transformations, both in the field of life and health, and the field of resources and the environment."

Special attention is given to active learning and to the development of scientific habits of mind. In general, science teaching must provide a coherent representation of the universe. Major goals in science teaching include discovering, observing, reasoning, experimenting, and constructing models and hypotheses. These general goals are of course adapted to the level of study and to the track chosen. Scientific methods and attitudes and intellectual honesty are also specifically defined in the physics program. Chemistry has a special role with respect to security, health, and environment in the education of citizens.

Major Changes in the Science Curriculum

In the biology and geology curricula, the last 10 years have seen more focus on teaching through problems concerning concepts. The general approach is more naturalist and less conceptual. At the upper secondary level, increased attention is given to cellular and molecular biology and global tectonics, taking into account both technical aspects and human implications. A systemic approach is used, studying functions in biology, and the circulation and transformation of material in geology.

The change in the grades at which the physical sciences are now taught has led to the development of new programs. The physics and chemistry curricula are aimed at providing knowledge and skills for effective interaction within a technological world, including developing a critical mind, examining new technologies, and rejecting any trivialization of science.

The introduction of computer science has changed the way experiments are used in teaching physics. Computer-assisted experiments are more and more common. Software makes it easier to teach some theoretical concepts. Computers have enabled physics instruction to be broadened by helping students to overcome underlying mathematical difficulties.

Computer science has also been introduced into the life and earth sciences curricula, in the form of computer-assisted experimentation, data banks, and software tools. In addition, new technologies and means are now being used, including satellite images and haploid cultures. Practical work has been modernized accordingly.

Current Issues in the Science Curriculum

What constitutes appropriate scientific training for the twenty-first century has been the subject of much debate. The need to teach scientific culture is elaborated in the official guidelines for science teaching: "Scientific culture, inscribed in human history, contributes to the general culture of everybody."

In biology, recent curricular changes were developed according to scientific logic with an augmented genetics program. Approaches to earth science have been broadened in areas such as the solar system, external geodynamics, oceanic and atmospheric phenomena, and aspects of energy. An important issue is the process of constant adaptation to the modern world.

Programs in physical sciences are now designed so that teaching is rooted in the student's environment and in the applications of modern technologies. In this way, the physical sciences aim to show students that the world is intelligible and that we can act on our knowledge. To fulfill these goals, the physical science program is built around a dominant theme chosen for its practical importance. There were some criticisms that the curriculum is at times too formal, that the programs are too broad, and that examinations stress the use of formulae at the expense of understanding. These opinions are being taken into account.

Science Textbooks

Life and Earth Sciences

There has been a marked increase in the number of pages in textbooks at all levels. Generally, textbooks consist of the text of the lesson supplemented by photographs, drawings, graphs, and numerical data, and followed by exercises and summaries. Lessons and exercises often focus on the applications of science. Historical perspectives are a new feature in some textbooks. While student autonomy in the use of textbooks is a general goal, it is difficult to attain.

Physical Sciences

In recent years, physics and chemistry textbooks have had additional exercises and applications included, at the expense of the space devoted to the main lesson. Textbooks currently in use consist of 40 percent lesson development divided equally between text and graphs or illustrations, 20 percent applications of the lesson including experiments, and 40 percent exercises. In spite of this design, textbooks are rarely used for learning lessons. Exercises contained in them, on the other hand, are used extensively.

Pedagogy

The curriculum guides for the new science programs state that:

Pedagogical processes should be neither dogmatic nor formal and should, as often as possible, be based on rigorous exploitation of observations and experimental activities. Science teaching should be based on knowledge that pupils are familiar with, building on these preliminary concepts. The teacher will progressively introduce corresponding knowledge, using modern technologies such as audio-visual aids and computers. Several types of activities should be proposed, including manipulations, observation and performing of experiments, looking for documents, and oral presentations. Teachers should use diverse methods, including individual work, cooperative groups, and collective discussions.

The approach to science teaching is in general quite inductive. The relative emphasis placed on teaching facts and laws as opposed to process skills depends on subject and grade. Integration across disciplines is not a goal, since curriculum content is specific to each field. Nevertheless, biology and geology are taught by the same teacher, as are physics and chemistry. References to different disciplines are frequent but not consistently used. There is a stronger emphasis on oral and written communication in life and earth science than in physics and chemistry.

6 EVALUATION POLICIES AND PRACTICES

To achieve the mission of the education system, to allow students to study in school as long as possible, to generate evolution in the role, recruitment, and education of teachers, to improve school life: all these tasks require considerable effort. This effort can only be approved if significant progress is achieved and if results and feedback are given to the nation. Two requirements illustrate the action that needs to be taken: evaluation and assessment of achievement. (Loi d'orientation, 1989)

This official policy statement indicates the importance and breadth of action in the field of evaluation. Endeavors in the field range from the national to the local level. The main goals are "the improvement of the educational system, verifying the implementation of national education goals, adapting them to the different publics, and implementing a permanent regulation of the system."

The General Inspectorate, together with other agencies, is in charge of evaluation at all levels and pub-lishes an annual report. The *Direction de l'Evaluation et de la Prospective* is "in charge of the evaluation of the education system. Thus, it designs, implements, and analyzes evaluations of student achievement, of schools and training units, and of educational policies and innovations."

At the regional level, specific evaluations in different areas are also implemented. In schools and classes, great importance is placed on regular assessment in order to adapt pedagogy and help pupils assess themselves. Some evaluations are performed on national samples to assess achievement in relation to the national curriculum. In mathematics and science, among other subjects, these evaluations have taken place at the end of Grades 7, 9, and 10 regularly over the last 15 years.

Since the *Loi d'orientation* of 1989, large-scale evaluations have been obligatory for all pupils. Evaluations in mathematics and French are administered annually at the beginning of Grades 3 and 6. The results of these evaluations are used to enable teachers to adapt their pedagogy to the class, as well as to provide feedback to pupils and parents. Results may be calculated at the school and regional level, and results of a national sample are published by the Ministry. Since 1992, the same kind of evaluation, covering several other subjects including science, has been implemented at the beginning of Grade 10.

These evaluations, implemented by the *Direction de l'Evaluation et de la Prospective*, are based on standardized items and conditions, designed by specialists in the field, inspectors, teachers, and researchers, together with Ministry representatives. Item banks are also published to give teachers more tools to use in direct assessment of pupils. Within schools, the most common form of assessment remains conventional exercises marked by grade, rather than the pedagogical coding systems used in national evaluations.

7 REFERENCES AND SOURCES FOR FURTHER READING

Common Sources

1 Husén T, N T Postlethwaite 1994 *International Encyclopedia of Education*. Pergamon Press, Oxford

2 Organisation for Economic Co-operation and Development 1992 *Education at a Glance: OECD Indicators*. Organisation for Economic Co-operation and Development, Paris

3 United Nations Educational, Scientific and Cultural Organization 1994 *Statistical Yearbook*. United Nations Educational, Scientific and Cultural Organization, Paris

4 World Bank 1993 *World Development Report*. Oxford University Press, New York

Other Sources

5 Direction de l'Evaluation et de la Prospective 1995 *L'Etat de l'Ecole, no. 5*.

6 Direction de l'Evaluation et de la Prospective 1995 *Repères et Références Statistiques*.

7 Ministère de l'Education nationale et CNDP 1985 *Collèges: Programmes et instructions*.

8 Ministère de l'Education nationale et CNDP 1995 *Programmes de l'école primaire*.

9 Ministère de l'Education nationale, de l'Enseignement supérieur et de la Recherche *Bulletin officiel de l'Education nationale: août 1989, no. spécial 4 (loi d'orientation sur l'éducation); juillet 1992 no. 31; no. hors série du 24 septembre 1992; juin 1994, no. spécial 6; juillet 1994, no. spécial 7; juillet 1994 no. spécial 8; septembre 1994, no. spécial 11; février 1995, no. spécial 3; juin 1995, no. spécial 11*

10 Journal officiel de la République Française, janvier 1996

11 Direction de l'Evaluation et de la Prospective mars 1994 *Education et Formations, no. 37*.

12 Direction de l'Evaluation et de la Prospective décembre 1994 *Les Dossiers d'Education et Formations, no. 48*.

13 Institut national d'Etudes démographiques mars 1995 *Population et Sociétés, no. 299*.

14 Ministère de l'Education nationale mars 1994 *Enseigner dans les Ecoles*.

Notes on some important publications about the education system

Repères et références statistiques and *l'Etat de l'école* are yearly publications. *Repères et références statistiques* (about 340 pages) gives general information and statistical data about many aspects of the education system (schools, pupils, teachers, budgets, etc.).the Direction de l'Evaluation et de la Prospective has since 1993 published a yearly *Géographie de l'Ecole*, which gives comparative information (text and graphs) about the different regions of France.

The review *Education et Formations* is published three or four times a year and consists of a number of articles on the entire range of educational subjects. An abstract of every article is translated into English at the end of the review.

The *Dossiers d'Education et Formations* are study reports published about once a month that provide detailed information about specific subjects.

About once a week, a *Note d'Information*, which is widely circulated by the Direction de l'Evaluation et de la Prospective, provides four to six pages of information about a specific subject.

Statistical References

Section	Statistic	Reference	Page	Table	Year of Statistic
Country Profile	6%	5	12	–	1992
Country Profile	7%	5	12	2	1994
Country Profile	88%	5	13	2	1994
Country Profile	57%, 21%	5	14	–	1994
Structure of the System	34 to 70%	5	20	–	1994
Structure of the System	50 to 85%	6	61	1	1994–95
Structure of the System	91%, 7%, 0.5%, 1.5%	6	65	1	1994–95
Structure of the System	86%	6	35	1	1994–95
Structure of the System	88%,	6	55	3.1; 1	1994–95
Structure of the System	33%	6	39	1	1994–95
Structure of the System	325, 619, 80%, 66%, 300 to 1500, 57%, 900	6	43	1:2	1994–95
Structure of the System	68%	6	99, 107	–	1994–95
Structure of the System	about 50%	6	103	2,4	1994–95
Schools in the System	27, 23, 24	6	35	3	1994–95
Schools in the System	25, 30, 26, 22, 21	6	39	3	1994–95
Certification of Teachers	27%, 41%	6	253	–	1994
Teacher Profile	42, 41	6	223	1,2	1993–94
Teacher Profile	76%	6	225	2	1993–94
Teacher Profile	56%	6	237	1	1993–94
Teacher Profile	50%, 65%, 44%	6	239	1	1993–94
French TIMSS Populations	47%	6	107	1	1994–95
French TIMSS Populations	28%, 22%	6	107+99	–	1994–95

Germany

Kurt Riquarts, Institute for Science Education at the University of Kiel

1 COUNTRY PROFILE

Germany lies in central Europe and shares land borders with nine countries. From the Alps to the North Sea, Germany measures 876 km, with an area of 357,000 km^2 and a post-unification population of 80 million. The overall population density is 227 persons/km^2: 263 in the former West Germany and 145 in the former East Germany. In spite of the high population density there are only three cities, Berlin, Hamburg, and Munich, that have more than one million inhabitants. Two-thirds of the population live in communities of fewer than 100,000 inhabitants.

Eight percent of the total population is made up non-German ethnic groups: almost two million Turks, 916,000 people from the former Yugoslavia, and 558,000 Italians. The population is growing at a rate of less than 1 percent. The rate of growth for the German population is negative, but positive with regard to immigrants.

Germany is a federal republic of 16 *Laender,* or states, with a democratically elected, federal first chamber of parliament, the *Bundestag.* The *Laender* are represented in the second chamber, the *Bundesrat.* Areas of government jurisdiction are divided between the federal government and the states. Education and cultural affairs fall in the domain of the states.

Germany, a member of both OECD and the G7 Economic Group of countries, is rated as a high income country by the World Bank. The per capita gross national product for West Germany was US$23,560 in 1993 with an average annual growth rate of more than two percent between 1980 and 1993. This figure is reduced to less than one percent when the former East Germany is included. The average annual rate of inflation from 1980 to 1991 was less than three percent in the former West Germany only; in 1992, after unification, the inflation rate increased to 4 percent.

In 1992, expenditure on education was about five percent of the country's gross domestic product, thus placing Germany at the lower end of all OECD countries. According to UNESCO data, Germany's literacy rate is greater than 95 percent.

2 THE EDUCATION SYSTEM

Governance and Decision Making

Each of the sixteen *Laender* have sole jurisdiction over educational policy within their geographical area. Their authority includes regulation of curriculum and time schedules, professional requirements, school buildings and equipment, and teacher recruitment.

The states organize the supervision of the school system in three tiers: the Ministry of Education, regional school boards, and county school administrations. The intermediate level, regional school boards, does not exist in the nine smaller states.

The *Laender* coordinate their educational policy through the Standing Conference of *Laender* Ministers of Education. Resolutions and recommendations of the Conference of Ministers of Education only become legally binding when they are adopted into state laws, decrees, or regulations of *Laender* authorities. The Conference of Ministers of Education also deals with all curricular problems and innovations requiring coordination between *Laender*, such as recognition of examinations, the education of foreign students, and environmental education.

The federal Ministry of Education and Science has a concurrent right to legislate for the nonformal vocational sector, as well as to outline legislation on general principles for the university system. The Ministry also has joint legislative functions in education planning, the promotion of scientific institutions, and research projects of national significance.

The intended curriculum in mathematics and the sciences, as for all subjects, is defined at the state level according to school type and grade. All syllabi include the philosophy and rationale for the teaching of the subject, as well as a description of the content to be taught.

Authors and publishers develop textbooks based on the required state curricula. These textbooks are inspected to see whether they comply with the legally prescribed, subject-oriented curricula. For this reason, textbooks in Germany are an accurate reflection of the intended curriculum. Teachers and the subject department heads select the textbooks to be used from the state's list of approved textbooks.

Structure of the System and Participation Rates

Public and Private Systems

Although the state does not have a monopoly on education, only about 6 percent of school-age children attend private schools. These must be accredited by the state and are supervised by it. The state, however, is also required to subsidize them.

Structure of Education

Figure 1 shows the structure of education in Germany and the approximate enrollment rates for each level. Compulsory schooling commences at the age of 6 and finishes at 18. Nine or 10 of these years, depending on the school system of the individual state, must be spent in full-time schooling, and the following years either in full-time schooling or part-time vocational schools in conjunction with a trade or apprenticeship program.

Kindergarten, for three- to six-year-olds, is not directly linked to the education system and attendance is voluntary. Primary school, *Grundschule*, is the lowest level of the education system attended by all pupils and comprises Grades 1 to 4 for students aged 6 to 10.

Secondary level I, for students aged 10 to 16, offers differentiated teaching in accordance with student ability, talent, and inclination. Students are placed accord-

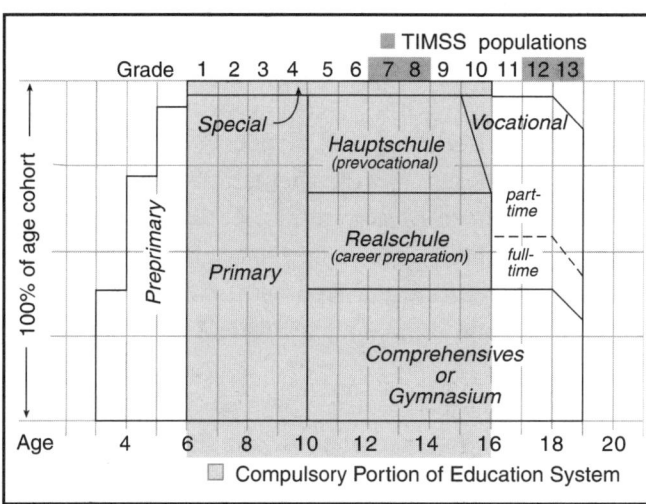

Figure 1: Structure of the Education System

ing to academic ability into one of the three systems, within which there is no streaming:

- *Hauptschule*, which provides a sound basis for subsequent vocational training;
- *Realschule*, which equips young people for subsequent careers in positions located between the purely theoretical and the purely practical;
- *Gymnasium*, which equips students for intellectual activity and prepares them for higher education.

Table 1 shows the distribution of students in secondary level I. The distribution of Grade 8 is representative of up and down grading within the three-fold system, which was stabilized by 1991.

Table 1: Distribution of Students in Secondary Level I

Secondary level I (10- to 16-year-old students)	Distribution of 14- to 15-year-old students (Grade 8) in 1991
Hauptschule (Grades 5–9/10)	28%
Realschule (Grades 5–10)	29%
Gymnasium (Grades 5–10)	30%
Comprehensive schools (not in all states) including Waldorf schools	9%
Special schools (various forms) for children with disabilities	4%

Secondary level II, for students aged 16 to 19, offers a three-year course qualifying students to enter university. Until the middle of the 1970s, the course was organized in terms of types of *Gymnasien* including classical, modern languages, mathematics, and science; this has been replaced by a system of basic and specialized courses combined with compulsory and optional ones. Many limitations are placed on students' choices to ensure that all students achieve a broad range of knowledge.

Secondary level II also encompasses full-time and part-time vocational education. The West German dual system of vocational education involves cooperative apprenticeships at two learning sites, the school and the workplace. Enterprise-based vocational training has two sponsors: the *Laender* governments, which establish and finance vocational schools, and the enterprises themselves, which finance and provide apprenticeships. Table 2 shows the distribution of students in secondary level II.

Table 2: Distribution of Students in Secondary Level II

Secondary level II (17- to 19-year-old students)	Distribution of 17- to 19-year-old students in 1993
Gymnasium and Comprehensive Schools (Grades 11 to 13)	31%
Full-time Vocational Education	16%
Part-time Vocational Education	53%

Funding the System

Schools are financed in three ways:

*   Personnel costs are paid by the states. Teachers, who are usually state civil servants, are assigned to schools in numbers that correspond to the number of pupils enrolled. The amount paid by the 16 states for teachers' salaries in 1991 was US$30 million, about 36 percent of their budgets.
*   Nonpersonnel costs such as building maintenance, equipment, laboratories, and libraries are paid by the county, although the state contributes to new construction and large investments such as new computers. The amount paid by the counties in 1991 was US$8 million, about 15 percent of their budgets.
*   In some states, parents pay for textbooks and learning materials. In the states governed by the Social Democratic party, textbook costs are usually paid by the county.

Enrollment in Mathematics and Science

In all of the 16 *Laender* in Germany, mathematics and the sciences are part of the core curriculum for Grades 1 to 10, and all students must study them. Mathematics is taught in all *Laender* and in all types of schools throughout these grades. In Grades 1 to 4, science is taught in all *Laender* as an integrated subject called *Sachunterricht*, which could be translated as "teaching about things" or "teaching about real objects". One-third of total instructional time is usually devoted to science-related topics.

In Grades 5 to 10, the sciences are taught separately as biology, chemistry, and physics. Only in the comprehensive systems is there a coordinated science program for Grades 5 to 8. The extent to which the three sciences are taught varies according to the type of school as well as to the *Laender*. Biology and physics are not normally taught in every grade, and chemistry teaching is usually begun later than the other two sciences.

Table 3, therefore, is included as an example of enrollments in mathematics and science. The data are drawn from the state of Northrhine-Westfalia, which has the highest population in Germany, and include figures from primary schools, *Gymnasien*, and *Gymnasien* and comprehensive schools.

Schools in the System

The School Year

The school year includes 38 weeks of instructional time, or between 190 and 220 days, depending on whether it is a five- or six-day school week. The school year ends with a six-week summer break that begins between mid-June and early August, based on a rotating system among the states. Two additional breaks include two weeks each at Christmas and Easter. In addition, other breaks are organized at the *Laender* level.

In Grades 1 to 4, a five-day school week is divided

Table 3: Mathematics and Science Enrollment (State of Northrhine-Westfalia)

Mathematics and Science Enrollment by Grade (Percent of Grade Cohort)

Grade	Modal Age	Mathematics	Science	Physics	Chemistry	Biology
1	6	100	100	–	–	–
2	7	100	100	–	–	–
3	8	100	100	–	–	–
4	9	100	100	–	–	–
5	10	100	–	–	–	100
6	11	100	–	100	–	100
7	12	100	–	–	100	100
8	13	100	–	100	–	100
9	14	100	–	100	100	100
10	15	100	–	100	100	–
11	16	100	–	55	50	95
12	17	100	–	30	20	50
13	18	25	–	30	20	50

into 20 to 25 class periods of 45 minutes each. The school day begins at 8:00 a.m. and finishes at 1:15 p.m. About 20 percent of instructional time is devoted to mathematics, and 5 to 10 percent to science-related topics. There is neither streaming nor a choice of courses at this level.

In Grades 5 to 10 the school week is divided into 30 to 35 class periods of 45 minutes each, either in a five-day school week (instruction in the mornings plus two or three afternoons) or in a six-day school week (instruction in the mornings, usually between 8:00 a.m. and 1:15 p.m.). About 13 percent of instructional time is devoted to mathematics, and 10 to 15 percent to the sciences. Students at this age are streamed into *Hauptschule*, *Realschule*, and *Gymnasium*, but within these divisions there is no streaming.

At secondary level II (Grades 11 to 13), an academic school week is divided into a minimum of 32 class periods, of which three periods or 9 percent are devoted to mathematics, and six periods or 19 percent to the sciences.

Class Size

The average class size varies among the different types of schools. In 1994 the average was 22 in primary school, 22 in *Hauptschule*, 25 in *Realschule*, and 26 in secondary level I *Gymnasien* and comprehensive schools.

Teacher Certification

Although instructors at preschools and kindergartens as well as instructors at general and vocational schools do not require university education, most teachers are trained at universities and institutes of higher education. Admission to training depends on possession of the *Abitur*, a qualification obtained through the secondary school leaving examination. Two training phases can be distinguished:

- Academic studies for a period of three to five years, during which preservice teachers receive scholarly training in educational and social studies, the subject(s) specialty, and in didactic studies and methods. This concludes with part I of the degree examinations.

- The introduction to school practice, usually taking 18 months. This phase comprises practical involvement in schools and complementary training at seminars. This concludes with part II of the degree examination.

There have been attempts to merge the two phases of training, but at present they remain separate.

Teachers in Germany are usually prepared for specific kinds of schools, either for primary schools, *Hauptschule*, *Realschule*, *Gymnasien*, or vocational schools. In most *Laender*, teacher education for primary schools and *Hauptschule* is completely integrated into universities. In some *Laender*, teacher education takes place at colleges of education, as was generally the case before the 1970s. Primary school and *Hauptschule* teachers must complete at least six semesters (eight in Bremen) at an institute of higher education followed by a preparatory period of one to three years. Their studies usually involve one or two subject specializations as well as general training. Teacher education for *Realschulen* includes the study of two main subjects over the same period at college or university, followed by 18 months of preparatory service. Teacher education for the *Gymnasium* entails studying two subjects for at least eight semesters. The degree courses are discipline oriented, and prospective teachers are not taught separately from other students. Interdisciplinary and vocational courses are rare at universities.

The technical teaching staff at vocational schools have completed training in the relevant skill, further training at a higher technical institution, and special courses designed to provide basic training in education. Teachers of general subjects and specialized theory at vocational schools must complete eight semesters at an institute of higher education and usually a year of relevant practical training as well. The probationary teaching period usually lasts 18 months. Teachers at special schools are trained in therapy at university institutes or colleges of education. This normally takes eight semesters followed by probationary training for at least 18 months. In most *Laender*, only candidates who have completed their training for positions in primary schools, *Hauptschulen*, *Realschulen*, or *Gymnasien* can receive training for appointments at special schools.

In all *Laender* there is ample opportunity for teachers to attend in-service development. This is intended to keep teachers up-to-date on the subjects they teach, as well as on broader fields of psychology and sociology in education and didactic methods. Courses are organized regionally or at the state level, often during school hours, and teachers are excused from school duties to attend. In recent years, correspondence courses for teachers have been instituted.

The greatest number of further education courses for teachers are offered by the state institutes for continuing and further education created in the 1970s. These institutes have full-time employees as well as part-time teachers for special courses. The foci for further education in science are procuring scientific knowledge and enabling an exchange of experience. The highest priority is, as studies have shown, "procuring knowledge" for teaching science, oriented toward the usual paradigm of the subject.

Teacher Profile

Teachers' salaries follow the civil service system: teachers at elementary schools are paid at the level of those who complete specialized colleges. All other salaries are comparable to those of other university graduates, such as judges or doctors in the health-care system. The difference in salary between a teacher and a principal is, in percentage, marginal. Of more importance is the automatic salary increase for civil servants: every two years they are awarded an automatic increase.

The age profile of teachers shows a massive overrepresentation of older teachers due to both an unfavorable age pyramid caused by the Second World War and the abundant hiring practices of the 1970s. Due to decreasing enrollment in the late 1980s and the limited financial resources available, very few new teachers have been hired in recent years.

From 1960 to 1990, the ratio of male to female teachers shifted in favor of female teachers, from 42 to 62 percent over the thirty-year period. This did not occur at the same rate in all types of schools; there was a clear increase in female teachers at elementary schools to about 67 percent, whereas at the *Gymnasien* there was only a slight increase to about 37 percent. The considerable increase in the percent of female teachers in 1992 (to 72 percent in elementary schools and 42 percent in *Gymnasien*) is due to the inclusion of the states from the former East Germany.

3 TIMSS POPULATIONS

Germany, IEA, and TIMSS

Germany participated in the First International Mathematics Study, in science and English as parts of the Six-Subject Survey, as well as in studies on civics education, classroom environment, writing, computer education, and reading literacy. Germany has been an active member of IEA since 1982.

German TIMSS Populations

Germany is participating in TIMSS at the Population 2 and 3 levels. Population 2 includes Grades 7 and 8. These adjacent grades have the largest number of thirteen-year-old students. Population 3 includes all students in their final year of upper secondary education, including vocational education. Since the final year of secondary education is Grade 12 in some *Laender* and Grade 13 in others, the sample was drawn from both Grades 12 and 13, as well as from the final year of vocational education. Mathematics and physics specialists were defined as those students enrolled in special subject-area courses. The inclusion of vocational students meant only about 10 percent of students in Population 3 were identified as physics specialists and 28 percent as mathematics specialists.

4 MATHEMATICS CURRICULUM AND PEDAGOGY

Goals for the Mathematics Curriculum

The curriculum for mathematics in Germany is laid down in syllabi for each state and for each of the different types of schools. They are prepared by teachers and school administrators (sometimes with the participation of parents and pupils) and approved by the single state's parliament. These syllabi advise teachers on aims, content, teaching approaches, and methods of assessment.

In general, the syllabi state that the general aims of mathematics education are to:

- provide fundamental knowledge and skills in important areas of mathematics;
- provide security in the techniques, algorithms, and concepts which are necessary for mastering everyday life in society;
- develop the ability to state facts mathematically and to interpret the contents of mathematical formulae; it should make possible the solving of non-mathematical or environmental phenomenon through mathematics;
- teach pupils to think critically and to question;

- give examples of mathematics as a cultural creation in its historical development and in its importance in the development of civilization;
- provide terms, methods, and ways of thinking that are useful in other subjects.

Major Changes in the Mathematics Curriculum

Over the past ten years there has been a continuing revision of syllabi in the various states, replacing unified differentiated syllabi with more specific syllabi for the different types of schools. Structural aspects in the teaching of mathematics have been reduced to conform with the ideas of the new mathematics, especially in the lower grades.

For primary and secondary level I there is now greater emphasis on applied mathematics, mostly in the form of additional word problems and "dressed" tasks. Explicit environmental concerns in connection with applied mathematics or social arithmetic are particularly common in comprehensive schools, *Hauptschulen* and *Realschulen*, where the teachers have developed supplemental material themselves.

At the primary level, pocket calculators and computers have played almost no role up to the present. The overwhelming majority of primary school teachers emphatically reject the use of calculators and computers in the classroom. In secondary schools, pocket calculators have been smoothly integrated, beginning in Grade 7. Computers are not a common tool, nor are they a subject of teaching. They are used as tools for calculations or simulations in secondary level II, as the subject of teaching in the newly defined "basic information technology education" in Grades 9 and 10, or are integrated into technology education.

Mathematics research has not influenced mathematics teaching except in some specialized secondary level II courses, such as statistics and probability, which were introduced into the curriculum in the late 1970s. Research in the psychology of mathematics education has not yet entered syllabi, textbooks, or classroom practice. Studies on the video analysis of classroom interaction have had a more direct impact on teachers; they have, to a certain degree, created a consciousness of teacher routines and lead to discussions among teachers.

Current Issues in the Mathematics Curriculum

Trends in the changes to the intended curriculum are more apparent in approach than in content. Thus, the new goals for mathematics teaching are to:

- present mathematics both as a theoretical study and as a tool for solving problems in the natural and social sciences.
- provide experience with fundamental mathematical ideas such as the idea of generalization, the need for proofs, structural aspects, algorithms, the idea of infinity, and deterministic versus stochastic thinking.
- use inductive and deductive reasoning, methods for proving, axiomatics, normalization, generalization and specification, and heuristic work.
- provide variation in argumentation and representation levels in all fields and aspects of mathematics teaching.
- teach historical aspects of mathematics.

Mathematics Textbooks

Textbooks that are approved by the local states reflect the state syllabus completely. The teacher or teachers in the mathematics department of a school may choose a textbook from the state-approved list. In general, teachers do not expect the students to use the textbook except as a source book for exercises. Mathematics teachers complain about textbooks and students' lack of interest in reading them, but students are not taught how to read and study with books. More photographs, pictures, illustrations, and graphs are used than previously, but they tend to be included to make the books more attractive and are usually not directly connected to the written text.

Textbooks are used differently in different types of schools: in *Gymnasien* they are used as a collection of tasks and exercises while in *Realschulen* and *Hauptschulen* they are used more directly as an explanatory tool and a collection of rules and phases to be memorized. Work sheets and self-developed materials are used in comprehensive schools. While there has been a decrease in the number of pages and topics, textbooks are still a collection of very different topics and themes within one book. At the same time, there has been an increase in the number of tasks and exercises

with less information, fewer explanations, and almost no review materials.

In *Realschulen*, *Hauptschulen*, and comprehensive schools textbooks lack theoretical consideration and presentation of proofs. At all levels, textbooks do not include enough projects, historical highlights, real applications, or information about mathematics in working life. At the secondary level II in *Gymnasien* and comprehensive schools there are theorems and proofs, but neither projects nor information on mathematics in careers.

Major topics in typical mathematics syllabi at the Population 1 level include:

- meaning, operations, properties of operations: whole numbers;
- equations and formulae;
- estimation and number sense;
- measurement units;
- common fractions;
- data representation;
- basic geometry.

At the Population 2 level in Gymnasium, mathematics syllabi include:

- percentage;
- number theory;
- pattern, relations and functions;
- equations, inequations, formulae;
- two-dimensional geometry: coordinate geometry, polygons and circles;
- congruence and similarity;
- binomial theorem;
- data representation and analysis;
- uncertainty and probability.

Hauptschule syllabi include the following:

- percentage;
- equations and formulae: functions, relations, and equations;
- congruence and similarity;
- two-dimensional geometry: polygons and circles;
- three-dimensional geometry;
- meaning, operations and properties of operations: whole numbers;
- proportionality concepts and problems;
- data representation and analysis.

At the Population 3 level in *Gymnasium*, major topics include:

- differentiation, integration, differential equations, partial differentiation;

- uncertainty and probability;
- vectors, linear algebra.

Pedagogy

The move from content-oriented towards more pedagogy-oriented didactic thinking fostered the development of innovative material, especially in primary mathematics. The main tendencies in mathematics pedagogy in Germany are:

- Careful analysis of topics with respect to their pedagogical significance. Applied mathematics is used as a way of illuminating real world structures that are created in part by mathematics. Consequently, applied mathematics is taught as a method for stimulating creative behavior.
- Detailed investigations of the principle of application; in particular, investigation of the prerequisites for genuinely carrying out applications by constructing mathematical models, based either on generally fixed predictive rules in the natural sciences or on negotiable conventions in the social sciences.
- Developing and testing projects that are easy to understand and are oriented to subject matter, often with an emphasis on regional matters. These projects serve as unities of meaning where the principle of application can be experienced, for example "packaging milk," "railway traffic between Aachen and Cologne," or the "jumbo jet."
- Preliminary experience, which cannot be explained by means of concepts, with fundamental ideas of stochastic chance experiments; expressing observations in everyday language; statistical investigations of the students' everyday world using methods of clear representation; suitable and interesting distributions, statistical correlation of two variables.

These tendencies are expressed in some syllabi, but are still far from classroom reality.

5 SCIENCE CURRICULUM AND PEDAGOGY

Goals for the Science Curriculum

In Germany, the curricula for the three sciences taught at the secondary level are laid down in syllabi for each state and for each of the different types of schools. They are prepared by teachers and school administrators, occasionally with the participation of parents and students, and approved by the single state's parliament. The syllabi advise teachers on aims, content, teaching approaches, and methods of assessment.

In general, the goals of the sciences are as follows. Biology:

- to impart knowledge and insights about the structure and essential processes of plants, animals, and human beings;
- to teach students that humans are subject to biological laws, that humans have structural characteristics common to all living creatures, and is thus part of the biosphere;
- to open the way to objective criticism;
- to enable students now and later to contribute to mastering current life situations and problems, such as avoiding damage to civilization, protecting the environment, and securing food production;
- to respect life.

Chemistry:

- to provide students with the basic knowledge and concepts required to understand their environment, which is to a great extent shaped by chemical processes;
- to make students capable of formulating, checking, and applying hypotheses when experimenting, developing model conceptions, and understanding chemical symbols;
- to create a fundamental understanding of the facts and procedures of chemistry in research and technology.

Physics:

- to learn to recognize and understand fundamental aspects of physics and technology phenomena in the world around them;
- to acquire the approach to thinking and working known as the scientific method, expressed in procedural terms;
- to learn that the application of physics knowledge has had and will continue to have consequences for technology and for society.

In each case these aims are more specifically determined by the type of school (see Section 4).

Major Changes in the Science Curriculum

Reforms of the curriculum do not take place at the national level. In the *Laender*, permanent reforms of the syllabi for the various subjects are ongoing. As studies have shown, the average life expectancy for a syllabus is seven years. This usually means that after implementation of a syllabus is completed the next revision commission is called into action.

There have been no major changes with regard to the content of the science curriculum, with the exception of the addition of biotechnology to the biology curriculum at secondary level II. It should be noted, however, that the opening of physics education to technical and technological aspects, and especially to moderate science-technology-society perspectives has increased.

Chaos and self-organizing systems are on their way to finding a place within the curriculum. In any case, there is a great deal going on in the area of research and development, and it is affecting teacher training.

The computer is, of course, becoming more and more important. Many new programs have been developed, including those from the teaching materials industry, yet the research boom has lost some of its impetus, since not all expectations have been fulfilled. The use of new technology has been included in the core curriculum, either as a separate unit usually taught in Grade 8 or incorporated into the mathematics syllabus.

Current Issues in the Science Curriculum

Under Germany's system (see Section 2) of opposition to formal change and openness to contextual renewal, science education is stable in that it has its traditional share of the timetable, but it is constantly being refined. Social demands regarding what should be taught in science classes have contributed to this process. The issues can be summarized under three main concerns:

- education for an elite versus education for all;
- a focus on practical knowledge for future jobs versus specialized academic studies; and
- a focus on integration versus separation.

Subject-matter didactics deal with these questions, mainly focusing on:

- a shift to a more integrated science education, or at least to a better coordination of the three single subjects;
- an awareness of the necessity of incorporating interests, preconceptions, learning abilities, and special needs (for example, girls in science) into instruction;
- a shift to pupil-centered teaching and learning strategies that take their demands more seriously, for example, real life contextualization of content or an examination of career paths.

Science Textbooks

Only those textbooks approved by the *Laender* and containing the complete state syllabus may be used in schools. Teachers from individual schools' science departments choose the textbooks from an approved list. The points mentioned in Section 4 apply equally to the use of science textbooks.

Science topics for the various TIMSS populations, taken from various original syllabi, are shown below:

Population 2: *Gymnasium*
Physics:
- light;
- electricity;
- magnetism;
- energy types, sources, conversions.

Biology:
- animals, animal reproduction;
- animal behavior;
- plants, fungi;
- biomes and ecosystems;
- pollution; conservation of land, water, and sea resources;
- human biology and health;
- evolution, speciation, diversity.

Chemistry:
- chemical changes, rate of changes;
- metabolism: exothermic, endothermic;
- segregation procedure;
- definitions: elements, connections, metals, non-metals, air.

Population 2: *Hauptschule*
Physics:
- light;
- electricity;
- magnetism;

energy types, sources, conversions.

Biology:
- animals, animal behavior;
- plants, fungi;
- reproduction;
- biomes and ecosystems;
- pollution; conservation of land, water, and sea resources;
- human biology and health;
- food production.

Chemistry:
- air, characteristics of gases;
- metals: iron, steel;
- concept of ions;
- acids, bases, salts;
- building materials: cement, lime, glass.

Population 3: *Gymnasium*

Physics:
- electricity;
- magnetism;
- mechanics (force, impulse, energy);
- wave phenomena;
- quantum theory;
- nuclear physics;
- relativity theory;
- kinetic theory.

Biology:
- cell biology;
- genetics;
- metabolic and developmental physiology;
- processing information and behavior;
- ecology and environmental protection;
- evolution theory.

Chemistry:
- organic chemistry;
- electrochemistry.

Pedagogy

The reform of the education system that took place at the end of the 1960s was to abolish all hindrances to equal opportunity within the system. The aim of the reform was to guarantee fundamental equal rights within education, including equality for girls and boys. Parents were to choose which secondary school in the three-fold system their children would attend. Coeducation was the basic principle that was to guarantee equal treatment of all students regardless of sex. Con-

tent and teaching methods among the three types of secondary schools were to become more similar.

Approaches to the teaching of the sciences, therefore, must become student-oriented, taking account of pupils' interests, learning abilities, and ideas. The special needs and interests of girls, science-technology-society perspectives, and the importance of project teaching are becoming increasingly recognized.

Additional changes in science pedagogy include the following:

- Findings from research on effective learning environments have fostered the trend towards interactive learning, involving process-focused student activities, communication about science methods, and an awareness of the importance of information processing skills.
- A shift in the teacher's role in the teaching-learning process can also be noted, changing the role from a disseminator of information into a facilitator of the learning environment.
- In-service training supports these changes by offering fewer content-based courses and more courses on teacher strategies and improvement of the teaching and learning environment.

6 EVALUATION POLICIES AND PRACTICES

At the national level there are no mandatory tests or examinations. Responsibility for testing lies with the schools, which must conform to strict state guidelines on evaluation practices for each type of school as well as for each particular subject.

At the end of each school year achievement in all subjects is evaluated to allow students to proceed to the next grade. The subject teacher performs these evaluations, which are based on written tests as well as achievements in the classroom. All subjects are valid, but German, mathematics, and the compulsory foreign languages carry more weight.

At the end of Grade 10, written tests and oral examinations are administered in two or three main subjects. The tests are designed by the subject teacher or the teachers in the department, but need approval by the state Ministry of Education before they are administered.

The school leaving examination at the end of Grade

13, or *Abitur*, follows the same procedures as above; only six of the sixteen states have a centralized examination. The *Abitur* is used to determine university entrance qualifications and is recognized by all states. The Standing Conference of the Ministers of Education has agreed on binding standards which set down the content to be tested in the *Abitur* examinations.

7 REFERENCES AND SOURCES FOR FURTHER READING

Common Sources

1   Husén T, N T Postlethwaite 1994 *International Encyclopedia of Education*. Pergamon Press, Oxford
2   Organisation for Economic Co-operation and Development 1995 *Education at a Glance: OECD Indicators*. Organisation for Economic Co-operation and Development, Paris
3   United Nations Educational, Scientific and Cultural Organization 1995 *Statistical Yearbook*. United Nations Educational, Scientific and Cultural Organization, Paris
4   World Bank 1995 *World Development Report 1995*. Oxford University Press, New York

Other Sources:

5   BMBW (Bundesministerium für Bildung und Wissenschaft) 1994 *Grund- und Strukturdaten [Basic and Structural Data]*. Ausgabe 1994-95 BMBW [and earlier editions], Bonn
6   Riquarts K 1996 *Framework for Science Education in Germany*. IPN, Kiel
7   *Bildung im Zahlenspiegel [Education in Numbers]*. 1993 Statistisches Bundesamt 1993 (and earlier editions), Wiesbaden
8   Arbeitsgruppe Bildungsbericht (am Max-Planck-Institut für Bildungsforschung)1994 *Das Bildungswesen in der Bundesrepublik Deutschland [The Educational System in the Federal Republic of Germany]*. Rowohlt, Reinbek bei Hamburg
9   Riquarts K et al. (eds.) 1990–94 *Naturwissenschaftliche Bildung in der Bundesrepublik Deutschland, 4 Bände [Science Education in the Federal Republic of Germany, 4 volumes]*. IPN, Kiel
Band 1: 1990 Bedingungen und Einflußgrößen naturwissenschaftlich-technischer Bildung [Conditions of Influences on Science and Technology Education]. IPN, Kiel
Band 2: 1994 Naturwissenschaftliche Bildung in öffentlichen und privaten Institutionen [Science Education in Public and Private Institutions]. IPN, Kiel
Band 3: 1991 Didaktiken naturwissenschaftlicher Fächer und naturwissenschaftsbezogener Lernbereiche [Didactics in Science Subjects and Science Related Areas of Learning]. IPN, Kiel
Band 4: 1992 Aktuelle Entwicklung und fachdidaktische Fragestellungen in der naturwissenschaftlichen Bildung [Current Developments and Subject-Oriented Questions Concerning Science Education]. IPN, Kiel

Statistical References[*]

Section	Statistic	Reference	Page	Table	Year of Statistic
Country Profile	8%, 2 million, 916,000, 558,000	5	13	–	1993
Country Profile	US$23,560, 2%, 4%	4	163	1	1993
Country Profile	95%	3	–	–	1991
Country Profile	5%	2	73	F01	1992
Structure of the System	28%, 29%, 30%, 9%, 4%	5	46–47	–	1993
Structure of the System	31%, 16%, 53%	5	23	–	1991
Structure of the System	US$30 million, 36%	5	276	–	1991
Structure of the System	US$8 million, 15%	5	276	–	1991
Schools in the System	38, 190–220	6	85	2	1995
Schools in the System	20%, 5–10%, 13%, 15%, 9%, 19%	6	87	3	1995
Schools in the System	23, 22, 25, 26	6	110	10	1992
Teacher Profile	42%, 62%, 37%, 72%, 42%	6	105	8	1992
Timss Populations	10%, 28%	6	65	–	1993

[*]All statistical data used in this article can be found in Riquarts (1996).

Greece

Georgia Kontogiannopoulou-Polydorides, University of Patras

Vasilis Koulaidis, University of Patras

George Stamelos, University of Patras

Joseph Solomon, University of Patras

1 COUNTRY PROFILE

Greece lies in southeastern Europe and occupies an area of roughly 135,000 km². Its physical geography is highly diverse, including mountainous regions and many small islands. Development in Greece in the later half of the twentieth century has been characterized by the development of urbanized areas, greater Athens being the largest with a population of 3.6 million. Rural areas consist of many small, and sometimes isolated communities.

The population of Greece is almost 11 million. The most common religion is Orthodox Christianity, but there are also small Roman Catholic and Muslim communities. The overwhelming majority of the population speaks and uses the Greek language, and Greek is used in schools. Schools specializing in the education of the Muslim community exist as well.

The Hellenic Republic, the official name of Greece, is a constitutional parliamentary democracy, with the people represented in an elected parliament called the *Koinovoulio*. Most executive power lies with the government and its head, the prime minister, while the president retains some limited constitutional powers.

According to OECD data, Greece is among the 25 wealthiest countries. The per capita gross national product in 1994 was nearly US$8000. Greece participates in the Council of Europe, and has been a member of the European Union since 1979. Public expenditure on education averaged nearly 8 percent of total spending between 1991 and 1994, or 4 percent of the gross national product. Expenditure on public and private education combined averaged more than 6 percent of the GNP over the same period.

Education has always been highly valued in Greek society. Since the creation of the Greek State in 1828, special efforts have been made to establish and expand the school network. The need for skilled personnel in the growing service sector laid the groundwork for the development of the current schooling system and the prevailing ideology of education.[1]

This framework, as well as a national preoccupation with classicism, helps to explain the special status of languages and mathematics in the Greek curriculum.[2] Subjects related to science have not traditionally been given much weight in the curriculum. Until recently, an upgraded status for these subjects has been at the forefront of all demands for curricular reform.

2 THE EDUCATION SYSTEM

Governance and Decision Making

The education system in Greece is highly centralized. Some initial steps have recently been taken toward decentralization, and administration of the education system now occupies two levels: central administration and partially decentralized services. The Ministry of Education, through local education districts, ensures that schools follow a centrally prescribed curriculum. All schools teach the same topics for each subject, and the same textbooks, teachers' guides, worksheets, and notebooks are used.

Central administration includes the Ministry of Education and Religious Affairs and the Pedagogical Institute, the function of which is to study and pro-

[1] See K Tsoukalas 1977 *Dependence and reproduction: The social role of educational mechanisms in Greece (1830–1922),* Themelio, Athens.

[2] See for instance C Noutsos 1979 *Secondary Curricula and Social Control,* Themelio, Athens.

pose curricular policy to the Ministry as well as to supervise the production of textbooks. The National Council of Education advises the government on questions of educational policy for all levels of education, issues concerning continuing education and in-service training, and areas such as the education of immigrants and special needs groups.

Decentralized services include agencies at the regional and school levels. The Ministry of Education exerts considerable influence over these agencies, however, and consequently their autonomy is limited.

Structure of the System and Participation Rates

The education system in Greece follows a 6-3-3 model, with the first nine grades being compulsory. Figure 1 shows the structure of the education system and the approximate enrollment rates at each level. The first level of education includes two years of non-compulsory preschool education and six years of compulsory primary school, called *Demotiko*. Children enter Grade 1 at the age of six. The official enrollment figure for the primary cohort is 97 percent, but it is possible that some groups, such as the Romany, may not have been included.

The second level of education consists in the three-year lower secondary school, *Gymnasio*, and the three-year upper secondary school, *Lykeio*. Eighty-five percent of the age cohort continue to the end of *Gymnasio* although it is part of compulsory education. There are three types of *Lykeia*: general, which has an academic

orientation; technical-vocational; and multilateral. Seventy-four percent of the age cohort graduate from *Lykeio*: 43 percent from general *Lykeia*, and 31 percent from multilateral and technical-vocational *Lykeia*.

A new type of postsecondary education, institutes of vocational training, has been recently developed. These institutes offer two-year courses for the training of technicians such as electricians and graphic artists, and other skilled personnel such as secretaries.

Higher education has a dual-track structure, including universities and technological education institutions. Universities offer four- to six-year courses depending on the field of study, while technological education institutions offer three-year courses and additional practical training. The technological educational institutions are similar in character to the British Polytechnics before they were converted into universities.

All students graduating from a *Lykeio* of any type may write entrance examinations to tertiary education, and between 80 and 85 percent do so each year. The number of candidates is normally double that of *Lykeio* graduates, since many candidates write the examinations repeatedly in order to improve their marks and their chances of acceptance. At least 30 percent of candidates succeed in entering university or a technological education institution. Many students not accepted into tertiary education choose to enter existing postsecondary institutes of vocational training or to study abroad.

Schools in the System

The School Year

The school year extends from September 1 to August 31 of the following year, but no teaching takes place between the end of June and the end of August in primary and secondary education. The teaching year is divided into three terms, followed by a summer vacation from the beginning of July to the end of August. All students attend school for between five and seven 45- or 50-minute periods per day from Monday to Friday.

Class Size

Education law has established that class size should not exceed 30. In actuality, for the school year 1989–1990, the average class size in primary school was closer

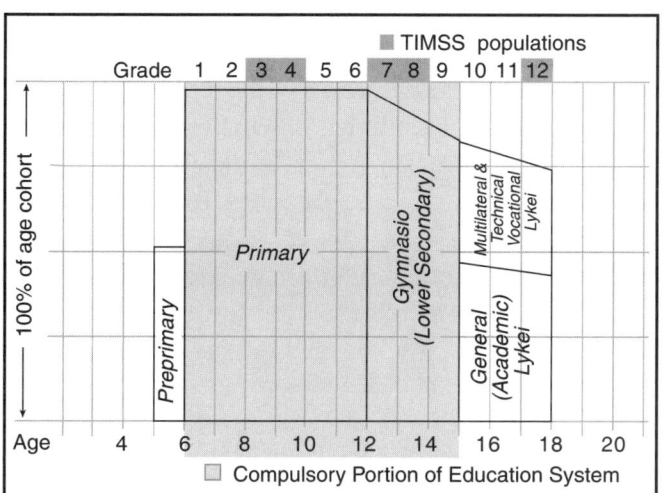

Figure 1: Structure of the Education System

to 20, owing to the many small classes in rural areas. In lower secondary schools the average class size was 19, while in general and multilateral *Lykeia* the corresponding figure was 17.

Mathematics and Science Programs

Streaming and setting are not permitted in Greek schools. Mathematics is a compulsory subject in all grades of primary and secondary school. Science subjects are compulsory for the last two grades of primary school and all grades of secondary school. In the first four grades of primary school, elements of science are taught within the environmental education program.

Teacher Certification

Initial Education

Primary teachers must have a four-year degree from a department of education in order to be certified. This is a recent development; the first departments of education within universities were created in 1985 and the first teachers with university degrees graduated in 1989. The majority of practicing primary school teachers have a degree from a two-year program at a teacher education college. Universities are in the process of retraining primary school teachers without university degrees, and almost 1000 teachers undergo this process each year. Secondary school teachers are required to hold a university degree in a specialty subject. They complete their studies in the relevant department within universities, but do very little, if any at all, pedagogical study.

Each year the Ministry of Education appoints teachers in numbers determined by the needs of schools. Prospective teachers at all levels must be enrolled on a waiting list, according to their specialization and the level at which they will teach. There are two waiting lists for primary education: the first for university graduates and the second for teacher education college graduates. For secondary education, there are 22 lists, one for each subject. There are many prospective teachers on the lists, and there is often a waiting period of several years for an appointment as a new teacher. Because of this wait, there has been discussion about reforming the list system, but no changes have yet been implemented.

Continuing Education and Sabbatical Leave

Teachers may continue their education at Ministry-operated in-service education centers, which offer refresher courses in pedagogy for both primary and secondary teachers. There are 17 such in-service centers throughout the country, and more than 1000 teachers attend each year. Currently, the in-service training center system is under reform.

Teachers who hold a university degree may also attend postgraduate courses at a university. By law, every teacher accepted for postgraduate study is entitled to a sabbatical leave with full salary.

Assessment of Teachers

Teacher assessment procedures are regulated by presidential decree. Its clauses have not been put in practice, however, because the decree is undergoing a reform.

Teacher Profile

In a recent statement, the Ministry of Education referred to the need for strengthening the social status of primary and secondary school teachers. Primary teachers, in particular, do not enjoy a high social status, and the majority come from fairly low socioeconomic strata. The fact that teacher education has recently been incorporated into university may modify this in the future. Secondary teachers have been required to hold a university degree since the formal education system was established. It is likely that this has influenced their social status despite the fact that their salaries do not differ from those of primary teachers.

Among primary school teachers, the gender split is virtually equal. This situation is expected to change in the future, since the number of female students in primary teacher education university departments significantly exceeds the number of males. In lower secondary schools, female teachers comprise the majority of teachers: 63 percent in 1990. In general and multilateral *Lykeia,* 52 percent of teachers are male, as are 65 percent in technical-vocational *Lykeia.*

 TIMSS POPULATIONS

Greece, IEA, and TIMSS

The Greek National Center (GRENAC) of IEA was officially founded in 1989, but has been involved in the research activities of IEA since 1987. GRENAC has participated in the following studies: Second International Mathematics Study (SIMS), Computers in Education Study, Reading Literacy Study, Third International Mathematics and Science Study (TIMSS), Preprimary Study, and the Civic Education Project. Recently, the Hellenic Coordinating Center of IEA was established at the University of Athens, coordinating all IEA activities in the country.

Greek TIMSS Populations

Population 1 includes students in Grades 3 and 4. These grades contain the largest proportion of students in the nine-year-old age cohort. At the time of testing, nearly 100 percent of nine-year-old students were in one of these two grades.

Population 2 includes students enrolled in Grades 7 and 8 in regular schools. These classes consist mainly of thirteen-year-old students. At the time of testing, nearly 100 percent of thirteen-year-old students were in one of these two grades. Students enrolled in special schools were excluded from Populations 1 and 2.

Population 3 includes students enrolled in the final year of the upper secondary school, *Lykeio*, who are specializing in mathematics and science.

4 MATHEMATICS CURRICULUM AND PEDAGOGY

Goals for the Mathematics Curriculum

The official curriculum in mathematics is constructed centrally by the Ministry of Education. Teams of specialists at the Pedagogical Institute define the content, produce new teacher guidebooks and textbooks, and suggest curricular changes to the Ministry. Schools are required to implement the national curriculum as it is given.

The aims of mathematics education for primary school are to develop mathematical thinking and to encourage children to use measurement and other relationships to understand their environment and solve real problems. For secondary schools, the goals of mathematics education are to:

- Enable students to develop rational, abstract, and critical thinking. Develop the ability to analyze and generalize. Enable students to apply principles and rules. Cultivate the ability to use mathematical proofs.
- Influence cognitive development and social traits: "Mathematics helps students to observe carefully, to concentrate, to insist, to take responsibility, to use their imagination in creative ways, to think and behave in a disciplined way, to cultivate a sense of beauty and morality, and to use critical thinking."
- Help students understand concepts as well as quantities, properties, and relationships, especially those that are relevant to understanding and solving problems related to the technical, economic, and social aspects of contemporary society.
- Help students develop the ability to communicate by using the language of mathematics clearly and accurately.
- Help students understand the role of mathematics in the development of knowledge.
- Prepare students to continue their studies at a higher level.

Major Changes and Current Issues in the Mathematics Curriculum

At the primary level, emphasis is given not only to the development of computational skills, but also to the acquisition of mathematical concepts. It should be noted that major curricular changes take place approximately once every decade.

At the lower secondary level the following changes have occurred during the last five years:

- The influence of modern mathematics is less obvious while additional emphasis is given to applications.
- Calculator use was introduced.
- Numerical skill development, including approxi-

mation and the calculation of errors is now introduced.

- The study of statistics is now included.
- Applications of mathematics in everyday life and in other subject areas are considered.

At the upper secondary school level, geometry has been compressed while statistics, combinations, and probability theory have been added.

Mathematics Textbooks

Textbooks are published under the supervision of the Pedagogical Institute. There is one textbook for each grade, and classroom teaching usually follows it closely. In recent years, there has been an increase in both the number of topics and the amount of the content covered by each topic. Textbooks emphasize explanations and exercises, with applications of mathematics included in a limited way. The historical background of mathematics is usually confined to a brief sketch, particularly in secondary school textbooks.

Primary and lower secondary school textbooks tend to include many pictures and graphs, while in upper secondary, illustrations are used to a more limited degree. Students at higher levels are expected to read and study from their textbooks, and to complete a heavy load of homework.

Pedagogy

Most sources of pedagogic innovation come from research projects within university education departments, the Pedagogical Institute, and groups of practicing teachers. By law, innovation may only be introduced into the curriculum by the Pedagogical Institute. In practice, however, the function of the Pedagogical Institute is bureaucratic and its actions confined to the implementation of Ministry decisions, rather than design of educational policy.

University departments may propose pedagogical innovations. However, the centralized character of the Greek education system and its uniform curricula mean that testing of innovations can only be done on a limited scale. Thus there are no established channels for the dissemination of research outcomes to practicing teachers. In-service teacher education centres, because of their links with faculties of education, provide the only channel for the dissemination of research.

Mathematics teaching is usually formal and textbook-based. Guidance in teaching methods comes from highly detailed and structured teachers' guides published by the Ministry. In upper secondary schools, students perform many computational exercises.

5 SCIENCE CURRICULUM AND PEDAGOGY

Goals for the Science Curriculum

The responsibility for construction of the science curriculum lies with the Pedagogical Institute. The last major reform concerning science-related subjects took place in 1983.

The goals for science in primary school are to:
- help students organize concepts and experiences in order to understand the environment;
- promote responsible attitudes and relationships between humanity and the biophysical environment;
- develop research skills for the enhancement of knowledge;
- familiarize students with group work

The goals of the secondary science curricula are to:
- familiarize students with natural phenomena and their underlying laws, and acquire the ability to interpret them;
- develop students' interest in scientific research;
- enhance students' abilities in observation and involvement in experiments.
- enable students to appreciate the importance of cooperation within the world's scientific and technical domains, both for the progress of science and the improvement of living conditions.

Major Changes in the Science Curriculum

The orientation of the science curriculum has shifted away from memorizing a body of knowledge, including definitions and typology, toward an understanding of concepts and their applications in everyday life. Observation and experiments are encouraged, but inadequate facilities and a lack of relevant training among teachers seriously limit their implementation. Science teaching tends to be subject-centered and traditional, with an emphasis on pencil-and-paper problem-solving skills and a general absence of practical work.

Current Issues in the Science Curriculum

The dynamics of change within the science curriculum must be seen in the light of several factors. A characteristic of modern Greek society is the willingness of the service sector to absorb new technology at a level far above the technical production capacity of the country. Similarly, some science teachers are anxious to utilize computers to their fullest extent. Fundamental changes within schools are hindered by the inertia of the whole educational service and a traditional approach to teaching. Yet, these hinderances coexist with an ability to exploit new technology.

It seems likely that new technologies could become more common in science teaching, although institutional inertia has slowed changes in the way science is taught in schools. The Ministry of Education is actively promoting the use of technology in schools, and has recently been seeking software development for science teaching in secondary schools. Over 75 percent of *Gymnasia* already have a network of 30 or more computers within their school, and this trend is expected to increase.

Science Textbooks

Since the middle of the 1980s, new textbooks have been produced for virtually all science subjects, in both primary and secondary education. As with mathematics, the volume of content has increased in the new books. Teaching is based mainly on textbooks published by the Ministry of Education.

Textbooks focus on both explanations and applications of science. In lower secondary school textbooks, there are simple exercises with numerical applications, while in upper secondary, more complex exercises and historical highlights are included. In primary and lower secondary school textbooks, many pictures and graphs are used, but they are less common in upper secondary school textbooks.

Pedagogy

Science teaching in schools at both the primary and secondary level is formal, teacher-centered, and textbook-based. The national curriculum recommends learning activities and gives detailed guidelines for their implementation. In practice, however, primary and lower secondary students rarely perform experiments. Experiments that take up more than 20 percent of available class time are supposed to be conducted by teachers as demonstrations. Research has shown that primary teachers, probably because of inadequate training, are hesitant to conduct experiments, so many students do not even see demonstrations.[3] In upper secondary schools, students spend a great deal of time on computational exercises and perform relatively little, if any at all, experimental work. It should be noted, however, that in technical and multilateral *Lykeia* there are technological subjects that involve laboratory work such as analytical chemistry and electronics.

6 EVALUATION POLICIES AND PRACTICES[4]

National Examinations

There is only one national examination, the centrally administered university and technological institute entrance examination. This examination is mandatory only for students wishing to enter an institution of higher education, and between 80 and 85 percent of upper secondary school graduates write it. The prerequisite is a leaving certificate from an upper secondary school or *Lykeio*. Formally, there is no distinction between certificates from general and technical-professional *Lykeio*, but in practice, 90 percent of the candidates who enter university have a certificate from the general *Lykeio*. Similarly, most graduates of technical-vocational *Lykeia* who wish to continue their studies enroll in a technological institute.

The entrance examination is an event that garners

---

[3] See Koulaidis 1995.

[4] For a detailed account, see G. Kontogiannopoulou-Polydorides, V. Koulaidis, and J. Solomon, (in press), Educational monitoring in Greece: W. J. Pelgrum and W. G. R. Stoel (eds), *Methods of Educational Monitoring in the European Union: Country reports and synthesis.*

considerable attention every year, and it is one of the few examinations that the public believes to be worthwhile. Research has shown that this examination system tends to favor students from middle-class family backgrounds; such students have a higher entrance-rate into departments that are in heavy demand.

School-based Examinations

Assessment of student performance is school-based. Summative assessment in the form of tests takes place at all levels beginning in Grade 5, and its purpose is to determine whether students can continue to the next grade level. It is estimated that all students in Grades 5 to 11 and about 80 percent of students in Grade 12 take the performance tests. Almost all test items are short-response or essay questions. Formative assessment in the form of oral tests or short written tests is used for diagnosing difficulties and providing feedback for everyday instruction.

CONCLUDING REMARKS: ANTICIPATED REFORMS

Mathematics and science teaching will be influenced by a series of reforms that are currently being planned by the Ministry of Education. As far as may be foreseen, the most important changes will concern:

- the organization of studies in upper secondary schools, with the possibility for final-year students to construct their own program, by selecting among a number of courses offered;
- the reorganization of the upper secondary school certificate with the introduction of external examinations at the end of the three years of study. This will also affect the mode of entrance in higher education institutes;
- the character and aims of the curriculum for primary and lower secondary schools, with the production of a relatively concise curriculum for the whole of compulsory education;
- the system of teacher preservice and in-service training and the existing system of teacher appointment in schools.

7 REFERENCES AND SOURCES FOR FURTHER READING

Common Sources:

1   Husén T, N T Postlethwaite 1994 *International Encyclopedia of Education*. Pergamon Press, Oxford
2   Organisation for Economic Co-operation and Development 1995 *Education at a Glance: OECD Indicators*. Organisation for Economic Co-operation and Development, Paris
3   United Nations Educational, Scientific and Cultural Organization 1995 *Statistical Yearbook*. United Nations Educational, Scientific and Cultural Organization, Paris
4   World Bank 1995 *World Development Report 1995*. Oxford University Press, New York

Other Sources:

5   Varnava-Skoura G, B Vasilou, S Georgakakos 1993 *Education in Greece. Quantitative Data 1960–1990*. Program EUROFORM Pedagogical Institute, Athens
6   1995 Teachers' profile must change. *Eleftherotypia* 6/9/95 [Editorial]
7   Kontogiannopoulou-Polydorides G, V Koulaidis, J Solomon, in press, Educational monitoring in Greece, in Pelgrum W J and W G R Stoel (eds.) *Methods of Educational Monitoring in the European Union: Country reports and synthesis*. Octo, University of Twente
8   Koulaidis V 1995 *EU Project: Science in Schools and the Future of Scientific Culture in Europe: Country Report*, EU XII DG
9   *Law 2327: National Council of Education* FEK, no. 156, 31/7/95
10 TIMSS/IEA 1992 *Mathematics Curriculum Expert Questionnaire*.
11 TIMSS/IEA 1992 *Science Curriculum Expert Questionnaire*.
12 Ministry of National Education and Religious Affairs 1995 *Educational Policy Review: Background Report to OECD on Education*, Organization of Textbook Publishing, Athens
13 Noutsos C 1979 *Secondary Curricula and Social Control*, Themelio, Athens

14 National Statistical Service of Greece 1990 *Annual Reports on Education*.

15 Pyrgotakis I E 1992 *Greek Teachers: An Empirical Approach to the Analysis of Their Work Conditions*. Grigoris, Athens

16 Stamelos G 1993 Basic trends in primary education in Greece: 1981–82, 1991–92. *Ta Ekpaideftika*. volume 29–30

17 Stamelos G 1994 Student movement and graduation in secondary education in Greece. *Ta Ekpaideftika*, volume 33

18 Tsoukalas K 1977 *Dependence and reproduction: The Social Role of Educational Mechanisms in Greece (1830–1922)*. Themelio, Athens

19 Tsountas K S, M G Chronopoulou 1995 *Secondary Education and Educational Personnel*. Ekfrasi, Athens

Statistical References

Section	Statistic	Reference	Page	Table	Year of Statistic
Country Profile	8%	8	66	1 (app. VI, B)	1991–94
Country Profile	4%	8	66	1 (app. VI, B)	1991–94
Country Profile	6%	8	75	–	1995
Structure of the System	97%	8	21	III-4	1991
Structure of the System	85%	13	99	1	1992
Structure of the System	43%	13	99	1	1992
Structure of the System	31%	13	99	1	1992
Structure of the System	74%	13	6	1	1992
Structure of the System	80%	4	6	–	1991
Structure of the System	85%	4	28–29	–	1991
Structure of the System	30%	11	38	–	1995
Schools in the System	class size	14	–	–	1989–90
Teacher Profile	50%	1	38	–	1991
Teacher Profile	50%	1	15	–	1991
Teacher Profile	63%, 52%, 65%	14	–	–	1990
Science Curriculum	75%	4	8	–	1993
Science Curriculum	20%	4	5	–	1993
Evaluation	90%	4	5	–	1993
Evaluation	80-85%	4	5	–	1993
Evaluation	80%	4	5	–	1993

Hong Kong

Frederick K. S. Leung, University of Hong Kong

Nancy W. Y. Law, University of Hong Kong

1 COUNTRY PROFILE

Hong Kong is situated on the southern coast of the Guangdong province of China. The territory consists of Hong Kong Island, Kowloon Peninsula on the mainland, and the area beyond the peninsula together with the surrounding islands, known as the New Territories. Hong Kong Island was ceded by China to Britain in 1842 after China's defeat in the first Opium War. In 1860, after the second Opium War, Kowloon Peninsula was added to the colony, and in 1898, the New Territories were leased to Great Britain for 99 years. This lease expires on July 1, 1997, when the whole colony will be restored to Chinese sovereignty.

Hong Kong has a total land area of 1074 km^2, consisting mainly of low-lying hills. Large areas have been reclaimed from the sea on the north shore of Hong Kong Island and around Kowloon to provide land for urban development.

Hong Kong has one of the highest population densities in the world, with 5404 persons/km^2. From 1980 to 1991, the average population growth was just over 1 percent per year. The number of residents emigrating from Hong Kong increased from an average of 20,000 in the early 1980s to 62,000 in 1990. The natural increase in population was 40,400 in 1991, while the number of legal immigrants from mainland China was about 26,800. About 98 percent of the population of 5,809,000 (1993 estimate) are Chinese, most of whom have their origins in Guangdong province. There are about 64,700 foreign nationals from English-speaking countries (UK, USA, Australia, and Canada), 80,200 Filipinos, 17,700 Indians, and 12,400 Japanese. English and Chinese are both official languages of the government, but the most commonly used language is Cantonese.

The British government did not attempt to develop Hong Kong's system of government into a parliamentary democracy until the 1980s. Before 1985, the Legislative Council consisted entirely of civil servants and members appointed by the governor. Since 1985, some members have been selected by indirect election. In 1991, 18 out of the 60 members were directly elected and by 1995, all members were either directly or indirectly elected.

From 1980 to 1991, Hong Kong enjoyed a high economic growth rate of an average of almost 6 percent per annum in per capita gross domestic product. Its economy is now ranked by UNESCO as high income, with an estimated per capita GDP of US$13,430 in 1991. Total government expenditure on education in 1991 was more than US$2 billion, the equivalent of 17 percent of total public expenditure for the fiscal year. This was significantly higher than many high-income countries, since the average for OECD countries was 12 percent for that year.

2 THE EDUCATION SYSTEM

Governance and Decision Making

Many advisory bodies, each with a high proportion of its membership appointed from outside the government, have been set up to advise the government on the planning, development, and management of the education system at all levels. The Education Commission is the highest advisory body on education, providing guidance on the overall objectives of education and the formulation of policy and priorities.

The Education Department is responsible for planning and providing schools; allocating pupils to places, supporting curriculum development, setting academic targets and related assessments, monitoring teaching standards, and administering funding of public sector schools and private institutions in receipt of public funds. The Curriculum Development Council, supported by the Curriculum Development Institute of the Education Department, is responsible for curriculum reviews. It publishes curriculum guides providing broad guidelines for developing subject syllabi at kindergarten, primary, and secondary levels, as well as specific subject syllabi at the various levels. The Curriculum Development Institute is also responsible for monitoring the quality of textbooks through review and publication of a list of recommended textbooks. Decisions about which textbooks to select from the list are made at the school level. Instructional techniques and other classroom processes are left to individual teachers and schools.

Structure of the System and Participation Rates

Structure of Education

The structure of the education system is shown in Figure 1, along with approximate enrollment rates for each level. Formal schooling begins at the age of six, and the first nine years of education are free, universal, and compulsory. Most children, however, begin preschool education in kindergartens, which are all privately run, at the age of three. Hong Kong follows the British system of six years of primary school, five years of secondary school, and two years of preuniversity, known as sixth form. At the age of six, all children enter Grade 1 and attend six years of primary education. At the end of compulsory education, Grade 9, most students stay on for two further years of secondary education leading to the first public examination, the Hong Kong Certificate of Education Examination. Between 37 and 38 percent of these students stay on for the two-year sixth-form course and take the Hong Kong Advanced Level and Advanced Supplementary Level Examinations, which are prerequisites for entrance into university. About 24 percent of the age cohort follow public-sector tertiary education courses at degree and sub-degree levels.

Public and Private Systems

Most children attend schools in the public sector. In 1994, private schools accounted for only 10 percent of primary student enrollment and 12 percent of secondary student enrollment. Only a very small proportion of schools in the public sector are directly operated by the government, while the rest are operated by non-profit organizations that receive public funding. Of schools in the aided sector of primary and secondary education, 1 percent follow the national curriculum of England and Wales. The rest of the aided schools and government schools follow the local curriculum laid down by the Curriculum Development Council. Some schools in the private sector follow non-local curricula; these are generally referred to as international schools. International schools account for approximately 2 percent of total primary enrollment and one percent of total secondary enrollment.

Technical and Vocational Study

Full and part-time technical education, industrial training, and vocational studies are provided through the statutory Vocational Training Council to postcompulsory education students as an alternative to further academic study. Craft-level courses are mainly for students who will leave school after Grade 9, and technician-level courses are offered to students who leave at Grade 11. Courses leading to certificates, diplomas, and higher qualifications are also available.

Language of Instruction

Nearly all primary schools in Hong Kong use Chinese as the medium of instruction. At the secondary and

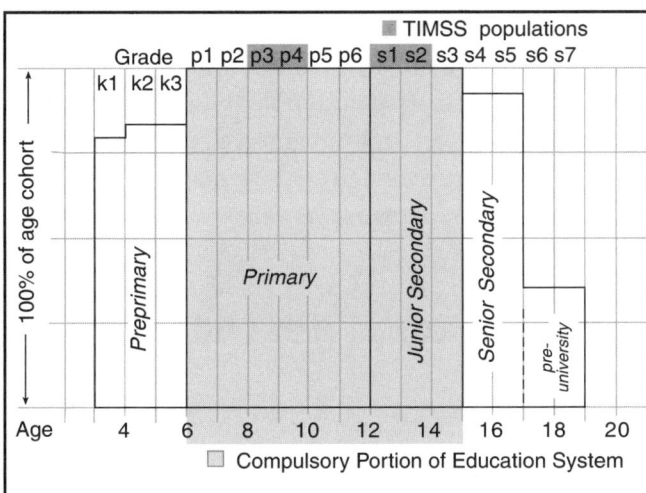

Figure 1: Structure of the Education System

sixth form levels, Grades 7 to 13, English is theoretically the medium of instruction in most schools, except in Chinese middle schools, which use Chinese. Public examinations can be taken in either English or Chinese. With the popularization of education, the English standard of most students has not been high enough for them to follow all instructions in English. Although English textbooks are still used in these schools and the majority of students take their public examinations in English, teaching is predominantly done in Chinese, switching to English only when technical terms are encountered.

Mathematics and Science Programs

Mathematics and science are compulsory for all students up to the end of Grade 9. A high proportion of both males and females continue to study mathematics in Grades 10 and 11, but by Grade 12 only 50 percent of males and 27 percent of females are enrolled in a mathematics course. In the sciences, however, enrollment for both males and females drops significantly beginning in Grade 10. The figures in Table 1 represent enrollment estimates based on numbers of students writing end-of-secondary-school examinations in mathematics, physics, engineering science, biology, statistics, and chemistry in 1994.

Table 1: Mathematics and Science Enrollment by Grade and Gender (percent of grade cohort)

Grade	Modal Age	Mathematics		Science	
		M	F	M	F
10	15	97	94	55	35
11	16	97	94	55	35
12	17	50	27	65	26

Schools in the System

The School Year

The school year in Hong Kong is about 200 days, beginning the first weekday in September and ending around the middle of July. There are approximately 90 days of holidays for most schools, including two weeks at Christmas, one week at Chinese New Year, one week at Easter, a seven-week summer vacation, and several other holidays.

Students in secondary schools and full-day primary schools attend school from Monday to Friday. Typically, these schools are in session 7 hours per day, 35 hours per week. Approximately 27 hours are spent on instruction, the rest being used for lunch, recess, assembly, and other activities. Some primary schools operate only in the morning or the afternoon. Students attend these schools Monday to Friday, plus one Saturday every fortnight. These schools are in session for an average of 27 hours per week, of which about 22 hours are spent on instruction and the remainder on other activities.

Class Time Allocated to Mathematics and Science

In secondary schools, between 12 and 15 percent of the timetable is devoted to mathematics, and about 6 percent to science. In primary school, the respective percentages are about 15 percent and 10 percent. Class size at the secondary level is typically 40 students per class, dropping to 30 per class by Grades 12 and 13. At the primary level, class size is usually between 30 and 35.

Streaming and Tracking in Mathematics and Science

Students follow the same curriculum up to Grade 9, but some schools offer remedial lessons in which low-ability students are grouped together for mathematics instruction. From Grade 10 onward, students choose between an arts and a science stream; some schools offer a commerce stream as well. In Grades 10 and 11, science-stream students usually take a course called additional mathematics as well as the general mathematics course taken by all students. They may also take physics, chemistry, and biology. Some arts stream students take biology or human biology; and many take geography, which includes earth science. In Grades 12 and 13, students further specialize in science, arts, or commerce subjects.

Certification of Teachers

Primary Level

Most primary school teachers are not university graduates. They are required to hold a Certificate in Education, which entails full-time attendance at a three-year course if they have completed Grade 11, or a two-year course if they have completed Grade 13. In 1992, the government decided to introduce posts for university graduates in primary schools, and the target is to have university graduates comprise 35 percent of the teaching force by the year 2007.

Until 1994, the Certificate in Education course was offered by Colleges of Education. In 1994, the Colleges were restructured to form the Hong Kong Institute of Education, a publicly funded tertiary institution independent of direct government intervention. It offers certificate courses as well as regular in-service courses for practicing teachers. The universities offer bachelor's degrees in primary education, as well as some specialized programs in primary school administration for senior teachers.

Not all teachers are trained in the subjects they are teaching, and the percentages of trained teachers in the public and private sectors differ drastically. Eighty-nine percent of mathematics teachers in the public sector, for example, are trained in the subject, but only 21 percent of those in the private sector are trained. The overall percentage is 84 percent. For science, 87 percent of public sector teachers are trained in the subject, and 19 percent of private sector teachers. The overall percentage is 83 percent.

Secondary Level

Those holding certificates in education, but who do not have a university degree may also teach in secondary schools, but only up to Grade 9. They should constitute no more than 30 percent of the teaching force in the secondary sector, but in 1994, the actual percentages were 34 in the public sector and 37 in the private sector. In public-sector schools, 88 percent of non-graduate and 70 percent of graduate teachers hold a recognized teaching qualification. In the private sector, 51 percent of nongraduate and 43 percent of graduate teachers hold a similar qualification. Most graduate teachers obtain their teaching qualification from one of the three universities in Hong Kong offering postgraduate diplomas or certificates in education. These are either one-year full-time courses, catering to recent graduates, or two-year part-time courses, catering to in-service teachers without initial teacher education. The overall figures for trained nongraduate and graduate teachers are 85 and 67 percent, respectively.

Sixty-five percent of graduate and 67 percent of nongraduate mathematics teachers in the public sector are trained in the subject area they are teaching. The respective figures for the private sector are 52 percent and 56 percent. The overall percentages for graduate and nongraduate teachers are 64 percent and 66 percent.

Science is taught mainly as a single subject in Grades 7 to 9, and 85 percent of graduate and 82 percent of nongraduate science teachers in the public sector are trained in the subject. The respective figures for the private sector are 66 percent and 63 percent. From Grade 10 onward, physics, chemistry, biology, and engineering science are taught as separate subjects. The overall percentages for trained graduate and non-graduate science teachers are 85 percent and 73 percent respectively.

The Education Department runs regular in-service courses for trained practicing primary and secondary school teachers. These courses are organized and run by the various divisions of the Education Department, offered by the former Colleges of Education, or contracted out to various tertiary institutions.

Teacher Profile

Teachers are paid on the same scale as civil servants, and although their salaries are in general lower than other professionals, they are considered quite well-paid. The salary scale for primary school teachers ranges from about US$19,000 to US$67,000 per year, and that for secondary teachers runs from about US$28,000 to US$76,000 per year. In contrast, the average salary for managerial and professional employees, excluding top management, is about US$42,000 per year.

There are more male teachers in mathematics and science than in other areas. For the general teacher population, about 75 percent of teachers in the primary sector are female. Furthermore, "female teachers were found to be highly concentrated in junior ranks" [12:66]. The average age of primary teachers is 39. In the secondary sector, female teachers constitute 50 percent of the teaching force. Secondary teachers are in general younger than primary teachers, with an average age of 35.

3 TIMSS POPULATIONS

Hong Kong, IEA, and TIMSS

Hong Kong has been member of IEA since its participation in the Second International Mathematics Study (SIMS), and since then has participated in the Second International Science Study (SISS), the Reading Literacy Study, and the Preprimary Study.

TIMSS Populations

Initially, Hong Kong intended to participate in all three populations. However, because of financial and personnel limitations, testing at the Population 3 level was not conducted.

Students at the Population 1 level were in Grades 3 and 4 and were about 8 to 10 years old at the time of testing. Those at the Population 2 level were in Grades 7 and 8 and were 12 to 14 years old at the time of testing.

4 MATHEMATICS CURRICULUM AND PEDAGOGY

Goals for the Mathematics Curriculum

The goal for mathematics education in both primary and secondary levels is "to enable pupils to understand and to appreciate relationships and patterns in both number and space in their everyday lives, and to be able to express them clearly and concisely" [14, 15].

The objectives for mathematics education are described in the curriculum guides, known as syllabi, published by the Curriculum Development Institute. For primary schools, the overall objectives are to:

- stimulate the interest of children in learning mathematics;
- foster mathematical thinking and creativity in children;
- teach basic mathematical concepts and computational skills, providing a foundation on which fur-

ther learning of mathematics and science can be built;
- develop pupils' ability to apply mathematics to problem-solving in daily life;
- encourage an appreciation of the order and pattern of number and shape.

For secondary schools, the overall objectives are to:
- continue the development of numeracy begun in primary school, including number work and the ability to cope with approximations, percentages, rates and ratios, and simple mensuration;
- prepare students to understand everyday applications outside the classroom, by teaching the fundamentals of statistics and probability, for example;
- provide a basis for further work in science and mathematics by teaching the use of mathematical symbolism and facility with its manipulation, by developing an ability to use the basic logical patterns and conventions of reasoning, and by introducing necessary tools such as the trigonometric ratios;
- introduce a general sense of the pattern and power of mathematics both as a tool and as part of our cultural heritage.

These objectives reflect an emphasis on treating mathematics more as a tool than a way of thought.

Major Changes in the Mathematics Curriculum

Recently there has not been any significant change in the Grade 1 to 11 curriculum. In Grades 12 and 13, two new advanced supplementary level subjects, Applied Mathematics and Mathematics and Statistics, were introduced in 1992 to broaden the sixth form curriculum. The former was designed for mathematically oriented students, and the latter for those not specializing in the physical sciences.

The impact of modern technology on the curriculum has been minimal, and the use of technological aids in the teaching of mathematics is rare. The only change has been the introduction, in 1990, of some new topics such as "approximation of simple equations in one unknown" and "improvement of accuracy by the method of bisection" at the certificate of education examination level.

The use of computers in the teaching and learning of mathematics is not common, nor are programmable and graphic calculators. Only pocket scientific calculators are allowed in the certificate of education examination. Usually, schools start allowing students to use calculators in Grade 8 or 9. Computers are available in the schools, but are used in computer science lessons and for school administration purposes, not for mathematics and science teaching.

Current Issues in the Mathematics Curriculum

Although compulsory education up to the age of 15 was achieved in 1978, most issues of debate pertain to the adjustment of the curriculum to meet the challenge of providing general education for all. For the mathematics curriculum, one impetus for change comes from the failure of the too-academic curriculum to meet the needs of the majority of students, especially the low achievers.

One suggestion for change was to restructure the mathematics curriculum "to consist of a core covering about 50 percent of the subject content and the rest being additional levels of learning. The core . . . represents the basic domains of learning that students at a certain level should achieve" [18]. Lower-ability students would then concentrate their efforts on the core. Another suggestion was to rearrange the present curriculum into modules, to allow students to take different combinations of modules according to their different abilities.

In 1990, the Education Commission suggested in its fourth report a target and target-related assessment scheme "based on criterion-referencing principles, in which students are assessed not against one another but against targets at progressive levels" [19]. This concept was later revised to a target-oriented curriculum, stressing task-based learning and assessment rather than the traditional content-based approach.

There have been heated debates on the target-oriented curriculum both inside and outside education circles. Many of the criticisms were on the haste with which it was implemented without proper piloting. There was severe opposition from the schools, particularly those with conservative teachers. But some opposed the curriculum on theoretical grounds. Among the opposing voices were some pointing out that this approach represents a typical Western philosophy which contradicts the Chinese culture in Hong Kong

[20]. The voices of opposition challenged the new curriculum's stress on child-centered learning and the criterion referencing principle, and accused it of promotion of individualism, a notion which has a negative connotation in the Chinese culture.

Another issue of concern is the place of the Additional Mathematics Curriculum. Originally, additional mathematics was meant for mathematically inclined secondary school students. Now it has become a subject taken by virtually all science students, making it a hidden prerequisite for A-level mathematics. Many students find the Additional Mathematics Curriculum too demanding, but lowering the standard of the curriculum will defeat its purpose. Debate on how to tackle the problem is still ongoing.

Mathematics Textbooks

Teaching is dominated by the textbooks used. Because of examination pressure, textbooks usually follow a standard definition-theorem-proof-example-exercise format, and not much emphasis is laid on the cultivation of interest, understanding of concepts, or the development of problem-solving abilities. Projects, historical highlights, applications of mathematics, and mathematics in careers are rare. Much space is devoted to worked examples and drill and practice exercises.

At the secondary and Sixth Form levels, public examinations can be taken in either English or Chinese. Since the majority of students take the examinations in English, most textbooks are written in English, although over 98 percent of the students are Chinese. Such a language barrier hinders students from reading the textbook on their own. English is theoretically the medium of instruction in most schools, except Chinese middle schools, which use Chinese. Nearly all primary schools in Hong Kong use Chinese as the medium of instruction. But since the popularization of education, the standard of English for most students is not high enough for them to be able to follow a class taught solely in English. So although English textbooks are still used in these schools and the majority of students take public examinations in English, teaching is predominantly done in Chinese, switching to English only when technical terms are encountered.

Pedagogy

Teaching is examination-driven and is teacher-centered. Recently, there have been suggestions of adopting a more investigative approach in the teaching and learning of mathematics, but the following description in the Hong Kong report of SIMS still applies today:

> Schools in Hong Kong have a reputation for reliance on teacher dominated instructional strategies. . . . There has developed a tendency for classes to be taught by lecture-style delivery, with little student participation, apart from note taking and completing assigned work. . . . The intense pressure of examinations, the expectations of parents, pupils and colleagues appear to encourage teachers to impart knowledge and instill in students a need to learn for a predominantly recall mode of performance [21:23].

5 | SCIENCE CURRICULUM AND PEDAGOGY

Goals for the Science Curriculum

Draft general curriculum guides for the primary and secondary curriculum were published by the Education Department in 1992, and brief references were made in both documents to the general aims of science education [14,15]. The emphases were on "increasing pupils' knowledge and understanding of the natural world and the world as modified by human beings, and with developing skills and competence associated with science as a process of inquiry" [15:15]. The aims of both primary and secondary science curricula are similar.

Official subject curriculum guides and examination syllabi have a great deal of influence on the textbooks used in schools. While most teachers do not consult curriculum guides in their everyday teaching, textbooks must follow these guides carefully in order to be included on the recommended book list. Teachers tend to follow textbooks rather closely in their teaching, particularly at the primary and lower secondary levels. At the senior secondary level, because teachers have to prepare their students for public examinations, many consult the examination syllabi as well. In fact, at this level, teaching is very much influenced by the examination syllabus and the styles of examination questions.

Curriculum guides and examination syllabi are determined by subject committees of the Curriculum Development Council of the Education Department and the Hong Kong Examinations Authority respectively. These subject committees include members from various science departments and faculties of education in tertiary institutions, as well as subject teachers recommended by principals.

Major Changes in the Science Curriculum

Beginning in 1995, primary science will be combined with two other subjects, health education and social studies, into the single subject of general studies. One of the reasons for this move was to allow an integrated treatment of common areas of concern in the three subjects. However, the actual syllabus for general studies is basically an amalgamation of the original syllabi. Whether schools make use of the opportunity for innovative integration and the incorporation of social, personal, and scientific perspectives in the analysis and solution of problems has yet to be seen.

At the junior secondary level, science is usually taught as a single subject, Integrated Science. This subject has seen little curricular change for more than a decade. However, recent suggestions have been made both at the school and subject committee level to explore ways of changing the curriculum to emphasize personal and social relevance, as well as to allow more investigative and problem-solving activities. This move for change represents an effort to modify the curriculum guide into a document that gives specific directives only on learning objectives. This would leave pedagogical decisions open to textbook writers and teachers, allowing them to cater to students of varying academic abilities. Explorations and debates are still continue, and the subject is still taught largely as it was a decade ago.

At the senior secondary level, the sciences are taught as individual subjects with physics, chemistry, and biology being the most common. Biology and chemistry have undergone a number of curricular changes directed at highlighting social and environmental issues relating to scientific and technological developments. Another change that has direct impact on the curriculum is the use of continuous school-based teacher assessment of practical work, in place of the single practical examination within the public examination. This encourages the incorporation of extended investigative work into the curriculum and thus the development of higher level scientific process skills. Physics, on the other hand, has remained essentially unchanged over the past decade.

Advances in information and communication technology over the past decade have had a tremendous impact on the workplace and the home. Unfortunately, this has not filtered through to the curriculum or teaching practice in science. Computers are rarely, if ever, used in public-sector schools in either science teaching or experimentation.

Current Issues in the Science Curriculum

Science curricula of the 1980s were academic in orientation and their structures were designed to reflect a scientist's view. However, secondary education is available to nearly everyone and there is a consequent need to cater to mixed abilities. The challenge is to make the science curriculum relevant and meaningful to the majority of students, while retaining the rigor of the scientific disciplines and preparing students for higher studies in science. Many science teachers fear that thematic, problem-oriented curriculum structures would seriously affect the rigor of the subject, leading to a shallow, fragmented understanding of the discipline. This fear is perhaps greatest among physics teachers.

The science curriculum has tended to be content-oriented and at senior levels driven by public examinations. Another challenge to science education in Hong Kong is whether the development of investigative and higher-order scientific process skills can become a major curriculum goal in practice, and not just a statement in the curriculum guides. Some subjects have tackled this problem by making use of the backwash effects of public examinations on curriculum; they use continuous school-based teacher assessment of practical skills in place of one-off, uniform, practical examinations. Whether this has been successful remains to be explored.

Science Textbooks

Textbooks have paramount influence on the implemented curriculum, particularly at the primary and junior secondary levels where there are no public examinations. Most teachers rely heavily on textbooks to determine teaching content and sequence, and many may not be aware of the official curriculum guides and syllabi. Generally, at all levels, each student has his or her own science textbook and frequently secondary students also have an experimental workbook.

Textbooks are published by commercial publishers and each school decides on the particular textbooks it wishes to use. However, all schools in the public sector must choose textbooks from a recommended list published by the Curriculum Development Institute of the Hong Kong Education Department. Textbooks have to be sent to the latter organization for scrutiny before they can appear on the recommended book list.

Textbooks generally adhere closely to the content and topic sequence of official syllabi. Students use them frequently, for study purposes as well as for exercises.

Pedagogy

Teaching in science classrooms tends to be teacher-centered and didactic. At the primary level, teaching is conducted by teacher presentations supplemented by television programs produced and broadcast by the Education Department. There are usually no laboratories in primary schools, but they are standard in secondary schools. These are generally used for teacher demonstrations and prescribed, closed-ended student experiments. Group work is not common except in laboratories, where groups of four or five students share a set of equipment. Cooperative learning is rarely organized, and investigations and open-ended projects rarely take place, even at the preuniversity level. Teaching frequently follows the textbook closely. Many teachers do not see themselves as having a role in school-based adaptation of the teaching curriculum, beyond minor addition or elimination of topics and teaching sequences. Consequently, integration of topics within the science curriculum rarely occurs. Inductive approaches to teaching science are commonly used, especially at the secondary level. The deductive approach is sometimes also used, especially in physics.

Activities focusing on oral and written communication are not common features in science classrooms although the science examination committees have tried to push for more attention to be paid to the development of these skills. The use of computers, videodisks, and other electronic media besides video cassettes is extremely rare.

6 EVALUATION POLICIES AND PRACTICES

Public examinations are principally conducted for school and university placement purposes. Competition in these examinations is intense, especially at the higher levels. With the broadening of the apex of the education pyramid, however, the intensity has lessened in recent years. The backwash effect of examinations on teaching is severe. Teachers tend to follow the examination syllabus closely in their teaching, and students concentrate their efforts on examination-related content only.

Within schools, internal examinations are organized two to four times a year. In addition, teachers usually organize their own class tests, although the practice differs from school to school.

The types of public assessment are as follows:

- an academic aptitude test consisting of multiple-choice test items, administered at the end of Grade 6 for place allocation in secondary schools;
- a mean eligibility rate for place allocation in senior secondary schools based on internal assessment performed during Grade 9;
- a certificate of education examination consisting of multiple-choice items and short- and long-answer questions, which is both a certification for secondary education and for place allocation in the sixth form, and is administered in Grade 11;
- the advanced and advanced supplementary level examinations consisting of short and long questions, which is administered in Grade 13 for place allocation in tertiary institutes.

7 REFERENCES AND SOURCES FOR FURTHER READING

Common Sources

1   Husén T, N T Postlethwaite 1994 *International Encyclopedia of Education*. Pergamon Press, Oxford
2   Organisation for Economic Co-operation and Development 1995 *Education at a Glance: OECD Indicators*. Organisation for Economic Co-operation and Development, Paris
3   United Nations Educational, Scientific and Cultural Organization 1995 *Statistical Yearbook*. United Nations Educational, Scientific and Cultural Organization, Paris
4   World Bank 1995 *World Development Report 1995*. Oxford University Press, New York

Other Sources:

5   Roberts, D (ed) 1993 *Hong Kong 1992*. Hong Kong, Government Printer
6   Bray M, P Ieong 1996 The growth and diversification of the international schools sector in Hong Kong. In: *International Education*, Vol. 26(1)
7   Cheng KM *et al* 1996 *Preparation of Students for Tertiary Education*. Hong Kong, UGC
8   Hong Kong Government 1995 *Enrollment Survey 1994*. Hong Kong, Government Printer
9   Hong Kong Examinations Authority 1994 *H.K.C.E.E. Annual Report*. Hong Kong, HKEA
10 Hong Kong Examinations Authority 1994 *H.K.A.L.E. Annual Report*. Hong Kong, HKEA
11 Education Commission 1992 *Report No. 5: The Teaching Profession*. Hong Kong, Government Printer
12 Hong Kong Government 1995 *Teacher Survey 1994*. Hong Kong: Government Printer
13 Hong Kong Government 1994 *1994 Report of Salaries and Employee Benefits Statistics*. Hong Kong, Census and Statistics Department
14 Hong Kong Curriculum Development Council 1992 *Guide to the Primary Curriculum (Draft)*. Hong Kong, Government Printer
15 Hong Kong Curriculum Development Council 1992 *Guide to the Secondary 1 to 5 Curriculum (Draft)*. Hong Kong, Government Printer

16 Curriculum Development Committee Hong Kong 1983 *Syllabus for Primary Schools: Mathematics.* Hong Kong, Government Printer

17 Curriculum Development Committee Hong Kong 1985 *Syllabus for Mathematics (Forms I-V).* Hong Kong, Government Printer

18 Curriculum Development Institute 1993 *Report of the Working Group on Support Services for Schools with Band 5 Students.* Hong Kong, Curriculum Development Institute

19 Education Commission 1990 *Report No. 4: The Curriculum and Behavioral Problems in Schools.* Hong Kong, Government Printer

20 Sheng Kung Hui Primary Schools Council 1994 *TOC—Why We Insist on School Autonomy.* Hong Kong [in Chinese]

21 Brimer A, P Griffin 1985 *Mathematics Achievement in Hong Kong Secondary Schools.* Hong Kong, Center of Asian Studies, University of Hong Kong

Statistical References

Section	Statistic	Reference	Page	Table	Year of Statistic
Country Profile	5,404/km^2	3	1–3	1.1	1993
Country Profile	over 1%	4	288-289	26	1980–91
Country Profile	20,000	5	366	–	early 1980
Country Profile	62,000	5	366	–	1990
Country Profile	40,400	5	368	–	1991
Country Profile	26,800	5	366	–	1991
Country Profile	5,809,000	3	1–3	1.1	1993
Country Profile	64,700; 80,200; 17,700; 12,400	6	1	–	1992
Country Profile	over 5%	4	163	1	1980–93
Country Profile	US$13,430	4	238–239	1	1991
Country Profile	US$2 billion, 17%	5	402	App. 8A	1991
Structure of the System	37–38%	7	8-9	8.28	1994
Structure of the System	10%	8	46	3.11	1994
Structure of the System	12%	8	72	4.8	1994
Structure of the System	1%	8	52,77	3.17,4.13	1994
Structure of the System	2%	8	53	3.18	1994
Structure of the System	1%	8	78	4.14	1994
Structure of the System	97, 94	9	494	–	1994
Structure of the System	55	9	494,497	–	1994

Continued

Statistical References continued

Section	Statistic	Reference	Page	Table	Year of Statistic
Structure of the System	35	9	494,499	–	1994
Structure of the System	50	10	482,487	–	1994
Structure of the System	27	10	482,488	–	1994
Structure of the System	65	10	485,488	–	1994
Structure of the System	26	10	484	–	1994
Teacher Certification	89%, 21%, 84%, 87%, 19%, 83%	12	77,78	5.13	1994
Teacher Certification	34%, 37%	12	103	–	1994
Teacher Certification	88%, 70%, 51%, 43%, 85%, 67%	12	115	–	1994
Teacher Certification	65%, 67%, 52%, 56%, 64%, 66%, 85%, 82%, 66%, 63%, 85%, 73%	12	124, 132	6.13	1994
Teacher Profile	US$42,000	13	17-22	6	1994
Teacher Profile	75%	12	66	–	1994
Teacher Profile	39	12	67	5.6	1994
Teacher Profile	50%	12	108	–	1994
Teacher Profile	35	12	109	6.5	1994

Hungary

Judit Krolopp, National Institute of Public Education, Center for Evaluation Studies

Péter Vári, National Institute of Public Education, Center for Evaluation Studies

1 COUNTRY PROFILE

Hungary lies in the Carpathian Basin of middle-eastern Europe, surrounded by seven countries: Austria, the Slovak Republic, Ukraine, Romania, Yugoslavia, Croatia, and Slovenia. It covers an area of 93,000 km² and has a population of over 10 million, a number that is annually decreasing. Twenty-five percent of the population lives in and around the capital, Budapest.

Poised between eastern and western Europe, Hungary has throughout history experienced a flow of different cultures through its borders. A Turkish occupation of 150 years and the long decline of the Austro-Hungarian Empire left their mark on Hungarian language, architecture, and culture. Today, Hungary's population is largely homogeneous, with minority groups of Croatian, German, Romany, Romanian, Serbian, Slovak, and Slovenian background composing just over 1 per cent of the total population.

After 40 years of communist rule, Hungary has for the past five years functioned as a republic governed by a parliamentary democracy. The government is headed by a prime minister who is chosen by the political party in power. Areas of government jurisdiction are divided between the central government and the local governments of Budapest and the 19 counties. Central government powers include jurisdiction over areas such as defense, external affairs, monetary and fiscal matters, social services, and energy.

The World Bank rates the Hungarian economy as upper-middle income, and its per capita gross domestic product in 1991 was US$2989. The per capita gross national product (at purchaser values) for 1993 was US$3350 with an average annual growth rate of 2 percent between 1980 and 1993. The average annual rate of inflation during this period was 13 percent.

In 1993, Hungary allocated 7 percent of total expenditure to education, approximately 7 percent of the year's GNP. In 1993, the distribution of university degrees as the highest level of education was as follows: 14 percent in the 25 to 34 age group, 14 percent in the 35 to 39 age group, 15 percent in the 40 to 49 age group, 11 percent in the 50 to 54 age group, and 15 percent in the 55 to 60 age group.

2 THE EDUCATION SYSTEM

Governance and Decision Making

The administration of education underwent a significant reorganization after the political changes of 1990, with local governments assuming responsibility for what was formerly the central government's exclusive domain. Since the reorganization, policies relating to teacher education, secondary school final examinations, and university entrance examinations have been set by the Ministry of Education. Within guidelines defined by the Ministry of Education, school boards now set local policy: the operation of schools, the implementation of curriculum, and the hiring of teachers.

Education in Hungary is in a transitional state. Before the reorganization of 1990, the Ministry of Education set a national curriculum and provided approved textbooks and teacher's guides. The last version of the national curriculum dates to 1978, with some modifications from 1981. A national core curriculum, under development since 1989, was introduced in September of 1995 and will be implemented over a period of three years. The core curriculum describes the content and basic goals of primary and secondary education, leaving some subject matter and the choice of pedagogical methods and textbooks to individual schools and teachers.

Structure of the System and Participation Rates

Structure of Education

Before the Second World War, the education system was heavily streamed and it was difficult for students to change tracks. After the Second World War the government established an eight-year general school for elementary education serving all students aged 6 to 14, in Grades 1 to 8. Secondary education, however, remained heavily tracked with three types of schools at the secondary level: a four-year academic secondary school (Grades 9 to 12), a four-year vocational secondary school (Grades 9 to 12), and a three-year trade school (Grades 9 to 11). After the elections in 1990, the new government allowed the school structure to become more diversified. Apart from the four-year academic secondary schools, eight- and six-year academic secondary schools have appeared, draining enrollment from Grades 5 to 7 of elementary schools. The current structure of the education system is shown in Figure 1, along with participation rates for the various levels.

Compulsory schooling begins at age 6 and ends at age 16, normally Grades 1 to 10. According to 1992 World Bank data, 87 percent the age cohort of 6- to 14-year-olds were enrolled in elementary school. Of secondary school age children, 81 percent were enrolled in secondary schools, and 15 percent of 18-year-olds were enrolled in tertiary institutions.

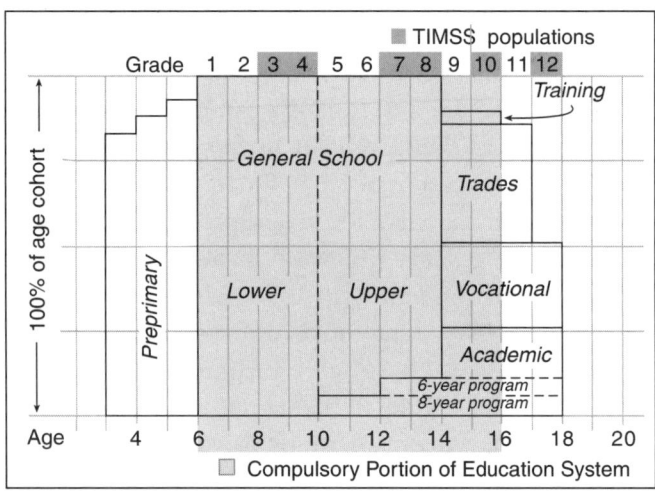

Figure 1: Structure of the Education System

Public and Private Systems

Most schools are funded by either the local or state government, but the number of schools funded by church or private foundations has been increasing. According to 1993–94 data, 97 percent of all secondary schools are funded by local governments, the state, or the church, while the remaining 3 percent are funded by foundations or enterprises. This represents a significant change from the days before the reforms of 1990, when state-funded schools were the only ones in existence.

Mathematics and Science Programs

Mathematics is compulsory throughout elementary school and each of the three types of secondary school: academic, vocational, and trade. Science is a compulsory subject from Grades 1 to 8. The pure science subjects, biology, chemistry, geography, and physics are taught only in academic secondary schools, where they are compulsory. Twenty-seven percent of all secondary school students are enrolled in these schools; 36 percent attend vocational and 37 percent trade secondary schools. At both trade and vocational schools, specialized areas of science are taught.

Schools in the System

The School Year

Students attend primary school for 185 instructional days per year and secondary schools for 198 days. At the school director's discretion, up to 10 of these days may be used as holidays or for other activities such as excursions or ski-breaks. In addition to these holidays, there are several national holidays, when schools are closed. The school year begins September 1, and lasts until the end of the second week in June. For students in the last year of secondary school, the school year ends before mid-April, allowing a four-week break to prepare for final examinations, which are held in May. There are two official holiday periods during the school year, from Christmas until the beginning of January, and a two-week period at the beginning of April.

Students attend school five days a week, from Monday to Friday. The number of 45-minute lessons per week increases from 20 to 38 from the lowest grade of primary school to the end of secondary school. The prescribed maximum number of lessons per day varies from four to seven, although these numbers are usually higher in practice.

Class Time Allocated to Mathematics and Science

Mathematics and science are compulsory for all students in Grades 1 to 8, and all students have the same number of science lessons. Beginning in secondary school, students may study elective courses in mathematics, science, and other subjects. Some secondary schools specialize in mathematics, and students enrolled in these schools may take between five and seven lessons of mathematics per week instead of the usual three or four. Mathematics and science classes, like other classes, typically have between 25 and 35 students.

Certification of Teachers

In Hungary, those who wish to become teachers follow one of three programs. Teachers' colleges offer a three-year course for those who plan to teach Grades 1 to 4 and a four-year program for those who plan to teach distinct subjects in Grades 5 through 8. Secondary school teachers take a five-year program offered by universities.

Pedagogy and teaching methodology are part of every course in the teacher education program. During the five-year program, preservice teachers study methodology for one year and visit primary school classes regularly for two and a half years. They also practice-teach in schools. Requirements for the four-year level course are similar, but with a greater emphasis on methodology.

In-service development and upgrading for teachers are not compulsory. There are, however, several courses each year that teachers may attend voluntarily. These courses meet for one three- or four-hour session each week for either one or two years, and the topics covered are either scientific or methodological. Lecturers and instructors are professors from universities and colleges.

Teacher Profile

Teaching is not a high-status position in Hungary. There is a national salary scale for teachers, and their salaries are based on their level of qualification and the number of years experience in education. Teachers are among the lowest-paid government employees, and their income lags far behind that of self-employed professionals.

3 TIMSS POPULATIONS

Hungary, IEA, and TIMSS

Hungary has been a member of IEA since 1968. Before joining TIMSS, Hungary participated in the Six-Subject Survey, the Second International Mathematics Study, the First International Science Study, the Second International Science Study, the Written Composition Study, the Classroom Environment Study, the Computers in Education Study, and the Reading Literacy Study.

Hungarian TIMSS Populations

The samples for Populations 1 and 2 were representative of the country as a whole. Populations 1 and 2 included all students enrolled in Grades 3 and 4, and Grades 7 and 8 of elementary school. Since the vast majority of students attend eight-year elementary schools, the Populations 1 and 2 samples were from the same schools.

For Population 3, the stratification was based on the school type, with a distinction made between secondary and vocational school students. Vocational schools span Grades 9 to 11, but the final year is spent on out-of-school practice, rendering the school-leaving population inaccessible. Students of the previous grade, Grade 10, were assessed instead. Population 3 thus covers Grade 12 of secondary schools and Grade 10 of vocational schools. For Population 3 only the literacy portion of the test was administered.

4 MATHEMATICS CURRICULUM AND PEDAGOGY

Goals for the Mathematics Curriculum

The goal of the mathematics curriculum is to serve the

needs of the other sciences, such as physics and chemistry. It also aims to provide insight into mathematics as a science and to develop students' thinking processes.

The national curriculum now used in the schools was written under the supervision of the Ministry of Education in 1978 and last modified in 1981. A national core curriculum has been under development since 1989, but the rapid pace of political and economic change delayed its formal adoption until September 1995. Schools have three years to implement the new curriculum. Until the three-year transitional phase ends, individual schools and teachers may choose whether they wish to use the newly edited textbooks or follow the old curriculum with accompanying textbooks and curriculum guides.

The 1978 mathematics curriculum contains about 75 to 80 percent basic and 20 to 25 percent supplementary material, with a time schedule to be followed for basic topics. Accompanying the curriculum is a detailed curriculum guide that covers methodological approaches. Basic material is compulsory for all students, while the extent to which the supplementary material is taught depends on the teacher as well as on the ability and interest of the students. This system provides scope for talented students to acquire a deeper knowledge of certain subject areas.

Major Changes in the Mathematics Curriculum

The 1978 mathematics curriculum was modified in 1986 for the primary level and in 1987 for the secondary level. Since the early 1990s, the Ministry of Education has approved a number of local curricula, all of which aim to satisfy needs that are no longer met by the old curriculum. These needs include an increased consideration of topics represented in university entrance exams. Local curricula have the same aim as the national core curriculum, which is to provide teachers with more freedom, within a loose content framework, to choose the textbooks and methods most suited to the both students and teachers.

Current Issues in the Mathematics Curriculum

The new national core curriculum is the most domi-

nant issue in education in Hungary today. It describes the minimum knowledge that must be acquired by every student by the end of compulsory education. The national core curriculum covers 10 core subject areas, determining approximately 50 percent of the subject matter and allowing teachers to select the remaining 50 percent. The content set out in the national core curriculum provides a framework for textbook writers to follow. Schools and teachers decide which textbook to use and what supplementary material to teach in addition to the prescribed topics. Students must purchase the selected textbooks.

Mathematics Textbooks

When textbook publishing was highly centralized, there was one mathematics textbook for each grade in each type of school. Since the end of the state's monopoly on textbook publishing, textbooks have been offered by four major and a number of minor publishers. There are now several textbooks available, particularly for the lower grades of elementary school and for secondary school. Because of the growing number of six- and eight-year secondary schools, there are more and more textbooks being written for Grades 7 to 12. Textbooks may be submitted to the Ministry of Education for evaluation and approval; approved textbooks will be heavily subsidized and will thus be marketed at a substantially lower price.

Mathematics is built up spirally, with the same areas taught in secondary and elementary schools at an increasingly complex level. Additional topics are introduced at higher levels. In mathematics textbooks, between four and eight main topics are broken down into subtopics with sample exercises. The textbook most often serves as background material for problem-solving based on collections of mathematics exercises.

Pedagogy

Mathematics is taught in a rather traditional fashion in Hungary. Primary mathematics is taught inductively, with some reliance on manipulative materials to develop mathematical thinking. The older the students get, the more mathematics classes lose this component, and by Grades 6 to 8 mathematics is presented as a deductive structure based on axioms and requiring abstract thinking. Because of the thematic structure used, it is difficult to integrate topics within the mathematics curriculum. This structure echoes that used in teacher preparation, where it is only at the end of the mathematics program that preservice teachers are examined in the subject as an integrated whole.

In primary and secondary schools, subject matter is based on open-ended projects and problem solving. Computers are not used in mathematics classes, while calculator use is allowed beginning in Grade 7 or 8 at the teacher's discretion. Activities are limited almost solely to written exercises and demonstrations at the blackboard.

5 SCIENCE CURRICULUM AND PEDAGOGY

Goals for the Science Curriculum

In Hungary, science is not taught as an integrated subject, but is broken down into physics, biology, chemistry, and geography. These are taught as separate subjects in primary school and academic secondary school. In vocational secondary school, science is generally taught as an integrated subject. In some schools, however, the nature of the program requires a deeper involvement in certain areas of the natural sciences, for example the teaching of biology in an agricultural vocational school. Science is not a compulsory subject in vocational schools; it is taught only when necessary to the nature of the program.

Major Changes in the Science Curriculum

The physics, biology, chemistry, and geography curricula were described in the 1978 national curriculum. In 1979, a separate curriculum was written for academic secondary schools, stating the grades and number of lessons in the sciences that were to be taught. Physics was compulsory in Grades 9 to 12, chemistry in Grades 9 and 10, biology in Grades 11 and 12, and geography in Grades 9 and 10. A modification of this curriculum in 1983 resulted in some changes to the number of years and the number of lessons in chemistry and biology. Chemistry became a compulsory subject in Grades 9 to 11 for two lessons per week instead of the previous two to four lessons (two lessons in Grade 9, four in Grade 10). Biology was taught in Grades 10 to 12, two lessons per week instead of the previous four to two lessons (four lessons in Grade 11 and two in Grade 12).

In 1983, there was a reduction in the quantity of subject matter taught. At the same time and as a result of political change, the ideological part of education, dialectical materialism, was deleted. The science curriculum has not changed since 1985, but six- and eight-year academic secondary schools use an alternative curricula approved by the Ministry.

Current Issues in the Science Curriculum

The national core curriculum describes the minimum knowledge that must be acquired by every student by the end of compulsory education. It covers 10 areas of knowledge over 10 years of education, and the requirements are grouped in blocks of 2 years. The curriculum stipulates about 50 percent of the subject matter, allowing individual schools and teachers to determine the remaining half. Less instructional time per week is devoted to science subjects than previously, resulting in a shift in emphasis between grades.

Science Textbooks

Science textbooks are in a rather chaotic state. The textbooks currently used in elementary schools (Grades 1 to 8) were written to comply with the 1978 curriculum, closely following the subject matter from lesson to lesson. During the latter half of the 1980s, several new elementary school textbooks were published, but many lacked a teachers' guide, with the result that teachers based their teaching on personal experience. The trend toward the disappearance of teacher's guides points to an aim of the national core curriculum: the loosening of centrally designed teaching to promote more freedom for teachers.

Similarly, secondary school textbooks are in disarray. Many have been published in manuscript form and sometimes one textbook covers subject matter for three grades. Publication of textbooks has been decentralized since 1990, but only those books approved by the Ministry are placed on the central textbook list and receive subsidies. Individual schools and teachers choose which textbooks they want to use. Students are required to purchase their own textbooks, the prices of which have become 10 to 20 times higher during recent years.

Pedagogy

With the exception of a few vocational secondary schools where science is taught as an integrated subject, the teaching of science is divided into four areas—physics, biology, chemistry, and geography—each area being taught as a distinct subject and an organic whole. Where two topics depend on each other, careful curriculum planning is needed in order that one may follow the other in the correct order. Learning about the structure of molecules, for example, is part of the chemistry curriculum and must precede cell biology.

The teaching of science tends to be inductive in the first few grades involving many manipulative exercises, experiments, and practical investigations. Teaching gradually becomes more deductive and reaches an axiomatic, academic level by secondary school.

Computers are seldom used in the classroom. Teachers must keep to a tight time schedule in order to cover the curriculum, so there is little use of videotapes, slides, or other instructional media. After Grades 7 and 8, student activities are almost exclusively written, except for oral presentations to the class.

6 EVALUATION POLICIES AND PRACTICES

Assessment in Hungary is conducted on three levels: at the national, school, and class levels. There is one national assessment, a matriculation examination for students finishing academic or vocational secondary school. The examination is constructed centrally and students write it concurrently all over the country. Mathematics is a compulsory part of the examination, in addition to Hungarian language and literature (written and oral), history (oral), and a foreign language (written and oral). The examination in mathematics is written, and those who fail can try again on an oral exam. Students completing academic secondary school must choose one or more subjects for the free-choice part of the examination. They may choose any of the subjects studied during secondary school, including any

of the science subjects. The final component of the matriculation is a lecture prepared and presented by students on a topic chosen from the subject under examination. Topics are based on the curriculum and are thus similar in all schools. The presentations are performed in the school and are evaluated by teachers.

Classroom-level assessment provides teachers with an opportunity for regular monitoring of student performance. Assessment may consist of a test covering the subject-matter dealt with during previous lessons, or it may take the form of an oral recitation of newly learned materials. At the end of each major topic there is usually a test to ensure students have a sufficient grasp of the topic. Tests and recitations are graded, forming the basis for the final grade in that subject. If a student fails a subject, there is a second examination on the subject matter for the whole year. If a student fails the second examination as well, he or she must repeat the year. Final examinations are given only in secondary school and not in vocational school.

7 REFERENCES AND SOURCES FOR FURTHER READING

Common Sources

1   Husén T, N T Postlethwaite 1994 *International Encyclopedia of Education*. Pergamon Press, Oxford
2   Organisation for Economic Co-operation and Development 1995 *Education at a Glance: OECD Indicators*. Organisation for Economic Co-operation and Development, Paris
3   United Nations Educational, Scientific and Cultural Organization 1995 *Statistical Yearbook*. United Nations Educational, Scientific and Cultural Organization, Paris
4   World Bank 1995 *World Development Report 1995*. Oxford University Press, New York

Other Sources

5   Ministry of Culture and Public Education 1993–94 *Statistical Report: Primary Level Education 1993–94*. Statistical Department of the Ministry of Culture and Public Education, Budapest
6   Ministry of Culture and Public Education 1993–94 *Statistical Report: Secondary Level Education 1993–*

94. Statistical Department of the Ministry of Culture and Public Education, Budapest

7 Central Statistical Institute 1994 *Statistical Yearbook 1994.* Central Statistical Institute, Budapest

Statistical References

Section	Statistic	Reference	Page	Table	Year of Statistic
Country Profile	93,000 km^2	7	13	1.1	1994
Country Profile	10 million	7	13	1.1	1994
Country Profile	25%	7	13	1.3	1994
Country Profile	1%	7	13	1.22	1994
Country Profile	US$2989.	4	238	1	1991
Country Profile	US$3350.	4	162	1	1993
Country Profile	1.2%	4	162	1	1993
Country Profile	13%	4	162	1	1980–93
Country Profile	7%	3	4–15	4.1	1993
Country Profile	7%	3	4–15	4.1	1993
Country Profile	14%	7	315	8.17	1994
Country Profile	14%	7	315	8.17	1994
Country Profile	15%	7	315	8.17	1994
Country Profile	11%	7	315	8.17	1994
Country Profile	15%	7	315	8.17	1994
Structure of the System	87%	4	217	28	1992
Structure of the System	81%	4	217	28	1992
Structure of the System	15%	4	217	28	1992
Structure of the System	97%	6	30	2	1993–94
Schools in the System	class time	7	328	35	1994

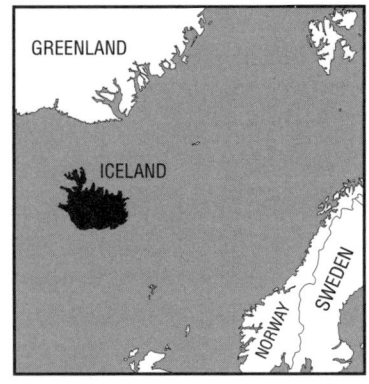

Iceland

Einar Gudmundsson, Institute for Educational Research

Anna Kristjansdóttir, University College of Education

Stefan Bergman, University College of Education

Thorlakur Karlsson, University of Iceland

1 COUNTRY PROFILE

Iceland lies in the north Atlantic, 287 km southeast of Greenland, 798 km northwest of Scotland, and 970 km west of Norway. It occupies an area of over 100,000 km² with approximately 260,000 inhabitants.[1] The first settlers, most of whom came from Norway, arrived in the ninth and tenth centuries. Today, there is virtually a single ethnic group and language in Iceland.

Most Icelanders live within 10 km of the coast, since the center of the island is uninhabitable. The population density is greatest in the capital region, where over 58 percent of the population lives. Communities with more than 200 inhabitants constitute about 91 percent of the population, while about 9 percent of the population lives in small communities and rural areas. The average growth rate of the population has been slightly greater than 1 percent per year since 1960. Immigration and emigration have remained at approximately the same level since 1990.

Iceland is governed by a democratically elected parliament, or *Althing*, founded in the year 930. The *Althing* creates laws and forms the government, which is headed by a prime minister.

The per capita gross national product in Iceland was US$23,300 according to the exchange rate current in 1994, but US$19,900 according to purchasing power parity. The GDP per capita in Iceland was US$22,600 according to current exchange rate in 1994, but US$19,300 according to PPP. Both the GNP and GDP have grown an average of less than 1 percent per year since 1990. The annual inflation rate in Iceland is one of the lowest among OECD countries, having ranged between 2 and 5 percent for the past several years.

---

[1] Census of December 1, 1994

Icelanders spent 13 percent of total public expenditure on education in 1992, or 5 percent of the GDP. About 45 percent of all 20-year-olds finished high school in 1993. Fifty-four percent of female 20-year-olds and almost 37 percent of male 20-year-olds finished high school that year. The adult literacy rate is about 99 percent.

2 THE EDUCATION SYSTEM

Governance and Decision Making

Parliament legislates for all educational levels in Iceland. The Ministry of Education is responsible for putting this legislation into practice in all educational establishments, with the exception of agricultural colleges, which fall under the aegis of the Ministry of Agriculture. A primary and lower secondary school act came into effect in 1995, and a new upper secondary school bill will become law in 1996. These provide for a greater degree of decentralization at both educational levels.

At the compulsory education stage, the Ministry of Education issues a national curriculum guide and monitors local education authorities in their fulfillment of responsibilities outlined in the primary and lower secondary school act. Local authorities finance the school system in their district, while a schools council administers the schools and ensures that all children of compulsory school age attend school. Local authorities must submit an annual report on their district to the Ministry of Education.

The Ministry of Education issues a national curriculum guide for primary and lower secondary schools, and another for the upper secondary schools. The state provides all students in primary and lower secondary school with textbooks. In general, one textbook exists in Icelandic for each subject for each grade; and, where a choice exists, the school or teacher decides which to use. No laws or directives apply to teaching methods at any level.

Structure of the System and Participation Rates

Structure of Education

Figure 1 shows the structure of the education system in Iceland and the approximate enrollment rates for each level. Approximately 75 percent of children between two and five years of age in Iceland attend preschool. A few preschools accept children under the age of two, and about 15 percent of children in this age group attend.

Education is compulsory in Iceland from Grades 1 to 10 of primary and lower secondary school. Virtually 100 percent of the 6 to 15 age group is enrolled in compulsory schooling.

About 85 percent of students aged 16 to 19 enroll in upper secondary schools offering academic courses or vocational courses. Students leaving academic upper secondary school write a matriculation examination, which is a qualification for entry into the University of Iceland and other tertiary-level institutions. Some schools also graduate students with vocational qualifications, and a handful of schools offer vocational training exclusively. Other upper secondary schools offer

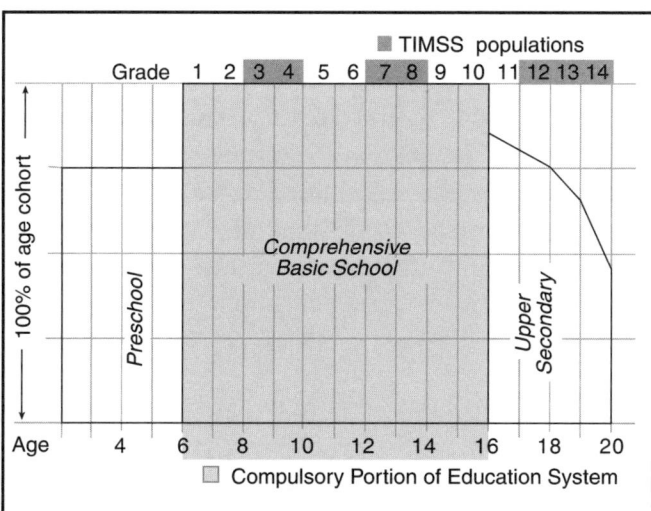

Figure 1: Structure of the Education System

training in seamanship, fish processing, agriculture, arts, and business.

Public and Private Systems

The majority of students attend state schools. Private schools do exist, however, at all levels. About 4 percent of children attend private preschools. Almost 2 percent of primary and lower secondary school students attend private schools, which constitute just under 3 percent of the schools at this level. One upper secondary school of the 65 that offer education at this level is privately operated; about five percent of upper secondary students attend this school.

Schools in the System

The School Year

Most primary and upper and lower secondary schools operate for nine months of the year, from September to the end of May. The academic year varies from one upper secondary school to another, but most begin in early September. The number of school days varies from year to year, but is generally 172 to 174 days in primary and lower secondary schools. In upper secondary schools, the objective is to provide no fewer than 150 teaching days, but there are at present as few as 130 in some cases. There are approximately 20 holidays in primary and secondary school which do not fall on weekends.

Instructional Times

Pupils attend school Monday to Friday. In primary and lower secondary school, a typical school day is from 8:00 a.m. to 3:00 or 4:00 p.m. In upper secondary schools, the school day begins between 8:00 and 9:00 a.m., and students are often in school until 4:00 or 5:00 p.m. Pupils in primary and lower secondary school attend between 25 and 34, 40-minute lessons per week. The new primary and lower secondary school act suggests that Grades 1 to 4 should have 30 lessons per week; Grade 5 to 7, 35 lessons; and Grades 8 to 10, 37 lessons. In academic courses at upper secondary school, students have an average of 36 lessons per week, while those in vocational courses have as many as 46 lessons.

Class Size

In primary and lower secondary schools, the average class size is 15 to 20 students, with larger classes in the capital area than in the regions. In upper secondary schools, the average class size varies between 18 and 20.

Grouping in Mathematics and Science

Officially, there is no streaming in the Icelandic school system, although it is done in some schools at the lower secondary level and in some private schools. Mathematics is studied in every year of compulsory schooling, and science in Grades 1 to 9. During the 10 years of compulsory schooling, students spend approximately 15 percent of teaching time on mathematics and about 6 percent on science. In upper secondary schools, the amount of study time devoted to mathematics and science varies greatly. Students majoring in science or physics in the academic courses of upper secondary school, for example, spend a great deal of time on mathematics. As a minimum, upper secondary students must spend at least 9 percent of their time on mathematics and 9 percent on science.

Certification of Mathematics and Science Teachers

Teachers in primary and secondary schools in Iceland must have one of two qualifications. They may complete a three-year course of study at the University College of Education, which qualifies them to teach in primary and lower secondary schools. These students major in a particular subject, such as mathematics or science. The college also offers a course for prospective teachers in vocational, technical, and fine art subjects in upper secondary schools. By the second route to certification, teachers may hold a B.A., B.Sc., or higher degree from a university, after which they take an additional year of teacher education at the University of Iceland. The University College of Education offers a master's program in special education and school management.

In-service education for teachers at the University College of Iceland is offered in a number of areas through summer courses, workshops, seminars, school-based development projects, and distance learning.

Teacher Profile

In 1994, the average salary of teachers in upper secondary schools was about 94 percent of the average salary of all graduate public employees. The average salary of teachers in primary and lower secondary schools was 74 percent of the same group. However, it should be pointed out that about 15 percent of teachers at primary and lower secondary level are not university educated. These teachers, who were qualified under an older system in which university was not compulsory, fill about 11 percent of teaching jobs. In this context, it should be noted that graduate public employees are lower paid than those in the private sector, and that teachers received a larger proportional salary increase in 1995 than most other employee groups.

The average age of teachers at primary and lower secondary level is 42 years, and 46 years in upper secondary schools. About 79 percent of teachers in primary and lower secondary schools are women and 21 percent men, while in upper secondary schools 41 percent are women and 59 percent men.

3 TIMSS POPULATIONS

Iceland, IEA, and TIMSS

Iceland's Institute for Educational Research has been affiliated with IEA since 1988. The first IEA project in which the Institute participated was the Reading Literacy Study. Iceland is now a participant in TIMSS, but not in other IEA research projects. In recent years, there has been growing interest in IEA's comparative studies, both among scholars and educational authorities. Results of such studies are thought to have an important role in educational policy-making, as well as in the evaluation of the effectiveness of the school system.

Icelandic TIMSS Populations

Iceland is participating in TIMSS at all three population levels. Population 1 consists of 8- and 9-year-old children, who were in Grades 3 and 4 of primary school. Population 2 includes 12- and 13-year-old children in Grades 7 and 8 of primary and lower secondary school.

Population 3 consists of all the students who were to graduate that year from an upper secondary school, and included students in the 17- to 20-year-old age group. This population was composed exclusively of generalists, and not mathematics or physics specialists.

Grades 3 and 4 of primary and lower secondary school include 99 percent of children aged 8 to 9, and Grades 7 and 8 include 99 percent of children aged 12 and 13. In upper secondary schools, an estimated 60 percent of each year graduates from a course of study, and this was the group from which Population 3 was drawn.

4 MATHEMATICS CURRICULUM AND PEDAGOGY

Goals for the Mathematics Curriculum

The Ministry of Education issues a curriculum guide for primary and lower secondary schools that deals in detail with individual subjects. General objectives are laid down for all subjects, the content is discussed, and there is a description of methods. The discussion of objectives emphasizes the importance of conceptual study and understanding of symbols, the importance of applying mathematics to everyday life, and the ability to build on knowledge at the upper secondary stage. Emphasis is also placed on problem-solving skills, logical reasoning, and the ability to use mathematics to describe phenomena.

The national curriculum guide for primary and lower secondary schools, issued in 1989, is well-known to school staff, and its overall objectives are taken into account when school curricula are drawn up. Its influence on actual teaching is, however, not great, since discussion of teaching methods in the curriculum guide is less clear than the discussion of content.

Legislation lays down guidelines on the role of the upper secondary school in preparing students for citizenship in a democracy, for the world of work, and for further study. The laws state that the Ministry of Education shall produce curricula, monitor teaching, and provide guidance on development work related to up-

per secondary schools. According to the curriculum guide, the objective of mathematics teaching is that the student be capable of understanding the concepts and symbols of mathematics and of applying them to the solution of various problems, both in everyday life and within many other fields of study.

The upper secondary school curriculum guide, last issued in 1991, has for the past decade been prepared by a group of upper secondary school teachers on behalf of the Ministry of Education. In addition to the objectives mentioned above, it also includes descriptions of the content at individual stages of study. The curriculum guide is familiar to teachers, and the study guidelines of individual schools are often closely linked to the discussion of content.

Major Changes in the Mathematics Curriculum

The national curriculum guide includes complex objectives, rather than the descriptions of the content of study for individual age groups that characterized previous curriculum guides. No research has been carried out on the usefulness of the curriculum guide in providing practical guidance to teachers.

The national curriculum guide for upper secondary schools has, for the past decade, covered all upper-secondary education after compulsory schooling. Prior to that time, there was no national curriculum. Its aims include standardizing and coordinating stages of study in different majors. The upper secondary schools themselves have played the major role in this process, and in fact made most of the decisions involved, with the agreement of the Ministry of Education.

Discussion has begun on revision of the mathematics curriculum at both school levels. It has been decided that work will be carried out in tandem, in order to ensure better coordination and a systematic approach to mathematical studies. It will probably deal with the influence of calculators and computers on mathematical studies, from the beginning of primary school until the tertiary level. In the current curriculum guide, guidance on the use of this technology is unclear, although the curriculum guide for primary and lower secondary schools recommends adoption of technology where applicable.

Current Issues in the Mathematics Curriculum

In recent years, there has been discussion of how the individual school sets its objectives. School syllabi have therefore been a focus of attention, but in many cases a gap remains between the provisions of the curriculum guide and the content of available textbooks. This gap needs to be bridged in a clearer and more systematic fashion.

Another urgent issue is the question of how using calculators influences the way people calculate. Since calculators came onto the market in the 1970s, education authorities have not prohibited their use in Icelandic schools. Individual schools or teachers have, however, done so. This inconsistent approach has led to the existence of a range of permissibility from a total ban at the primary level to almost unchecked use of calculators in higher grades. There has been a growing debate in recent years on how calculators may be used to improve student numeracy and skills in applying mathematics to everyday life.

Computer use is increasing in Icelandic education, from the first grades to the tertiary stage. However, in both primary and secondary schools, computer use is often inadequately connected to individual subjects. A considerable gap exists between teachers who have made progress and those who do not know how to apply the technology in teaching. Progress in this field has been hindered by the fact that the authorities have not formulated a clear policy on the application of information technology in education; this situation is currently being addressed.

Education policy has favored a minimum of separation of pupils into groups, while those with special needs receive supplementary instruction either within the class or elsewhere. In recent years, a few schools at the lower-secondary level have begun to stream pupils by ability. There is an ongoing debate on alternative methods of taking different abilities into account; this is closely linked to the debate on the quality of evaluation as a part of the teaching process. Considerable interest is shown in improving assessment of the quality of study and teaching.

The training of mathematics teachers is also an urgent issue. There has been a shortage of teachers in Icelandic schools for several years, and there has been a serious lack of trained mathematics teachers in lower and upper secondary schools. In lower secondary schools, for instance, only one mathematics teacher in five majored in mathematics during teacher education. In upper secondary schools, the shortage of teachers with the required training in the subject is severe. No decision has yet been made on how to deal with this problem.

Mathematics Textbooks

Teaching materials for primary and lower secondary schools are largely published by the National Center for Educational Materials, a state institution that operates under special legislation. In 1984, the Center took over responsibility for deciding what material should be published, and in what form. Study materials are allocated to schools on a quota system. In general, the youngest pupils are given materials to keep, while they are loaned to the older pupils. There are no regulations obliging a school to use a certain set of study material in teaching.

The National Center for Educational Materials offers both materials that have been produced and tested in Iceland, and translations of materials from other countries. Almost all teaching material used in Grades 1 to 7 was produced in the 1970s in Iceland. This material is not influenced by the reactionary tendencies which were then common in other countries. In lower secondary schools, original Icelandic material and material translated from Swedish are both used. They are based on very different approaches to mathematical study, and differ greatly in terms of how the content is handled and the method of study to be used.

A large variety of books are in use in upper secondary schools. These include books that have been produced in Iceland, translated books, and photocopied booklets produced by individual schools or teachers. At this educational level, students buy their own textbooks. The presentation of material varies somewhat among the available books, but there is no fundamental difference in the way material is organized, the content, or the quantity of text and problems.

Pedagogy

At the primary and lower secondary level, discussion of innovations and research results takes place within courses for teachers, within schools' development projects, in periodicals published by professional bodies, at teachers' congresses, and on special days set aside for teachers' activities. Issues that are regarded as important, and on which emphasis is placed, include:

- the effects of technology and the schools' reaction to them;
- the use of manipulative materials to reinforce conceptual skills;
- methods of including topics based on students' environment, experience, and ideas;
- the importance of verbal and written communication in mathematical study;
- greater literacy in the applications of mathematics, as used in the media and elsewhere.

The drawing up of syllabi has also received considerable attention, as has the question of evaluation. Organized discussion of teachers' professional development is also under way.

There is growing emphasis on the need to develop a joint debate of teachers in primary and secondary schools. However, since the working environments of the two differ considerably, and since each group has its own professional organization, there is much that divides the two groups.

5 SCIENCE CURRICULUM AND PEDAGOGY

Goals for the Science Curriculum

General objectives for science teaching in Icelandic schools are included in the national curriculum guides for primary and secondary schools. According to the guides, science is part of general education, a preparation for further study and participation in a democratic society, and practical training for a number of professions. The primary and lower secondary school curriculum guide (1989) lays down the principal objectives and content of study at this level, while the individual school sets out teaching methods, textbooks, methods of evaluation in a syllabus adapted to the school environment.

The upper secondary school curriculum guide (1990) includes little detailed discussion of the objectives of science teaching. The curriculum guide describes the objectives and content of individual courses,

defines major options, and standardizes the content and standards of core courses. The curriculum guide is written by Ministry of Education specialists and staff, school directors, and working parties on behalf of professional teachers' organizations. Revision of the curriculum guide is an ongoing process.

There is little monitoring of how schools put the science curriculum into practice. In primary and lower secondary schools, tests have been carried out on a sampling basis in order to assess pupils' progress, but this has not been done in upper secondary schools.

Major Changes in the Science Curriculum

Changes in science teaching described in the 1989 curriculum guide concern both the relative importance of academic and other values, attitudes to teaching, and methods of teaching. The principal change is that science teaching now begins in the first grades of primary school, where it had been an ill-defined aspect of social studies. Biology, chemistry, and physics have been merged into a single subject, natural science. Steps have been taken to integrate the disciplines, but individual schools determine to what degree. Geology remains a part of the social studies and geography curriculum. The objectives for natural science were broadened, and more emphasis placed on the development of the individual, conceptual study, comprehension of the natural environment and humanity's place in it, development of ethical standards, and respect for life and the environment. There is also greater emphasis on subject matter from everyday life, environmental education, and the interaction of technology and science.

Schools are given more scope to choose the emphasis of study, and to adapt educational activity to the school's own environment. At the lower secondary level, the subjects of study are classified into nine groups, including: natural phenomena; the substances in our environment; forms, utilization, and sources of energy; technology and science; shared characteristics of living things; animals; plants; ecology; and the body and health. Pupils study subjects from all groups, as chosen by teachers and pupils.

In the national curriculum guide for upper secondary schools, objectives of knowledge and methodological skills are emphasized, with little attention paid to the social aspects of science teaching. Environmental education receives some attention within individual disciplines. Computer science is organized in the form of a special course for all students on computer technology and applications.

As a general policy, computers are to be used broadly at all grade levels in primary and lower secondary schools. They are primarily used for word processing or for communication, but also to some extent for data processing, data gathering, and measurement, especially in physics teaching. Computers have a growing influence on science teaching at this level, but it does not seem to be especially systematic. Computer equipment and use vary between schools, and there is a lack of suitable software and training for teachers. Pupils have access to computers in computer rooms, physics classrooms, the school library, and increasingly their own classrooms.

At the upper secondary level, the curriculum guide refers very infrequently to computer use in the sciences. In biology, computer use has been limited to teaching programs, and to some extent data processing. In chemistry, there has also been little use. Computer use is greatest in physics, where they have been used for data processing and measurement in many upper secondary schools. A teaching-aid workshop in physics was established in 1990, and this has encouraged the use of measuring instruments in connection with computers.

Current Issues in the Science Curriculum

The curriculum guide for primary and lower secondary schools is now under review, providing the opportunity to introduce new policies, experience, and ideas on science education. The government has made a statement on the need for more education in the natural sciences, technology, and the environment at all educational levels, and for the reorganization of science education in schools. Much still remains unclear on the pursuit of these objectives.

Since the beginning of 1996, primary and lower secondary schools have been operated by local authorities, whose influence on the schools' development will increase. The state will, however, retain responsibility for policy-making and monitoring, and plans to increase this beyond the present level. The government now places greater emphasis on technology, and admits that

this has been a hitherto-neglected aspect of education. It is remarkable that a society with such a high level of technology as Iceland has not, up to the present, given this field more prominence in the general education system.

Much remains to be done in revision of study material, retraining for teachers, and other means conducive to the development of science education. Funding for the production of study material in the sciences is lacking. Experience shows that it has been a lengthy process to put national curriculum guide policies into practice, and the revision of study material is an especially time-consuming process.

The computerization of schools will continue, and technological media will influence teaching in the near future. Progress in the acquisition of technology and its use in schools has been rapid, but there has been criticism of the fact that the use of computers is not well defined in schooling, nor is the connection with education and learning.

Science Textbooks

The slow renewal of textbooks has had a negative influence on science teaching in Icelandic primary and lower secondary schools, particularly in physics. Most textbooks are published by a state institution, the National Center for Educational Materials, which decides which books to publish and allocates them to schools. Recently published books on science for Grades 1 to 7 are lavishly illustrated, with varied projects, experiments, and discussion topics. They place greater emphasis on integrated science subjects, the balance of nature and the environment, and on conceptual study. The student can learn directly from the book, allowing classroom teaching to concentrate on projects and discussion.

Teachers require a great deal of their textbooks, and many of them follow the book strictly. They are not compelled to use specific books, but they generally have a limited choice. All textbooks at the primary and lower-secondary school level are written and published in Icelandic. Since the 1970s, study material in biology at lower-secondary level has been translated and adapted from other languages.

Upper secondary textbooks are either original Icelandic books or foreign books translated and published by Icelandic publishers. Books in other languages are used in some disciplines, especially physics and biology. Publishers decide which of these books to publish, and do not require official approval. The textbooks are chosen by the teacher and purchased by students.

Pedagogy

A new national curriculum guide was published in 1989 and work is underway on a revised guide. The new guide will be based both on the experiences of recent years and new policies in science and technology. The following pedagogical issues will receive some attention:

- The curriculum guide should continue to emphasize varied methods of teaching, but the approach to practical teaching should be reviewed. Teachers are professionally responsible for choosing teaching methods, and choose these themselves or in collaboration with students. The emphasis upon environmental education in science teaching has led to greater variety of methods, and increased the part played by project work.

- The curriculum guide should accurately define the emphases and content of material, especially at the primary and lower secondary level. Newer textbooks in science encourage the development of varied methods by emphasizing research, experiments, field work, group work and discussion, personal experience, expression, and conceptual study.

- Icelandic schools have adopted computers and other technology to a considerable degree. Computerization has not, however, had an impact on science teaching as of yet. Computer use should be more systematic than hitherto.

- In recent years constructivism has had a growing influence on teaching and on the form of study material, and this trend may be expected to continue. Teacher education has lent support to this development.

Developments in science teaching in upper secondary schools have been slow since the period of great change in the 1970s. Much remains unclear about development in the future, but there are discernible signs of the influence of computers, environmental education, and constructivism.

6 EVALUATION POLICIES AND PRACTICES

Evaluation in the Icelandic school system consists primarily of examinations and tests written and organized by teachers. In Grade 10, compulsory national examinations are held in four subjects: mathematics, Icelandic, Danish, and English. No national examinations are held in science, physics, chemistry, or biology in either primary or secondary school, and all examinations in upper secondary school are school based.

National Examinations

The purpose of the Grade 10 national examination is to provide information on pupils' knowledge at the end of compulsory education in the four subjects tested. When combined with the school's assessment of the pupil, the examinations provide upper secondary schools with information on the pupil's level of achievement. Upper secondary schools base their admissions on examination results and stream students accordingly. Icelandic law provides all students with the right to enter upper secondary school after 10 years' education in primary and lower secondary school, regardless of their academic standing.

In Grade 10, preparation for the national examinations dominates teaching in the four subjects. Thus the examinations have considerable influence upon the instruction given in most schools. The national examination in mathematics is criterion referenced and covers the study material for Grades 7 to 10. Students write answers to the problems given in the mathematics examination, showing their calculations; multiple-choice questions are not used.

The new primary and lower secondary school act requires that more national examinations be held. In addition to those in Grade 10, there will also be national examinations in Grades 4 and 7. No decision has been made on which subjects will be examined, but mathematics is likely to be included. There is some doubt as to whether examinations will be held in science, physics, chemistry, or biology in these grades. Similarly, it is likely that national examinations will be introduced in upper secondary school in the coming years.

School-Level Examinations

The national curriculum guide provides for varied evaluation in mathematics, particularly in the choice of methods of evaluation, projects used to evaluate knowledge and skills, and types of work. In science, evaluation in primary and lower secondary school must cover knowledge, understanding, skill, and viewpoints within the various fields of science. Achievement should therefore be evaluated not only by written examination but also by consideration of work, skill, and attitude, by both continuous assessment and evaluation of various projects. In practice, knowledge is largely evaluated by means of written examinations. The most common forms are essay or short-answer-type questions. Multiple-choice questions are rare in class examinations.

There is little discussion of evaluation in mathematics and science in upper secondary schools in official rulings, but evaluation of various different kinds is assumed. This may consist of a single examination at the end of a stage of study, or continuous assessment of work.

In mathematics, traditional class examinations are the principal form of evaluation in upper secondary schools. In science, written examinations are the most common form, but grades are given for the written examinations plus practical work. The most common forms of science examination in upper secondary school are essays and short-answer questions.

There are no standardized achievement tests in mathematics and science for primary or secondary schools, but interest in such examinations has grown over the past three years. Work is now in progress on standardized achievement tests in mathematics for Grades 2 to 10 of primary and lower secondary schools.

7 REFERENCES AND SOURCES FOR FURTHER READING

Common Sources

1   Husén T, N T Postlethwaite 1994 *International Encyclopedia of Education*. Pergamon Press, Oxford
2   Organisation for Economic Co-operation and Development 1995 *Education at a Glance: OECD Indicators*. Organisation for Economic Co-operation and Development, Paris
3   United Nations Educational, Scientific and Cultural Organization 1995 *Statistical Yearbook*. United Nations Educational, Scientific and Cultural Organization, Paris
4   World Bank 1995 *World Development Report 1995*. Oxford University Press, New York

Other Sources

5   Benediktsson R 1993 *A alfjódavettvangi í kennslustofunni. [The international classroom]*. Uppeldi og menntun, 2, 131–14
6   Bergmann S 1995 Introduction of modern environmental education in a small, developed society. In MacDermott F (ed.) *Proceedings of The Conference on the Exchange of Promising Experiences in Environmental Education in Great Britain and the Nordic Countries.* (pp. 39–51) Bradford, University of Bradford
7   Bergmann S 1994 Entwicklungstendenzen des Biologieunterrichts an den Schulen Islands. [Trends in biology teaching in Icelandic schools.] In H Entrich and Staeck L (eds.) *Biologische Bildung in einem Europa des politischen Umbruchs.* (pp. 167–171) Berlin, Heuchtturm-Verlag
8   Delury G E (ed.) 1987 *World Encyclopaedia of Political Systems and Parties* (2nd ed) New York, Facts on File Publications
9   Elley, W B 1992 *How in the World Do Students Read?* The Hague, IEA
10 Frumvarp til laga um framhaldsskóla. 1995 [Bill on upper secondary schools.]
11 Gestsson V 1995 Margmidlun fyrir grunnskóla. [Multimedia for comprehensive basic schools.] *Ny menntamál, 13* (1), 14–16
12 Gudjónsson H 1988 *Almenn efnafrædi. [General chemistry.]* Reykjavík, Mál og menning
13 Gudjónsson H 1990 *Almenn efnafrædi II. [General chemistry II.]* Reykjavík, Mál og menning
14 Gunnlaugsdóttir M 1995 Personal contact at Ministry of Culture and Education (The data were gathered from schools at the authors' request)
15 Hansen F B, O J Proppé 1992 Teacher education in Iceland. In F. Buchberger (ed.) *ATEE-Guide to Institutions of Teacher Education in Europe* (pp. 156–167). Bruxelles, Association for Teacher Education in Europe

16 Jónasson Jón Torfi 1990 *Menntun og skólastarf á Íslandi í 25 ár 1985–2010.* [25 Years of Education and Schooling in Iceland 1985–2010.] Reykjavík, Author

17 Jónsdóttir K (in press) *Icelandic National Dossier.* Reykjavík, Ministry of Culture and Education

18 Jónsson, A H, M Olafsdóttir 1993 Stada líffrædikennslu á framhaldsskólastigi ári 1993. [Biology teaching in upper secondary schools in 1993.] *Rádstefna um líffrædikennslu 24–25 Sept. 1993.* (pp. 33–37) Reykjavík, Samlíf-Samtök líffrædikennara, Líffrædifélag Íslands

19 Kjararannsóknarnefnd opinberra starfsmanna. 1995, April *Fréttarit KOS.* [KOS Newsletter.] Reykjavík, Author

20 Kjartansson, H 1993 Stefna í náttúrufrædikennslu í grunnskólum. [Science teaching policy in comprehensive basic schools.] *Rádstefna um líffrædikennslu 24–25 Sept. 1993* (pp. 15–19). Reykjavík: Samlíf-samtök líffrædikennara, Líffrædifélag Íslands

21 Lög um grunnskóla nr. 66/1995. [Act on comprehensive basic schools, no. 66/1995.]

22 Lög um framhaldsskóla nr. 57/1988. [Act on upper secondary schools, no. 57/1988.]

23 Macdonald, MA 1993 *Science Curriculum Materials In Compulsory Schools in Iceland. Status Report C. The Status and Future of Science Education in Iceland (age 6–15).* Reykjavík, The Research Centre of the University College of Education

24 Macdonald, MA, MA Garner 1991 *Science in Iceland: From Curriculum to Classroom in the 1990s.* A paper presented at the Third Nordic Conference on Science and Technology Education held in Hanko, Norway 5–8 August 1991

25 Menntamálaráduneytid 1989 *Adalámskrá grunnskóla.* [National Curriculum Guide for comprehensive basic schools.] Reykjavík, Author

26 Menntamálaráduneytid 1990 *Námskrá handa framhaldsskólum: Námsbrautir og áfangalysingar.* [National Curriculum Guide for upper secondary schools: Tracks and course descriptions.] Reykjavík, Author

27 Menntamálaráduneytid 1991 *Til nyrrar aldar. Framkvæmdaáætlun menntamálaráduneytisins í skólamálum til ársins 2000.* [Towards a New Century: Ministry of Education and Culture Action Plan for Education to the Year 2000]. Reykjavík, Author

28 Menntamálaráduneytid 1994 *Nefnd um mótun menntastefnu.* [Educational Policy Committee.] Reykjavík, Author

29 Ministry for Foreign Affairs 1995 *Gender and Equality in Iceland: National Report to the Fourth United Nations World Conference on Women in Beijing 1995.* Reykjavík, Author

30 Ministry of Culture and Education. 1994 *Fjöldi nemenda í grunnskólum 1994–1995.* [Number of students in Icelandic comprehensive basic schools 1994–1995.] Reykjavík, Author

31 Olafsson T 1994 Statusrapport om anvendelse af EDB i naturfag i grundskolelæreruddannesle i Island. [Status report on the use of EDB in science in teacher training in Iceland.] In U Vasström (ed.) *Lärarnas lärare i naturorienterande ämnen år 2000* (pp. 78–79) Helsingborg, Nordisk Ministerråd

32 Sigurgeirsson I 1991 *The Role, Use and Impact of Curriculum Materials in Intermediate Level Icelandic Classrooms.* University of Sussex

33 The Statistical Bureau of Iceland 1994 *Landshagir: Statistical Abstract of Iceland 1994.* Reykjavik, Author

34 Stefánsdóttir L, S Myrdal 1993) Fjarkennsla um tölvunet. [Distance learning through computer nets.] *Uppeldi og menntun, 2,* 121–130

35 Sveinsson G 1995 Personal contact at the National Economic Institute of Iceland

36 Thórsdóttir AB 1995 Personal contact at the Governments Personnel Office. (The data were gathered from teachers' payroll records, where each teacher's birthday appears)

Statistical References

Section	Statistic	Reference	Page	Table	Year of Statistic
Country Profile	287 km, 798 km 970 km	33	19	1.3	1995
Country Profile	103,000	33	19	1.3	1995
Country Profile	264,919	33	26	2.1	1993
Country Profile	58%*	33	29	2.3	1993
Country Profile	91%*, 9%*	33	35	2.6	1993
Country Profile	1.25%*	33	26	2.1	1960–1993
Country Profile	1-1.5 percent*	33	46	2.13	1986–1993
Country Profile	$23,300 $19,900	35	NA	NA	1994
Country Profile	$22,600 $19,300	35	–	–	1994
Country Profile	0.5%	35	–	–	1990–1994
Country Profile	1.5%, 4.1%, 3.7 %	35	–	–	1992–1994
Country Profile	13%, 5%	33	223	15.4	1992
Country Profile	45%, 54%, 37%	33	274	18.7	1992–1993
Country Profile	99%	9	9	2.1	1991
Structure of the System	75%, 15%	17	Chapter 3	3.12	1992
Structure of the System	100%	17	Chapter 4	4.14	1994–1995
Structure of the System	85%, 74%, 66%	17	Chapter 5	5.5	1993–1994
Structure of the System	45%	33	274	18.7	1992–1993
Structure of the System	4%	17	Chapter 2	2.6.1	1995
Structure of the System	2%	17	Chapter 2	2.6.2	1995
Structure of the System	3%	17	Chapter 4	4.14	1994–1995
Structure of the System	One of 65	17	Chapter 5	5.5	1993–1994
Schools in the System	172–174	17	Chapter 2	2.10.1.2	1994–1995
Schools in the System	20	17	Chapter 2	2.10.1.3	1994–1995
Schools in the System	150, 130	17	Chapter 2	2.10.1.3	1994–1995
Schools in the System	25, 34	17	Chapter 2	2.10.2.2	1994–1995
Schools in the System	30, 35, 37	17	Chapter 2	2.10.2.2	–
Schools in the System	36, 46	17	Chapter 2	2.10.2.3	1994–1995
Schools in the System	15–20*	30	–	–	1994–1995
Schools in the System	18–20	17	Chapter 5	5.3.1.3	1994–1995
Schools in the System	15%, 6%	17	Chapter 4	4.7	1994–1995
Schools in the System	9%, 9%	17	Chapter 5	5.3.1.6	1994–1995
Teacher Profile	94%*, 74%*	19	21–24	2.6	1994
Teacher Profile	15%*, 11%*	17	Chapter 4	4.14	1994–1995
Teacher Profile	42, 46	36	–	–	1995
Teacher Profile	79%, 21%, 41%, 59%	29	3	5.2, 5.3	1993

* Computed from numbers in the reference.

Islamic Republic of Iran

Ali Reza Kiamanesh, Center for Educational Research

Fatemeh Faghihi, Center for Educational Research

1 COUNTRY PROFILE

The Islamic Republic of Iran is a mountainous country with an area of 1.6 million km² located in southwest Asia. Known as Persia until 1935, Iran is bounded by the central Asian countries of Armenia, Azerbaijan, and Turkmenistan, and the Caspian Sea to the north, Afghanistan and Pakistan to the east, the Oman Sea and the Persian Gulf to the south, and Iraq and Turkey to the west.

Iran's population in 1995 was almost 60 million. Thirty-four million, or 58 percent, are urban residents. About 99 percent of the population are Muslims, most belonging to the official religion, the Twelve Imamiyyah Shia sect. There are minority groups of Christians, Jews, and Zoroastrians, whose religions are officially recognized and whose followers may participate in the country's political, economic, and social affairs. Religious minority groups have their own representatives in the Islamic Consultative Assembly.

The official language and script of Iran is Persian, but the Turkish, Kurdish, Arabic, Lori, Gilani, Mazandarani, and Baluchi languages are used in some regions. Iranian culture and civilization date back two thousand years when Aryan tribes of Medes, Parthians, and Persians entered the land, which was later given the name Iran, or land of the Aryans. In the middle of the seventh century, the majority of Iranians accepted the religion of Islam.

In the wake of the Islamic Revolution in Iran of 1979, begun under the leadership of Imam Khomeini, a new political and ideological culture and identity were introduced. One year after the Islamic Revolution, the Literacy Movement was established to promote child and adult literacy. By 1995, fifteen years after the establishment of the Literacy Movement, more than 79 percent of Iranians over the age of six were literate.

Public expenditure on education has increased significantly in recent years, as the figures in Table 1 demonstrate.

Table 1: Public Expenditure on Education, 1993 to 1995

Year	1993	1994	1995
Public Expenditure on Education, in billion dollars	US$1.5	US$1.9	US$2
Gross National Product, in billion dollars	US$31	US$38.6	US$45.8
Expenditure on Education, Percent of Budget	21	20	23
Expenditure on Education, Percent of GNP	5	5	5

2 THE EDUCATION SYSTEM

Governance and Decision Making

The Ministry of Education creates and implements rules and policies pertaining to education, after policies are approved by the Supreme Council of Education, a legislative body with authority over preuniversity education. Within each province a general directorate operates under the Ministry of Education's jurisdiction. Each directorate contains several district offices, the number depending on the density of the region's population.

The preparation of textbooks is a continuous activity aimed at meeting specific objectives for each level of education. In 1994–95, the Ministry of Education published close to 171 million volumes of school textbooks covering over 1000 topics. The Organization for Research and Educational Planning, an affiliate of the Ministry of Education, has two subordinate bureaus that prepare textbooks: the Curriculum Development and Textbook Compilation Bureau, and the Textbook Publication and Distribution Bureau. Curriculum committees within the Curriculum Development and Textbook Compilation Bureau, consisting of educators, curriculum specialists, subject-matter specialists, teachers, administrators, and university professors, determine the specific objectives of education for Grades 1 to 12. These goals, when approved by the Supreme Council of Education, form the draft curricula for primary, junior secondary, and secondary education.

Structure of the System and Participation Rates

Structure of Education

Iran's education system follows a 5-3-4 model. The structure of the system is shown in Figure 1, along with the approximate enrollment rates for each level. Preprimary education is available as a one-year program to prepare five-year-olds for the primary school environment. General education is offered at the primary level for children aged 6 to 10, and at the junior secondary level for students aged 11 to 13. At the primary level, 96 percent of the age cohort is enrolled in school, while at the junior secondary level the figure is 70 percent. Primary and junior secondary school,

Grades 1 to 8, form the compulsory years of education.

Secondary education for students aged 14 to 17 is divided into two streams: academic and vocational. Academic secondary education teaches theoretical courses for students wishing to pursue higher education or obtain employment. There are three major areas of study: mathematics and physics, experimental sciences, and literature and the humanities. Vocational education provides skill development and training for students wishing to pursue specific technical careers. It is divided into three areas of study: technical, agricultural, and vocational. The secondary system is currently undergoing a fundamental reorganization. When this reorganization is complete, secondary school will be three years instead of four, although there will be an additional year for students writing university entrance examinations.

Forty-four percent of the age cohort attends some form of secondary education. In 1994–95 in academic secondary schools, 45 percent of students were female and 55 percent male. In vocational schools in the same year, 21 percent were female and 79 percent male.

Public and Private Schools

Nongovernment organizations are able to establish private schools in Iran, provided they follow the same curricula and extracurricular programs as public schools. Ninety-eight percent of primary schools were public in 1994–95, 94 percent of junior secondary schools were public, and 93 percent of academic secondary schools were public. There are no private vocational or technical schools.

Mathematics and Science Enrollment

Science is compulsory until the end of Grade 11. In 1994–95, 51 percent of males and 52 percent of females in Grade 12 were enrolled in a science course. Mathematics is compulsory until the end of Grade 10. In 1994–95, 52 percent of males and 49 percent of females in Grade 11 were enrolled in a mathematics course. Among Grade 12 students the figures were 22 percent of males and 10 percent of females. In 1994–95, Grade 12 was in the process of being phased out; during that year, some students were enrolled in Grade 12, others left school at the end of Grade 11, and the rest proceeded to a preuniversity year. Low enrollment figures in Grade 12 mathematics were a result of this process.

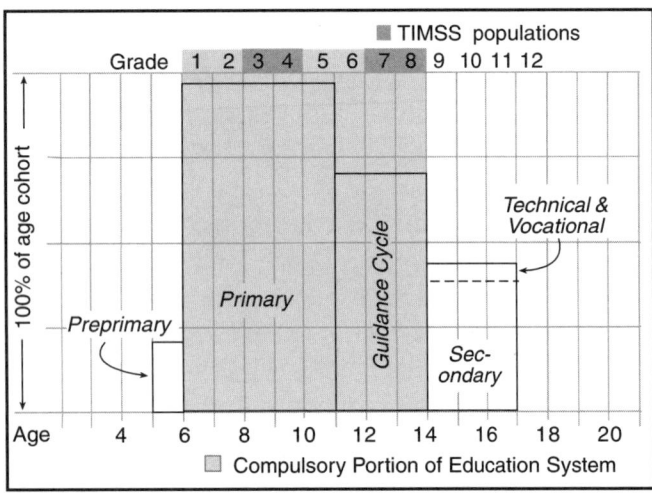

Figure 1: Structure of the Education System

Schools in the System

The School Year

The annual school calendar for all three levels of education includes 200 instructional days between September 23 and June 21 of the next year. There is a summer vacation between June 22 and September 22, and a New Year's holiday of 13 days, from March 21 to April 2. In addition, there are several national holidays throughout the year. Students attend school 6 days per week, from Saturday to Thursday.

Class Time Allocated to Mathematics and Science

In Grades 1 to 5, students attend school for 24 to 28 hours per week, depending on the grade level. Between 18 and 21 percent of school hours are devoted to mathematics and between 11 and 13 percent to science. In Grades 6 to 8, students attend school for 30 to 33 hours per week, spending 12 to 17 percent of class time on mathematics and 12 to 13 percent on science. In Grades 9 to 12 students attend school for 32 to 36 hours per week. Depending on their grade level and area of study, students spend between 9 and 30 percent of class time on mathematics and between 6 and 44 percent on science. The average class size in primary education was approximately 29 in 1994–95. In the same year, class size in both junior and academic secondary was about 32.

Streaming and Tracking

There is no streaming or tracking in Grades 1 to 8. Secondary education is divided into academic and vocational streams. There are different admission criteria for the two branches, as determined by the Supreme Council of Education. The objective in preparing these criteria is to provide conditions that will enable students to enter the branch appropriate to their ability. An important goal is to ensure that there are sufficient graduates for the needs and priorities of the country. In general, Grade 8 students must achieve a minimum grade in designated courses for admission to secondary school. Each branch of secondary education specifies different designated courses; mathematics is required for entering the Mathematics-Physics branch, and science for entering the Experimental Science branch.

Certification of Mathematics and Science Teachers

Teachers may be certified in one of two ways: at a teacher education center or a university. Teacher education centers provide certification for primary and junior secondary teachers. These centers are post-secondary institutes that select students from secondary schools through a national entrance examination, with graduates receiving an associate diploma in primary and junior secondary education. There are 12 fields of study within these centers: mathematics, experimental sciences, physical education, English, technical and vocational education, social sciences, primary education, Persian literature, counseling, Islamic ethics and Arabic language, art, and special education. Preservice teachers specialize in one of these areas, and generally teach only that area. About 20 percent of these courses is devoted to pedagogical training.

University teacher education programs produce specialist teachers for both academic and vocational secondary schools. After four years, students are awarded a bachelor's degree in their area of study. About 18 percent of these courses is devoted to pedagogical training. In addition to the seven universities specializing in teacher education, several other universities and higher education institutes provide teacher education programs. Preservice teachers in these programs take pedagogical courses along with courses in their specialty subjects.

Within the Ministry of Education is a directorate of in-service programs, which develops short-term and long-term courses for all Ministry personnel, including teachers. The education content of the short-term programs is divided into three categories: general, special, and skill development. General programs consist of courses such as Islamic or political knowledge. Special programs concentrate on science topics to update teachers on specialized knowledge. Skill development programs stress teaching methods, recent changes in textbooks, and appropriate instructional aids. Long-term courses lead to higher degrees and are offered by in-service higher education centers, teacher education centers, and universities.

Teacher Profile

Teaching is a respected profession in Iran, and teachers occupy a middle position both in regard to salary and professional status. In the primary school system in 1994–95, 53 percent of teachers were female and 47 percent male. At higher levels of education, however, the proportion of males increased. In junior secondary schools in the same year, 45 percent of the students were female and 55 percent were male. In academic secondary schools, 40 percent were female and 60 percent were male, while in vocational schools, 15 percent were female and 85 percent male. Among specialist teachers at the secondary level, about 13 percent are mathematics and 19 percent science specialists.

3 TIMSS POPULATIONS

Iran, IEA, and TIMSS

Iran became associated with IEA and TIMSS in 1991. In addition to TIMSS, Iran is participating in the Language Education Study.

Iranian TIMSS Populations

Iran is participating in TIMSS at the Populations 1 and 2 levels. Population 1 consists of all students enrolled in Grades 3 and 4, which contain the largest number of students in the nine-year-old age range at the time of testing. Population 2 consists of all students enrolled in Grades 7 and 8, which contain the largest number of students in the thirteen-year-old age range at the time of testing.

4 MATHEMATICS CURRICULUM AND PEDAGOGY

Goals for the Mathematics Curriculum

There is a national curriculum in mathematics that all the schools are required to implement. Goals are prepared by a mathematics council and approved by the Supreme Council of Education. The goals for mathematics education in Grades 1 to 8 are to:

- develop a systematic way of thinking and inferring, as well as a systematic approach to using acquired knowledge in conclusions and abstractions;
- develop the ability to do regular, daily numerical calculations;
- develop the ability to do simple mental calculations including numerical and measurement estimation;
- familiarize students with the aspects of mathematics that relate to other subjects;
- develop the ability to solve problems;
- develop an understanding of the mathematical concept of each problem, and the ability to express it in a mathematical framework.

The goals for mathematics education in Grades 9 to 12 are defined in four general categories:

- mathematics as a component of the study of nature and the universe;
- mathematics as an essential element in developing cognitive abilities;
- mathematics as a part of the world of the future;
- mathematics as an element in the cultural development of society.

These categories are further described in specific, detailed goal statements.

Major Changes in the Mathematics Curriculum

At present, there is no curriculum guide and teachers refer to textbooks for curricular changes. New editions of the five volumes of primary mathematics textbooks have recently been published, with the principal change being a concept of numbers based on new theories of learning. Multiplication and division of common fractions, for example, have been omitted from the Grade 4 curriculum, and division of decimal fractions from Grade 5. Examples and exercises from real life situations have been incorporated into all new books.

In the junior secondary curriculum, new textbooks have been published for all three grades, with the following curricular changes incorporated:

- Integers have been considered in an algebraic structure.
- Elements of analytic geometry have been included.
- Geometric transformations such as symmetry have been used to introduce shapes such as cones and spheres.

At the secondary level, applications of mathematics are widely considered in the new editions of both academic and vocational textbooks. A new textbook for computer studies based on algorithmic thinking has been published for secondary schools. Computational mathematics and algorithms are considered important at all levels, and more emphasis is now being placed on the application of definitions.

Current Issues in the Mathematics Curriculum

Evaluations of both the elementary and junior secondary school curricula have been conducted and revisions are planned. New mathematics curricula have been developed to accommodate the shift to a three-year secondary school system, and mathematics and computer textbooks will be issued to conform with the new curricula. New computer textbooks are in progress. In compiling the new mathematics textbooks, some mathematical structures such as Boolean algebra, group algebra, rings, and fields have been omitted from the curriculum, while complex numbers, graph theory, and limits and derivatives of logarithmic functions have been added. Moreover, some concepts such as conic sections, integrals, and congruence were shifted from Grade 12 to Grade 11, the final grade of the new system.

Mathematics Textbooks

Textbooks are prepared and published by the Ministry of Education, and all schools must use them. In elementary and junior secondary schools, teachers are expected to follow the textbook page by page to ensure that they adhere to the curriculum. At these levels, textbooks contain a workbook section in which students solve problems and do exercises. Most textbooks contain many graphs and illustrations.

Pedagogy

Pedagogical practices are determined by District Offices of Education or within the school and may be summarized as follows:

- Equal emphasis is placed on teaching concepts as well as rules and computational procedures. Rules and computational techniques are more strongly

emphasized at the primary level, however, as is concept development at the secondary level.
- Cooperative learning is not a common procedure in mathematics classes at any level.
- Mathematics is usually integrated across topics within mathematics, particularly at the lower levels.
- Calculators are not used in mathematics classes. Computers are not intended for use in mathematics classes, and are not usually used.
- Both deductive and inductive approaches are used in teaching mathematics.
- Manipulative materials are frequently used at the primary level, but their use is restricted at upper levels.
- Investigations and open-ended projects are used occasionally, but are not emphasized in the curriculum.
- Oral and written communication are not emphasized in mathematics education.

5 SCIENCE CURRICULUM AND PEDAGOGY

Goals for the Science Curriculum

The Curriculum Development Bureau, which consists of several groups of experts, is responsible for all curriculum development. All the bureau's decisions concerning curriculum structure and content are incorporated into official curricular documents and must be implemented by schools.

The goals of science education are categorized in four domains: cognitive, skill learning, psychomotor, and affective. The goals in the skill-learning domain, which are applicable to all branches of science are to:
- develop the student's interest in and positive attitude toward science, to benefit both students and society;
- develop the student's ability to learn how to learn, and to acquire self-learning habits and lifelong education skills;
- develop the student's ability to communicate, to present opinions and hear those of others, and to communicate orally and in written and graphic form;

- familiarize students with scientific methods and engage their curiosity and perseverance in undertaking scientific and experimental tasks;
- develop problem-solving skills;
- develop the student's ability to accommodate continuous change in science, technology, and professions;
- develop concepts and generalizations from a variety of facts and observations.

Major Changes in the Science Curriculum

As in other countries, computer technology is rapidly increasing at all levels of education. Science curricula at the primary and junior secondary levels now place more emphasis on skill learning and performance aspects.

Current Issues in the Science Curriculum

Five years ago, a single science curriculum committee was created and began formulating new science education goals and curriculum frameworks for both the primary and secondary levels. Formerly, specialist groups within the Bureau of Curriculum Development formulated curriculum frameworks for the individual subjects in their areas of specialty. A draft version of the Grade 1 science textbook is being prepared by the science curriculum committee, along with a workbook and teacher's guide. When the new textbook is issued, information sessions will be held to inform teachers about changes to the curriculum.

The curriculum was updated because new topics in science and technology needed to be incorporated into textbooks, and because teaching methods were old and inefficient. Updating teaching methods is an essential component of any revision of the science curriculum. The most recent revision of the science curriculum shifted its emphasis from content- to process-centered teaching methods.

Science Textbooks

The Ministry of Education writes and publishes all textbooks, including science textbooks. Primary textbooks are paid for by the government, and junior secondary and secondary textbooks are government subsidized so that students may purchase them at a reduced rate. Students have their own science textbooks, from which they read and do exercises.

Science teachers rely so heavily on textbooks in classroom teaching that many consider the textbook to be the program of study. The domination of textbooks over classroom process is such that when a new book is being compiled, teachers become anxious and demand in-service training to cope with the changes. In developing the recent changes to the curriculum, the curriculum committee considered the use of different approaches, including explanations, exercises, and historical highlights. These have now been incorporated into secondary science textbooks and the draft versions of the primary textbooks, and the new books are interesting and easy to read.

Pedagogy

Pedagogical practices are determined by District Offices of Education or within the school, and may be summarized as follows:

- Concept development takes place through the teaching of skills at all levels of education.
- Cooperative learning is sometimes used in science education.
- Science is usually integrated across topics within science, particularly at the lower levels of education.
- Computers have many applications in science education, and are used more frequently than calculators.
- Both deductive and inductive approaches are used in science education, with more emphasis on deductive.
- Investigations, laboratories, and open-ended projects are strongly recommended in the curriculum.
- Activities focusing on oral and written communication are used in science classes.

6   EVALUATION POLICIES AND PRACTICES

Since curriculum development in Iran is centralized, there is also a centralized assessment system. Final assessments are designed for the last grade of each level: Grade 5 of primary, Grade 8 of junior secondary, and Grade 12 of secondary. The first two examinations are administered on a provincial basis, while the Grade 12 or school leaving examination is administered nationally. The goal of these examinations is to evaluate students' knowledge and ability to succeed at the next level of education.

At the end of each grade, written and oral examinations are administered in all subjects. In addition, there are written examinations in all subjects every three months. The instruments used in internal exams are designed by teachers and there are no regulations governing them, but most teachers use essay-type and open-ended questions. Experience indicates that the types of assessment and questions used tend to affect teachers' mode of instruction in the classroom, with some emphasizing memorization and others assessment of skills.

CONCLUDING REMARKS

Iran participated in TIMSS to achieve a number of goals, all of which are rooted in the fact that mathematics and science courses are the fundamental bases of technological and scientific development in any society. Furthermore, economic and scientific improvement depends to a large extent on the abilities of students in these two subject areas. Iran, like other countries, expects the school curriculum to provide students with the necessary and appropriate content by using correct methodology, by raising student awareness, and by creating a positive attitude among them.

It is hoped that participation in a large-scale study—especially since it is Iran's first active participation in such a study since the revolution—will provide the country with experience in carrying out a broad and extended project. Furthermore, it will create an opportunity for specialists and authorities in mathematics and science education to provide answers to the questions posed by the study regarding the intended, implemented, and attained curricula, and the relationships between curricula and social and educational contexts.

Such an experience will also provide an opportunity for specialists to evaluate the nation's education system and identify its weaknesses and inadequacies, as well as to compare the present education situation in Iran with that of other countries. This will also al-

low researchers to identify the roots of social differences, and, if necessary, call for revision to curricular structure and content, teacher education and methodology, teaching-learning processes, and the examination and evaluation system.

7 REFERENCES AND SOURCES FOR FURTHER READING

Common Sources

1 Husén T, N T Postlethwaite 1994 *International Encyclopedia of Education*. Pergamon Press, Oxford
2 Organisation for Economic Co-operation and Development 1995 *Education at a Glance: OECD Indicators*. Organisation for Economic Co-operation and Development, Paris
3 United Nations Educational, Scientific and Cultural Organization 1995 *Statistical Yearbook*. United Nations Educational, Scientific and Cultural Organization, Paris
4 World Bank 1995 *World Development Report 1995*. Oxford University Press, New York

Other Sources

5 Ministry of Education 1993 *Education in the Islamic Republic of Iran*
6 Bureau of Coordination of Plans and Development Planning 1995 *Educational Statistics*
7 Educational Statistics: Ministry of Education, June 1995.

Statistical References

Section	Statistic	Reference	Page	Table	Year of Statistic
Country Profile	56 million	1	1–7	1.1	1991
Country Profile	74%	5	137	–	1991
Country Profile	table	5	174–176	8.1, 8.2	1990–93
Structure of the System	109, 84, 44	5	69	3.6	1992–93
Structure of the System	45,55	5	43, 50	3.3, 3.4	1990–93
Structure of the System	21, 79	5	56, 194	3.5, 8.4	1990–93
Structure of the System	99, 98, 96	5	37, 43, 50	3.2, 3.3, 3.4	1992–93
Schools in the System	29, 32	6	2–4	–	1994–95
Teacher Profile	15, 85	5	56, 194	3.5, 8.4	1990–93
Teacher Profile	47, 53, 55, 45, 60, 40, 13, 19	6	b–l	–	1994–95

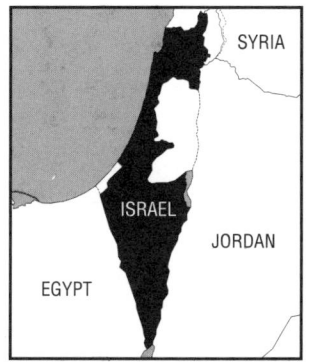

Israel

Pinchas Tamir, The Hebrew University of Jerusalem

1 COUNTRY PROFILE

Israel is a small country in the Middle East, located at the eastern end of the Mediterranean. It has a total area of 21,000 km², excluding occupied territories.

Israel is an ancient land of many cultures and ethnic groups, and is known as the Holy Land to three major religions: Judaism, Christianity, and Islam. The Jewish nation that inhabited the country during the Biblical era lived in exile for 2000 years. Descendants of these people began to return at the end of the nineteenth century and established the State of Israel in 1948. From a population of 600,000 Jews in 1948, Israel has grown into a nation of 5.5 million people.

Much of the country's population growth has been the result of large-scale immigration of Jews, including those from European countries during World War II, from Arab countries in the 1950s and 1960s, and more recently from the former Soviet Republics. Immigration into Israel continues, but in lessening numbers, while there is a constant movement of Israeli citizens to other countries, particularly the United States. Great difficulties emerge when such diverse communities need to integrate into one society, study at the same schools, and use the same language. It is interesting to note that the Hebrew language, which had been dormant for 2000 years, has been revived in the last 100 years and has become the official and dominant language of Israel. The use of Hebrew as a unifying language has contributed substantially to the national identity.

About 80 percent of Israeli citizens are Jews, 14 percent Moslems, 3 percent Christian, and 2 percent Druze. Arab, Druze, and Beduin children are educated in their own languages at separate schools. The majority of Arab Israeli citizens live in the central and northern parts of the country. Eighty percent of the Jewish population and 38 percent of non-Jews live in the cities of Tel Aviv, Jerusalem, Haifa, Beersheba, and Nazareth, and the rest live in small towns and villages spread between Metulla in the north and Eilat in the south.

Israel is governed by a democratically elected parliament, known as the *Knesset,* and an elected prime minister. A president, elected by the *Knesset,* acts as head of state, and fulfills representative duties. The prime minister forms the cabinet and together they carry out many of the governing functions.

Between 1990 and 1995 there was a considerable rise in monies allocated to education: a more than 6 percent increase in 1990 and a more than 9 percent increase in 1995. The education budget increased from approximately US$2 billion to US$4.5 billion during these years. The literacy rate in Israel among those with more than four years of schooling was 94 percent in 1993 for the Jewish population and 83 percent for the Arab population.

2 THE EDUCATION SYSTEM

Governance and Decision Making

The intended curriculum is determined centrally by the Ministry of Education, but schools often include locally determined curriculum material that reflects the special needs and resources of the local community. Textbooks are usually selected by heads of subject departments in collaboration with classroom teachers. Teachers, with the guidance of department heads, determine the pedagogical methods used in the classroom. For most subjects, there is no provision for a textbook, but a large number of topic booklets are available.

Structure of the System and Participation Rates

Structure of Education

Most children remain at home until the age of 3, when they enroll in private kindergartens. Compulsory education begins at age 5 and ends at 16. Elementary school includes Grades 1 to 6 for students aged 6 to 11. Junior secondary school includes Grades 7 to 9 for students aged 12 to 14, and senior secondary school includes Grades 10 to 12 for students aged 15 to 17. Until 1968 elementary school included Grades 1 to 8, and high school, Grades 9 to 12. Today, about half of all schools still follow the old 8-4 model, while the remaining half have changed to the new 6-3-3 model. The general structure of the education system is shown in Figure 1, along with approximate enrollment rates for each level.

The percent of students attending school in 1993 was as follows:

- Jewish: boys, 89 percent; girls, 97 percent; all, 93 percent of the cohort at all ages.
- Arab: boys, 70 percent; girls, 71 percent; all, 70 percent of the cohort at all ages.

The relatively low enrollment of Jewish boys reflects the fact that many boys decide to enter the work force at the age of 16. Among the Arab communities, both boys and girls are likely to leave school at 16 in order to begin working.

School Types

The education system in Israel comprises several types of senior secondary schools. In 1994, in the Hebrew system, the percent of students attending each type of school was as follows: 56 percent academic, 41 percent technological, and 3 percent agricultural. Among Arab students, the figures were 75 percent academic, 24 percent technological, and 1 percent agricultural. About 3 percent of the total number of students study in special education classes. In all school types, the curriculum is designed by the Ministry of Education, which is also responsible for the matriculation examinations, or *bagrut*. Practically all students in academic schools write the *bagrut* examination. Kibbutz schools, like most comprehensive schools, follow the academic curriculum. Students who are slow learners are usually directed to technical or agricultural studies.

Schools in the System

The School Year

Elementary schools are in session from September 1 to June 30, with 220 instructional days per year. The secondary school year is about 210 days long, ending June 20. Major long holidays in Hebrew schools include two weeks in October, one week in December, and two weeks in April. In the Arab system there are two weeks of winter holiday in December and two weeks of spring holiday in April.

Schools are open six days per week, except Arab elementary schools which are open five days per week. Hebrew schools close on Saturday, Arab Moslem schools close on Friday, and Christian schools close on Sunday. The number of hours of classes per day increases gradually as students progress through the system. Students in the early years of elementary school attend four 45-minute class periods daily, while secondary students attend seven class periods between 8:00 a.m. and 3:00 p.m.

Class Size

In 1994, the average class size in Hebrew schools was 28 students in elementary, 30 in junior secondary, and 28 in senior secondary schools, with 50 percent of the total number of classes larger than 30 students. In Arab schools the corresponding figures were 31, 33, and 32, with 68 percent of the total number of classes larger than 30 students.

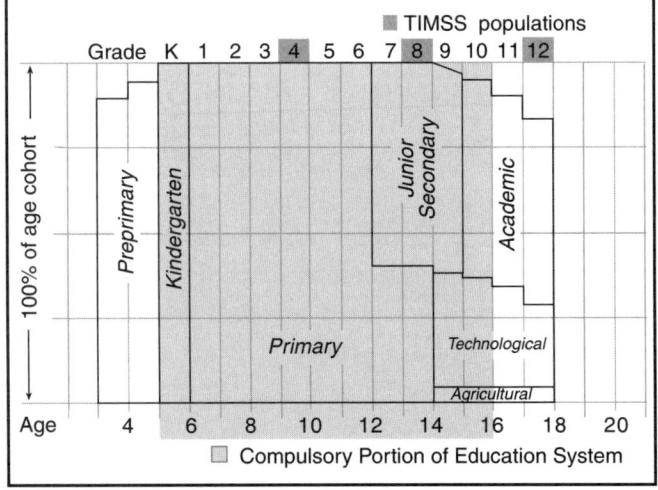

Figure 1: Structure of the Education System

Streaming and Tracking

Schools are not tracked and there is no official policy on within-school tracking. Primary school classes generally include a wide ability-range of students. In mathematics, all students study together until the end of Grade 6. In Grades 7 and 8, many schools divide students into two or three groups that follow the same curriculum at different levels of complexity. In Grades 8 and 9, all students take mathematics and science; streaming takes the form of allocating students to three broad levels on the basis of achievement tests and student records from lower grades, and students study toward a matriculation examination at one of the three levels. Students in Grades 10, 11, and 12 choose from a wide range of options, and setting in mathematics and science sometimes occurs at these levels. In Grades 11 and 12 students may specialize in certain subjects, such as history or biology. Some subjects, such as English and mathematics, have several levels of matriculation examinations, allowing students to choose or to be directed to the level that is most appropriate for them. Students in the lowest level of mathematics in junior secondary school usually choose a vocational track in senior secondary, studying some mathematics, but not necessarily complying with the official curriculum. These students do not take a mathematics matriculation examination.

Teacher Certification

The Ministry of Education issues teaching permits to graduates of teachers' colleges and of teacher education programs within universities. The teacher education institution awards a teaching certificate, which is required to obtain a permit to teach in schools. All early-age and elementary teachers study at teachers' colleges, whereas all secondary school teachers are prepared in universities and must complete at least a bachelor's degree in the discipline they choose to teach.

The Ministry of Education has established special departments within the Ministry for preservice and in-service teacher education and has invested a large proportion of its budget in the improvement of teaching and the promotion of the social and professional status of teachers. Most recently, a national effort focused on science, mathematics, and technology, known as "Tomorrow 98" has received wide recognition and support. "Tomorrow 98" initiates and maintains activities that promote and facilitate school science and mathematics. Evaluation of these in-service courses has shown considerable success in many projects, especially in terms of teacher attitudes and pedagogical content knowledge.

Teacher Profile

In 1993, the average age of Jewish elementary school teachers was 39, and of secondary school teachers, 42. The corresponding figures among Arab teachers were 36 and 36.

In 1993, the percent of college and university graduates, mainly holding bachelor's degrees, was 24 percent among Jewish elementary teachers and 63 percent among Jewish secondary teachers, compared with 15 percent and 56 percent respectively among Arab teachers.

In 1993, Jewish teachers taught for an average of 22 hours per week, while Arab teachers taught 24 hours per week. Ninety percent of elementary teachers and 70 percent of secondary teachers are female. In secondary school, subjects are taught by specialist teachers. Among junior secondary school teachers, the percent of females in 1995 was biology, 86 percent; chemistry, 77 percent; physics, 63 percent; computers, 67 percent; electronics, 15 percent; and mathematics, 76 percent. Among senior secondary specialist teachers, the percent of females in 1995 were: biology, 79 percent; chemistry, 80 percent; mathematics, 66 percent; physics, 40 percent; computers, 67 percent; and electronics, 18 percent.

Teachers' salaries in Israel are neither high nor low, compared with other occupations or with other countries. Elementary school teachers, until recently, were not required to have academic degrees and their salaries are considerably lower than those of secondary teachers. Secondary teachers have two advantages over elementary teachers: a full-time elementary school work week requires 30 periods of teaching, while the secondary week is only 24 periods per week; and elementary teachers begin their summer vacation on July 1, while secondary teachers start on June 20.

Secondary teachers study their major subject at university and have very few required courses. Many secondary teachers enjoy the status of researchers, since the course contents are identical for researchers and teachers.

3 TIMSS POPULATIONS

Israel, IEA, and TIMSS

Israel was one of the first countries to join IEA. Prior to TIMSS, Israel participated in the Six-Subject Survey, the Second International Mathematics Study, the Second International Science Study, the Computer Education Study, and the Written Composition Study. In addition to IEA publications, all of these studies published local reports in Hebrew. Various findings from SISS were published in professional science education journals in Israel.

Israeli TIMSS Populations

The majority of schools in Israel are Jewish, with teaching conducted in Hebrew. Eighteen percent are Arab schools, with teaching in Arabic. Population 1 included all nine-year-old pupils in Hebrew-medium schools, the majority of whom were enrolled in Grade 4 at the time of testing. Approximately 70,000 nine-year-olds from about 1100 elementary schools participated in TIMSS at the Population 1 level.

Population 2 included all 13-year-olds enrolled in Grade 8 in Hebrew junior secondary state schools at the time of testing. The cohort amounted to approximately 67,500 and was divided into two types: 340 old-model elementary schools with Grades 1 to 8 (about 40 percent of the cohort), and 315 new-model junior secondary schools with Grades 7, 8, and 9.

Population 3 included 17-year-olds enrolled in Grade 12. In 1995–96, when the study was conducted, the state school population consisted of approximately 63,000 students in about 530 Hebrew academic, technological, and agricultural schools. Of these schools, 125 were randomly sampled.

4 MATHEMATICS CURRICULUM AND PEDAGOGY

Goals for the Mathematics Curriculum

The general goals of mathematics education are to:
- know the basic facts and terms in mathematics;
- become familiar with the basic principles of the language of mathematics, which is increasingly used in the sciences and social sciences;
- develop an awareness of the potential of mathematics as an instrument for creating models of situations in natural and social sciences as well as in daily life;
- develop the ability to reach conclusions from simple mathematical models with regard to the situations for which they were created;
- develop skills in logical thinking, such as hypothesizing, generalizing, analyzing, criticizing, and reaching conclusions;
- develop the ability to use calculus;
- develop a positive attitude towards mathematics.

Mathematics is divided into a number of topics: numbers, including whole numbers; fractions and decimals; estimation and number sense; other numbers, and number concepts; measurement; geometry; proportion; functions, relations, and equations; data representation; probability and statistics; analysis; validation; and structure. The emphasis on each subject varies from basic to advanced according to course level through the 12 years of schooling. Mathematics studies in secondary school are based on a three-level program that prepares students for the matriculation examinations, or *bagrut*: a basic course and two levels of advanced courses.

Major Changes in the Mathematics Curriculum

There have been no major changes to the elementary school curriculum in a number of years. A new curriculum guide for Grades 7 to 9 was published in 1990, and several significant changes were included.

- Most students do not study geometry after Grade 9.
- Students of different abilities learn the same topics at different levels of difficulty.

In the early 1980s, a new curriculum was introduced for secondary school. Some schools adopted the entire curriculum, and some combined sections of it with the previous one. The senior secondary curriculum includes the following changes.

- Only students in advanced mathematics programs study geometry in high school.
- The concept of "function" receives a great deal of attention, following a unified approach and using a variety of contexts.
- Calculus teaching is now begun in Grade 10 instead of Grade 12.
- New topics such as vectors and linear programming have been introduced.
- Over the next five years, computers will be introduced as tools for learning mathematics.

Current Issues in the Mathematics Curriculum

The mathematics curriculum is determined by a team of experts and used by most teachers as prescribed. The relevance of the mathematics curriculum to the life of students is considerably less acute than in science because most teachers and students view mathematics as a tool for other subjects. Several issues relating to the mathematics curriculum and instruction may be identified:

- The use of calculators and computers. Many educators fear that students who grow accustomed to using calculators and computers will no longer be able to make calculations by themselves. In spite of this, several computer courses have been developed and used in trials. The 1993 curriculum for secondary schools encourages the use of both calculators and computers.
- The place of statistics in the mathematics curriculum.
- The teaching of algorithms as opposed to the promotion of an understanding of underlying mathematical concepts.
- The anxiety over mathematics that many students experience. By offering three different levels of matriculation examination, students are given the op-

portunity to choose the program in which they can succeed, thus reducing their anxiety about the study of mathematics.
- The question of whether mathematics should be a compulsory subject in the external matriculation examination.

Mathematics Textbooks

Some of the newer textbooks, particularly those developed at the Examination Development Center or in academic institutions, have improved markedly in recent years. Virtually all new textbooks contain all curricular topics together with time allocations, and they usually include a teachers' guide as well. More explanations and examples are provided than in previous textbooks. The Ministry of Education must approve all textbooks in use.

Pedagogy

The teaching of mathematics has always been based on problem solving and, hence, there is no need for reform stressing high-level cognitive ability. Students in most mathematics classes are engaged in problem-solving activities, and it is recommended that teachers continue this practice.

5 SCIENCE CURRICULUM AND PEDAGOGY

Goals for the Science Curriculum

The intended curriculum is developed centrally by subject-specific curriculum committees. These committees are nominated by the country's chief subject-specific inspectors. Members of the biology committee, for example, include three highly regarded university professors of biology, three biology teachers, three leading science educators who have been active in the high school biology project, and the country's chief biology inspector. Among its other responsibilities, the committee designs and approves the curriculum guides.

The curriculum guide specifies goals and topics, suggests time allocations for topics, and lists activities such as laboratory investigations. It does not, however, prescribe learning materials such as textbooks. Most teachers follow the curriculum guides closely, especially in senior secondary school, while preparing students for matriculation examinations. In certain subjects, such as biology, the guide requires teachers to choose the topics they teach. Matriculation examinations in these subjects are designed to allow students to choose some topics, so that they are tested on materials they have had an opportunity to study.

Major Changes in the Science Curriculum

During the last 10 years, substantial reforms in the teaching of science at all levels have been proposed. Most of these reforms have been based on a document developed by a committee headed by the president of Weizmann Institute of Science. These reforms have been designated, "Tomorrow 98." The goals of the reform are not different from many that can be found associated with similar reforms elsewhere. For example, the content goals formulated for Grades 7, 8, and 9 are to:

- develop scientific and technological literacy in selected topics, maintaining some depth;
- provide a scientific and technological base for continued study in these areas;
- use educational technology;
- promote an understanding of the relationships among science, technology, and society.

Goals in the cognitive domain are:

- develop scientific reasoning skills;
- develop technological reasoning skills;
- develop self-knowledge through individual projects;
- treat students' misconceptions appropriately.

Goals in the affective domain are:

- develop positive attitudes toward science and technology;
- develop a capacity for critical thinking;
- develop positive attitudes toward experiments;
- recognize humanity's responsibility to preserve the environment.

In addition to the curriculum changes described above, some more general trends may be observed. Fewer students are enrolling in technological schools and are instead choosing academic schools. This means that fewer students are enrolling in technological and

vocational schools. At the same time, the use of microcomputers had been slowly increasing. Along with a continuing emphasis on the separate structure of the sciences, the curriculum stresses the opportunity for integrated study of the disciplines. The integrated approach is believed to increase the probability that what students learn in school will be used in later life.

Current Issues in the Science Curriculum

The relevance of science to the lives of students continues to be a major issue. A shift toward school-based development of curriculum and materials is increasing. Curriculum specialists are changing their roles, visiting schools, and assisting in the development of school-based materials instead of designing inflexible programs.

Self-learning is recognized as an important goal of education that may be accomplished through the integration of activities and experiences within the learning process. The study of secondary school biology is an example of this, since several of its activities are directed toward the development of self-learning. Such activities include performing individual research projects in ecology, for which structured, whole-class experiences are inappropriate. An important component of self-learning is self-assessment, which may be seen in the use of meta-learning devices such as concept mapping, or the justification of a choice in some multiple-choice items. Finally, more and better use of homework may be a means of promoting meaningful learning.

Science Textbooks

All curriculum projects produce textbooks, which are usually paperbacks covering theme-specific topics prescribed by the subject curriculum committees of the Ministry of Education. In the sciences, several small books per course are published, allowing them to be updated one at a time without great expense. In addition, the use of several smaller books allows teachers some flexibility in structuring the course. There is a new textbook on evolutionary theory, for example, which is a topic teachers may choose to teach or pass over.

Textbook writing and publication are not controlled by the Ministry of Education, but a textbook may not be used in school unless it has Ministry approval. Textbooks are usually chosen by teachers or head teachers, and students buy their own textbooks at all levels.

A special organization, known as the Israel Science Teaching Center, was established in 1967 and has been in operation ever since. A team from each of the universities produces teaching materials for each grade and for each science or mathematics curriculum, using new concepts and methods based on both academic and practical approaches. Although these learning materials are subsidized by the Ministry of Education, teachers are free to use them or not, as they choose.

Pedagogy

In many ways science teaching in Israel may be looked upon as a large-scale educational laboratory. Pedagogical studies, curriculum trials, and other innovations are common in many schools, particularly those close to a university.

Each discipline is autonomous regarding curriculum. Although chemistry, physics, and biology claim to teach science by inquiry, with a strong emphasis on laboratories, the investigative nature of the courses is not born out in assessment. Each discipline uses a different assessment scheme in its matriculation examination, which in turn has a profound effect on teaching. Thus, the physics examination requires students to perform an already-familiar investigation, biology includes an inquiry-oriented practical examination, and chemistry requires no practical examination. Chemistry laboratories therefore tend to be used very little. Chemistry and physics students learn by rote, while biology students devote several hours per week to serious practical work. Biology teaching uses a variety of learning experiences, including outdoor activities, individual ecology projects, and concept mapping.

Another pertinent pedagogical decision is associated with the use of living organisms. Children enjoy studying living animals, but there is a danger of hurting the animals. This conflict has been the focus of considerable research, resulting in specific recommendations that allow schools to study animals while ensuring they are not endangered.

The most recent development in the science curriculum is the emphasis on science, technology, and society. This is true for all levels and all sciences.

6 EVALUATION POLICIES AND PRACTICES

A matriculation examination in mathematics is administered in Grade 11 or 12. The examination includes questions and problems that require complete solutions or proofs. All students are examined in mathematics, but science subjects are elective. Results in all subjects are included in the matriculation certificate required for entering institutions of higher learning. Matriculation examinations are developed by *ad hoc* committees for each subject, with each committee including two professors in the discipline, two teachers, one science educator, and an assessment specialist. The assessment tends to drive all educational activities, and to convey the goals of the course to students and teachers better than any other communication. After the examination, the tests are published and become learning material.

Occasionally, there are national or regional external examinations offered at various grade levels. These are often multiple choice in form. The examinations help to evaluate the program and serve as entrance or placement tests to the appropriate level junior or senior secondary school. The purpose of these evaluations is to provide feedback on learning to students, teachers, and researchers.

Traditionally, assessment through paper-and-pencil tests has been the major, often the only, measure that counted. Student achievement in such tests reflects the level of learning and can serve as a measure of the quality of learning. However, there is other useful information that can be obtained. The critical role of prior knowledge in enhancing or disrupting learning can be examined through the use of diagnostic assessment.

In Israel, 50 percent of a student's final grade is determined by the teacher and the rest by weighted average of all components, since it is believed that external examination is not as valid a measure of achievement. Furthermore, educational leaders in Israel are now advocating more school-based activities of all types. There is some discussion about replacing certain external examinations, for example, those at the end of secondary school, with internal measures administered by teachers to their own students.

There has been substantial improvement in some external examinations. The old practical examination in physics is gradually being replaced by one similar to the biology model, in which an unfamiliar task is assigned. Innovative test formats have been developed, including one that asks for a written description of the association between two concepts, and another that requires justification for multiple-choice answers.

7 REFERENCES AND SOURCES FOR FURTHER READING

Common Sources

1 Husén T, N T Postlethwaite 1994 *International Encyclopedia of Education*. Pergamon Press, Oxford

2 Organisation for Economic Co-operation and Development 1995 *Education at a Glance: OECD Indicators*. Organisation for Economic Co-operation and Development, Paris

3 United Nations Educational, Scientific and Cultural Organization 1995 *Statistical Yearbook*. United Nations Educational, Scientific and Cultural Organization, Paris

4 World Bank 1995 *World Development Report 1995*. Oxford University Press, New York

Other Sources

5 Fullan M, A Pomfret 1977 Research on curriculum and instruction implementation, *Review of Educational Research*, 47:335-397

6 Ministry of Education and Culture 1993 *Tomorrow 98: Science and Technology in Secondary Schools*. Maleh Hahamisha (in Hebrew)

7 Shprinzak D, E Bar, D Lewy-Mazalos 1995 *The Educational System Through Figures*. Ministry of Education, Culture and Sport, Sept. of Economics and Statistics (in Hebrew), Jerusalem

8 Central Bureau of Statistics, Ministry of Education, Culture and Sport 1995 *Educational Institutions 1993/94*, Series of Education and Culture Statistics, No. 235 (in Hebrew)

9 Central Bureau of Statistics 1995 *Education and Culture, Selected Data* [reprint from *Statistical Abstract of Israel, 1995*, No. 46] (in Hebrew)

10 Central Bureau of Statisitcs 1995 *High School Survey 1995*. Ministry of Education and Culture, Jerusalem

11 Central Bureau of Statistics 1995 *Israel Statistical Yearbook 1995*. Central Bureau of Statistics, Jerusalem

12 Tamir P et al. 1988 *Science Teaching in Israel in the 1980s*. The Science Teaching Center, Hebrew University of Jerusalem, Israel (in Hebrew)

13 Ministry of Education, Culture and Sport 1995 School Files for Grades 4, 8, and 12 of the Hebrew Public School System. Computation Unit, Jerusalem (in Hebrew)

Statistical References

Section	Statistic	Reference	Page	Table	Year of Statistic
Country Profile	600,000, 5.5 million	7	7	–	1995
Country Profile	80%, 14%, 3%, 2%	11	3	46	1995
Country Profile	80%, 38%	11	–	–	1995
Country Profile	6.4%, 9.4%	7	67	–	1995
Country Profile	US$2 billion, US$4.5 billion	7	60, 61	–	1995
Country Profile	94%, 83%	7	97	–	1993
Structure of the System	89%, 97%, 93%	7	86	–	1993
Structure of the System	70%, 71%, 70%	7	86	–	1993
Structure of the System	56%, 41%, 3%	8	22	5	1994
Structure of the System	75%, 24%, 1%	8	33	12	1994
Structure of the System	3%	7	11, 47	–	1995
Schools in the System	28, 30, 28, 31, 33, 32	8	42	b	1994
Schools in the System	50%, 68%	8	44	c	1994
Teacher Profile	39, 42, 22, 90%	9	661	22.31	1993
Teacher Profile	36, 36, 24, 70%	9	662	22.31	1993
Teacher Profile	24%, 15%	9	663	22.31	1993
Teacher Profile	63%	9	664	22.32	1993
Teacher Profile	56%	9	665	22.33	1993
Teacher Profile	86%, 77%, 76%, 63%, 67%, 15%, 79%, 80%, 66%, 40%, 67%, 18%	10	–	–	1995
TIMSS Populations	18%	9	663	–	1995
TIMSS Populations	70,000, 1100, 2350	13a	48	–	1995
TIMSS Populations	67,500. 340, 40%, 315	13b	16, 33	–	1995
TIMSS Populations	63,000, 530	13c	22	–	1995
TIMSS Populations	125	13c	22	–	1995

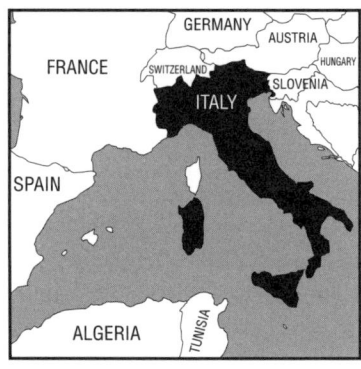

Italy

Anna Maria Caputo, Centro Europeo dell'Educazione

 ## COUNTRY PROFILE

Italy occupies a peninsula in southern Europe that extends into the Mediterranean Sea. Along the Alpine range in the north of the country, it shares a frontier with France to the west, Switzerland and Austria to the north, and Slovenia to the east. Two small independent countries lie in the Italian peninsula: the Republic of San Marino (60 km²) and Vatican City (0.4 km²). Italy has an area of 301,302 km²

As of the 1991 census, Italy had a resident population of more than 56 million inhabitants, with an average density of 188 persons/km². The population has remained relatively stable over the last 10 years. The past 40 years have seen changes in the distribution of the population, with medium-sized towns growing in number, and greater concentration of people living in urban areas. The number of inhabitants in urban areas has increased from 84 to 94 percent of the total population, while the percent in rural areas has fallen from 16 to 6 percent.

In terms of language and tradition, Italy is one of the most homogeneous nations in Europe, since it was one of the first nations in Europe to achieve linguistic, cultural, and political unity through the expansion of the Roman Republic. Today, there are ethnic and linguistic minorities in Italy, including French, German, Ladin, Slovenian, Greek, Albanian, Catalan, Serbo-Croat, and other groups, but they total less than 6 percent of the population and are concentrated in well-defined geographical areas. There are three regions[1] with special laws establishing the language of a minority group as an official one alongside Italian.

Italy is a democratic republic whose constitution of 1846 confers legislative powers and control of the government to parliament. Parliament consists of two houses, the chamber of deputies and the senate, both of which are elected for a five-year term by direct universal suffrage. The president, as head of state, is elected by both houses in a plenary session. The government is composed of the prime minister and a council of ministers. The constitution provides for a series of regional, provincial, and municipal authorities to be responsible for governing at a local level.

Italy is a founding member of the OECD and is also a member of the G7 Economic Group of Countries. The World Bank ranks Italy among the leading developed countries. The per capita gross national product in 1993 was US$19,840, with an average annual growth rate of 2 percent between 1980 and 1993 and an inflation rate of about 9 percent during the same period.

Education receives about 10 percent of public spending. This amount is less than the 12 percent average for other OECD countries. In 1992, expenditure on education accounted for 5 percent of the GDP compared to the OECD average of 6 percent, 5 percent for the G7 countries, and 6 percent for the European Union. Expenditure per student in primary education in state schools was US$4,050, compared to US$3,410 for OECD countries in 1992. Expenditure per student in secondary education in state schools was US$4,700, while the OECD average was US$4,760. According to UNESCO statistics, Italy has a literacy rate in excess of 95 percent.

---

[1] Val d'Aosta: Franco-Provencal group (70,000 people); Alto Adige: German group (287,503 people) and Ladin group (18,434 people); Friuli: Slovenian group (53,193 people).

2 THE EDUCATION SYSTEM

Governance and Decision Making

The Italian constitution establishes certain fundamental principles in education, including the duty of the State to guarantee a network of teaching institutions open to all without distinction; the right of private individuals to establish schools with no ties to the state; the duty of parents to provide at least eight years of education for their children; and the provision of free compulsory education in state schools. The administration of the education system in Italy has traditionally been centralized, although some authority is given to local governments.

The Ministry of Education supervises and coordinates all educational activities provided by state and private institutions. It studies and promotes education, oversees all educational institutions, promotes changes in curricula and programs, and provides in-service training of personnel. It administers the budget, staff recruitment, and staff mobility. Finally, it controls and administers the automation and mechanization of services. The Ministry of Education operates through directors, inspectors, and services that are responsible for various tasks, and is represented at the local level by regional education superintendents and provincial directors of education. These authorities implement national policies and directives, maintain contacts with other local bodies, and coordinate the services offered to the public.

The official intended curricula are defined by the Ministry of Education and may be adapted by schools to accommodate local contexts. Textbooks are chosen by schools on teachers' suggestions, with the input of parents and students on class committees. Educational activities and methods are chosen by individual teachers, who may program their work jointly with other teachers at the school level. Teaching methods may be influenced by in-service courses and educational support materials.

A number of participatory committees have been operating in Italian schools since 1974 and, within the limits of national legislation, are entrusted with several functions:

- Class committees in secondary schools and inter-class committees in primary schools, both composed of teachers, parents, and in some cases students, make proposals to the teachers' committee regarding educational and teaching activities.
- The teachers' committee, which is composed of teachers, is involved in such tasks as preparing the teaching program and educational initiatives of the school, selecting textbooks and educational support material, promoting experimental initiatives, and proposing forms of in-service teacher education.
- The school committee, which is composed of representatives of teachers, parents, nonteaching staff, the principal, and students in senior high schools is responsible for the estimated and final budgets and for school organization and planning.

Structure of the System and Participation Rates

Structure of Education

The education system is divided into four levels: preprimary schools, primary schools, secondary schools, and universities. Secondary schools are divided into two levels: level I, denoting middle or junior high schools, and level II, denoting senior high schools. In addition, there are vocational training institutes providing direct job training to young people aged 14 to 18. Figure 1 shows the structure of the education system and the approximate enrollment rates for all levels. Compulsory education begins at the age of 6, lasts

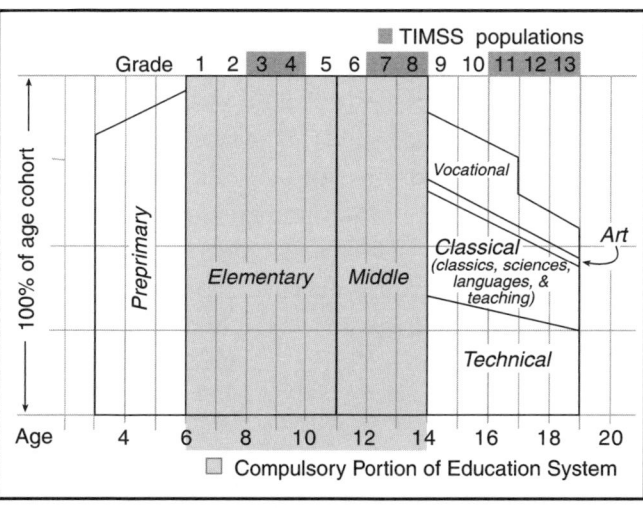

Figure 1: Structure of the Education System

for 8 years, and includes primary and junior secondary school. The minimum leaving age is 14.

Primary School

Primary school attendance is compulsory for children between the ages of 6 and 11[2]. In 1993–1994, 8 percent of primary school children attended nonstate schools, most of which were operated by Roman Catholic religious organizations. Primary schools are made up of five grades divided into two cycles according to developmental level. The first cycle lasts two years and teaches basic skills, while the second cycle gradually introduces pupils to concepts. Classes are made up of no more than 25 pupils, with a limit of 20 pupils for classes containing a student with a disability. The law provides for the presence of three teachers for every two classes, or four teachers for every three classes.

Junior Secondary School

Junior secondary school, or *scuola media*, is free, compulsory, and lasts three years. The entry requirement is the primary school leaving certificate, which is normally obtained at the age of 11. Children aged between 11 and 14 years[3] normally attend junior secondary schools, with more than 4 percent attending non-state schools.

Senior Secondary Education

After finishing compulsory education and passing the junior secondary school leaving examination, students may attend senior secondary school for an additional three, four, or five years. Senior secondary education is not free of charge; students pay a fee to the state and also to the school itself. All schools offering postcompulsory instruction are part of senior secondary education and fall into one of the following categories:

- classical schools, including the *Liceo Classico*, which prepares humanities students for university; the *Liceo Scientifico*, which prepares mathematics and science students for university; the *Istituto Magistrale* for primary teacher education; the *Scuola Magistrale* for preprimary teacher education; and

---

[2] In 1993–94 the participation rate was 101.5 percent because of delays, anticipations and repeating; the population for age ranges was estimated.
[3] In 1993–94 the participation rate was 107.5 percent because of delays, anticipations and repeating; the population for age ranges was estimated.

the *Liceo Linguistico* which prepares language students for university;

- art schools, including the *Liceo Artistico* and the *Istituti d'Arte*, which train students in the visual arts and lead to university or fine arts academies;
- technical schools, which prepare students for professional, technical, or administrative occupations in the agricultural, industrial, or commercial sector, and which give access to university;
- vocational schools, which train students to become qualified first-level technicians, and which may lead to university.

Some figures on senior secondary education from the school year 1993–1994 are given below:

- 89 percent of junior secondary school pupils enter senior secondary education;
- 72 percent of young people aged 14 to 19 continue secondary education;
- 70 percent of senior secondary school-leavers go to university;
- 9 percent of senior secondary school students attend non-state schools.

Schools in the System

The School Year

The school year runs from the beginning of September to the end of June and provides for at least 200 instructional days. The Ministry of Education establishes the dates for teaching, examinations, and holidays. There is a two-week break each year at the end of December, a one-week break around Easter, and five days for national and local holidays.

Instructional Times

Primary schools are normally in session for 27 hours per week, rising to 30 in the second year with the inclusion of a foreign language. Classes run from Monday to Friday or Saturday morning, including one or two afternoons according to parental choice. Whatever the timetable, teachers must devote at least three hours to mathematics and two hours to science per week.

Junior secondary students attend classes in various subjects for 30 hours per week, for five hours a day from Monday to Saturday. Twenty percent of the timetable is devoted to mathematics and sciences. By parental request, school hours may rise to 36 or 40 hours per week, Monday to Friday, if there are enough pupils to make up one or more classes. The extra hours are used for complementary activities or for an extension of the regular school subjects. The hours include lunch break, recess, and open-air activities.

Senior secondary students attend school from Monday to Saturday for a number of hours that varies according to school and class type. In the classical school group, times range from 25 to 34 hours per week; in technical schools from 30 to 40 hours; 40 hours in vocational schools; 39 to 44 hours in the art group. The time devoted to mathematics and sciences varies from 0 to 20 percent, according to school and class type.

Class Size

Since 1993, the Ministry of Education has gradually increased class size to 20 students, due to policies of financial restraint. This involves the following changes:

- The number of preprimary school sections has been increased. At present, the average class size is 23. The number of pupils per section ranges from 14 to 28 and does not exceed 20 in classes that include a child with a disability.
- The number of primary school classes has been reduced. At present, the average class size is 17 and the number of pupils per section ranges from 10 to 25, not exceeding 20 in classes that include a child with a disability.
- The number of junior secondary classes has been reduced. At present, the average class size is 20 and the number of pupils per section ranges from 15 to 28, not exceeding 20 in classes that include a child with a disability.
- The number of senior secondary classes has been reduced. At present, the average class size is 22 and the number of pupils per section ranges from 15 to 29.

Grouping of Students

Schools are not tracked as there is no official policy in Italy for in-school tracking. It is common practice for classes to be comprised of students with varying abilities. Since the 1970s, school policy has been to include children with disabilities in normal classes in preschools and compulsory schools, with the consequent closure of special classes for the disabled. The new policy involves special teachers who promote the integration and learning of children with disabilities into the classroom. The integration process for these students is irreversible; the question is no longer whether or not to integrate, but in what way, and how to program educational and other classroom activities in the presence of pupils with disabilities. In the late 1980s, school policy also provided for students with disabilities to attend normal senior secondary classes.

Certification of Teachers

Teaching in Italy requires the following qualifications:

- Secondary school teaching requires a university degree in a discipline involving four or five years of study. Specific pedagogical training is not necessary.
- Primary school teaching requires a primary school teaching diploma, which entails four years of secondary education at the *Istituto Magistrale*.
- Preprimary school teaching requires a preprimary school teaching diploma, which entails three years of secondary education at the *Scuola Magistrale*.
- Technical and practical teaching in secondary schools requires a technical secondary school diploma, which entails five years of secondary education.

Primary school teachers may teach all subjects, while secondary school teachers may teach only subjects in which they were certified by state examination. For example, teachers can teach mathematics and the sciences in junior high schools with any science degree (physics, mathematics, natural sciences, or biology), while they can teach mathematics in senior high schools only if they have a degree in mathematics, astronomy, physics, engineering, or statistics.

Initial training takes place during the first teaching year. Training activities take place inside school and in seminars. The training activities focus on teaching, communication, methodological, and interpersonal skills; educational psychology; and aspects of law and administration. Themes such as student guidance, student integration, health education, and prevention of drug addiction are also dealt with. Each teacher must devote 40 hours to these activities, grouped into periods at the beginning and the end of school year.

In-service teacher training is a right and duty of teachers that allows them to update their knowledge and teaching methods, as well as to participate in research work and educational innovation. By contract, teachers must devote 100 hours over three years to this. The national in-service training plan is implemented at national, regional, provincial, and local levels, coordinating all initiatives, and begins with needs expressed by schools. The plan was established by union organizations and the Ministry of Education. In-service training is also offered by teachers' associations, university faculties and departments, and by private institutions. Updating courses include foreign language teaching, intercultural education, and experimentation and implementation of new curricula.

Teacher Profile

Women constitute the majority of teachers in Italy, especially at the lower grade levels: 100 percent of preprimary teachers are women, as well as 93 percent of primary, 71 percent of junior secondary, and 54 percent of senior secondary teachers. At the senior secondary level, women constitute 62 percent of humanities teachers, 55 percent of mathematics and science teachers, and 44 percent of technical applied subject teachers. In junior secondary schools, there are no significant differences across school subject groups: women teachers constitute up to 70 percent of mathematics and science teachers.

Most teachers in Italy are in the 40 to 49 age group, except for preprimary school teachers who are, on average, younger. The social status of teachers is considered to be average, even though teaching has lost much of its professional prestige. The starting salary of primary school teachers is US$18,161, just over the per capita GDP of US$17,482. Junior secondary teachers have a starting salary of US$19,708, well above the per capita GDP. As teachers' careers unfold, however, their salaries do not keep pace with other professions. At the end of their working lives, usually 40 years of teaching, the salary of primary school teachers is US$27,852, while that of junior secondary teachers is US$30,327.

3 TIMSS POPULATIONS

Italy, IEA, and TIMSS

Italy has been a member of IEA since 1968, and the *Centro Europeo Dell'Educazione*, in Frascati, Rome, has been the national center for IEA coordination since 1979. Before TIMSS, Italy took part in the Six-Subject Survey, the Second International Science Study, the Computer Education Study, the Written Composition Study, the Reading Literacy Study, and the Preprimary Project, which is still in progress.

Italian TIMSS Population

Population 1 is defined as all pupils enrolled in Grades 3 and 4 of state primary schools. These grades contained the largest proportion of nine-year-old pupils at the time of testing, May 1995. At the mid-point of the testing year, 96 percent of the children in the two grades were nine-year-olds.

Population 2 is defined as all pupils enrolled in the second and third year of state junior secondary schools. These years contained the largest proportion of 13-year-old pupils at the time of testing, May 1995. At the mid-point of the testing year, 88 percent of the pupils in the two grades were 13-year-olds.

Population 3 is defined as all students enrolled in the final year of senior secondary schools, regardless of whether they study mathematics, science, or physics. Students attending specialized courses in mathematics or physics were considered as subsets of the general population. This definition identified approximately 20 percent of Population 3 students as mathematics specialists and 16 percent as mathematics and physics specialists.

4 MATHEMATICS CURRICULUM AND PEDAGOGY

Goals for the Mathematics Curriculum

Mathematics curricula are defined for each school type and grade by the Ministry of Education. The official curriculum contains general and specific goals, guidelines for the number of hours of instruction per subject (for secondary education only), teaching contents and indicators, as well as learning objectives. A commission set up by the Ministry of Education, consisting of teachers, university lecturers, school principals, inspectors, Ministry personnel, and education experts, defines the curriculum.

Primary School

The current official curricula were approved in 1985 and became operational in the 1990–91 school year. The new curricula state that the goals of mathematics are to develop concepts, methods, and attitudes that will allow students to order, quantify, and measure real events and phenomena. The official intended curricula concentrates on theme-based learning objectives that define the content to be taught. Each theme is accompanied by suggestions of learning experiences that promote the achievement of the objectives. The themes proposed are problems, arithmetic, geometry and measurement, logic, probability, statistics and computer education.

Junior Secondary School

The current official intended curriculum has been in place since 1979–80. The goals are to develop students' abilities in logic, abstract concepts, and deduction, as well as a scientific mentality to deal with problems. In this context, mathematics education aims to:

- stimulate pupils' intuitive capacity;
- lead pupils to check the validity of their intuitions and deductions with increasingly organized forms of reasoning;
- prompt pupils to communicate in increasingly accurate and clear language, by means of symbols or graphs that facilitate thought processes;
- guide the capacity for synthesis, promoting the progressive clarification of concepts and thus recognizing similarities in different situations, in order to arrive at a unified view of fundamental ideas;
- begin a process of awareness and mastery of calculus.

The list of objectives accentuates the idea of progression of knowledge and operating skills.

The intended curriculum is divided into seven broad themes, including geometry, numerical sets, mathematics of certainty, mathematics of probability, problems and equations, the method of coordinates, geometric transformations, correspondence, and structural similarities. The teacher establishes which teaching path to follow over the three-year period.

Senior Secondary Schools

The structure of senior secondary schools has remained unchanged for decades, with a clear distinction between preparation for university education and preparation for the job market. Each type of school has its own curriculum, including objectives, contents and methods. A new curricula was proposed by the Brocca Commission and accepted by the Ministry. The goals for mathematics education are the same for all school types and vary according to the two-year or three-year secondary school cycles. The curricula already provide for the long-awaited extension of compulsory education by two years, and also for the teaching of mathematics in the three-year senior secondary school cycle.

In the first two-year cycle of senior secondary education, mathematics promotes the development of intuitive and logical skills; the ability to use heuristic procedures; abstraction and concept formation; the capacity for inductive and deductive reasoning; the development of analytical and synthetic attitudes; language precision; and the capacity for coherent and logical reasoning. A detailed list of learning objectives is also provided at the end of the two-year cycle. Contents are grouped under five broad themes: two-dimensional and three-dimensional geometry, numerical sets and calculus, relations and functions, probability and statistics, logic, and computer science.

In the following three-year cycle, mathematics develops:

- knowledge at higher levels of abstraction and normalization;
- the capacity to grasp the distinct characters of various kinds of language, including historical, natural, formal, and artificial;
- the capacity to use mathematical methods, tools, and models in different contexts;
- the attitude to critically reexamine and logically order knowledge gained.

Major Changes in the Mathematics Curriculum

Primary Schools

The word mathematics has replaced the term arithmetic. Primary mathematics does not limit itself to counting sums, but develops abilities in reasoning, spatial geometric intuition, and statistical thought that will be taught in greater detail during junior secondary education. New themes have been added to the primary curriculum, such as logic, statistics, probability, and computer education.

While very basic computer education requiring pencil and paper is by now firmly established, computers still have limited use, and only about one-quarter of schools have them for educational purposes. Teachers use computer education as a tool for analyzing problems. They propose activities with materials such as charts or games with colors, arrows, or blocks to introduce content such as problems, data algorithms, language, communications, systems, and models. The use of calculators is discouraged by teachers, who wish to emphasize the importance of mental calculating abilities.

Junior Secondary Schools

Mathematics teachers are now free to organize the subject according to personal preference. Changes in content include the use of Cartesian coordinates, geometric transformations, and the elements of probability and statistics. One important item missing from the curriculum, which was written in 1979, is computer education. Teachers have tried to combat this problem by introducing flow charts for problem solving and by using computers in mathematically significant contexts. Fifty percent of schools at this level are provided with computers. Calculator use is tolerated, but not encouraged.

Senior Secondary Schools

Several new elements have been introduced into the curriculum of the first two years, including probability and statistics, logic, and computer science. In the following three years, the greatest change lies in the teaching of mathematics up to the final year. Computers are now used throughout senior secondary education. Nearly all senior secondary schools have computers and an increasing number of mathematics teachers are using them as a teaching tool. Calculators are allowed and in many cases encouraged.

Current Issues in the Mathematics Curriculum

At the primary level, the next few years will be devoted to making curricular objectives, contents, and methodologies operational. Teachers will be encouraged to develop the connections and integration between mathematical contents and other school subjects.

The most pressing issue concerning senior secondary schools is the potential raising of the minimum school leaving age. Teachers who are not involved in experimental projects will continue to change methodology and content taught according to Brocca Commission guidelines.

Mathematics Textbooks

By law, textbooks are provided in primary and secondary education. Teachers consider them as a resource for themselves and a working instrument for students. Today's textbooks tend to make better use of color, illustrations, photographs, cartoons, and drawings, and generally have better page formats and less densely printed text. Textbooks for primary schools must be compiled by publishers according to Ministry of Education regulations on methodology and content.

There is little consistency in the proportion of space mathematics textbooks devote to explanation and exercises. Some put great emphasis on exercises and worked examples, but provide few explanations or extended activities. Others give ample space to explanations and detailed analysis with the goal of stimulating reading and study. However, in general, textbooks for primary and junior secondary schools do include some material on classroom and outside activities. Textbooks for the first two-year cycle in senior secondary schools concentrate on curricular topics, with the introduction of a few new topics such as computer science.

Pedagogy

Mathematics pedagogy at all levels involves experience, representation, and formalization. Choosing an experience to stimulate students' curiosity is the task of the teacher, as is that of integrating students' reflections to deeper mathematical developments. Learning to communicate in speech and in writing about mathematics is a fundamental skill at all grade levels.

The starting point for the mathematics curriculum is an evaluation, performed by the teacher, of the student's knowledge and strategies for solving problems. At the junior secondary level, for example, teachers must build on knowledge pupils have gained at the primary level, referring to concepts and information needed for developing new themes and problems.

Mathematics education is inductive at the primary and junior secondary levels, while at the senior secondary level the deductive component takes on greater importance. In order to achieve the curricular objectives for primary school, it is important to provide pupils with an adequate manipulative and representative base. Each pupil must be allowed to use common or structured materials that provide adequate models of mathematical concepts. The curricula also stress a gradual movement away from the use of physical materials to the use of corresponding mental images in order to carry out and interpret tasks. Similarly, in secondary education, programming languages, problem-solving algorithms, and the operational aspects of computers may be used as a different approach to abstraction.

5 SCIENCE CURRICULUM AND PEDAGOGY

Goals for the Science Curriculum

Primary Schools

The science curriculum includes goals such as developing cognitive curiosity, creativity in formulating hypotheses, and attention to relations between events. Older students concentrate on increasing mastery of research techniques, autonomy in judgment, and conscious respect for the environment. The curriculum requires students to acquire a basic grounding in five broad areas, including living things, health, the Earth, natural resource management, and materials and their characteristics. Instead of presenting a detailed list of contents, the curriculum suggests a series of activities to carry out with pupils.

Junior Secondary Schools

The teaching of experimental sciences encourages students to know the structures and mechanisms of nature, discover the importance of formulating hypotheses, and identify interactions between the physical and biological worlds. The curriculum is divided into five broad themes, grouping content and aims to meet the

requirements of a basic grounding in science. The themes include: matter; the Earth and the solar system; the structure, functions, and development of living things; the environment; and scientific progress and society. The themes and contents are integrated with suggested teaching approaches. The teacher establishes the teaching program for the three years, taking into account students' interests and maturity along with the school's sociocultural and territorial contexts.

Senior Secondary School

All school types have at least one science teacher. In the experimental curricula, vocational and technical-industrial schools for electronics, telecommunications, and mechanics, the subjects taught include physics, earth science in the first year, and biology in the second. In some vocational and technical curricula, specialist courses such as industrial chemistry and health physics are taught. In the *Liceo Scientifico,* all students of the three-year cycle attend a three-year specialist physics course.

Major Changes in the Science Curriculum

Primary Schools

The official curriculum treats science as a subject distinct from geography and history, with which it was combined in the previous curriculum. In addition, science is now taught in Grades 1 and 2; formerly, science education began in Grade 3. The goals and content have been coordinated with those of the junior secondary science curriculum, creating a science education process that develops gradually throughout compulsory education.

Junior Secondary Schools

The experimental science curriculum allows teachers to define the syllabus to reflect students' interests and maturity and the sociocultural and territorial contexts of the school. Opportunity is given to teachers to teach the biological aspects of sexuality, although it is considered advisable to treat this theme with the involvement of all teachers and students' families.

Senior Secondary School

The new official curriculum presents several new elements. It lists common goals for science education for all school types in the first two-year cycle. In addition, it introduces laboratory experimentation for all sciences, and computer and calculator use in physics laboratory activities.

Current Issues in the Science Curriculum

Science education is seen as essential for all students, whether as a basic component of compulsory education, or as preparation for higher level studies or qualified employment. Implementing the new primary school curriculum will involve the development of technological competence, the use of laboratories, and activities allowing integration with other school subjects. Following the curricular guidelines in junior secondary schools will lead to developing a series of important social themes in health education, including drug use, AIDS, sex education, smoking, and nutrition. Similarly, environmental education and pollution-related issues will become increasingly important.

Senior secondary schools will see an increasing use of computers for simulations, data collection and processing, and word processing. Moreover, there will be a greater use of multimedia supports to meet the lack of sophisticated laboratory equipment.

Science Textbooks

For many years there have been criticisms that textbook authors expect phenomena to be taught and explained using words alone. Textbooks now portray explanations and invite pupils to check the statements made by participating in suggested activities. Textbooks suggest not only contents but also methods and procedures, to some extent replacing the teacher as the source of information.

Many textbooks, even at the junior secondary level, are too concerned with covering all topics thoroughly. As a result, most are too long for the reading capacity of most students, and lively, easy-to-read books remain a rarity. However, science textbooks do use color, photographs, illustrations, graphs, and cartoons to make the books more interesting to students. Their prices are quite high, as are the prices for books which include software for simulations.

Pedagogy

At the primary level, science topics are developed from simple problems, with the aim of increasing knowledge organically. The topics are dealt with mainly through practical activities, sometimes in a laboratory, or through environmental exploration activities. Practical work facilitates group discussion, detailed analysis, and information gathering using books or audiovisual materials. Teachers choose which topics to teach, and certain topics are revisited over the five years at increasing levels of complexity.

The approach in junior secondary schools begins with problem identification from direct observation of facts, phenomena, and environments. The teacher's task is to guide pupils in observations, class discussions, hypotheses, and experimentation to encourage written and spoken communication. Methodologies at the senior secondary level share a common experimental approach, but differ in content and emphasis across the various sectors.

6 EVALUATION POLICIES AND PRACTICES

Primary Schools

Pupils are assessed throughout the school year through data gathered by teachers. Teachers use various evaluation techniques: tests, questionnaires, written and oral reports, group activities, systematic observations, and questioning. At regular intervals, pupils are assessed jointly by their teachers, and the evaluation is recorded in a school report delivered to pupils' families every four months. The school reports show the pupil's initial profile and progress to date. At the end of Grade 5, pupils write an examination before proceeding to junior secondary school. The examination consists of two written tests on Italian language and mathematics, and an interdisciplinary oral. The written tests are defined at the school level.

Junior Secondary Schools

Teachers make systematic observations of each pupil's learning progress and level of maturity. On the basis of these observations, the class committee writes a performance evaluation in each subject every three months. From these analytical evaluations and other elements, such as participation in school life, the class committee arrives at an overall evaluation, which determines admission to the next grade. At the end of junior secondary school, pupils write an examination in Italian, mathematics, and a foreign language, as well as a multidisciplinary oral. The test papers are selected at the school level and must offer students opportunity to demonstrate their abilities. All elements of the examinations are analyzed to produce an overall evaluation, which determines whether the student obtains the school leaving certificate.

Senior Secondary School

Assessment is performed twice or three times per year, as decided by the teachers' committee. The assessments consist of a discussion and joint evaluation by the teachers of each student. An evaluation that takes both marks and participation into account is made for each subject. In the final assessment of the year, students must obtain a mark of at least 6 out of 10 to pass. Students who obtain less than the pass mark in any subject must repeat the year.

At the end of the senior secondary school courses lasting four or more years, students who have had positive evaluations write a final examination known as the *esame di maturità*. It consists of two written papers and an oral, with an examining board appointed by the Ministry of Education. The two written papers are chosen by the Ministry, while the oral is based on two school subjects and the two written papers. The board of examiners arrive at a joint evaluation of the written papers and oral. The results are certified by a diploma, which allows access to university or other forms of higher education.

Students of three-year courses must pass written final examinations in order to obtain a diploma demonstrating their vocational qualifications. The diploma may be used to access the fourth year of a five-year experimental course of the same type, or to enter the job market.

7 REFERENCES AND SOURCES FOR FURTHER READING

Common Sources

1 Husén T, N T Postlethwaite 1994 *International Encyclopedia of Education*. Pergamon Press, Oxford

2 Organisation for Economic Co-operation and Development 1995 *Education at a Glance: OECD Indicators*. Organisation for Economic Co-operation and Development, Paris

3 United Nations Educational, Scientific and Cultural Organization 1995 *Statistical Yearbook*. United Nations Educational, Scientific and Cultural Organization, Paris

4 World Bank 1995 *World Development Report 1995*. Oxford University Press, New York

Other Sources

5 Organisation de Coopération et de Developppment Economiques 1995 *Regards sur l'Education: Les indicateurs de l'OCDE*. Organisation de Coopération et de Developppment Economiques, Paris

6 Sistema Statistico Nazionale Istituto Nazionale di Statistica [National Statisitcal System National Statistical Institute] 1993 *Annuario Statistico Italiano edizione 1992. [Italian Statistical Yearbook 1992 edition]*. ISTAT, Roma

7 Sistema Statistico Nazionale Istituto Nazionale di Statistica [National Statisitcal System National Statistical Institute] 1995 *Annuario Statistico Italiano 1994. [Italian Statistical Yearbook 1994]*. ISTAT, Roma

8 Sistema Statistico Nazionale Istituto Nazionale di Statistica [National Statisitcal System National Statistical Institute] 1995 *Rapporto Annuale, La Situazione del Paese 1994. [Annual Report, State of Italy 1994]*. ISTAT, Roma

9 Ministero dell'Interno [Ministry of the Interior] 1994 *Primo Rapporto Sullo Stato delle Minoranze in Italia–1994. [The First Report on the State of Minority Groups in Italy]*. Ministero dell'Interno, Roma

10 Censis 1993 *27° Rapporto Sulla Situazione Sociale del Paese 1993. Con il Patrocinio del CNEL [The 27th Report on the Social State of Italy. Funded by CNEL]*. Franco Angeli, Roma

11 Censis 1994 *28° Rapporto Sulla Situazione Sociale*

del Paese 1994. Con il Patrocinio del CNEL. Franco Angeli, Roma

12 Istituto Geografico de Agostini [De Agostini Geographic Institute] 1994 *Calendario Atlante de Agostini 1995. [De Agostini Geographic Abstract 1995].* Istituto Geografico de Agostini, Novara

13 Ministero della Pubblica Istruzione [Ministry of Education] 1985 *Programmi Didattici per la Scuola Primaria. [Official Intended Curricula for Primary School].* Istituto Poligrafico e Zecca dello Stato, Roma

14 Ministero della Pubblica Istruzione [Ministry of Education] 1979 *Programmi, orari di insegnamento e prove di esame per la scuola media statale. [Official Intended Curricula, Timetable, and Examinations for Junior Secondary School].* Gazzetta Ufficiale n. 50 (s.o.), Istituto Poligrafico e Zecca dello Stato, Roma

15 Ministero della Pubblica Istruzione [Ministry of Education] 1992 *L'Istruzione in Italia. [Education in Italy].* Ministero dell Pubblica Istruzione, Roma

16 1991 *Piani di Studio della Scuola Secondaria Superiore e Programmi dei Primi Due Anni. Le proposte della Commissione Brocca. [Structure of Senior Secondary School and Curricula of the First Two-year Secondary School Cycle. The Proposals of the Brocca Commission].* Studi e Documenti degli Annali della Pubblica Istruzione, 56 [Education Ministry Studies and Documents Series no. 56] Le Monnier, Firenze

17 1992 *Piani di Studio della Scuola Secondaria Superiore e Programmi dei Trienni. Le proposte della Commissione Brocca. [Structure of Senior Secondary School and Curricula of the Three-year Secondary School Cycle. The proposals of the Brocca Commission].* Studi e Documenti degli Annali della Pubblica Istruzione, 56/90 [Education Ministry Studies and Documents Series no. 56/90] Le Monnier, Firenze

18 Cavalli A 1991 L'insegnante depresso. Un ceto professionale di fronte ai valori che cambiano. [Depressed teacher: A professional status confronted with values that are changing]. In *il Mulino.* XL(335): 465–472

19 de Lillo A 1991 Insegnanti di classe. Le origini sociali di chi lavora nella scuola. [Teachers' Social Origins]. *il Mulino* XL(335): 473–481

20 Martinelli A 1991 Immagine di una professione. [Image of a Profession]. In *il Mulino* XL(335): 482–503.

21 *Sistema educativo e mercato del lavoro nel contesto internazionale. [Education System and Labour Market in International Framework].* ISTAT, Roma

22 Sistema Staistico Nazionale Istituto Nazionale di Statistica [National Statistical System National Statistical Institute] 1995 *Italian Statistical Abstract 1995.* ISTAT, Roma

Statistical References

Section	Statistic	Reference	Page	Table	Year of Statistic
Country Profile	60km^2	12	119	–	1994
Country Profile	0.4km^2	12	119	–	1994
Country Profile	301,302km^2	6	31	1.1.	1991
Country Profile	56 million	7	57	2.1	1991
Country Profile	188/km^2	7, 6	57, 31	CEDE elaboration from 2.1 & 1.1	1991
Country Profile	84%–94%, 16%–6%	8	114	CEDE elaboration from table 1	1951–1991
Country Profile	less than 6%, 70,000, 287,503, 18,434, 53,193	9 9	31, 61, 73, 94, 100, 114, 188, 205, 236, 253, 273, 355, 360	CEDE elaboration from numerical data	1991
Country Profile	US$19840, 2%, 9%,	4	163	1	1993, 1980–1993
Country Profile	10%, 12%	5	122	F13	1992
Country Profile	5%, 6%, 5%, 6%	5	76	F01	1992
Country Profile	US$4050., US$3410.	5	89	F03 (2)	1992
Country Profile	US$4700. US$4760.	5	90	F03 (3)	1992
Country Profile	95%	4	163	1	1990
Structure of the System	8%	11	140	4	1993–94
Structure of the System	101.5%	11	183	20	1993–94
Structure of the System	45 (4%)	11	140	4	1993–94
Structure of the System	107.5%	11	184	22	1993–94
Structure of the System	89%	11	184	22	1993–94
Structure of the System	72%, 70%	11	185	24	1993–94
Structure of the System	9%	11	140	4	1993–94
Schools in the System	23, 17, 20, 22	11	150	6	1993–94
Teacher Profile	100%, 93%, 71%, 54%	5	194	P36(B)	1992
Teacher Profile	62%, 55%, 44%, 70%	CEDE & Ministry database	–	–	–
Teacher Profile	40, US$18,161, US$17,482, US$27,852	5	188	P35(A)	1992
Teacher Profile	40, US$19,708, I US$17,482, US$30,927	5	189	P35(B)	1992
Italian TIMSS Populations	96%, 88%, 20%, 16%	CEDE & Ministry database	–	–	–
The Mathematics Curriculum	50%	CEDE & Ministry database	–	–	–

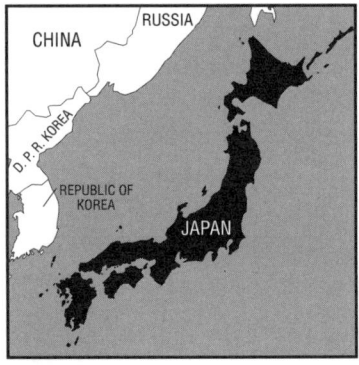

Japan

Masao Miyake, National Institute for Educational Research

Eizo Nagasaki, National Institute for Educational Research

1 COUNTRY PROFILE

The Japanese archipelago lies east of continental East Asia, separated from it by the Sea of Japan. There are four main islands, including Hokkaido in the north, and Honshu, Shikoku, and Kyushu moving south-west, as well as many small islands. The total land area is 377,801 km².

Nearly all of the 124 million residents of the archipelago are Japanese, and 77 percent are urban dwellers. The population grew 6 percent between 1980 and 1992, and has now achieved a density of 330 persons/km². The size of the 4 to 14 age group is gradually shrinking, while the 65 and above group is gradually increasing.

According to the constitution of 1947, Japan's government is based on a tripartite division of powers: legislative, administrative, and judicial. The legislative branch, known as the *Diet*, consists of democratically elected representatives in two houses, the House of Councilors and the House of Representatives. The prime minister is elected indirectly, by a vote in the *Diet*.

In 1993 Japan had one of the highest gross national products per capita, at US$31,490 in 1993. Japanese economic growth averaged 3 percent per annum between 1980 and 1993, a rate that was among the world's highest. The rate of inflation for the same period was less than 2 percent, one of the lowest in the world. In 1990, nationwide spending on education accounted for 17 percent of public expenditure, an amount equivalent to 6 percent of the GNP.

2 THE EDUCATION SYSTEM

Governance and Decision Making

The Ministry of Education, Science, Sport, and Culture is the administrative body responsible for school education, and all educational activities come under its supervision. Local bodies establish and maintain virtually all elementary and lower secondary schools, and are responsible to a prefectural or municipal board of education. The Ministry of Education supervises and subsidizes local boards of education.

The Ministry of Education prepares and distributes a course of study that forms a standard curriculum that all textbooks must follow. Groups of cities or towns usually combine to form adoption areas for the selection of textbooks for compulsory education. Municipal boards of education in adoption areas decide which textbooks should be selected on the basis of suggestions from school teachers. At the upper secondary level, textbooks are selected by individual schools from those authorized by the Ministry of Education.

Structure of the System and Participation Rates

Structure of Education

Education in Japan follows a 6-3-3 pattern, and includes primary education in elementary schools, secondary education in lower and upper secondary schools, and higher education in colleges and universities. Preschool education consists of kindergartens and nursery schools, and more than half of all five-year-old children are enrolled in kindergarten. Figure 1 shows the

Figure 1: Structure of the Education System

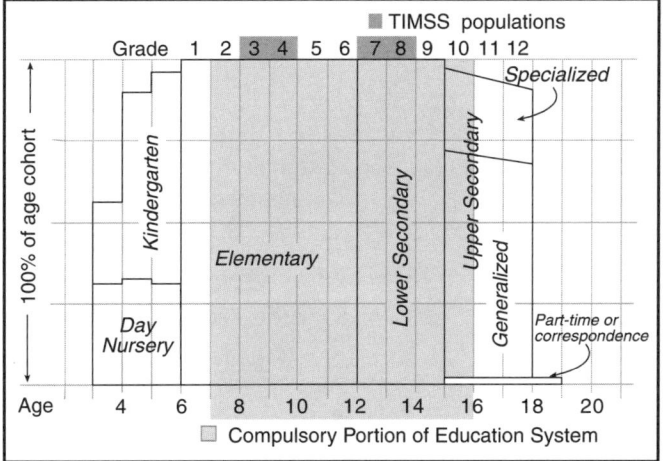

structure of the education system and the approximate enrollment rates for each level.

Participation Rates

Compulsory education consists of six years of elementary and three years of lower secondary school; virtually all children between the ages of 6 and 15 are enrolled in school. In 1992, 96 percent of the age cohort entered upper secondary school, 89 percent graduated from upper secondary school, and 33 percent entered university.

School Types

In upper secondary schools, education can be full time, part time, or by correspondence. Full-time students complete upper secondary school in three years, while part-time and correspondence students take at least four years. There are two streams at the upper secondary school level, general and specialized. About 74 percent of all students in upper secondary schools were enrolled in general courses in 1993. Specialized courses provide vocational and other education for students who have selected a future career. Courses include agriculture, industry, commerce, fishery, home economics, nursing, science and mathematics, physical education, music, art, and English.

Public and Private Systems

There are both public and private institutions at all levels of the academic hierarchy. The federal government bears most of the expenses of national schools, while municipal and prefectural schools are supported locally, with some assistance from the federal government. As a rule, private schools are self-supporting through tuition, donations, and business income. National and prefectural governments do, however, provide financial assistance toward maintaining and improving private schools. Eighty percent of kindergarten students are enrolled in private schools, while 1 percent of elementary students, 4 percent of lower secondary students, and 29 percent of upper secondary students are enrolled in private schools.

Schools in the System

The School Year

The school year begins April 1 and ends March 31. School holidays include national holidays, second and fourth Saturdays, and three long vacations. Most elementary and secondary schools are in session for 35 weeks, or 190 days. Most schools use a three-term school year: April 6 to July 20, September 1 to December 25, and January 8 to March 25.

Public school students attend school from Monday to Friday, and two to three Saturdays per month. Most elementary students attend six hours per day from Monday to Friday and four hours per day on Saturday. Most secondary students attend seven hours per day from Monday to Friday and four hours per day on Saturday. School days from Monday to Friday include approximately two hours for lunch, assemblies, and other activities. In elementary and lower secondary schools, total class time and percent of class time per week spent on mathematics and science are noted in Table 1.

Table 1: Class Time Spent on Mathematics and Science

Grade	1	2	3	4	5	6	7	8	9
Class time in hours	19	20	21	22	22	22	25	25	25
Mathematics	16%	19%	18%	17%	17%	17%	10%	13%	13%
Science	—	—	11%	10%	10%	10%	10%	10%	10%

(In Grades 1 and 2, life and environmental studies are taught instead of science and social studies.)

Class Size

In 1994, the average elementary class size was 29 students. During the same year, lower secondary schools averaged 34 students per class. The maximum class size for public upper secondary schools is 40 students.

Streaming and Tracking in Mathematics and Science

In public elementary and lower secondary schools, there is no official policy on within-school streaming, and schools are not tracked. From elementary to the end of junior secondary school, a compulsory program of mathematics and science is taught to all students in mixed-ability classes. Schools may offer optional courses beginning in Grade 9, which interested students may take. In upper secondary level, schools offer courses geared to differing abilities and interests, and students are placed in tracks according to their entrance examination achievement. In Grades 11 and 12, several optional courses in mathematics and science are offered.

Certification of Teachers of Mathematics and Science

Boards of education at the prefectural level certify successful teaching candidates. Prospective elementary and secondary teachers must have a four-year bachelor's degree, with several courses in education theory and pedagogy as part of their degree. There are two-year college courses in education, but graduates are awarded an associate bachelor's degree and may teach only at the elementary or junior secondary level. Seventy-two percent of elementary teachers, 88 percent of lower secondary mathematics teachers, and 94 percent of lower secondary science teachers have four-year bachelor's degrees.

All prefectural boards of education offer compulsory and noncompulsory, professional development programs in both subject-matter and school management. Compulsory induction training for beginning teachers has also been implemented by all prefectural boards since 1989. In addition to these opportunities, teachers may participate in study meetings at school, local research groups, university courses, and national associations for mathematics and science educators. Teachers wishing to become principals, vice-principals, or supervisors must take examinations offered by prefectural boards of education.

Teacher Profile

Teachers enjoy an economic status similar to other professions, but the rate of increase in salary level is somewhat low. Average starting monthly salaries and monthly salaries at the age of 41 for several professions requiring university education are given in Table 2.

Table 2: Professional Salaries

Profession	Starting monthly salary	Monthly salary at 41 years of age
Upper secondary teacher	US$1980	US$4240
University teacher	US$1990	US$4530
Clerk	US$1770	US$4310
Engineer	US$1830	US$4310

(In these salaries, several allowances are included, such as dependency, commuting, and housing allowances.)

In 1992, the average age of elementary teachers was 40, of lower secondary teachers 39, and of upper secondary teachers 42. Most elementary school teachers are female, but at all other levels most teachers, particularly teachers of mathematics and science, are male.

Table 3: Percent of Male and Female Teachers

Level	Male	Female
Elementary school	41%	59%
Lower secondary school	64%	36%
Mathematics	77%	23%
Science	85%	15%
Upper secondary school	79%	21%
Mathematics	91%	9%
Science	92%	8%

(Figures for teachers of mathematics and science are estimates.)

3 TIMSS POPULATIONS

Japan, IEA, and TIMSS

Japan has been a member of IEA since 1961. Prior to joining TIMSS, Japan participated in the First International Mathematics Study, the First International Science Study, the Second International Mathematics Study, the Second International Science Study, and the Computer Education Study.

Japanese TIMSS Populations

Japan is participating in TIMSS at the Population 1 and Population 2 levels. Population 1 includes all students enrolled in Grades 3 and 4 at the time of testing. These grades contained half the students in the nine-year-old age cohort at the time of testing, and at the midpoint of the testing year, virtually all nine-year-old students were in one of these grades. Population 2 includes all students enrolled in Grades 7 and 8. These grades contained half the students in the thirteen-year-old age cohort at the time of testing. At the midpoint of the testing year, virtually all thirteen-year-olds were in one of these grades.

4 MATHEMATICS CURRICULUM AND PEDAGOGY

Goals for the Mathematics Curriculum

Within the Ministry of Education is a curriculum council that advises the Ministry on the development of the intended curriculum. On the council's advice, the Ministry compiles national courses of study and curriculum guides in which goals and content are explained. Provincial boards of education determine an official curriculum that must accord with the national course of study. Schools are required to implement the official curriculum as given.

In elementary school, the goals are to help children develop the ability to consider daily life phenomena insightfully and logically, and to acquire fundamental knowledge and skills regarding numbers, quantities, and geometric figures. The curriculum is intended to foster positive attitudes toward mathematics and a willingness to make use of it in everyday life. In secondary school, the goals are to help students deepen their understanding of the basic concepts, principles, and rules concerning numbers, quantities, and figures. Students should learn to represent phenomena mathematically and appreciate a mathematical way of viewing and thinking. The curriculum is intended to foster a willingness to apply mathematics to everyday life.

Major Changes in the Mathematics Curriculum

The national courses of study were revised in 1989 and implemented beginning in April 1992 for Grades 1 to 6; April 1993 for Grades 7 to 9; and April 1994 for Grades 10 to 12. The fundamentals of the revision are to emphasize basic mathematics; to develop an appreciation of a mathematical way of thinking; to foster students' attitudes toward the use of mathematics; to enrich situations in which students can be treated more individually; and to encourage logical thinking. In the intended mathematics curriculum, computers are to be used as teaching aids beginning in elementary school, and are to be widely used in lower and upper secondary school mathematics. Some optional courses in the use of computers in mathematics will be offered in the upper secondary curriculum. Calculators are to be used from Grade 5 onward.

Current Issues in the Mathematics Curriculum

The Second International Mathematics Study and other national surveys revealed that Japanese students seemed to dislike mathematics. This attitude appears to be growing among students. National associations for mathematicians and mathematics educators have declared mathematics to be at risk and have taken steps to counteract the negative trend. In the latest revision of the national course of study, the usefulness of mathematics was strongly emphasized. It still seems, however, that mathematics is somewhat isolated from context. Calculators and computers were strongly recommended in the intended curriculum, but it appears that they are not being used effectively in the classroom.

Whole-class teaching is the typical style, and differentiation of the mathematics curriculum does not take place until Grade 11. Until then, teachers must work with students of all ability levels in the same classroom.

In 1995, a new curriculum council was established whose major theme was to introduce a five-day school week. How to cope with the reduction of total class time is a major issue in all subjects.

Mathematics Textbooks

Textbook writing teams consist of mathematics educators, mathematicians, and teachers. The books are approved by the Ministry of Education and published by commercial publishers. Local boards of education in consultation with teacher representatives decide which textbooks from the approved list will be used. The boards pay for textbooks used in compulsory education, while in upper secondary school, parents pay for textbooks.

Teachers at all levels are expected to use textbooks in implementing the intended curriculum. Teachers commonly use textbooks for reviewing, consolidating basic knowledge and skills, and summarizing rules. In addition, upper secondary school teachers use textbooks to introduce new topics and explain processes by working examples. Students use textbooks for reading, highlighting important information, and doing exercises.

Generally, mathematics textbooks follow a problem-solving approach consisting of the five stages: inquiry, explanation, examples, exercises, and applications. More real-life topics and visual elements such as illustrations are gradually being included in textbooks, particularly at the elementary and lower secondary levels.

Pedagogy

The curriculum guide emphasizes a balance between teaching rules and computational techniques, and developing concepts. Teachers, however, tend to stress rules and techniques in their teaching. Mathematics is usually integrated across topics within mathematics, but tends to be taught as a separate subject. This trend becomes increasingly strong in secondary school mathematics. Calculators are recommended for use beginning in Grade 5, but teachers rarely use them; this is a particular problem in secondary school mathematics.

Computers are also recommended for use at all grades, but, as of yet, they have not been used a great deal.

Inductive approaches are usually used in elementary school and lower secondary school mathematics, while formal deductive approaches are introduced gradually, beginning in Grade 8. Whether formal inductive approaches are appropriate for use in Grade 8 is a major issue. Manipulative materials are used in elementary school mathematics, but are rarely used in secondary school mathematics. The use of investigative and open-ended projects is strongly recommended, particularly in the intended curriculum of lower secondary mathematics. At present, teachers agree that projects are important, but are unsure of how to approach them. Communication in mathematics as a new theme has not been given much attention, since in the elementary curriculum it has been described as a discussion method, and a teacher-centered style is dominant in secondary schools.

5 SCIENCE CURRICULUM AND PEDAGOGY

Goals for the Science Curriculum

The goals and contents of the science curriculum are based on the course of study prepared by the Ministry of Education. Textbooks are prepared in accordance with the course of study and science classes are conducted using textbooks. Science teaching begins in Grade 3 and is a required subject throughout compulsory education and the first year of upper secondary school. The course of study lists the overall goals of science education as well as the goals and contents of the classes to be taught each year.

According to the course of study, the goals of elementary school science education are "to become familiar with and observe nature, and to conduct experiments, thereby fostering problem-solving capabilities and a love of nature, as well as gaining an understanding of natural objects and phenomena, and promoting scientific ways of seeing and thinking." The goals in secondary school are "to enhance interest in nature, conduct observations and experiments, foster scientific investigative abilities and attitudes, as well as to deepen the understanding of natural objects and phenomena, and to promote scientific ways of seeing and thinking."

Major Changes in the Science Curriculum

In keeping with societal changes and the needs of each age group, the course of study has produced age-appropriate goals and contents that correspond with the qualities, capabilities, and knowledge that students will require for the future. The fifth revision of the course of study since the end of the Second World War was implemented in elementary schools in 1992, in lower secondary schools in 1993, and in upper secondary schools in 1994. The main changes were as follows. In elementary schools, science was eliminated in Grades 1 and 2 , and replaced by life and environmental studies. The development of problem-solving abilities was clarified and emphasis was placed on experimentation, observation, and direct experience. In lower secondary schools, besides an increased focus on experimentation and observation, emphasis was placed on investigative activities with direct relevance to specific objects, phenomena, and daily life. The course of study also acknowledges and encourages discussion of advances that have been made in science and technology. In upper secondary schools, new science courses were introduced in order to allow students some flexibility in selecting courses according to their interests and abilities. Emphasis is placed on learning the scientific method while developing thinking, decision-making, and expressive abilities.

Current Issues in the Science Curriculum

The most pressing issues are the introduction of the five-day school week and its accompanying reduction of science class time, as well as the reorganization of science content in the new course of study. Other issues include how to conduct science education using computers, how to keep science education in step with advances in science and technology, and how to teach classes that deal with environmental problems caused by the progress of science. Yet another issue is whether to introduce information education, or the gathering and utilization of information, as a new subject.

Students' growing aversion toward and disinclination to study science, as well as the tendency of students to study nonscientific fields, are very serious problems for a nation built on science and technology. Also of concern is the problem of how to promote national scientific literacy, given that after graduation, students seem to forget the scientific concepts that they are supposed to have acquired.

Science Textbooks

Textbook writing teams consist of science educators, scientists, and teachers. The books are written to conform with the course of study, approved by the Ministry of Education, and published by commercial publishers. Local boards of education in consultation with teacher representatives decide which textbooks from the approved list will be used. The boards pay for textbooks used in compulsory education, while in upper secondary school, parents pay for textbooks.

Throughout elementary and lower secondary school, almost all science classes are conducted using textbooks. Students read textbooks, follow them in conducting experiments and observations, and solve the practice problems in them. Elementary school science textbooks include the following topics: living things and their environment, matter and energy, the earth and space. Lower secondary school textbooks include the following: objectives and phenomena relating to matter and energy, living creatures, and natural objects and phenomena surrounding them. In upper secondary schools, integrated science, biology, chemistry, earth science, and physics are offered.

Pedagogy

Traditionally, science education has been centered on acquiring knowledge and concepts. The new course of study seeks to achieve a balance between the traditional approach, and an approach that emphasizes inquiry and process. Almost all classes, however, tend to focus on the transmission of knowledge from teacher to student. Science education is supposed to focus on experimentation and observation, but with the exception of elementary science teaching, it cannot be said that this is happening. There is so much content to be covered that little time can be spared for laboratory work.

Computers are used very little in elementary science classes. In lower secondary and upper secondary science, however, computers are used effectively in the course of conducting observations and experiments, for information searches, processing data from experiments, and measuring in experiments. Traditional science education dwelt on pure science, but it has come to include familiar objects, phenomena, and content relating to daily life. The proposed effect is to stimulate an interest in nature, intellectual excitement, curiosity, and a desire to explore the world.

Almost no approaches emphasizing the relationship between science, technology, and society have been introduced. Scientific thinking, decision making, and expressive ability are emphasized, while written and oral communication in science classes are stressed for the first time.

6 EVALUATION POLICIES AND PRACTICES

National Level

Officially, there is no system of external evaluation. The three levels of entrance examinations, however, tend to play the same role as external evaluation. The first examination is run by individual national and private lower secondary schools, and takes place at the end of Grade 6 to qualify students for entry into these schools. A relatively small number of students take this examination, while all other students enter public lower secondary schools automatically. The second takes place at the end of Grade 9, before entry to upper secondary school. Virtually the entire 15-year-old cohort takes this examination, which is run by individual prefectures. Examinations for national and private schools are run by individual schools. The third examination takes place in Grade 12, and about one-third of the age cohort takes it. For national and private universities, there are two stages to this process: an examination given by the center for university entrance examination and an examination given by each university. There is a strong and growing belief that these examinations exert a powerful negative influence on the implemented and attained curriculum of these school levels.

There is no regular testing of the intended curriculum. From time to time, The Ministry of Education does administer nation-wide surveys to evaluate the intended curriculum as part of the preparations for the next revision. The latest survey conducted was administered to Grades 5 to 9 over the last three years. Most instruments are short-answer problems or extended problems in paper-and-pencil tests.

Local Level

Some local boards of education implement surveys to evaluate the intended curriculum locally. The purpose of these examinations is not to rank or evaluate schools, but to provide feedback to schools on their performance. Most boards of education employ inspectors to examine and provide feedback.

Classroom Level

Classroom evaluation is conducted to foster students and improve teaching. Alternative evaluation methods are recommended, but paper-and-pencil tests are still the main method used. The results of evaluation may be used for class placement within schools. Recently, the Ministry issued a revised framework for classroom evaluation, known as the permanent cumulative record. There are two major points to note in this framework. Emphasis on the recording system has changed from a norm-referenced one to a criterion-referenced one, and more emphasis is put on evaluating of ways of mathematical and scientific thinking and the affective domain.

7 REFERENCES AND SOURCES FOR FURTHER READING

Common Sources

1 Husén T, N T Postlethwaite 1994 *International Encyclopedia of Education*. Pergamon Press, Oxford
2 Organisation for Economic Co-operation and Development 1995 *Education at a Glance: OECD Indicators*. Organisation for Economic Co-operation and Development, Paris
3 United Nations Educational, Scientific and Cultural Organization 1995 *Statistical Yearbook*. United Nations Educational, Scientific and Cultural Organization, Paris

4 World Bank 1995 *World Development Report 1995.* Oxford University Press, New York

Other Sources

5 Japan Textbook Research Center 1995 *Report on The Basic Research on the Function of Textbooks as Learning Materials.* Tokyo (Japanese)

6 Ministry of Education, Science, Sports and Culture 1989. *Course of Study for Elementary School, Course of Study for Lower Secondary School, Course of Study for Upper Secondary School.* Tokyo (Japanese)

7 Ministry of Education, Science, Sports and Culture, 1994 *Education in Japan 1994: A Graphic Presentation.* Tokyo

8 Ministry of Education, Science and Culture 1980, *Japan's Modern Educational System: A History of the First Hundred Years.* Tokyo

9 Ministry of Education, Science, Sports and Culture 1995. *Statistical Abstract of Education, Science and Culture* 1995 edition. Tokyo

10 Ministry of Education, Science, Sports, and Culture 1995, *Statistics in Education.* Tokyo (Japanese)

11 Ministry of Education, Science, Sports, and Culture 1994, *Statistical Survey Report on School Teachers.* Tokyo. (Japanese)

12 National Institute for Educational Research 1990 *Basic Facts and Figures about the Educational System in Japan.* Tokyo

13 National Personnel Authority 1995, *White Paper on Government Employees.* Tokyo (Japanese)

14 *World Education Report 1993* by UNESCO (Japanese version)

Statistical References

Section	Statistic	Reference	Page	Table	Year of Statistic
Country Profile	377,801 km^2, 124 million, 6%, 330/km^2	3	1–5	1.1	1993
Country Profile	77%	14	118	1	1993
Country Profile	US$31,490, 3.4%, <2% (1.5%)	4	163	–	1993
Country Profile	17%, 6%	7	50–51	–	1990
Structure of the System	96%, 33%	7	18	–	1992
Structure of the System	89%	10	–	–	1994
Structure of the System	74%	7	14	–	1993
Structure of the System	80%, 1%	7	22	–	1992
Teacher Profile	US$1980–US$4310	13	265–268	2-7-2-8	1994
Teacher Profile	41%, 59% 64%, 36% 79%, 21%	11	5	1	1992

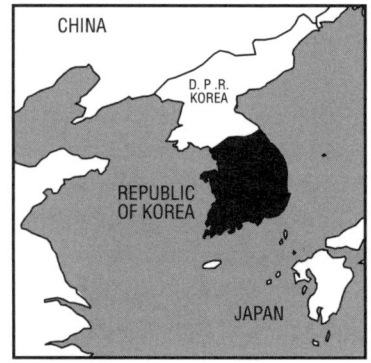

Republic of Korea

JinGyu Kim, National Board of Educational Evaluation

1 COUNTRY PROFILE

Korea, a peninsula located on an eastern spur of the Asian continent, occupies a total land area of 220,843 km². The Republic of Korea, commonly known as South Korea, comprises 99,016 km² of the peninsula. South Korea's population is estimated at 43 million, with a density averaging 441 persons/km². The population is growing at a rate of less than 1 percent per year. Population growth has declined in recent years. The Korean people are ethnically homogeneous, and are distinct in both language and culture from other Asian peoples.

Korea's history spans 5000 years. Today, it is a democratic nation with a government composed of three branches: an executive branch, guided by the president; a unicameral legislature, called the National Assembly; and a judicial branch. The president is elected every five years, but representatives in the National Assembly are elected for a four-year term. The country is divided into 15 administrative units: Seoul, the capital city; five municipalities; and nine provinces. In the past, the president appointed the governors and mayors of local governments. Since July 1995, however, municipalities and provinces have elected their mayors, governors, and representative bodies to govern local affairs.

Korea is an increasingly industrialized nation with a growing economy ranked as upper middle income by UNESCO. A series of five-year economic development plans established by the central government regulates the economy. As of 1993, the per capita gross national product stood at US$7660, a figure that grew at an average annual rate of 8 percent between 1980 and 1991. Expenditure on education in 1994 was 23 percent of net government expenditure and almost 4 percent of the GNP. According to UNESCO data, Korea's literacy rate is greater than 95 percent.

2 THE EDUCATION SYSTEM

Governance and Decision Making

In South Korea, there are three tiers of administration in the education system: the Ministry of Education, and local offices of education at the provincial and county levels. The Ministry of Education is responsible for forming policies relating to education, publishing and approving textbooks, directing subordinate agencies in planning and policy implementation, and providing financial support for national universities. Each local office of education has a board that is elected by the local parliament and that makes decisions regarding educational matters in the area. In 1991, there were 15 regional offices of education in six municipal cities and nine provinces, while at the county level there were 180 local offices of education. All colleges and universities are directly administered by the Ministry of Education. Superintendents, as heads of the regional offices of education, are responsible for directing all secondary schools located in their own area. All elementary schools and middle schools are responsible to each local office of education.

Korea's schools are required to follow a national curriculum set by the Ministry of Education. Since the Republic of Korea was established in 1948, the curriculum has been reformed six times, in seven- to ten-year cycles, in order to meet new educational needs. The sixth curriculum was implemented in March, 1995. The Ministry of Education produces national curriculum guides outlining intended learning outcomes, the content to be taught by grade level and subject, and the time allocation for each subject. Textbooks are divided into three types: those authorized by the Ministry, those inspected for quality and approved by the

Ministry, and those recognized by the Ministry as relevant and usable in school. Decisions about instructional methods and other classroom processes are made by teachers and schools.

Structure of the System and Participation Rates

Structure of Education

The Education Law of 1949 established a school system of the 6-3-3 type, and this is still used in Korea. Figure 1 shows the structure of the education system and the approximate enrollment rates for each level. Grades 1 to 12 are divided into elementary (Grades 1 to 6), middle (Grades 7 to 9), and high (Grades 10 to 12) schools. Secondary schools include both middle and high schools. As of 1995, Korea had approximately 20,000 education institutions. Among them were 8960 three-year kindergartens for children aged three to five, which are typically not part of the public school system. The proportions of pupils aged three, four, and five are 10, 28, and 44 percent respectively. The total proportion of the age cohort attending three-year kindergarten is 27 percent.

All children between the ages of 6 and 11 receive free compulsory education. There are nearly four million students enrolled in almost 6000 primary schools under the guidance of 138,000 teachers. Virtually 100 percent of the cohort was enrolled in elementary school. Until Grade 6, students are taught all subjects by one teacher, although special teachers may be available for physical education, music, and art.

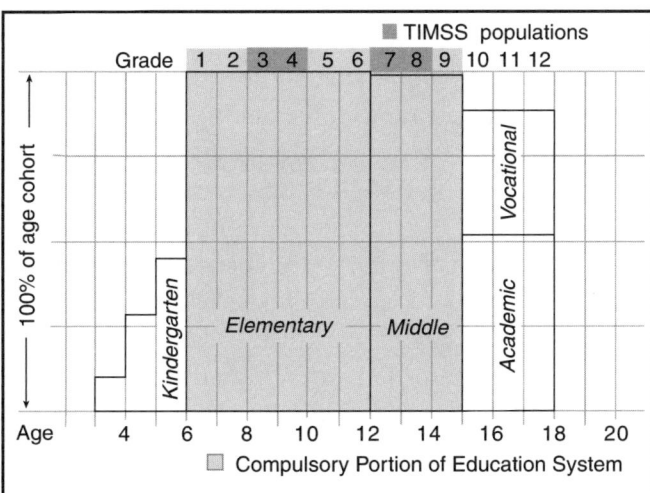

Figure 1: Structure of the Education System

All candidates from primary school are accepted into middle school. Candidates are assigned to one of the middle schools in their residential district, through a computerized placement program. Almost 2.5 million students are registered in middle schools, virtually 100 percent of the age cohort. The gender split among teachers at this level is virtually even. Compulsory education for middle school students was first introduced in 1985 in agricultural and fishing villages. Middle school students in towns and cities pay their own tuition and fees.

High schools are divided into general or academic schools and vocational schools. In 1995, there were 1.2 million students and 56,000 teachers in over 1000 general high schools. Twenty-two percent of these teachers were women. There are 900,000 students and 44,000 teachers in 762 vocational high schools. Twenty-five percent of these teachers were women. The proportion of students registered in academic and vocational tracks in high schools are 52 percent and 38 percent respectively. The remaining 10 percent of the cohort is not enrolled in education at this level. Approximately 99 percent of middle school graduates go on to high school.

Public and Private Systems

Most schools in Korea are part of the public school system, but there are a number of private schools, run by religious, social, or private organizations, which receive some funding from the government. In 1994, nearly 2 percent of elementary school children, 24 percent of middle school students, and 63 percent of high school students were enrolled in private schools.

Mathematics and Science Enrollment

All students study a compulsory mathematics program until the end of Grade 12, and a compulsory science program until the end of Grade 11. Thirty-two percent of males enrolled in Grade 12 take one or more science course, as do 26 percent of females.

Schools in the System

The School Year

The Korean school year runs from the first day of March to the last day of February, and includes more than 220 instructional days for elementary, middle, and high schools. Instructional days include 204 academic activity days and over 16 noninstructional days such as entrance ceremonies, festivals, school excursions, athletic meetings, and commencement. There are three vacation periods during the school year: a summer and a winter vacation, each of five weeks' duration, and a one-week spring vacation at the end of school year.

Students attend school from Monday to Saturday. At the primary level, schools are typically in session for between 24 and 31 hours each week, while secondary schools are in session for 34 hours. These times include one to two hours daily for lunch, exercise, and other activities. In 1995 the average elementary class had 37 students. The average class sizes for middle schools and high schools were 49 and 48 students, respectively.

Streaming and Tracking

There is no official policy on within-school tracking in South Korea, and primary and secondary school classes are composed of students with a wide range of abilities. All students study a compulsory program of mathematics and science up to the end of Grade 11. General mathematics and science courses are usually offered up to Grade 10, while in Grades 11 and 12 elective courses are offered for sciences, humanities, and vocational majors.

The Mathematics Program

Secondary school mathematics is divided into four courses: General Mathematics, Mathematics I, Mathematics II, and Applied Mathematics. General mathematics courses are usually offered in Grade 10 for all secondary students. Mathematics I is a basic course offered for general, academic secondary school students. Mathematics II is an advanced course offered for science majors in general secondary school. Applied Mathematics is offered for vocational secondary school students who have taken a general mathematics course.

The Science Program

Secondary school science courses are divided into three levels, General Science, Science I, and Science II. General Science courses are usually offered in Grade 10 for general secondary school students and in Grades 10 and 11 for vocational secondary school students. In Grades 11 and 12, additional courses (Science I and II)

are offered in general secondary schools. Science I includes Physics I, Chemistry I, Biology I, and Earth Science I for humanities majors in general secondary schools. Science II includes Physics II, Chemistry II, Biology II, and Earth Science II for science majors in general secondary schools.

Teacher Certification

Teachers follow one of two certification programs: a four-year degree program in an education college or a teaching certificate program offered at a general college or university. Graduates of national and private colleges must take a public screening test overseen by regional education offices before they may begin teaching. The 1990 reform of the teacher appointment system abolished the priority privilege for national education college graduates. Formerly, education college graduates were chosen for teaching positions before general and private education college graduates.

After taking 50 courses, 150 credit hours in a four-year degree program, national teacher's college graduates receive the primary school teaching certificate. Certificates for primary school teachers are given for all subjects taught at primary school. All national teachers' college students are required to take five credit hours in mathematics instruction and science instruction, respectively. There are seven course enrichment programs for mathematics and science education majors in national teachers' colleges.

Secondary science and mathematics teachers take as many as 20 courses in an area of specialty: 7 or more courses in educational theory, two courses in specific mathematics and science teaching methods, and a four-week student-teaching practice.

In-service development is offered in four categories: qualification studies for promotion, general studies to broaden knowledge of education theory and practice, adjustment training for new recruits to become acquainted with the profession, refresher training for in-service teachers to perform their teaching duties better, and two-year studies for selected teachers to obtain a master's degree from the Korean National University of Education. The promotion training program lasts 30 days (180 hours) or longer, and general training lasts 10 days (60 hours) or longer. All in-service courses are necessary for advancement except a two-year studies program.

Teacher Profile

Social and economic rankings of the professions show that the status of teachers is lower than that of other professionals in Korea. Teachers' social ranking was twentieth for elementary teachers, eighteenth for secondary teachers, and fourth for professors among 26 professions. The rankings in economic status were twentieth for elementary teachers, twenty-second for secondary teachers, and tenth for professors among 26 professions. Employees of major industrial companies earn considerably more than teachers. The annual average salary for first year teachers was US$18,100 in 1992. Teachers with 10 years of experience earned US$25,300 in the same year, compared to US$28,000 to US$38,000 earned by clerical staff in major industrial companies.

In 1994, the average ages of teachers were 42, 39, and 40 for elementary, middle, and high school teachers respectively. Forty-four percent of elementary teachers and 63 percent of secondary teachers were under 40. The proportion of female teachers is increasing. In 1985, 43 percent of elementary teachers, 39 percent of middle school teachers, and 20 percent of high school teachers were women. In 1995, 56 percent of elementary teachers, 50 percent of middle school teachers, and 24 percent of high school teachers were women. In general, elementary school teachers teach all subjects in the curriculum. The proportion of female mathematics teachers in middle and high schools are 45 percent and 17 percent, respectively. The proportions of female science teachers in middle and high schools are 43 percent and 21 percent, respectively.

3 TIMSS POPULATIONS

Korea, IEA, and TIMSS

Korea has been a member of IEA since 1982. Korea participated in the Second International Science Study and the Classroom Environment Study, and is now participating in TIMSS, the Second Civic Education Study, and the Language Education Study.

Korean TIMSS Populations

Population 1 includes all students enrolled in Grades 3 and 4 in regular elementary schools. These grades contained the largest proportion of nine-year-olds at the time of testing. Population 2 includes all students enrolled in Grades 7 and 8 in regular middle schools. These grades contained the largest proportion of students aged thirteen at the time of testing. South Korea did not participate at the Population 3 level. Many students in their final year of secondary school are overworked and psychologically burdened in preparing for college entrance examinations.

4 MATHEMATICS CURRICULUM AND PEDAGOGY

Goals for the Mathematics Curriculum

Korea has a national mathematics curriculum that schools are required to implement. The rationale behind curriculum structure and content is presented in a curriculum guide that states that the general objectives of mathematics education are to help students to:

- Understand the basic concepts, principles, and rules governing numerical quantities and geometric figures through the use of mathematical approaches in observing and organizing the various phenomena faced in everyday life;
- Solve various problems logically by practicing basic mathematical skills and applying these skills to everyday life;
- Enhance the abilities and attitudes needed to apply mathematical knowledge and skills to the solving of various problems.

Major Changes in the Mathematics Curriculum

The emphasis of the mathematics curriculum has changed over the past 10 years. Applications of mathematics in real-world settings are now stressed. The curriculum now includes problem solving, basic mathematics skills for everyday life, mathematical activities in the classroom, the development of a mathematical attitude and motivation, and relevant aspects of mathematical content.

There is no official policy on the use of electronic aids such as calculators and computers in mathematics classrooms, and the use of electronic aids depends solely on the teacher's interest. The use of such devices in mathematics, however, will be unavoidable in the future, when mathematics teachers will use calculators and computers to enhance students' problem-solving abilities in the new curriculum. Applied mathematics courses offered for vocational secondary students include the use of calculators and computers. These students use calculators and computers as a learning tool to understand mathematical concepts and to perform effective problem-solving activities in mathematics classes.

Current Issues in the Mathematics Curriculum

Although the focus of the mathematics curriculum has shifted, changes in teaching practice have been slow. Mathematics instruction has not been found to be as effective as intended, and it has been suggested that students be exposed to a variety of teaching methods in their mathematics classes. However, mathematics teachers prefer to use the lecture method because of both the quantity of material to be taught and the number of students in their classrooms. The concern of both students and parents regarding passing the college entrance examinations intensifies the problem. To help students cope with the examinations, most mathematics teachers emphasize short-term recall rather than open-ended exploratory activities, and the stated objectives of the courses are often ignored at the classroom level.

Mathematics Textbooks

In elementary schools, students use a textbook and a workbook published by the Ministry of Education for each grade level. The workbook focuses on computational drills and the practice of practical applications. During the past 10 years mathematics textbooks have changed significantly in format and approach, including more pictures and diagrams and using color more frequently. The textbooks and workbooks are developed either by consultants or by external agencies working under contract to the Ministry of Education. The research and development agencies are typically research institutions or university departments. The authors are usually composed of classroom teachers, education specialists, and professors.

In 1987, the policy on secondary school textbooks was changed from a government authorized to a government approved system. The Ministry approves five textbook series, produced by private publishers, for each subject, and a committee of mathematics teachers within each school chooses which series will be used. Beginning in 1996, the number of approved textbooks was no longer limited to five series for each subject. The Ministry of Education now approves many textbooks, after reviewing their relevance to national curriculum guidelines and standards. Secondary school mathematics textbooks are usually developed by university professors with the participation of mathematics teachers.

Pedagogy

Open-ended questions are increasingly used in external texts such as national assessments and college entrance examinations, and these are stressed in the classroom. Mathematics teachers try to emphasize concrete manipulative activities and thinking processes so that students discover principles and rules, and solve the problems embedded in these discoveries. Instructional methods for problem solving such as individual exploration, small group activities, peer tutoring, and discussion are expected at all levels. To facilitate learning, homework is assigned with the objective of reviewing ideas learned, reinforcing skills, and consolidating learning at the end of a unit.

Although calculators have become a part of daily life in Korea, students are not allowed to use them in mathematics instruction. The only exception is in applied mathematics courses offered for vocational high school students. Perhaps mathematics teachers believe that the use of calculators may cause a decline in students' computational skills. These skills are considered one of the main components of mathematics. Students are not asked to deal with numbers larger than three digits in mathematics classes, so the performing of complex and tedious calculations is not required. Computer science, computer programming, and their applications do not belong to the mathematics curriculum and are taught as separate subjects.

Educational television has been an important part of classroom instruction for the past 10 years. Educational broadcasting programs are aired for more than seven hours every day, and most teachers take advantage of various programs to improve the quality of their instruction.

5 SCIENCE CURRICULUM AND PEDAGOGY

Goals for the Science Curriculum

Goals for the science curriculum are contained in the national curriculum published by the Ministry of Education. This publication advises teachers on aims, contents, teaching methods, and methods of assessment. The curriculum guide states the general goals of science education:

- to help students understand scientific knowledge and methods;
- to help students develop scientific thinking skills and creative problem solving abilities;
- to help students enhance their interest in and curiosity about natural phenomena.

Major Changes in the Science Curriculum

The interaction of science, technology, and society has been emphasized throughout the last decade, and a new unit on "environment and natural resources" was introduced into the secondary school earth science curriculum in 1989. The unit includes human and natural environment, natural resources, and the earth's future. Environmental education was introduced as an elective subject in the middle school curriculum in 1995, and environmental science has been taught as an elective in secondary school since 1996. A new, integrated science curriculum, General Science, has been taught in Grade 10 since 1996. It covers the basic concepts of four subfields of science: physics, chemistry, biology, and earth science, and includes scientific inquiry, materials, forces, energy, life, earth, environment,

integration within the sciences, and issues surrounding the impact of science and technology on society. Science-technology-society issues and environmental concerns reflect the fact that there is a strong social desire to broaden the base of expertise in technology. The availability and use of audio, video, and software materials will increase significantly in the future.

Current Issues in the Science Curriculum

The relationship between science, technology, and society was recently added to the existing science objectives, and instructional materials dealing with science-technology-society issues should be developed in the near future. Many educators and specialists recommended that the science curricula concentrate on technology and information systems to prepare students for the future needs of society.

Science education in Korea has two purposes: to have students learn science as part of an academic background, and to develop a society of well-informed people. As of 1995, only 17 percent of high school graduates took science education for academic purposes, and most students have little interest in studying the difficult content of the science curriculum. The science curriculum should therefore maintain a balance between the two purposes of science education.

There are some gaps between the intended and implemented curricula at the secondary level. The integration of topics is regarded as important and the curriculum is structured to facilitate this, but science instruction for middle school students is taught separately by science teachers with different specialties, such as physics, chemistry, earth science, and biology. Most secondary science teachers are not graduates of the general science education program, but come from discipline-oriented science education programs such as physics education, chemistry education, biology education, and earth science education. Because of the academic background of middle school science teachers, discipline-oriented science instruction has been strongly emphasized and the integration of science topics has largely been ignored in their classrooms. Nevertheless, the Ministry of Education has expanded the integrated science curriculum in Grade 10. By the end 1996, 140,000 science teachers will have participated in the in-service training program to prepare for the new curriculum At the senior levels, Grades 11 and 12, the integration of science topics is not an issue, because the curriculum is discipline-oriented.

Science Textbooks

Korea's national curriculum describes general objectives and topics by grade, but it is the textbooks that define the specific topics, their order, and their relative weight. Comments in Section 4 on the authors of mathematics textbooks also apply to science textbooks. In elementary school, students use a single textbook and workbook for each grade, published by the Ministry of Education. Secondary school science teachers select one of the five science textbook series published by private corporations and approved by the Ministry. The textbooks and workbooks are developed either by consultants or by external agencies working under contract to the Ministry of Education. The research and development agencies are typically research institutions or university departments. The authors are usually composed of classroom teachers, education specialists, and professors.

There is no change in the extent to which teachers are expected to use textbooks, and the amount and types of material included in new textbooks are similar to the textbooks of the past 10 years. The greatest change in textbooks is the new primary school science workbook, which has been used in conjunction with the regular textbook since 1989 and focuses on hands-on activities not found in the regular textbook.

Pedagogy

Teaching methods used in science classes emphasize creativity, problem solving, and investigative processes. Hands-on activities such as laboratory investigations are often performed in science classrooms. Small-group and cooperative learning strategies are being used more often in elementary schools, while in secondary schools science teachers use the traditional lecture method combined with group discussions. Large class sizes and a general shortage of laboratory equipment in secondary schools mean that most students are not able to spend the number of hours in the laboratory required by the science curriculum.

The curriculum recommends that secondary school science deal selectively with following topics: liquid crystal; alloys; super plastics; ceramics; super-conductive materials; light speed; optical fiber; the principle of optical communication; electronic circuits; semiconductors; earth satellites and spaceships; exploration of the solar system and space; gravity applications; biotechnology; making new kinds of life by breeding; monoclonal antibodies; artificial organs; biotechnology; and industry.

6 EVALUATION POLICIES AND PRACTICES

National Examinations

Korean students must write examinations to enter both high school and college. At the end of Grade 9, 98 percent of students write the high school entrance examination. Since 1974, a high school entrance examination system has been employed in 14 major cities where secondary school education has been standardized in both school facilities and finances. In these areas, candidates for vocational secondary schools are given entrance tests before candidates for academic secondary schools. Following the examinations, the candidates who pass are assigned to secondary schools in their residential district by computer lottery.

The President's Council for Educational Reform announced on May 31, 1995 that the secondary school entrance examination in 14 major cities will be abolished in 1997, and candidates who choose two secondary schools among all secondary schools in their residential districts will be assigned one of them by computer lottery. A new comprehensive report card will be used as the major criterion for selecting secondary school applicants. The report card includes information on overall performance and activities, including grade point averages by subject for the most recent three years, school attendance rate, records of individual affective characteristics, extracurricular activities, public service, and other records. The established secondary school examinations in the remaining cities and rural areas are managed by individual schools, and will continue to be administered for the time being.

The national college entrance examination, formally called the College Scholastic Ability Test, is given annually to all high school graduates who wish to enter college or university. The test attempts to measure applicants in terms of general scholastic ability. There are four sections in the test: verbal, mathematical, science and social inquiry, and English as a foreign language. The National Board of Educational Evaluation is responsible for developing and administering the College Scholastic Ability Test.

Nearly all of these examinations use a multiple-choice format, with the remainder using a short-answer format. Both assessments determine which students will be selected for admission to high schools and colleges or universities. These examinations have definite effects on the implementation of curriculum, particularly in the final years of secondary school. Most secondary teachers, for example, emphasize lectures instead of laboratory activities in order to help students pass the college entrance examinations.

Korean students write a third type of external assessment. The National Board of Educational Evaluation, operated by the Ministry of Education, has since 1989 annually administered academic achievement tests nationwide at the elementary, middle, and high school levels. The goal of this assessment is to monitor students' educational progress and the effectiveness of the system at the national level. The national assessments are performed annually in Grades 4, 5, and 6 of primary school, Grades 7 and 8 of middle school, and Grades 10 and 11 of secondary school. The national tests for primary school students cover Korean language, mathematics, science, and social studies. Middle school students take the national tests in the above four subjects and English. At the secondary level, there are national tests for general secondary school students in Korean language, Korean geography, general mathematics, general science, and English.

Local Assessments

Periodic large-scale assessments are undertaken by regional offices of education. These assessments usually include tests in all subjects at the local level. The goal of these assessments is to identify any serious discrepancies in achievement between rural and urban areas. There are no typical local assessments in Korea. Two of the 15 regional offices of education perform annual tests in all subjects at all grades levels. Seven regional offices of education administer local tests in three or five subjects in each grade in primary, middle, and high schools. In the remaining six regional offices of educa-

tion, students are not required to participate in local assessments.

Assessment at the Classroom Level

There are two main purposes for assessment at the classroom level, formative and summative. Formative assessment is used to provide feedback to students, to diagnose weaknesses in the teaching program, and to aid in the planning of future lessons. Summative assessment is used to report student achievement to parents and to monitor long-term progress. A wide variety of assessment procedures are used at the classroom level, including performance assessments, interviews, and journals as well as more traditional methods.

7 REFERENCES AND SOURCES FOR FURTHER READING

Common Sources

1 Husén T, N T Postlethwaite 1994 *International Encyclopedia of Education.* Pergamon Press, Oxford
2 Organisation for Economic Co-operation and Development 1995 *Education at a Glance: OECD Indicators.* Organisation for Economic Co-operation and Development, Paris
3 United Nations Educational, Scientific and Cultural Organization 1995 *Statistical Yearbook.* United Nations Educational, Scientific and Cultural Organization, Paris
4 World Bank 1996 *World Development Report.* Oxford University Press, New York

Other Sources

5 Ministry of Education 1993 *Education in Korea 1993–94.* Ministry of Education, Seoul
6 Ministry of Education (Translated by Kim J-H, Y-R Lee, S-G Noh) 1993 *The Sixth National Science Curriculum of the Republic of Korea.* Korean Educational Development Institute, Seoul
7 National Board of Educational Evaluation 1989 *Entrance Examination Policy for Colleges and Universities in Korea.* National Board of Educational Evaluation, Seoul

8 National Board of Educational Evaluation *Guidelines for the College Scholastic Ability Test.* National Board of Educational Evaluation, Seoul

9 National Board of Educational Evaluation *Handbook of Education Statistics.* National Board of Educational Evaluation, Seoul

10 National Board of Educational Evaluation 1995 *Statistical Yearbook of Education 1995.* National Board of Educational Evaluation, Seoul

11 Research Foundation 1986 *1986–87 Study in Korea.* Korea Research Foundation, Seoul

12 Presidential Council for Educational Reform 1995 *Educational Reform Strategies toward an Establishment of New Education System in Korea.* Presidential Council for Educational Reform, Seoul

13 Yoon J I, et al. 1994 *An International Comparative Study on Teachers' Socioeconomic Status and Views of Teaching Profession between Korea and Japan.* Seoul National University, Seoul

14 Korean Federation of Teachers' Associations 1995 *A Study on Strategies to Improve Teachers' Salaries, Benefits, and Welfare.* Policy Research Report vol. 62 Seoul, Korea

Statistical References

Section	Statistic	Reference	Page	Table	Year of Statistic
Country Profile	US$7660, 96%	4	163	1	1993, 1990
Country Profile	8%	4	163	1	1980–1993
Country Profile	23%	9	48	29	1994
Structure of the System	20,000, 8960, 6000, 138,000, 2.5 million	10	24	1-1	1995
Structure of the System	10%, 28%, 44%, 27%	10	39	2-4	1995
Structure of the System	1.2 million, 56,000, 1000, 900,000, 44,000, 762	10	24	1-1	1995
Structure of the System	2%, 24%, 63%	10	26–27	1-2	1994
Structure of the System	32%, 26%	10	210	5-1	1995
Schools in the System	37, 49, 48	9	18	3	1995
Teacher Profile	20th, 18th, 4th	13	34	3-5	1994
Teacher Profile	20th, 22nd, 10th	13	36	3-6	1994
Teacher Profile	US$18,100 US$25,300 US$28,000 US$38,000	14	48–50	4	1992
Teacher Profile	42, 39, 40	9	21	6	1995
Teacher Profile	44%	10	100–101	3-18	1995
Teacher Profile	63%	10	174–175 280–281	4-22, 6-28	1995
Teacher Profile	43%, 39%, 20%	9	21	6	1995
Teacher Profile	56%, 50%, 24%	9	22	7	1995
Teacher Profile	45%, 17%, 43%, 21%	informal	national	survey	1995
The Science Curriculum	17%	10	254	6-17	1995
The Science Curriculum	140,000	5	132		1993
Evaluation	97%–98%	9	38	20	1995

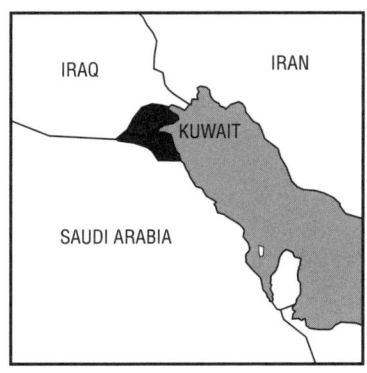

Kuwait

Mansour G. Hussein, Ministry of Education

1 COUNTRY PROFILE

The state of Kuwait is located to the northwest of the Arabian Gulf. To the north and northwest it borders the Republic of Iraq, and to the south and southwest the Kingdom of Saudi Arabia. Kuwait is divided into five provinces, namely the Capital, Hawally, Farwaniya, Al-Ahmadi, and Al-Jahra. The total population of the country is approximately 1.7 million, of whom almost one million are foreign nationals.

Kuwait has a written constitution which combines aspects of both presidential and parliamentary systems of government. Legislative authority is vested in both the Ameer as head of state and the National Assembly, which is composed of 50 members elected by the people. The prime minister is the head of government, and executive power is exercised through the ministers.

Kuwait ranks as one of the highest per capita income countries in the world. In 1993, the GDP was US$22 billion, with a national income of US$23 billion. Kuwait provides free education at all levels. In 1993, spending on education totaled US$1.6 billion, or 6 percent of the GNP and approximately 13 percent of total government expenditure.

2 THE EDUCATION SYSTEM

Governance and Decision Making

Kuwait's five provinces each have an Educational District that ensures the implementation of Ministry policy. The Board of Under-Secretaries, which reports to the central Ministry of Education, determines all educational policy. All decisions concerning curriculum and textbook selection are made centrally by special committees consisting of representatives from Kuwait University, and the Public Authority for Applied Education, General Supervision, Teachers, and Curriculum Unit.

Structure of the System and Participation Rates

The structure of the education system and the approximate enrollment rates for each level are shown in Figure 1.

Public Education

Modern education in Kuwait began with the establishment of the first public school in 1912. The First Education Council was formed in 1936 and public education became a national priority. Originally, the education system was confined to two stages, six years of primary and two years of secondary schooling. By 1956, this was expanded to two years of kindergarten, four

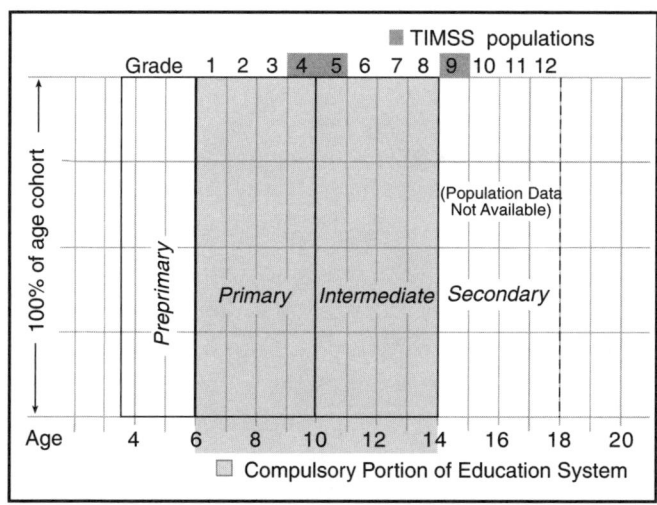

Figure 1: Structure of the Education System

years of primary, four years of intermediate, and four years of secondary schooling, and this system is still in place today. Public education is free at all stages, but only Grades 1 to 8, ages 6 to 14, are compulsory. Sixty-nine percent of Kuwaiti schools are public.

Several institutions of higher learning have been established, including an industrial college, a teacher education institute, and commercial and technical schools. Finally, in 1966, Kuwait University was founded, with faculties of science, arts, education, and a girls' college.

To keep pace with modern educational developments, the Ministry of Education adopted the unit-credit system for secondary schools in 1978–79. The two-term system was introduced in 1984–85, but it is the Ministry's intention to convert all secondary schools to the unit-credit system in the near future.

Private Education

Arab schools, which constitute 16 percent of all schools in Kuwait, operate according to the systems and syllabi of the Ministry of Education. Students follow the same curriculum and write the General Secondary School Certificate examinations along with public school students.

Foreign schools operate according to the systems and syllabi of foreign countries, the most commonly used syllabi are English, American, French, Indian, Pakistani, and Iranian. There are also several foreign embassy schools for the children of the diplomatic corps based in Kuwait, including a German and a Japanese school. Approximately 15 percent of Kuwaiti schools are designated foreign.

Schools in the System

The School Year

The school year begins in the middle of September for public schools at all stages, and ends in late May or early June. In 1995–96, primary schools were open for approximately 185 instructional days between mid-September and mid-May, while intermediate and secondary schools were open slightly longer. There are 14 holy days in the school year, and schools are closed on these days.

Students attend school for five days per week, from Saturday to Wednesday. The school day lasts for six hours, from 7:30 in the morning to 1:30 in the afternoon, and students attend school for 30 hours per week.

Class Size

According to the Ministry of Education *Statistical Diary* for 1994–95, the average class sizes in the three stages were 30, 31, and 26 respectively. All schools are single sex.

Class Time Allocated to Mathematics and Science

Table 1 shows the percent of class time allocated to mathematics and science in elementary school. In secondary school, the percent varies depending on whether the student is enrolled in the arts or the science stream, ranging between 3 percent and 20 percent. The figures for secondary school are shown in Table 2.

Table 1: Percent of Class Time Allocated to Mathematics and Science (Elementary)

Course	Grades 1–2	Grade 3	Grades 4–8
Mathematics	16	13	13
Science	6	10	10

Streaming and Tracking

Mathematics and science are both compulsory subjects from Grade 1 until the end of Grade 11. Grades 11 and 12, however, are divided into an arts stream and a sciences stream, and students choose which stream they wish to study. Arts students study religion, Arabic, English, French, humanities, and basic mathematics and science in Grade 11. No mathematics or science courses are offered in the Grade 12 arts stream, and graduates from this program may not be admitted to the colleges of science, engineering, or medicine at university. Students in the science stream study religion, Arabic, English, mathematics, physics, chemistry, biology, and geology. Fewer females than males are enrolled in the sciences division, and the Ministry of Education has established a program to encourage girls to enroll in secondary school science. All Grade 10 girls attend school-run discussions on the role of women in society and the need for women with science diplomas in industry.

Table 3 shows the percent of males and females enrolled in a mathematics or a science subject in Grades 11 and 12.

Table 2: Percent of Class Time Allocated to Mathematics and Science (Secondary)

Course Name	Grade 9	Grade 10	Grade 11 (Arts)	Grade 12 (Arts)	Grade 11 (Science)	Grade 12 (Science)
Mathematics	16	16	3	–	19	19
Integrated Science	9	–	–	–	–	–
Biology	–	6	–	–	6	10
Geology	–	–	–	–	6	–
Physics	–	6	–	–	13	16
Chemistry	–	6	–	–	10	13
Integrated Science	–	–	3	–	–	–

Table 3: Percent of Grade Cohort Enrolled in Mathematics and Science

Grade	Age	Mathematics		Science	
		M	F	M	F
11	16	49	40	49	40
12	17	52	43	52	43

Certification of Mathematics and Science Teachers

Teachers in Kuwait must follow one of two certification programs: a four-year university course, completing a Bachelor of Arts or Science in Education degree; or a four-year college course, completing a Bachelor of Education.

Preservice preparation is performed in two educational establishments, the College of Basic Education and Kuwait University. The College of Basic Education is part of the Public Authority for Applied Education and Training. Preservice teachers study for four years, during which the concentration is on preparing students theoretically and practically to become competent, qualified teachers. Many elementary school teachers are prepared at the college. Preservice teachers enrolled in Kuwait University study for four years, concentrating more on theoretical than practical studies. Teachers graduating from Kuwait University may teach at the level they have been trained for.

In-service education and upgrading are performed at two establishments, the Ministry of Education training center, and Kuwait University. The training center offers programs for newly recruited teachers, teachers transferring from stage to stage, teachers applying a new syllabus, new developments, or changes in the syllabus, and teachers teaching a new subject or course. Teachers wishing to obtain a Diploma of Education or other academic qualification may apply for a study leave and take courses at Kuwait University. Both theoretical and practical aspects of education may be studied.

Teacher Profile

Teachers are well-paid compared with other government employees. A recently graduated teacher earns US$1575. Married teachers earn US$1839, as well as earning an additional US$150 for each child in the family.

Most teachers begin their career at the age of 24 and retire at 50, and the average age at all levels is approximately 37. Because men are able to earn more money in other professions, not many choose teaching as a profession. As a result, there overwhelmingly more women than men at all levels of the education system. Table 4 shows the percent of female and male mathematics and science teachers, among both Kuwaiti and non-Kuwaiti groups.

Table 4: Percent of Female and Male Mathematics and Science Teachers

	Gender	Mathematics	Science
Kuwaiti teachers	female	92	93
	male	8	7
Non-Kuwaiti teachers	female	37	40
	male	63	60

3 TIMSS POPULATIONS

Kuwait, IEA, and TIMSS

Kuwait has been a member of IEA since 1991. A special committee was formed during the 1991–92 scholastic year to administer the TIMSS test.

Kuwaiti TIMSS Populations

Population 1 includes all intermediate schools in Kuwait, and a total of 4500 students were tested. One Grade 5 class from each of the 150 schools was chosen at random to participate in TIMSS. Grade 5 contains the largest proportion of students in the nine-year-old age cohort at the time of testing.

Population 2 includes all secondary schools in Kuwait, and a total of 2000 students were tested. One Grade 9 class from each of the 69 schools was chosen at random to participate. Grade 9 contained the largest proportion of students in the thirteen-year-old age cohort at the time of testing.

4 MATHEMATICS CURRICULUM AND PEDAGOGY

Goals for the Mathematics Curriculum

The Ministry of Education develops the curriculum, textbooks, and teachers' guides centrally, and schools are required to implement the official curriculum as given. Since 1981, Kuwait, with the other Gulf States, began to implement a unified core curriculum in mathematics. The curriculum guide states the general aims of mathematics education are to enable students to:
- understand the quantitative aspects of their environment;
- understand economic and social activities;
- study mathematics at a higher level;
- follow scientific and psychological advances in society and study other branches of knowledge;
- acquire mathematical knowledge and skill;
- acquire methods of mathematical thinking in order to solve mathematical problems in daily life;
- appreciate the beauty and consistency of mathematical structure;
- appreciate the role of mathematics in scientific and technical development;
- appreciate the role of Islamic civilization in the development of mathematics.

Major Changes in the Mathematics Curriculum

Until the late 1960s, developments in education were strongly influenced by the Egyptian model, and it was only in 1972, as a result of a general conference on curriculum, that a distinctively Kuwaiti system began to emerge. The conference membership included educational specialists from the Ministry of Education, as well as other government and private institutions. Through the conference, general goals of education were outlined for the first time, as well as specific objectives for each subject at each grade. Other outcomes included the encouragement of local textbook writing and publication, and the development of a system of continuous evaluation.

Since its initial development, the mathematics curriculum has undergone an extensive evaluation process resulting in several changes. The emphasis on abstraction in mathematics, which was an important trend in the 1970s, has been lessened; the relationship between mathematics and the sciences was brought into focus; and mathematics teaching and methodology were restructured. In addition, a commitment has been made to write and publish textbooks locally, and to define the goals and objectives for each stage of education.

In addition to the 1972 conference, Kuwait sought the expertise and cooperation of many regional and international agencies in its curriculum development. This endeavor resulted in several cooperative projects, including the UNESCO Project, which worked with several Arab universities and ministries of education for the development of mathematics education in the Arab states. Modern mathematics was taught on a trial basis in two Kuwaiti schools in 1970. By 1973, the new approach was generalized throughout the country.

The steering committee of the UNESCO Project later became a permanent committee for mathematics development. Membership in the committee includes mathematics experts from the Ministry of Education, head teachers and subject teachers from all levels, mathematicians and mathematics educators from Kuwait University, and mathematicians from the Teacher Training College and the Educational Research Center.

In 1981, the Gulf Arab States Educational Research Center, which has its headquarters in Kuwait, began a project for developing unified mathematics syllabi in the Gulf States. It involved:

- identifying the current status of mathematics syllabi through surveys and other research.
- defining the major features of the development process, including setting objectives for all study areas and levels and formulating these educational objectives into behavioral ones.
- setting content material for all study areas.
- submitting the proposed content to member states for comments and revision.
- developing a format for new mathematics textbooks, including teachers' manuals and study guides.
- proposing content material and methods of instruction.

In order to finalize the implementation schemes and curricular development, several conferences and studies on different aspects of the project have been held. Much work remains to be done, however.

Current Issues in the Mathematics Curriculum

After the full implementation of the unified curriculum in both the primary and intermediate stages, evaluating Grades 1 to 8 is considered to be the most immediate concern of the educational authorities. The impact on student and teacher performance must be assessed, as well as the influence on other subjects.

After the introduction of computers in both intermediate and secondary schools, the role of the teacher is expected to change. It is believed that computer use will change the student-teacher relationship, allowing students to become less dependent on teachers. The resulting impact on student discipline will be examined carefully.

The existence of two systems of secondary education, the two-term system and the unit-credit system, has been of great concern to education authorities for the last 15 years. It was expected that after a five-year period of trialling the unit-credit system, it would be approved for all schools. This has not yet occurred, however.

Evaluation in secondary education, particularly Grade 12, is also of great concern for education authorities at both the Ministry of Education and the Ministry of Higher Education. There are complaints about examination and evaluation standards, since many secondary school students graduate with high grades, only to fail at university.

Mathematics Textbooks

Mathematics textbooks are written to complement the unified curriculum prepared by the Gulf Arab States Educational Research Center. According to an agreement between ministers of education in the Gulf Cooperation Countries, a representative committee of all GCC countries was formed for the purpose of writing all textbooks. The first stage of writing took place in 1985–86 for Grade 1 textbooks, and these entered general use in 1987–88. Grade 12 textbooks, the last to be written, entered general use in 1995–96.

Pedagogy

Teachers are encouraged to use interactive learning, with an emphasis on making better use of community involvement. Inductive approaches are used at primary and intermediate stages, with more emphasis on deductive methods in secondary school.

Students do not usually use calculators at the primary and intermediate stages, but they are strongly encouraged to use them in secondary school. Computers are widely used in secondary schools, while intermediate schools have begun to introduce the use of computers to all students. Primary students do not use computers. All secondary schools have at least two computer laboratories, with 20 to 25 computers in each laboratory.

In addition to computers and calculators, a variety of instructional technologies are used in schools and have become a significant factor in instruction. All schools are supplied with videos and televisions, and the use of educational programs has became an important part of teaching in all schools.

5 SCIENCE CURRICULUM AND PEDAGOGY

Goals for the Science Curriculum

The official curriculum, along with textbooks and teachers guides, is determined centrally by the Ministry of Education and schools are required to implement it as given. In 1981, Kuwait and the other Gulf States began to implement a unified core science curriculum.

The goals of science education are to:

- acquire scientific facts and concepts;
- develop intellectual curiosity, objectivity, decision-making skills, a critical mind, honesty, humility, self-reliance, a love of teamwork, and an appreciation of the value of work;
- develop skills in identifying problems, gathering data, proposing theories, testing and experimenting, observing, interpreting results, and organizing and sorting items;
- develop scientific and practical skills in gathering samples from the environment, experimenting, and using of tools, apparatus, and natural resources;
- acquire scientific attitudes and interests;
- acquire an appreciation of scientists' efforts and their role in the progress of humanity and science.

Major Changes to the Science Curriculum

The General Curriculum Conference of 1972 affected not only the mathematics curriculum, but the science curriculum as well. Changes in mathematics occurred more rapidly as a consequence of the UNESCO and ALESCO projects, but renewal of the science curriculum followed soon after.

The National Committee for Science Development included science experts from the Ministry of Education, head teachers, science teachers from all three stages of schooling, scientists and science educators from Kuwait University, and scientists from the Teacher Training College and the Educational Research Center. The committee formed a number of resolutions concerning curricular reform, science teacher education, and the development of a comprehensive science program for all grade levels. The resolutions were approved by the Ministry of Education, and the process began with the writing of new textbooks for all stages. By the end of 1984, new textbooks had been issued for all grades, and the time allocated to science teaching increased significantly as the new curriculum was implemented.

In the new curriculum, physics topics were expanded to include sound, heat, electrostatics, and magnetism. A modified version of the theory of evolution was introduced in biology, and basic atomic structure became an integral part of chemistry. Geology was introduced as a separate subject. In the natural sciences, health education assumed a prominent role in Grades 1 to 4, while in Grades 5 to 8, an integrated science program replaced the traditional division of topics. The secondary school curriculum was extensively revised. Topics such as metabolism, physiology, and evolution were introduced into the biology curriculum, and the organic chemistry curriculum was almost completely rewritten.

In 1983, the Gulf Arab States Educational Research Center initiated a project for developing science syllabi in the Gulf States. It involved:

- identifying the current status of science syllabi through surveys and other research;
- defining of the major features of the development process, including setting objectives for all study areas and levels and formulating these educational objectives into behavioral ones;
- setting content material for all study areas;
- submitting the proposed content to member states for comments and revision;
- developing a format for new science textbooks, including teachers' textbooks and study guides;
- proposing content material and methods of instruction.

In order to finalize the implementation schemes and curricular development, several conferences and studies on different aspects of the project have been held. Much work remains to be done, however.

Current Issues in the Science Curriculum

After the Iraqi invasion in August 1990 and the subsequent liberation, Kuwaiti authorities in general and the Ministry of Education in particular began to take a different attitude toward science. Some of the resulting

important issues pertaining to science education are:
- science, technology, and society problems;
- environmental education;
- the importance of applied science versus theoretical science;
- science teaching in the computer era.

Science Textbooks

Currently, new science textbooks are being written to complement the unified curriculum designed by the Gulf Arab States Educational Research Center. Education ministers from the Gulf Cooperation Countries agreed in their annual conference that a representative committee of all countries should be formed to write the textbooks. As with mathematics, a separate committee was formed for each stage of the process. Textbook writing began in 1986 with Grade 1, and continued through several trials and evaluations before final approval and distribution in 1989. The Grade 12 textbooks are expected to be completed and in general use by 1997–98.

Pedagogy

There are many similarities between the teaching approaches used in both mathematics and science. The use of calculators is discouraged in both primary and intermediate schools, while all students use calculators beginning in Grade 9. Computers are used less frequently in science classes than in mathematics.

Both the curriculum and teachers' guides stress that cooperative learning is to be encouraged. Oral communication has become more important, and students are expected to communicate about scientific methods, results, and concepts. To a certain extent, laboratory work is considered important in primary school, and it increases in significance each year. In secondary school, 25 percent of the final grade is given for laboratory work in biology, chemistry, and physics.

6 EVALUATION POLICIES AND PRACTICES

Assessment policies in Kuwait are determined by policymakers within the Ministry of Education. The Department of Evaluation and Assessment within the Ministry corresponds with educational boards. Each educational board also has an internal division that monitors examinations in schools within their jurisdiction.

The goals of assessments are:
- to identify student achievement with reference to the intended curriculum;
- to evaluate the appropriateness of the curricula for the intended age group;
- to promote students according to their achievement.

National Examinations

At the end of Grade 12, the Ministry of Education holds a national examination in all subjects. The value of these examinations is 75 percent of the total grade, while the remaining 25 percent is determined by performance on internal examinations and class work. The minimum passing grade in each subject is 50 percent, and students who do not achieve this mark must retake the examination after three months. The rules allow students to rewrite up to three subjects, and students who fail more than three subjects must repeat the year.

School-based Assessment

Examination rules in Kuwait require that all students attain a certain level of knowledge before they can be promoted from one grade to the next. Students are tested through two types of examinations, both of which are administered by teachers within schools:
- Daily or weekly examinations, held with or without prior warning to students. These examinations make up 50 percent of the total grade.
- Examinations held at the end of the term and the school year. These are usually one-hour examinations in Grades 5 to 8, and up to two hours in Grades 9 to 12. These examinations make up the remaining 50 percent of the total grade.

The types of instruments used include oral examinations, homework checks, and short quizzes as well as the longer, written tests described. Grades obtained during the year through both short and long examinations determine whether the student will be promoted. It should be noted that there are no end-of-year examinations in Grades 1 to 4, although there is ongoing assessment.

7 REFERENCES AND SOURCES FOR FURTHER READING

COMMON Sources

1 Husén T, N T Postlethwaite 1994 *International Encyclopedia of Education*. Pergamon Press, Oxford
2 Organisation for Economic Co-operation and Development 1995 *Education at a Glance: OECD Indicators*. Organisation for Economic Co-operation and Development, Paris
3 United Nations Educational, Scientific and Cultural Organization 1995 *Statistical Yearbook*. United Nations Educational, Scientific and Cultural Organization, Paris
4 World Bank 1995 *World Development Report 1995*. Oxford University Press, New York

Additional Sources

5 Gulf Arab States Educational Research Center 1984 *GASERC's Scientific Efforts in the Domain of Developing Mathematical Education in Arab Gulf States*. Arab Bureau of Education for the Gulf States
6 Hussein M 1987 *The Mathematical Attainment of 13-year-old Students in Kuwait*. University of Southampton, England
7 Ministry of Education 1995 *Statistical Diary for 1994–95*. Planning Department, Kuwait
8 Ministry of Education 1995 *The Statistical Book of Schools for 1994–95*. Planning Department, Kuwait
9 Ministry of Education 1995 *The General Statistics of Government Schools for 1994–95*. Planning Department, Kuwait
10 Ministry of Planning 1995 *The Statistical Review*. Central Statistical Office, Kuwait
11 Ministry of Education 1994 *The Development of Education*. Kuwait
12 Ministry of Education 1994 *Annual Report*. Department of Evaluation and Assessment, Kuwait
13 Ministry of Education 1994 *Annual Report*. Department of Curriculum and School Syllabus, Kuwait
14 Ministry of Education 1994 *Educational Expenditure and Student Cost*. Department of Finance, Kuwait

Statistical References

Section	Statistic	Reference	Page	Table	Year
Country Profile	1.7 million, 1 million	10	2	2	1995
Country Profile	US$22 billion US$23 billion	10	9	17	1990–93
Country Profile	US$1.6 billion	7	41	26	1993–94
Country Profile	6%, 13%	3	4–12	4.1	1993
Schools in the System	30, 31, 29	7	26	–	1995
Schools in the System	Tables 1 & 2	13	13	5	1995
Schools in the System	Table 3	7	8	4	1995
Teacher Profile	US$1575, US$1839, US$150	14	16	1	1995
Teacher Profile	Table 4	9	35	7	1995
Mathematics Textbooks	textbook information	5	12	4	1988
Science Textbooks	textbook information	5	32	7	1988

Latvia

Andrejs Geske, University of Latvia

1 COUNTRY PROFILE

Latvia is located in northeastern Europe, bordering Estonia, the Russian Federation, Belarus, and Lithuania. Latvia covers 64,000 km², with a coastline extending 475 km along the eastern coast of the Baltic Sea and the Gulf of Riga. Latvia's present population is about 2.5 million, just under 30 percent of whom live in the capital city, Riga. Seventy percent of the population lives in urban centers since ineffective agriculture practices and poor socioeconomic conditions have depopulated the countryside.

Latvians are descendants of five ancient tribes: Cours, Latgalls, Zemgaels or Semigallians, Selonians, and Livs from the Finno-Ugric group. From the seventh to eleventh centuries, the Latvian tribes struggled against the Scandinavians and Russians. In the middle of the twelfth century, German expansion reached the Baltic territories and Latvia fell under foreign occupation for nearly seven centuries. The Republic of Latvia was formed as an independent, sovereign state in 1918 and in the short period of time before the Soviet occupation in 1940 it achieved remarkable success in industry, agriculture, culture, and education. Independence was reestablished in 1991.

Between 1940 and 1952, almost half a million Latvian citizens were deported. In 1944, rapid immigration from the Soviet Union began, principally from Russia, Belarus, and Ukraine. In 1937, 77 percent of all inhabitants were Latvian and 9 percent were Russian. By 1989 only 52 percent were Latvian while 34 percent were Russian. Since 1992, Russians, Ukrainians, and Belarussians have gradually been leaving Latvia to be repatriated in their own countries.

Latvia is a democratic, parliamentary republic with legislative power held by a 100-member parliament.

The parliament is based on proportional representation and is elected by general suffrage for a period of three years. The Cabinet of Ministers is the state body with executive power, and the president does not bear political responsibility for his or her actions. The constitution first adopted in 1922 was reinstated in 1993.

During the first period of Latvia's independence, from 1918 to 1940, great progress was made in social welfare, raising the standard of living, as well as in developing education, culture, and science. During the postwar period, however, Latvia's industries and economy were dominated by the former Soviet Union. Soviet industrialization policies neglected the vital interests of the Latvian nation and most production was exported to the Soviet Union. The forced development of manufacturing industry, imposed during the postwar period and determined by the interests of the former Soviet Union, began to decline in the mid-1980s with the decrease in military industrial production.

Latvia's per capita gross domestic product in 1994 was US$1400, with a small annual growth rate. The 1994 government budget included 13 percent for education. Although this percentage is rather large compared to other countries, the education system in Latvia has suffered great pecuniary embarrassment because past budgets have been so small. Educational expenditure totaled 4 percent of the GDP in 1994, and 6 percent in 1995.

2 THE EDUCATION SYSTEM

Governance and Decision Making

The system of education in Latvia is based on the Education Act of 1991, a document that is still undergoing revision and improvement. The education standard sets out aims, tasks, content, results to be achieved, and forms of tests. With the introduction of the education standard in primary schools, teachers are able to structure their teaching according to programs they prepare.

General education schools are subordinate to the Ministry of Education and Science and to local authorities, with the Ministry determining education standards for all subjects in the course of study. The standards are published in the form of a regulatory document stating the objectives and tasks of the course of study, the curriculum, and the required results in terms of knowledge, abilities, and skills. The Ministry is responsible for school accreditation, and determines the maximum number of classes for each level, the compulsory subjects for each level, and the content of final examinations.

All teachers' salaries come from the Ministry of Education budget. Local authorities are responsible for hiring and paying school staff and the material resources for schools. Because of this, there tends to be a great difference in resources such as libraries, computers, and audiovisual equipment available in different areas. The majority of local authorities are not interested in the educational process, having many more pressing responsibilities.

Until 1991, there was only one textbook per subject for each grade, and its use was compulsory. There are now several books available for each subject and teachers are free to choose the one that seems most suitable. Textbooks that have been approved by the Ministry of Education are somewhat less expensive because of government subsidies. Teachers determine the pedagogical methods used in the classroom as well as the division of material among lessons.

Structure of the System and Participation Rates

Structure of Education

The structure of the education system and the approximate enrollment rates for each level are shown in Figure 1. Preschool is the first stage of the education system. Kindergartens are established and maintained by local governments, businesses, and other organizations, with a curriculum approved by the Ministry of Education. In 1990, 48 percent of children in the one to six

age cohort were enrolled in a kindergarten, but by 1994 the figure dropped to 34 percent.

General Education

The Education Act of the Republic of Latvia states that compulsory education begins at 6 or 7 and continues until age 15 or the completion of basic school. General education is made up of three stages: elementary, basic, and secondary. Elementary school covers Grades 1 to 4 for students aged 7 to 11. Basic school covers Grades 5 to 9 for students aged 11 to 15. Subjects studied at these levels include Latvian, mathematics, a foreign language, drawing, gymnastics, music, and civics. Students at this level attend approximately four to six hours of class per day. At the end of Grade 9, all students write examinations in mathematics and Latvian language and literature in order to receive a graduation certificate.

Secondary Education

Secondary school covers Grades 10 to 12 for students aged 16 to 18. Compulsory subjects include Latvian language and literature, mathematics, a foreign language, world history, Latvian history, and physical education. Optional subjects include the study of a second foreign language, economics, geography, computer science, physics, chemistry, biology, music, nature and society, and others.

Vocational Education

First level vocational education includes vocational, agricultural, and craftsmen's schools, which prepare students for independent technical work in various

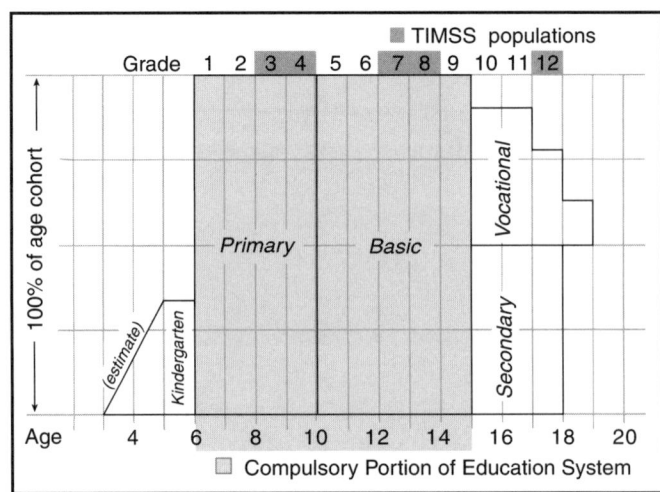

Figure 1: Structure of the Education System

Table 1: Enrollment in Science Classes

Grade	1	2	3	4	5	6	7	8	9	10	11	12
Science	100	100	100	100	-	-	-	-	-	-	-	-
Physics	-	-	-	-	-	-	-	100	100	80	80	80
Biology	-	-	-	-	-	100	100	100	100	80	80	80
Chemisty	-	-	-	-	-	-	-	100	100	70	70	70

fields. The length of study ranges between two and four years. Both theory and practice in the chosen craft are taught, together with elements of general education. During the last year of study, it is possible to enroll in additional courses from the regular secondary school curriculum in addition to the craft courses.

Second level vocational schools or secondary special education establishments include technical colleges, art and music schools, and agricultural schools. The length of studies ranges between four and five years following basic school, or two to three years following secondary school.

Mathematics and Science Enrollment

Mathematics is compulsory from Grade 1 to Grade 12. Science as a single, integrated subject is taught in Grades 1 to 4. Table 1 shows the enrollment of students in science classes from Grades 1 to 12.

Schools in the System

The School Year

The school year in Latvia begins September 1 and ends at the beginning of June. Grades 9 and 12, as the end of the respective education levels, are allocated extra time in June for external examinations. Holiday periods include two weeks at Christmas, one week in October, and another week in March.

Students attend school from Monday to Friday. Students have between 22 and 27, 40-minute instructional periods per week in elementary school, between 32 and 38 in basic school, and up to 40 in secondary school. Breaks between lessons are 10 to 20 minutes long.

Mathematics and Science Classes

Mathematics and science are compulsory up to the end of basic school. Mathematics is a compulsory subject in all grades of secondary school as well, but students may choose a basic or advanced course. Students in Grade 10 and above who wish to study science may choose basic or advanced physics, chemistry, or biology, which are offered throughout secondary school.

Certification of Mathematics and Science Teachers

Teacher education programs have changed several times in last 30 years and preservice teachers are now educated very differently than in the past. Teachers now acquire their education in one of the following education establishments in Latvia.

Universities

There are 12 faculties within the University of Latvia that provide the necessary qualifications to become a teacher. Within faculties, students concentrate on specialist subjects, and pedagogical studies are considered less important. Daugavpils Pedagogical University is a higher education establishment with six faculties concentrating on pedagogy. It prepares teachers for various levels, from elementary to basic school. After four years of study, students receive a bachelor's degree; those who remain for six years receive a master's degree. Some years ago the duration of study was five years for all candidates, and graduates received a university degree.

Higher Education Schools

Students completing a program at one of these schools receive a bachelor's or master's degree. A great deal of attention is given to pedagogy in these schools. Both primary and secondary school mathematics and natural science teachers in secondary school must have a university degree, or a degree or diploma from a higher educational school.

Education Colleges

Education colleges are secondary special education establishments and graduates are permitted to teach in elementary and basic schools. These schools are in the process of being transformed into higher education schools.

Unfinished Higher Education

After three years of study at any of the preceding institutions, students receive a teacher's qualification. They are permitted to teach in elementary and basic schools.

Teacher Profile

The socioeconomic status of teachers in Latvia tends to be very low. The average salary in 1994 was US$150 per month, while the average salary of teachers was only US$90 to US$125 per month. The average age of teachers is quite high, and there are not enough young people entering the profession. Between 90 and 95 percent of all teachers are female.

3 | TIMSS POPULATIONS

Latvia, IEA, and TIMSS

Latvia became a member of IEA in 1993. All IEA studies are carried out by the National Research Center in Latvia, located at the University of Latvia, Faculty of Pedagogy and Psychology. In addition to TIMSS, Latvia took part in the Computers in Education Study and the Language Education Study.

Latvian TIMSS Populations

Latvia is participating in TIMSS at all population levels. Grades 3 and 4 are being tested at the Population 1 level, and Grades 7 and 8 at the Population 2 level. For Population 3, only physics specialists, all Grade 12 students enrolled in advanced physics, were tested.

4 | MATHEMATICS CURRICULUM AND PEDAGOGY

Goals for the Mathematics Curriculum

The national standards for mathematical education consists of four parts: goals of instruction, content at standard levels, criteria of knowledge and skills, and forms and methods of evaluation.

The main goals for mathematics education in Grades 1 to 4 are:
- to develop mathematical thinking and abilities according to the individual capabilities of each student;
- to learn the mathematics necessary for everyday life as well as for further studies.

The main goals for mathematics education in Grades 5 to 9 are:
- to develop the use of mathematical methods in studying the world;
- to develop general mental abilities.

Goals of the math curriculum for Grades 10 to 12 are:
- to learn the mathematics necessary for practical purposes;
- to develop knowledge of the culture of mathematics;
- to develop mental abilities.

Major Changes in the Mathematics Curriculum

The mathematics curriculum has changed very little in recent years, although issues connected with everyday life are now being given more attention in elementary and primary schools. More time is spent at all levels on combinations, graph theory, and probability and statistics. This process will continue, particularly in advanced courses at the secondary level.

Current Issues in the Mathematics Curriculum

Many teachers are having difficulty with the new and

complex system of evaluation, which is used for all subjects. In the past, evaluation was based on a five-point system, in which two points were given for an unsatisfactory performance, three for satisfactory, four for good, and five for excellent. The new system involves a ten-point scale, but the principles are quite different from the five-point one. Under the old system, the mark was decreased according to errors made. Now, the mark is formed by summarizing positive achievements. The system is intended to promote the creative and unified use of skills and knowledge in unfamiliar situations.

Mathematics Textbooks

The number of available textbooks has increased in recent years, and teachers are sometime able to choose between three or four different books. Until 1991, there was only one book for each subject in each grade. Unfortunately, not all teachers appreciate the new books, for a variety of reasons. The new books are in some cases not an improvement over the old ones, and many teachers are simply too set in their teaching habits to be open to change.

Textbooks for the first grades of elementary school are brightly colored, but those for the middle grades are still black-and-white. The number of pictures and diagrams has increased. The new textbooks are designed so that students can use them for reading, study, and exercises without the teacher's assistance, whereas the old books were considered a supplement to the teacher's lecture.

Textbooks are sold on the open market and parents usually buy them for their children. School libraries have limited funds and cannot buy enough books for all students in all subjects. Usually, children whose financial resources are limited may borrow books from the school library. In general, textbooks are used for every subject.

Pedagogy

Mathematics is taught as a separate subject beginning in Grade 1. In elementary school usually there is one teacher for all subjects, including mathematics. From Grades 5 to 9 there is one qualified teacher for every subject, including mathematics. Unfortunately, there is a shortage of qualified teachers in all subjects, espe-

cially foreign languages and computer science.

Latvia is moving toward integrated education, and several integrated subject projects are being studied. At present, however, there are only a few elementary schools that include mathematics integrated with other subjects as part of these studies.

5 SCIENCE CURRICULUM AND PEDAGOGY

Goals for the Science Curriculum

The education standard determines the aims and tasks for elementary, basic, and secondary school in every subject. In elementary school, the main goals are:
- to develop an interest in nature;
- to provide knowledge of natural objects and phenomena;
- to create an understanding of the relationships between man and nature;
- to develop an understanding of the ethics involved in relationships between man and nature.

The main goals for basic schools include:
- to develop an understanding of the laws that govern nature;
- to promote sustainable development that does not damage the natural world;
- to prepare students for independent and rational action.

Goals for science subjects in the secondary curriculum are linked to mastering the content for each subject.

Major Changes in the Science Curriculum

During the 1980s, secondary education was compulsory in the Soviet Union and Latvia. Since Latvia became independent, however, this level of comprehensive education could not be sustained. Because many children now complete their education at the end of Grade 9, it became necessary to change the natural science program. Under the previous structure, the content to be learned at various levels was divided between primary and secondary curricula. The curricula have now been restructured in a cyclic pattern, so that students acquire the principles of all subjects in primary school, and the knowledge is deepened and increased in secondary school. These changes are particularly important since students may now choose not to take any natural science subjects in secondary school; the cyclic structure ensures that all students have some background in science when they finish school.

All courses include practical exercises. The majority of content includes subject-related issues, but some social issues are discussed as well. New information technologies have had very little influence on content and teaching methods because schools have not had the financial resources to purchase them.

Current Issues in the Science Curriculum

Science is not a mandatory subject in secondary schools, and many students choose not to take natural science. Educators and others are concerned about this development, since many colleges and universities require candidates to write a natural science entrance examination. By the age of 15, many students are cutting themselves off from the possibility of higher education by abandoning science. There is much discussion and research underway about creating an integrated, compulsory natural science course for those students who are not studying any other science courses.

Science Textbooks

General comments made in Section 4 about mathematics textbooks also apply to science textbooks.

At the Population 1 level, the major topics in typical science textbooks include time, natural phenomena, celestial bodies, animate nature, relationships between objects in nature, natural landscapes, the human body and hygiene, geography of Latvia, and geography of the Earth.

At the Population 2 level, textbooks include planning and mapping, geology, biology, principles of chemistry, inorganic substances (oxygen, hydrogen, and water), characteristics of inorganic substances (metals, metalloids, oxides, bases, acids, and salts), bodies and substances, motion and forces, light, and heat.

At the Population 3 level, textbooks include mechanics, molecular physics, electrodynamics, electromagnetic oscillations and waves, optics, principles of

the theory of relativity, quantum physics, and the constitution of the universe.

Pedagogy

Public awareness about education, its goals, and results has changed during the past few years. Formerly, knowledge and skills were the goals of education, but there is a growing awareness of positive attitudes as an expected result. While knowledge and skills may be estimated through conventional testing, attitude cannot. Equally, it is difficult to evaluate teachers' contribution to this area of education.

Specialists in the different branches of science teach natural science subjects from Grade 5 onward, just as they do mathematics. In many cases this means that there is little integration among science subjects. Not all students, for example, understand that an atom is the same in chemistry as in physics. Some primary schools are experimenting with an integrated natural science course, but most schools are still teaching the sciences as separate courses. A serious problem is the lack of trained teachers for the integrated natural science program. At present, none of the higher education establishments train teachers for this program. Many educators are reluctant to teach integrated courses because they fear that the level of complexity and the level of learning will decrease. Notwithstanding the successes in the implementation of a cross-curricular and interdisciplinary approach to natural science, educators must consider a number of issues in the comprehensive development of education:

- the lack of system-level strategies on supplementary concepts for natural science education in comprehensive schools;
- the lack of resources, including curricula, programs, textbooks, and teacher aids, including commonly agreed-on terminology in Latvian as basic requirements for encouraging and including natural science in all fields of education;
- the insufficient level of interdisciplinary and interactive preservice and especially in-service programs for environmental and administrative decision-makers, teachers, and other specialists.

6 EVALUATION POLICIES AND PRACTICES

Students must pass a national examination in mathematics at the end of Grades 9 and 12, which are the final years of basic and secondary school. Students have three to four hours to solve between 6 and 10 written problems. The examination is set by the Ministry of Education and all students write it simultaneously.

At the end of secondary school, students write several compulsory examinations in addition to the national examination. Mathematics, Latvian, and a foreign language are compulsory, and students choose another subject such as natural science. Usually these are oral examinations and include theory, problems, and practical tasks. The Ministry develops some 20 to 30 examination questions in every subject and publishes them three months before the examination. Each school selects the questions to be included in the examination, changing no more than 10 percent of them.

Diagnostic tests set by the Ministry are held once or twice per year in every subject for every grade. The same tests are written by all students of the same grade, and they are administered at approximately the same time. The duration of tests for elementary school pupils usually is one class period, or 40 to 45 minutes, but for basic and secondary school students two class periods is usual. The goal of testing is to help the teacher determine the weak and strong points of every student and the class as a whole, and to evaluate the knowledge of students according to the Educational Standard.

Teachers regularly estimate students' knowledge and skills during the school year. Students in Grades 1 to 3 are not evaluated, but beginning in Grade 4 students are given a term report summarizing their progress twice a year. Knowledge, skills, attitudes, and the dynamic of growth are taken into account. If their marks are unsatisfactory in several subjects, students must repeat the grade.

7 REFERENCES AND SOURCES FOR FURTHER READING

Common Sources

1   Husén T, N T Postlethwaite 1994 *International Encyclopedia of Education*. Pergamon Press, Oxford
2   Organisation for Economic Co-operation and Development 1995 *Education at a Glance: OECD Indicators*. Organisation for Economic Co-operation and Development, Paris
3   United Nations Educational, Scientific and Cultural Organization 1995 *Statistical Yearbook*. United Nations Educational, Scientific and Cultural Organization, Paris
4   World Bank 1995 *World Development Report 1995*. Oxford University Press, New York

Other Sources

5   Central Statistical Bureau of Latvia 1995 *Latvia in Figures. Collection of Statistical Data*. Central Statistical Bureau of Latvia, Riga
6   Ministry of Education and Science, Center for Education Information 1995 *Summary of Main Statistical Data of Schools at the Beginning of the School Year 1994–95*. Ministry of Education and Science, Riga
7   Ministry of Education and Science 1994 *Education in Latvia*. Ministry of Education and Science, Riga
All statistical data are taken from sources [5], [6], and [7].

Lithuania

Algirdas Zabulionis, University of Vilnius

1 | COUNTRY PROFILE

Lithuania is the largest of the Baltic states, occupying an area of 65,000 km². In 1995, the population was 3.7 million with a density of 57 persons/km². Lithuania's ethnic composition is relatively homogeneous, with 81 percent Lithuanians, 8 percent Russians, and 7 percent Poles. The remaining population is composed of Belarussians and people from other parts of the former Soviet Union. The capital is Vilnius, which has a population of 575,000.

Independence was first proclaimed in 1918 and restored in 1990, although Lithuania had a long and rich history of independence in the Middle Ages. Lithuania is now governed by a democratically elected parliament, the *Seimas,* headed by a prime minister who is the leader of the political party in power. Government jurisdiction is divided between the federal and municipal levels. Currently, the municipal level is undergoing significant reform with the creation of 10 regional governments.

Lithuania's economy is ranked by the World Bank as middle income. In 1993, the per capita gross domestic product was US$695, a figure that grew to US$803 in 1994. In 1994, registered unemployment was 6 percent and the average monthly salary US$125 per month. The 1994 federal budget allotted 22 percent for education, or approximately 5 percent of the GDP, and included financing for all levels of secondary and higher education. Funding for general secondary schools was 10 percent of the federal budget.

In 1994, 119 of every 1000 people over the age of 15 had a university-level education, 507 out of every 1000 had upper secondary or college-type education, and 176 per 1000 had basic school education. For every 10,000 inhabitants, there were 1447 students in general secondary schools, 133 students at vocational schools, 65 students in college-type schools, and 145 students in university level educational institutions in 1995–96.

2 | THE EDUCATION SYSTEM

New Developments

The secondary education system in Lithuania began to change with the development of a national school in 1988, well before independence was regained in 1990. After the restoration of statehood, the development of new education legislation became a priority for the *Seimas.* New education laws provide for substantial changes in the objectives, content, and structure of education. The objectives of the new education laws serve diverse, but related requirements: the necessity for citizens to acquire knowledge and understanding of the principles of a democratic, pluralistic society; the need to accept humanism and tolerance as basic values in life; the benefit of developing skills in independent decision making; and the acquisition of professional abilities. These objectives require substantial change in teaching methods, the creation of new textbooks, and the introduction of structural changes to create a more flexible and adaptive system of secondary education. In 1992, the government published a document entitled *General Concepts of Education in Lithuania*[1], which prescribes further reforms to the system. The last steps in educational reform were the publication of a draft general curriculum for Lithuanian secondary schools, containing a discussion of the concept of a modern school, syllabi for all subjects taught in school, and new education standards.

---

[1] Ministry of Education and Science of the Republic of Lithuania 1994 *General Concept of Education in Lithuania.* Publishing Center of Ministry of Education and Science, Vilnius

Governance and Decision Making

The *Seimas* defines the basic principles, structure, and objectives of education in Lithuania, while the Ministry of Education and Science and its specialized institutions devise and implement education policy. The Ministry is directly responsible for administration and financing of vocational and special schools. Municipalities are responsible for administration and financing of general education schools. Nevertheless, teacher's salaries and the assignment of teaching and auxiliary staff are determined by Ministry regulations.

The curriculum is defined centrally by the Ministry of Education and Science. Curriculum guides outline intended learning outcomes by grade and subject, including the rationale for teaching, a description of the content to be taught, and time allocations for each subject. Previously, teachers were issued a textbook series that was developed centrally for all Soviet Union countries, with small adaptations for national cases. This situation is changing slowly. Some translated textbooks are still in use, mainly in mathematics and science classes, as the principal teaching resource. Teachers may now include locally determined curriculum materials that reflect the needs and resources of their community.

Structure of the System and Participation Rates

Preschool

Public schooling in Lithuania generally begins with preschool for children aged one to six, and approximately one-third of the cohort attends. Parents are allowed to choose whether their children begin Grade 1 at the age of six or seven. Unfortunately, the starting age appears to depend not only on child development but also on the socioeconomic status of the family and the proximity of schools. At present, less than 20 percent of students in Grade 1 are under 7.

General Education

The public, general education system is divided into three levels: primary, basic, and upper secondary schools. Primary education consists of Grades 1 through 4 for children aged 6 or 7 to 10 or 11. Basic school offers Grades 5 through 9, and upper secondary school provides Grades 10 through 12. Education is compulsory for all students up to the age of 16. Beginning in 1998, basic school will be extended through Grade 10 and secondary school will consist of only Grades 11 and 12.

Primary and basic schooling follow a curriculum that allows little flexibility for students of differing abilities. This changes in Grade 9, however, with the introduction of a parallel, four-year gymnasium along with regular secondary education. The gymnasium is a new type of educational institution, developed since the education reforms began, and offering general education at a more advanced level. Traditionally, gymnasia are split into two profiles: humanities, and mathematics and science. The number of gymnasia is increasing, with about six thousand students or 10 percent of the cohort attending this type of secondary school in 1994. Basic or secondary school students who do not adapt to or who are for social reasons unable to attend general schools may attend a youth school, or *jaunimo mokykla,* for one or two years. Students leaving youth school are able to reenter either general or vocational school. Youth schools are geared toward providing the individual with basic knowledge about the world, elementary work skills, and general education.

Vocational education is an important part of the system. Students are accepted for two to four years of vocational schooling after graduation from basic school. Some are able to obtain a general secondary education diploma at the same time.

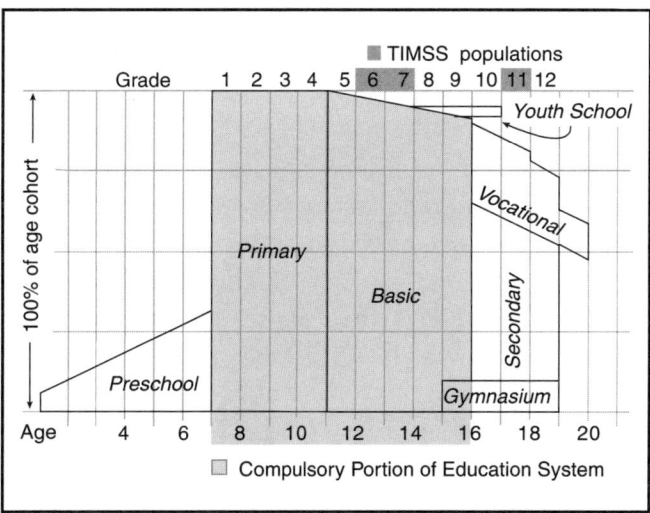

Figure 1: Structure of the Education System

Private Schools

The education law of 1991 authorized the creation of alternative, privately owned educational institutions. In 1996, there are only about 20 of such nonstate preschools or secondary schools and 16 colleges. Nonstate schools that conform to the state curriculum receive some state support, but enrollment in nonstate schools is still limited.

Participation Rates

There were more than half a million students at the beginning of 1995–96. They were learning in more than two thousand general secondary schools (836 primary, 592 basic, and 695 secondary schools), 106 vocational schools, 67 colleges, and 15 higher schools and universities. Because parents choose the entrance age to primary school, it is difficult to assess the percent of age cohort attending school. In 1995, 36 percent of children aged 3 to 6 attended preprimary education, with significant differences among rural and urban areas: 11 percent in rural regions compared to 50 percent in urban areas. In the same year, 38,000 students graduated from basic secondary school, corresponding to 72 percent of the 15-year-old age cohort. About two-thirds of them continued their education in upper secondary school, and one-third in vocational school or college. In the same year, 20,000 students, or 38 percent of the 18-year-old age cohort, graduated from upper secondary school. About 12,000 of them entered university, college, or vocational school. In 1995, the total number of students in youth school was 2500.

Schools in the System

The School Year

The school year in Lithuania begins September 1 and lasts 195 school days. Schools close for a week at the beginning of November, for two weeks at the end of December, and one week in March or April. In addition to these holidays, there are several national or local holidays on which schools are closed. The school year is divided into three terms of 12 to 14 weeks each, and students attend from Monday to Friday. Primary schools are in session up to 22 lessons per week, junior secondary schools for 28 lessons per week, and upper secondary schools for 32 lessons per week. The time allocated for one lesson is 45 minutes, with the exception of Grade 1, in which lessons are 35 minutes.

Class Size

The recommended class size for primary school is 24 students, but up to 30 students are allowed for the upper grades. In reality, class sizes vary from 13 in rural areas to 22 in cities. The average student-to-teacher ratio in 1995 was 17 to 1 in primary school and 10 to 1 in secondary school; these figures are significantly lower than in 1990, when they were 22 to 1 and 13 to 1. In primary school, students traditionally attend class in self-contained classrooms and are taught by one teacher, although specialist teachers for music, arts, or foreign languages may be available. In the upper grades, students are typically taught by single-subject teachers.

Language of Instruction

For the majority of students, the language of instruction is Lithuanian. There are a number of schools for linguistic minorities, and some schools offer separate classes with instruction in native languages. In 1995, 86 percent of students had Lithuanian as their primary language of instruction, 11 percent had Russian, and the remaining 3 percent, Polish. In recent years, the number of classes taught in Russian has declined, while the number of Polish classes has increased. These changes are connected to the departure of a portion of the Russian-speaking population; after this, a segment of Polish children chose to attend classes in Polish instead of Russian. There are also a few schools with instruction in Yiddish and Belarussian for those minorities.

Streaming and Tracking

There is no official policy on within-school tracking, and schools are not tracked at the primary and basic school levels. However, grouping by ability does occur in some schools. Students with high marks may specialize in some subjects or take enhanced or accelerated programs. This changes in upper secondary schools and gymnasia, which offer differentiated programs.

Gender Differences

Gender differences among students at the different levels of schooling are quite significant. In 1995, the proportion of girls in school changed from 50 percent in basic secondary school to 61 percent in upper secondary school. This disproportion increases in university, where more than 80 percent of students in pedagogical universities are female.

Certification of Teachers

Teacher education in Lithuania is carried out in universities and pedagogical colleges. Qualification and further education for teachers are provided by the Teachers In-Service Training Institute and in-service training centers. Education law allows those who have acquired college-level pedagogical education to teach. Those who have college-level, nonpedagogical education must acquire pedagogical qualifications through appropriate studies, the duration of which cannot be less than one year.

The qualification category or level of professional preparation is determined through certification. There are five main qualification categories: junior teacher, teacher, senior teacher, teacher-methodologist, and teacher-expert. Every five years, teachers must be recertified, either to confirm their category or to move to a new, higher category. The certification process was launched three years ago, and only about 25 percent of teachers have undergone the process. In general, teachers teach all subjects in primary school, and one or possibly two in basic and upper secondary school. Mathematics teachers, for example, are prepared to teach this subject in Grades 5 through 12.

Teacher Profile

Teachers in Lithuania unfortunately have quite a low socioeconomic status and relatively low salaries. The average salary of teachers is 10 percent lower than corresponding average monthly salaries in the public sector. As a consequence, there is a shortage of teachers in many subjects, particularly foreign languages.

The general characteristics of secondary school teachers in 1995 were as follows:
- 85 percent had a university-level education;
- 12 percent graduated from college;
- 87 percent had been teaching for more than four years;
- 14 percent were male, a decrease from 17 percent in 1990.

3 TIMSS POPULATIONS

Lithuania, IEA, and TIMSS

The first contacts with IEA were established in 1991. Lithuania joined TIMSS in 1992, one year before the country became formally associated with IEA. Lithuania is represented in IEA by the Center for Information and Prognosis, which is affiliated with the Ministry of Education and Science. Until recently, TIMSS was the only IEA study conducted in Lithuania, but the first steps have been taken toward joining the Civic Education Study.

Lithuanian TIMSS Populations

Lithuania is participating in TIMSS at the Populations 2 and 3 levels only. Population 2 includes all students enrolled in Grades 7 and 8 of basic school with Lithuanian as the language of instruction. Population 2 represents about 85 percent of the target population.

Population 3 generalists consist of two groups. The first includes all students in their final year of education, Grade 12, in upper secondary schools with Lithuanian as the language of instruction. The second includes all students in the final year of vocational and college-type schools with Lithuanian as the language of instruction. This population excludes students who graduated from secondary school before entering a vocational or college-type school. The Population 3 mathematics specialists were defined as all students in their final grade in schools offering an enhanced curriculum in mathematics, and in the mathematics and science gymnasia. There were fewer than one thousand of these students, and they all participated in the study.

4 MATHEMATICS CURRICULUM AND PEDAGOGY

Goals for the Mathematics Curriculum

Education in the former Soviet Union was highly centralized. Curricula for all subjects were prepared by subject-matter committees in the Education Ministry of the Soviet Union and then officially approved. The national versions in each of the republics were allowed only slight deviations. The enforced use of a single set of textbooks for the entire former Soviet Union further enforced the lack of local or national variation. New curricula are under development; in the interim, the Ministry of Education publishes a temporary curriculum each year.

The national curriculum for mathematics defines traditional goals in mathematics education, including providing students with the basic mathematical knowledge, skills, and abilities for use in everyday life and future careers. In addition, the curriculum aims to develop logical thinking and to foster and develop mathematical talent.

Major Changes in the Mathematics Curriculum

The reforms begun since 1990 include new general curricula and educational standards. The Ministry of Education publishes a new, temporary curriculum each year, which usually contains some changes from the previous year. These curricula and lesson plans include the distribution of subject lessons per week and are compulsory for all schools. Schools are allowed to organize the teaching process only within the framework prescribed by these two documents, for example, choosing three, four, or five mathematics lessons per week. The lack of diversity in textbooks has severely limited this freedom.

The changes in the education system as a whole are linked with changes in the general structure of the curriculum; it no longer prescribes in detail how many lessons teachers must give in each subtopic, but allows some flexibility in planning instructional times. Secondary school curricula allow even more freedom, in that schools may determine the number of mathematics lessons per week according to school traditions and student needs.

The former mathematics curriculum was based on a strongly academic tradition. Beginning in primary school, students were taught set theory and mathematical logic. Equations and inequalities were taught in Grades 6 to 8. Elements of linear algebra and calculus were taught in upper secondary school. The consequence of this was a large difference between the in-

tended and implemented curricula, since the subject was too difficult for most students. Reform in mathematics education is moving the curriculum toward real-life mathematics, stressing skills and abilities rather than knowledge. Some changes are related to the impact of new technologies on education, especially the use of calculators and computers. Significant use of computers during mathematics classes is still in the future, however. There is a definite shift away from continuous mathematics toward discrete subjects, including increased attention to combinations, graph theory, and numerical methods. Calculators are used beginning in Grade 5.

Current Issues in the Mathematics Curriculum

The old curriculum was prescriptive, while the new one gives much more freedom to the class teacher. To a certain degree, teachers are writing their own programs for mathematics, using the general curriculum as an guideline. To assist teachers in this task and to guarantee some comparability in mathematics education, standards for school mathematics are under development. The standards will describe the basic requirements in mathematical ability to be achieved by the end of primary, junior, and senior secondary schools. These requirements will describe different levels of achievement: minimal, mathematical literacy, and advanced.

Mathematics Textbooks

In 1990, the greater part of textbooks used in Lithuanian schools had been translated from Russian. Any original textbooks that were available were for primary schools. The problem of creating new textbooks and teaching materials is therefore one of the most pressing issues of education reform in Lithuania. The process of generating new textbooks and teaching materials has begun; textbooks for primary school and some grades of junior secondary school are already in use. Unfortunately, most mathematics textbooks are still at the planning stage.

Pedagogy

The methods used for mathematics teaching are the

traditional chalk-and-blackboard type, stressing teaching rather than learning. Computers and other new technologies have had little impact as of yet. Some attempts have been made to include real-life applications in school mathematics, to connect mathematics across the curriculum, and to define the mathematical literacy needed for every member of society.

5 SCIENCE CURRICULUM AND PEDAGOGY

Everything written above about mathematics education as it pertained to general changes in the education system is valid for science education as well. This includes the decentralization process, the freedom given to teachers and students, and the problems with standards in education.

Goals for the Science Curriculum

The main goals in the current national curricula are straightforward, and most begin with words *to help, to stimulate, to develop,* and *to show.* Beginning in Grade 1, students learn about the earth, humanity, animals, and nature. This integrated science course lasts throughout primary school, ending in Grade 5 with a course on the nature of Lithuania. Students study geography beginning in Grade 6, in an integrated course that includes earth science and the study of other countries. Biology is taught beginning in Grade 7, life science in Grades 7 and 8, human biology in Grade 9, and the environment in Grade 10. Physics and chemistry are taught as separate subjects in Grades 8 to 12.

Major Changes in the Science Curriculum

The reform in science education began in 1989, when new courses in integrated science and civics education were developed for primary school. The plan was to introduce integrated science for Grades 5 to 7, leaving the single-subject science courses, physics, chemistry, biology, and geography to be taught in the upper grades. The number of topics was reduced, keeping the focus on scientific literacy, especially as related to the environment, human biology and health, and new technologies. A significant change in the science curriculum involves a shift from the formal teaching of science knowledge to an approach that caters to the needs of real life. Such an approach enables students to practice and understand the methods and value of science.

Current Issues in the Science Curriculum

The old curriculum was prescriptive, while the new one gives much more freedom to the class teacher. To a certain degree, teachers write their own programs for science, using the general curriculum as an guideline. To assist teachers in this task and to guarantee some comparability in science education, standards for school science are under development. The standards will describe the basic requirements in scientific ability to be achieved by the end of primary, basic, and upper secondary schools. These requirements will describe different levels of achievement: minimal, scientific literacy, and advanced. The new curriculum also places a strong emphasis on environmental and ecological issues.

Science Textbooks

The process of writing a new generation of national textbooks in science has begun, and at the same time several textbooks from Scandinavian countries have been translated and adapted for Lithuanian schools. This has allowed the introduction of high quality, modern textbooks into the classroom, while providing adequate time for authors to develop national books of the same high caliber.

Pedagogy

There are no special national characteristics in science pedagogy. The change to an integrated science program has required intensive secondary school teacher training activities. Some attempts have been made to stress learning by doing and other active teaching methods, but these do not have much influence on science teaching in general.

6 EVALUATION POLICIES AND PRACTICES

Students must pass final examinations in order to graduate from basic and upper secondary school. Students write between five and seven examinations: compulsory examinations at the end of secondary school in native language and literature, mathematics, and foreign languages, as well as one subject from each elective block, including humanities, science, and social sciences. Students choose whether to write the advanced or basic level at the time of the examination. Upon passing their final examinations, students receive a secondary school diploma *brandos atestatas* and may apply to university, college, or vocational school.

Recently, there have been no national assessments of student achievement in mathematics and science. Some information on student ability in this area is routinely gathered through the regular examination system, that is, in basic and upper secondary school leaving examinations and university entrance examinations. A more systematic approach will be undertaken after the establishment of the National Center for Examinations; it is expected that students will write external examinations designed and administered by the Center in 1998.

7 REFERENCES AND SOURCES FOR FURTHER READING

Common Sources

1 Husén T, N T Postlethwaite 1994 *International Encyclopedia of Education*. Pergamon Press, Oxford
2 Organisation for Economic Co-operation and Development 1995 *Education at a Glance: OECD Indicators*. Organisation for Economic Co-operation and Development, Paris
3 United Nations Educational, Scientific and Cultural Organization 1995 *Statistical Yearbook*. United Nations Educational, Scientific and Cultural Organization, Paris
4 World Bank 1995 *World Development Report 1995*. Oxford University Press, New York

Other Sources

5 Lithuanian Department of Statistics 1995 *Lietuvos statistikos metrastis [Statistical Yearbook of Lithuania 1994–95]*. Lithuanian Department of Statistics Publishing Center, Vilnius
6 Lithuanian Department of Statistics 1995 *Lietuvos moksleivija ir studentija [The Students of Lithuania, in Lithuanian]*. Lithuanian Department of Statistics Publishing Center, Vilnius
7 Lithuanian Department of Statistics *1996 Vaiku ugdymas svietimo istaigose [Education in Schools, in Lithuanian]*. Lithuanian Department of Statistics Publishing Center, Vilnius
8 Ministry of Education and Science of the Republic of Lithuania 1994 *General Concept of Education in Lithuania*. Publishing Center of Ministry of Education and Science, Vilnius
9 Ministry of Education and Science of the Republic of Lithuania 1994 *Lietuvos bendrojo lavinimo mokyklos bendrosios programos [General Curriculum for Lithuanian Secondary School, in Lithuanian]*. Publishing Center of Ministry of Education and Science, Vilnius

Statistical References

Section	Statistic	Reference	Page	Table	Year of Statistic
Country Profile	65,000	5	12	1.1	1995
Country Profile	3.7 million	5	36	3.4	1995
Country Profile	57	5	36	3.4	1995
Country Profile	81%	5	39	3.8	1995
Country Profile	8%	5	39	3.8	1995
Country Profile	7%	5	39	3.8	1995
Country Profile	575,000	5	37	3.7	1995
Country Profile	22%	5	143	9.5	1995
Country Profile	5%	7	21	-	1994
Country Profile	10%	7	21	-	1994
Country Profile	119	7	7	–	1994
Country Profile	507	7	7	–	1994
Country Profile	176	7	7	–	1994
Country Profile	1,447	6	88	–	1995
Country Profile	133	6	88	–	1995
Country Profile	65	6	89	–	1995
Country Profile	145	6	90	–	1995
Structure of the System	20	7	34	–	1995
Structure of the System	16	7	48	–	1995
Structure of the System	2,000	6	34	–	1995
Structure of the System	836	6	34	–	1995
Structure of the System	592	6	34	–	1995
Structure of the System	695	6	34	–	1995
Structure of the System	106	6	10	–	1995
Structure of the System	67	6	10	–	1995
Structure of the System	15	6	10	–	1995
Structure of the System	36%	6	22	–	1995
Structure of the System	11%	6	22	–	1995
Structure of the System	50%	6	22	–	1995

Continued

Statistical References continued

Section	Statistic	Reference	Page	Table	Year of Statistic
Structure of the System	38,000	6	10	–	1995
Structure of the System	72%	6	45	–	1995
Structure of the System	20,000	6	10	–	1995
Structure of the System	38%	6	45	–	1995
Structure of the System	12,000	6	71	–	1995
Structure of the System	2500	6	34	–	1995
Schools in the System	13	6	37	–	1995
Schools in the System	22	6	37	–	1995
Schools in the System	17:1	6	45	–	1995
Schools in the System	10:1	6	45	–	1995
Schools in the System	22:1	6	45	–	1995
Schools in the System	13:1	6	45	–	1995
Schools in the System	86%	6	16	-	1995
Schools in the System	11%	6	16	–	1995
Schools in the System	3	6	16	–	1995
Schools in the System	50%	6	40	–	1995
Schools in the System	61%	6	40	–	1995
Teacher Profile	85%	6	43	–	1995
Teacher Profile	12%	6	43	–	1995
Teacher Profile	87%	6	40	–	1995
Teacher Profile	14%	6	40	–	1995

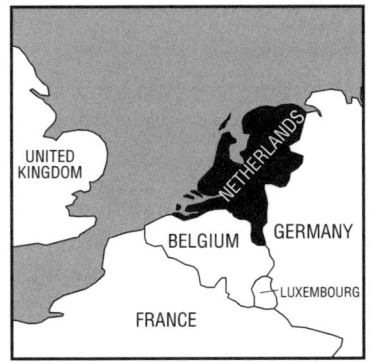

The Netherlands

Wilmad Kuiper, University of Twente

Anja Knuver, University of Twente

1 COUNTRY PROFILE

The Netherlands is a member state of the European Union. The topography is flat and a significant part of the country consists of river delta and lands reclaimed from the sea. About 27 percent of the country is below sea level and about 60 percent of the population lives on these lands. The Netherlands is bordered by the North Sea to the north and west, Germany to the east, and Belgium to the south. The total land area is about 41,000 km², and the country is divided into 12 provinces. With a population of 15 million in 1992 and a density of 372 persons/km², the Netherlands is one of the most densely populated countries in the world. Between 1981 and 1992, the total population increased by 7 percent, with an average annual growth rate of 0.6 percent. Since the mid-1960s there has been a decrease in the birth rate.

In the Netherlands, there is separation of church and state, and there is no state religion. Dutch society is becoming multiethnic to an increasing extent through an influx of people originating from Mediterranean countries and immigrants from former Dutch territories overseas. Many immigrants have settled in the large cities in the western part of the country. In 1992, about 7 percent of the population consisted of immigrants with non-Dutch nationality, and a further 2 percent had Dutch citizenship.

The Netherlands is a constitutional monarchy governed by a democratically elected parliament through a multiparty system. The Constitution provides that members of the Dutch Lower Chamber, the provincial councils, and the municipal councils be elected directly. Members of the Dutch Upper Chamber are elected by the provincial councils.

The economy of the Netherlands is ranked as high-income by the World Bank. The per capita GNP for 1991 was US$18,780, with an average annual growth rate of approximately 2 percent between 1980 and 1991. The average annual rate of inflation during this period was also 2 percent. Education expenditure in 1991 was 11 percent of total government expenditure and 6 percent of the gross domestic product. According to World Bank figures for 1993, approximately 64 percent of the labour force were employed in the service, transportation, or public sectors, 32 percent in industry, 20 percent in construction and manufacturing, and 4 percent in agriculture. According to UNESCO data, the Netherlands' literacy rate is greater than 95 percent.

2 THE EDUCATION SYSTEM

Governance and Decision Making

A distinctive feature of the Dutch education system is that it is based on a constitutional principle of freedom of education, including the freedom to found schools based on ideological or religious principles. As a result, there is a wide variety of schools in the Netherlands. There are two main categories: public schools, constituting 27 percent in 1993, and private schools, constituting 73 percent. Among the latter group are 30 percent Roman Catholic, 24 percent Protestant, and 19 percent nondenominational schools.

The Dutch education system combines decentralized administration and school management with centralized education policy. The Minister of Education, Culture, and Science controls education through centralized legislation with regard to the provisions of the Constitution. Indirect control occurs by means of regulations concerning the allocation of budget and other resources to schools. Based on the Constitution, public and private schools are funded on an equal basis. There is direct control through centrally imposed qualitative and quantitative standards that must be met by schools. In particular, these standards include the subjects that are compulsory by law in a particular type of school, minimum teaching time to be spent on subjects, binding regulations with regard to the objectives and content of schoolleaving examinations, and standards of competence to be met by teaching staff.

The intended curriculum for mathematics and the sciences is defined at the national level in terms of legislated core objectives for primary and lower secondary schools and legislated national examination programs at upper secondary schools. Generally, these curricular documents include a rationale for the teaching of mathematics and the sciences, a description of goals and objectives, and time allocations for each subject. The documents are developed by committees consisting of teachers, principals, and mathematics and science experts from the National Institute for Curriculum Development, the National Institute for Educational Measurement, university departments, and the Freudenthal Institute in the case of mathematics. Based on these documents, educational publishers develop and supply textbooks and other curricular support materials. Decisions about textbook choice, instructional techniques and other classroom processes are left to schools and individual teachers.

Structure of the System and Participation Rates

Structure of Education

The Dutch education system is divided into three levels: primary, secondary (subdivided into lower and upper levels), and tertiary. The 1985 Primary Education Act integrated kindergarten and primary school, creating a new style of primary education that starts at the age of four and covers eight grades. Figure 1 shows the structure of the education system and the approximate enrollment rates for each level. Education is compulsory from ages 5 to 16, but 95 percent of children begin school at the age of 4. Up to the end of Grade 8, students typically attend school in self-contained classrooms and are taught by one teacher, although specialist teachers may be available for physical education,

Figure 1: Structure of the Education System

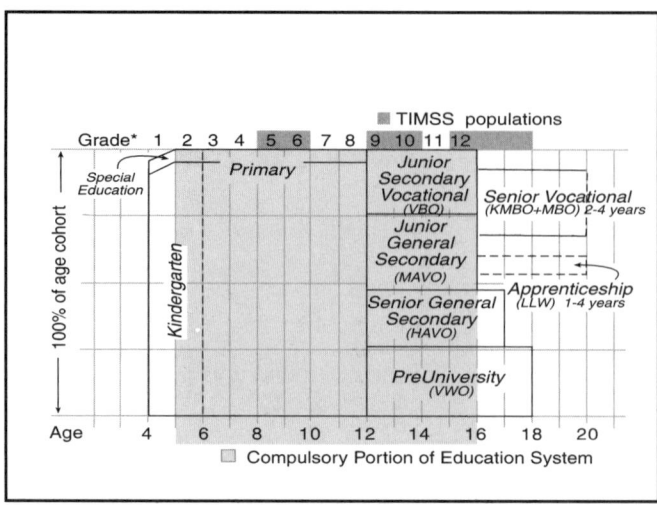

music, and art.[1] Special education schools at both primary and secondary levels are geared to children with different educational needs. About 6 percent of primary school students attend special schools.

Streaming and Tracking

Secondary education caters to students aged 12 to 17 or 18 years. Students may follow one of four main ability tracks:

• Junior secondary vocational or prevocational education, known as VBO. This is a four-year program specializing in technical, home economics, commercial, trade, and agricultural studies. It prepares students for the two-year short senior secondary vocational education known as KMBO, the three- or four-year long senior secondary vocational education known as MBO, apprenticeship programs, and the labor market.

• Junior general secondary education, known as MAVO. This program is also four years long, and prepares students for short or long senior secondary vocational education, apprenticeship programs, and the labor market.

• Senior general secondary education, known as HAVO. This is a five-year program preparing students for higher vocational education.

• Preuniversity education, known as VWO. It is a six-year course preparing students for university and higher vocational education.

At the end of primary education, students and their

---

[1] In TIMSS, the equivalent of the Dutch Grade 8 is Grade 6.

parents are advised by their classroom teacher which ability track to choose in secondary education. This advice is based on student achievement, especially in language, reading, and mathematics; in most cases, the rating of student achievement is underpinned by results of the standardized test administered at the end of primary school. Parents have the right to decide which track their child will attend, but secondary schools tend to refuse students who are advised to go to a lower track.

From August 1993 onwards, a common core curriculum was implemented in the first three grades of VBO, MAVO, HAVO, and VWO and called basic education. The core curriculum includes 15 subjects, among which are mathematics, combined physics and chemistry, biology, and geography and earth science.

Participation Rates

In 1993, there were 8139 primary schools. MAVO, HAVO, and VWO are offered in 1008 secondary schools, and VBO in 548. About 24 percent of the age cohort were enrolled in VBO, about 28 percent in MAVO, about 22 percent in HAVO, and about 26 percent in VWO. Virtually 100 percent of the cohort were enrolled in formal schooling during the compulsory years with mathematics and the science disciplines as compulsory subjects. In the last grade of MAVO, the last two grades of HAVO, and the last two grades of VWO, mathematics, physics, chemistry, biology, and geography are elective. Table 1 shows the percent of each grade cohort enrolled in mathematics and the various science disciplines as electives in 1993–1994. Unfortunately, statistics on the breakdown by gender have not been available since 1990.

Table 1: Mathematics and Science Enrollment

Track:	MAVO	HAVO	VWO
Grade:	4	4–5	5–6
Modal Age:	15	15–16	16–17
Mathematics	70	A: 54 B: 30	A: 62 B: 45
Physics	34	29	46
Chemistry	40	28	37
Biology	48	34	35
Geography	39	27	33

(Percent of grade cohort 1993–1994 in all electives, by track and grade)

Schools in the System

The School Year

For the first three years of primary education, children must be taught for a total of 2240 hours, with a minimum of 480 hours in each school year. For the rest of their primary school career, they must receive at least 1000 hours of schooling per year. The maximum teaching time allowed in one day is five-and-a-half hours. For both primary and secondary schools, the school year encompasses about 40 weeks. Classes begin in August or September and continue until June or July. The summer vacation between school years is six weeks long for primary schools and seven weeks for secondary schools. In addition, schools close for one week in mid-October, two weeks at Christmas, one week at the end of February, and for several days in April and May.

Students attend school from Monday to Friday. Primary schools are closed on Wednesday afternoons as well. A typical school day for a primary school student extends from 8:30 a.m. until 3:00 p.m. every day except Wednesday, when students leave at noon. Primary schools allow a 15-minute break in the morning and a 60- to 75-minute break for lunch. Secondary school students often start at 8:15 a.m. and continue classes until 2:00 or 3:00 p.m.

Percent of Class Time Spent on Mathematics and Science

In primary education, five hours per week are typically spent on mathematics and one hour on science instruction. At the secondary level, depending on track and grade, three to four 50-minute periods are spent on mathematics instruction, and two to four periods on instruction in each of the science disciplines.

Class Size

The average class size in primary education was 24 students in 1994. Class size varies a great deal in primary education: in 1994, 25 percent of classes had 20 students or fewer and 16 percent of classes had more than 30 students. School size is an important factor in class size, class sizes being relatively small in small schools. The average class size in secondary education is 25 students.

Certification of Teachers of Mathematics and Science

Education of primary teachers requires four years of full-time study at a higher vocational teacher education institution. At least 35 weeks of teaching and practical experience are required for certification. A minimum of 20 of these weeks are spent on practical experience in primary schools; the other 15 weeks are spent on teaching experience at the institution. The diploma certifies teachers for any subject and any grade in both primary and special education. In case of the latter, an additional two years of part-time study are recommended.

Secondary school teachers may be certified by one of two routes. The first entails four years of formal education at a higher vocational teacher education institution, leading to a second degree certification for teaching one subject at lower secondary level. Sixty percent of the training program is spent on content, while the remaining part focuses on the development of professional teaching skills. The second route entails a one-year teacher education program for candidates who have completed a university degree in a subject, leading to a first degree certification to teach that subject in upper and lower secondary schools. Fifty percent of the program is spent on practice teaching, and the other 50 percent on institutional activities including pedagogy.

Until recently, in-service training was offered and provided by teacher education institutions; counseling services were offered and provided by educational support institutions. Since 1994, schools have received a government-funded budget for in-service training and counseling services. In this paradigm, teacher education institutions, educational counseling institutions, and private companies offer their services in the area of both in-service training and counseling. However, a very recent study indicates that on average, half of each school's budget goes unused. Most schools do not have a straightforward in-service policy, and most teachers are unresponsive to in-service training and counseling possibilities.

Teacher Profile

In 1993–94, the mean age of Dutch primary school teachers was about 40; for lower secondary teachers the mean age was about 43. Fourteen percent of teachers in primary education and only 8 percent of teachers in secondary education are under the age of 30. Thirteen percent and 23 percent of teachers are above 50 in primary and secondary schools respectively. As to gender, 67 percent of primary school teachers are women. In lower secondary education, the majority of teachers, 68 percent, are male.

The attractiveness of the teaching profession has been a subject of concern over the past five years. A Committee on the Future of Teaching Profession studied the role, position, and appreciation of teaching profession and developed a plan of improvement. Observed problems include low mobility of teachers, job size and burden, absenteeism due to illness and incapacitation, relatively low remuneration and status of the profession, and limited professional development. The plan resulted in a policy with the following main components: measures for turning schools into professional labor organizations, relaxation of qualification requirements, and improvement of teacher education.

3 TIMSS POPULATIONS

The Netherlands, IEA, and TIMSS

The Netherlands has been associated with IEA from its inception. Prior to joining TIMSS, the Netherlands participated in the Six-Subject Survey, the Second International Mathematics Study, the Second International Science Study, the Computers in Education Study (internationally coordinated by the University of Twente in the Netherlands), and the Reading Literacy Study.

The Netherlands TIMSS Populations

Population 1 in the Netherlands includes all students enrolled in Grades 5 and 6 of primary education.[2] These grades contain the largest proportion of students in the nine-year-old age cohort. Population 2 includes all students enrolled in the first two grades of VBO, MAVO, HAVE, VWO. These grades contain the largest propor-

---

[2] In TIMSS, the equivalents of the Dutch Grades 5 and 6 are Grades 3 and 4.

tion of students in the thirteen-year-old age cohort. In TIMSS, these grades are designated as Grades 7 and 8. Population 3 includes all students in the final grade of VWO, the final grade of HAVO, and the second grade of MBO and KMBO. The apprenticeship system was left out of the sample due to the complexity of the system and lack of an adequate sampling frame. Mathematics and science literacy were the only areas assessed.

MATHEMATICS CURRICULUM AND PEDAGOGY

Goals for the Mathematics Curriculum

The intended curriculum in mathematics for primary schools is described in terms of core objectives for mathematics. The document states that mathematics education is aimed at teaching students to:

- understand the relationship between mathematics education and daily life;
- acquire basic mathematical skills, understand simple mathematical language, and apply both in practical situations;
- reflect on their own mathematical activities and check the accuracy of the results of those activities;
- recognize and trace simple relationships, rules, patterns, and structures;
- describe their own methods of investigation and reasoning in their own words and be able to apply those methods.

Core objectives have been formulated for five content domains. These include basic skills (number facts, mental calculation, estimation, column arithmetic); ratio, proportion, and percent; fractions, decimals, and numbers; measurement; and geometry.

The intended curriculum for mathematics education in lower secondary education is defined in terms of core objectives for basic education, which have been developed concurrently with the new national examination program for mathematics for VBO and MAVO. The new Mathematics 12 to 16 curriculum differs dramatically from the previous more formal and abstract curriculum from before 1993. The principal characteristics of the new curriculum are:

- mathematics content that appeals to students and that is explicitly linked to real-life situations that challenge students to "mathematize" and to construct their own solutions in a creative and meaningful way;
- an emphasis on reasoning, problem solving, and inquiry;
- more coherence across mathematical topics and between mathematics and other subjects.

The intended curricula for mathematics education for the last three grades of VWO and the last two grades of HAVO have been redesigned with regard to structure and content. They have been divided into two subjects—mathematics A and B—both of which lay more emphasis on applications and problem-solving skills. Mathematics A has a general goal of understanding and solving problems derived from real life, and prepares for higher education studies in which mathematics does not play a major role. Mathematics B is more formal and abstract, and prepares for scientific and technical studies.

Major Changes in the Mathematics Curriculum

In primary education, a realistic approach to mathematics education has gained popularity over the mechanistic approach. Previously, learning content was atomized in meaningless small parts and students were offered fixed solution procedures and trained by individual exercises. Realistic mathematics education, on the other hand, uses context problems, visual models, individual constructions, interactive education, and other methods. In realistic mathematics education, the teacher has a very important role, relating problem-solving strategies to the standard solution by visual schemes and models. In such a model, the teacher must have detailed knowledge of students' problem-solving strategies. Most teachers in the Netherlands use textbooks based on realistic mathematics education for their mathematics lessons. However, research reveals that it is very difficult for them to implement all the didactic characteristics of this approach.

Changes in emphasis in the new intended curriculum for mathematics in lower secondary education show a desire to make mathematics more accessible, interesting, relevant, meaningful and, taking into account student ability level, less formal and abstract.

The curriculum emphasizes learning by doing, particularly in the use of well-chosen concrete materials and in solving mathematical and nonmathematical problems for which no standardized solutions are available. Logical thinking, reasoning, and reflecting are also emphasized. The principal domains are arithmetic, measurement, and estimation; algebra, relations, graphs, and functions; geometry; statistics and probability; and integrated mathematical activities.

Major changes to the mathematics curriculum for upper secondary education involved the implementation of Mathematics A and B as electives in upper HAVO and VWO. The principal content areas of Mathematics A are:
- In HAVO: tables, graphs, and formulae; discrete mathematics; and statistics and probability.
- In VWO: applied analysis; applied algebra; probability and statistics; and computer science.

The principal content areas of Mathematics B are:
- In HAVO: applied analysis and 3-D geometry.
- In VWO: analysis and geometry.

In both Mathematics A and B, students are required to work with calculators and computers.

Calculator use in mathematics classrooms has increased during the past 10 years, both in primary and in secondary education. Computers have, as yet, limited use in the classroom although they are encouraged. Electronic aids such as graphing calculators and computers are perceived as essential tools in upper secondary mathematics education.

Current Issues in the Mathematics Curriculum

Changes in the intended mathematics curricula for primary and lower secondary education have to do with content, and teaching and learning approaches. Concurrently, new methods of assessment are being encouraged. Changes at the classroom level are facilitated by the introduction of textbooks emphasizing problem solving and real-world applications, and by various opportunities for teachers for in-service training and counseling services. It is reasonable to expect that it will take some years before these changes will be put into practice. Issues focused on by curriculum experts in primary education include applications, mental arithmetic, flexible computation, and estimation.

A substantial reform of the last two grades of HAVO and the last three grades of VWO is being prepared. The reform will be implemented from 1998 onwards, with the first renewed national examinations taking place in 2000 for HAVO and in 2001 for VWO. The reform is aimed at better meeting the heterogeneous needs of students and at improving the transition from HAVO and VWO to higher education. It entails:
- greater emphasis on stimulating students' self-motivation;
- a strong focus on acquisition of general skills;
- a shift toward a more facilitative teaching approach;
- an attempt to make content more relevant and useful;
- the replacement of students' current right of free composition of electives packages in upper VWO and HAVO by four prestructured profiles preparing for higher education.

Mathematics Textbooks

In primary and secondary schools, the main resource for each grade and each level is a textbook, but textbooks are frequently supplemented by additional, teacher-produced materials and tests. Mathematics textbooks provide a source of problems and exercises, and in secondary education they act as reference books, setting out theorems, rules, definitions, procedures, notation, conventions, and explanations.

Textbooks are developed by teams of mathematics education experts and distributed by educational publishers. Legislated core objectives and national examination programs are influential frames of reference for revising existing textbooks and developing new ones. Schools and teachers are free to choose the textbooks they prefer, and each student has his or her own copy of the textbook. In primary education, textbooks are paid for by schools, while in secondary education they are paid for by parents.

About 80 percent of primary schools use a mathematics textbook based on the realistic approach to mathematics education. Since 1993, new textbooks have gradually been implemented in lower secondary education, and these reflect the intentions of the Mathematics 12 to 16 curriculum. Textbooks vary in the way and the extent to which these curricular intentions are specified, and, of course, teachers vary in the extent to which they are able and willing to use modern textbooks as intended. Most textbooks for primary and lower secondary mathematics education are richly illustrated and offer possibilities for differentiation among high- and low-ability students.

Pedagogy

A new focus on contextualization, personal and social relevance, and problem solving changes the role of the mathematics teacher from that of transmitter of knowledge to one of facilitator of knowledge. Students, instead of being receivers of ready-made mathematics, are participants in a process of constructing and applying meaningful mathematical knowledge and tools. In addition to this profound shift in teaching behavior, curricular reform also involves the use of new, innovative curriculum materials, and a commitment to the aims and characteristics of this innovation from teachers. Change must occur in teaching practice along these three dimensions in order for it to be effective. As previously stated, it will take several years to implement such a complex, multidimensional reform. In the area of in-service education, there is a need for training in translating the philosophy and characteristics of realistic mathematics education to every day pedagogical concerns, including the use of innovative curriculum materials.

5 SCIENCE CURRICULUM AND PEDAGOGY

Goals for the Science Curriculum

The intended curriculum in science for primary schools is described in terms of core objectives. This curriculum has been in effect since August 1993 and describes the knowledge, insight, and skills to be acquired. The core objectives state that elementary science is aimed at teaching students to:
- develop insights into the physical world, to which human life is inextricably bound;
- gain pleasure in exploring nature with a critical and inquiry-oriented attitude;
- develop a responsible attitude toward a healthy living environment.

The core objectives for elementary science are spread across five domains: the human body, plants and animals, materials and phenomena in nature and technology, the environment, and science process skills. Specific skills include observing and measuring, cooperative learning, formulating questions, designing and executing a study, processing, and concluding.

Lower secondary science is split into combined physics and chemistry, biology, and geography and earth science. Basic characteristics of these new intended curricula are:
- the selection of science content appealing to students, the use of real-life contexts for learning science so that students apply science in a creative and meaningful way;
- an emphasis on activity-based student engagement, aimed at the acquisition of a rich variety of investigative, social, and communication skills such as conducting experiments, organizing and representing data, applying concepts, using the computer, and expressing an opinion;
- more coherence across science topics, and between science and other subjects including mathematics;
- about 30 percent of instructional time is designated for practical and field work.

Major Changes in the Science Curriculum

Changes in primary science education have to do with an approach characterized by:
- a broad and integrated conception of the content of elementary science, including biology, physics, chemistry, earth science, environmental, and technology education;
- the importance given to the immediate surroundings of the school as a source for science teaching and learning;
- an emphasis on the importance of sensory experiences and hands-on activities with concrete objects and organisms;
- a conception of the teacher's role as a facilitator of student learning rather than an instructor.

In some primary schools, science education is approached in a more traditional way than is proposed in the core objectives. The changes require teachers to use integrated education, learning by doing, learning by discovery and exploration, and extramural activities as part of a broad program for students to learn less from books and lectures and more from direct experience with environments, objects, and experiments.

Changes in emphasis in the new intended curricula for physics and chemistry, biology, and geography in basic education can be characterized as a shift in emphasis from a structural, receptive, and quantitative approach to a more context- and activity-based and qualitative approach. According to science education experts, this shift is important in promoting the formation of meaningful concepts, in facilitating the applicability of conceptions to realistic settings, and in making instruction more attractive and more motivating for students. However, from a descriptive study from 1993 on the context- and activity-based quality of actual teaching practice in physics, chemistry, and biology education, it appears there still is a considerable gap between the ideal curriculum and what actually happens in classrooms.

In upper secondary education, national examination programs for physics, chemistry, and biology as elective subjects in VWO and HAVO have been implemented over the past 10 years. Although contexts are described in some of the programs, they are still constructed around the traditional structure of the various disciplines. New elements in the physics program for VWO include biophysics, physical information, and the descriptive linkage between topics and context concepts. In 1995, the content part of biophysics (blood circulation) was moved to biology. A general complaint is that the physics and chemistry programs are overcrowded.

Calculators have increased in use in classrooms during the past 10 years. Computers have, as yet, limited usage in the classroom, although their use is encouraged. Electronic aids such as graphing calculators and computers are perceived as essential tools in upper secondary education including the school-leaving examination.

Current Issues in the Science Curriculum

Important issues in science education in primary schools include the following:
- The daily practice of science teaching in primary schools lacks a systematic and clear conceptual structure; topics are randomly selected from a huge pile of possibilities, and teachers do not guide students in a consistent way. Students have little op-

portunity to work with concrete materials.
- Assessment of science performance both nationally and within the classroom is given very little attention. This clearly has to do with the absence of a clear structure in science education. There is, however, a trend to include science in national assessment.
- In primary education language, reading, and mathematics are considered the most important subjects. There is little outside pressure on primary schools to give much attention to science, since secondary schools do not expect students to have attained a specific level in science.

Science education, particularly modern integrated science education, must be promoted within primary education; a clear structure of the knowledge and skills students should acquire must be developed. At the moment, technology and environmental education, which are related to science, are receiving a great deal of attention.

A substantial overall reform of the last two grades of HAVO and the last three grades of VWO is currently being prepared. The reform is expected to be implemented in 1998 and is aimed at better meeting the heterogeneous needs of students and at improving the transition from HAVO and VWO to higher education. It builds on the innovations currently taking place in lower secondary schools and entails:
- an emphasis on stimulating students' self-motivation, including independent learning and responsibility for their own learning;
- a strong focus on the acquisition of general skills, including learning to learn, problem-solving skills, investigative skills, and reporting skills;
- a shift towards a more facilitative teaching approach;
- an attempt to make content more relevant and useful;
- the replacement of the system of free composition by students of elective packages in upper VWO and HAVO by four prestructured profiles preparing for higher education.

Science Textbooks

The principal resource in science teaching is the standard textbook for each grade and, in secondary education, for each science discipline and track. Frequently, textbooks are supplemented by the use of teacher-produced materials and topic-specific booklets in primary education and teacher-made tests in secondary education. Textbooks are developed by teams of teachers and other science education experts and distributed by educational publishers. As in mathematics, there is an extensive supply of textbooks, but in each subject two or three textbook series are in widespread use.

Legislated core objectives and national examination programs are influential frames of reference for revising existing textbooks and developing new ones. As an illustration, in lower secondary education new textbooks implemented gradually since 1993 reflect the activity and context-based intentions of the core objectives for combined physics and chemistry, biology, and geography. Of course, teachers vary in the extent to which they are able and willing to use innovative textbooks as intended. Schools are free to choose the textbooks most suitable for their students and teachers, and each student has his or her own textbook.

Pedagogy

A shift in emphasis toward a more context and activity-based approach to science education and toward more independent learning by students demands a shift in the role of the teacher from that of transmitter of knowledge to one of facilitator of learning. Students, instead of being passive receivers of ready-made science, are participants in a contextually embedded process of constructing and applying meaningful scientific knowledge and tools. Curricular reform also involves the use of new, innovative curriculum materials as intended, and a real commitment from teachers to the aims and characteristics of this innovation. Change must occur in all three dimensions in order for it to be effective. As previously stated, it is expected that it will take several years to implement such a complex reform. In the area of in-service training, there is a need for educating teachers in translating the philosophy and characteristics of context- and activity-based science education to actual teaching practice. Innovative curriculum materials can make a valuable contribution to effective in-service training, and support in their use will be an important component of the implementation of the reform.

6 EVALUATION POLICIES AND PRACTICES

The National Institute for Educational Measurement (CITO) develops tests for standardized evaluation of student performance in primary education. Primary schools are free to join one or more of these assessments; for primary education, there are no centralized compulsory examinations or tests.

The majority of schools (71 percent in 1996) participate in a standardized test that is administered at the end of primary school. This test, developed by CITO, measures the performance of students in the last grade of primary education (age 12) and helps teachers, students, and parents in choosing the appropriate level of secondary education. The test includes about 250 multiple-choice problems on Dutch language, mathematics, and information processing. In 1996, a set of items on natural and cultural science, including areas such as geography, history, biology, physics, and social science were included for the first time. The test is administered by classroom teachers on three fixed days in February. The CITO test is the most widely used and most important standardized primary education test in the Netherlands; much educational research is based on student results.

In addition to the end-of-primary test, there are several other assessments performed by the CITO:

- The preparatory end-of-primary test, for students in Grade 7.[3] Sixty-three percent of schools participated in this test in 1995. The assessment determines whether students have mastered basic skills in Dutch language, mathematics, and information processing. Results are used by teachers to adapt instruction to achievement levels.

- The National Assessment of Educational Achievement (PPON), which is a periodic assessment of language and mathematics performed every five years at various levels in primary school. The purpose of the assessment is to inform education policy makers on the strengths and weaknesses of the system and to initiate discussions on future developments in education.

- The Student Monitoring System (LVS), which schools may participate in regularly in order to as-

---

[3] In TIMSS, the equivalent of the Dutch Grade 7 is Grade 5.

sess performance in mathematics and language, including reading and listening comprehension. The information is kept in a student tracking system.

The National Institute for Educational Measurement is also responsible for the development of tests administered at the end of basic education. This summative assessment is aimed at:

- providing diagnostic information to schools, teachers, and students about the extent to which schools have successfully taught basic education to students of various ability levels;
- providing monitoring information to the national government and test developers about the quality of basic education.

The assessment entails the administration of paper and pencil tests, including open-ended and multiple-choice items, laboratory work, and an investigative component. Schools must administer at least one of those tests per subject and code student answers using centrally provided coding rubrics. The first series of tests were administered in 1995, resulting in complaints from schools about the workload and the unclear status of the assessment and its results.

National examinations are administered in the final grade of each track. Examinations cover six to seven subjects. In all cases, Dutch, English, and French or German are compulsory. Mathematics and the science disciplines are electives. Results of these examinations are used as certification for further study or as a prerequisite for entry into certain occupations. All national examinations consist of two complementary parts: a central and a school examination. In the central examinations for mathematics and the various science disciplines, it is primarily theoretical knowledge that is assessed. In school examinations, the focus is on assessment of practical skills and on alternative forms of assessment, oral examinations, preparation of papers, practical tasks, and computer use. There is a trend to include predominantly contextualized open-ended questions (short answer and extended response) in central examination. Currently, at the central examination for physics and chemistry VWO, the use of a source-booklet with physics, chemistry, and biology-tables is allowed and assessed. Use of regular calculators is allowed, and it has been proposed that each student have a graphing calculator at his or her disposal during the central mathematics examination.

7 REFERENCES AND SOURCES FOR FURTHER READING

Common Sources

1   Husén T, N T Postlethwaite 1994 *International Encyclopedia of Education*. Pergamon Press, Oxford
2   Organisation for Economic Co-operation and Development 1995 *Education at a Glance: OECD Indicators*. Organisation for Economic Co-operation and Development, Paris
3   United Nations Educational, Scientific and Cultural Organization 1995 *Statistical Yearbook*. United Nations Educational, Scientific and Cultural Organization, Paris
4   World Bank 1995 *World Development Report 1995*. Oxford University Press, New York

Other Sources

5   Bleijerveld K 1993 *De natuur, waaronder biologie. Handreiking kerndoelen basisonderwijs*. SLO, Enschede
6   Berg E van den (in preparation) *The Effects of In-service Education on Curriculum Implementation*. University of Twente, Enschede
7   Bokhove J 1985 Een nadere beschouwing van de uitkomsten van het peilingsonderzoek rekenen. *School & Begeleiding*. 2: 4-7
8   CBS 1995 *Zakboek Onderwijsstatistieken 1994/1995*. Heerlen, CBS
9   CBS 1995 *Kwartaalschrift Onderwijsstatistieken 1995-II*. Herleen, CBS
10 CBS 1994 *Kwartaalschrift Onderwijsstatistieken 1994-II*. Herleen, CBS
11 CBS 1994 *Kwartaalschrift Onderwijsstatistieken 1994-III*. Herleen, CBS
12 Commissie Ontwikkeling Wiskundeonderwijs 1992 *Trajectenboek Wiskunde 12–16*. Rijksuniversiteit Utrecht/SLO, Utrecht/Enschede
13 Commissie Toekomst Leraarsschap 1993 *Het gedroomde koninkrijk*.
14 Dodde N L, J M G Leune 1995 *Het Nederlandse Schoolsysteem*. Wolters-Noordhoff, Groningen
15 Fullan M G 1991 *The New Meaning of Educational Change*. Cassell Educational Limited, London
16 Gravemeijer K, M van den Heuvel-Panhuizen, G

Donselaar, N Ruesing, L Streefland, W Vermeulen, E te Woerd 1993 *Methoden in het reken-wiskundeonderwijs, een rijke context voor vergelijkend onderzoek.* Freudenthal Instituut, Utrecht

17 Heuvel-Panhuizen M van den 1996 *Assessment and Realistic Mathematics Education.* CD-B Press, Centre for Science and Mathematics Education, Utrecht (thesis)

18 Howson G.A. 1995 *Mathematics Textbooks: A Comparative Study of Grade 8 Texts.* TIMSS Monograph No. 3. Pacific Educational Press, Vancouver

19 Inspecitie van het onderwijs 1995 *Groepsgrootte in het basisonderwijs.*

20 Kamer-Peters T 1992 *National Science Education. A Broad Outline.* SLO, Enschede

21 Kievit R.J. de, A L M Brouwers, H v d Meche, H Schalk, J Schipper 1991 *Op weg naar basisvorming. Natuur- en scheikunde.* SLO, Enschede

22 Kok D, M Meeder, M Wijers, J van Dormolen 1992 *Wiskunde 12–16: een boek voor docenten.* Freudenthal Instituut, Utrecht

23 Kraemer J M 1995 Beleidsvoorwaarden voor een voortgezette onderwijsontwikkeling. In: Dolk M (ed.) 1995 *Vijfentwintig jaar ontwikkeling reken-wiskundeonderwijs. Verleden, heden, toekomst.* NVORWO, Utrecht

24 Kremers E J J 1994 *Toetsen ter afsluiting van de basisvorming: een actueel overzicht.* Handboek Basisvorming, IV.2E

25 Kuiper W A J M 1993 *Curriculumvernieuwing en lespraktijk.* Universiteit Twente, Enschede (thesis).

26 Letschert J 1994 *Primary Education in the Netherlands.* SLO, Enschede

27 OECD (in preparation) *OECD Education Statistic.*

28 Process Management Basisvorming 1993 *Kerndoelen Basisvorming.* Info-reeks Basisvorming, deel 6. Almere

29 Rengerink J, H ter Heege 1993 *Rekenen/wiskunde. Handreiking kerndoelen basisonderwijs.* SLO, Enschede

30 Schuring H 1991 The Netherlands. In: Howson G A (ed.) 1991 *National Curricula in Mathematics.* The Mathematical Association, pp. 169-179

31 Timmermans P, H Schalk, B Waas 1992 *Op weg naar basisvorming. Biologie.* SLO, Enschede

32 Treffers A 1987 *Three Dimensions. A Model of Goal and Theory Description in Mathematics Instruction— the Wiskobas project.* Reidel Publishing Company, Dordrecht

33 Vakontwikkelgroep Algemene Natuurwetenschappen 1995 *Advies Examenprogramma Algemene Natuurwetenschappen HAVO/VWO.* SLO, Enschede

34 Vakontwikkelgroep Biologie, Natuurkunde en Scheikunde 1995 *Advies Examenprogramma Biologie, Natuurkunde en Scheikunde HAVO/VWO.* SLO, Enschede

35 Vakontwikkelgroep Wiskunde 1995 *Advies Examenprogramma Wiskunde HAVO/VWO.* SLO, Enschede

36 Vuyk E J 1994 The Netherlands: System of education. In: Husén T, N T Postlethwaite (eds.) 1994 *The International Encyclopedia of Education. Second Edition.* Pergamon Press, Oxford, pp.4067–4077

37 Voogt J M, W D J Vlas 1995 *Aansluiting HAVO-HBO: vereiste en gewenste kwalificaties.* OCTO, Enschede

38 Weerden J van (ed.) 1993 *Balans van het wereldoriëntatie-onderwijs aan het einde van de basisschool. Uitkomsten van de eerste peiling wereldoriëntatie einde basisonderwijs.* The National Institute for Educational Measurement, Arnhem

Statistical References

Section	Statistic	Reference	Page	Table	Year of Statistic
Country Profile	US$20,950	4	163	1	1993
Governance	27 %, 73 %, 30%, 24%, 19 %	14	57	2.6	1993–94
Structure of the System	8139	10	4	1	1993
Structure of the System	1008 548	11	4	1	1993
Structure of the System	Table 1 enrollment figures	9	9	1	1993–94
Schools in the System	class size 24	16	–	–	1994
Teacher Pofile	figures on age and gender	27	–	–	1993–94

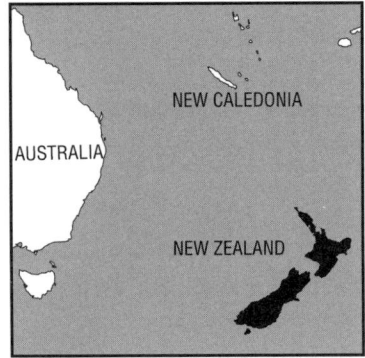

New Zealand

Robert A. Garden

1 COUNTRY PROFILE

New Zealand lies in the southern Pacific Ocean, 2000 km east of Australia. It is made up of the North and South Islands and a number of smaller islands, with a total land area of 268,000 km². The indigenous Maori people make up 13 percent of New Zealand's population of 3.5 million, with the remainder being principally of Caucasian descent. The first non-Maori to settle in New Zealand were from the United Kingdom, and for many years almost all immigrants were from there. Between 1986 and 1991, the total population increased by 4 percent, while the proportion in the 5- to 15-year-old population remained fairly constant. During the same period the proportion of Maori children in this age group increased from 19 percent to 20 percent, and the number of children of Pacific Island ethnic origin increased from 3 percent to 7 percent.

Although New Zealand's population density is relatively low, fewer than 13 persons/km², more than three-quarters of the people live in urban centers of more than 10,000. Immigration and emigration levels have kept pace in recent years, averaging between 1 and 2 percent of the total population respectively. In recent years, north Asia, the United Kingdom, and the Pacific Islands have contributed the largest numbers of permanent immigrants. Special provision has been made to cater to students who enter school without having learned English.

New Zealand is governed by a democratically elected parliament, the structure and processes of which are based on the Westminster system. Its economy is ranked as high income by UNESCO, but it had a low per capita GNP and per capita GDP during the 1980s. The average annual growth rate for gross domestic product from 1980 to 1991 was less than 1 percent, a figure that was one of the lowest among the high-income group of countries. New Zealand ranks among high-income economy countries in percent of total government expenditure allocated to education, 15 percent in 1993. Education expenditure in 1994 was 16 percent of net government expenditure and 6 percent of GDP. According to UNESCO data, New Zealand's literacy rate is greater than 95 percent.

2 THE EDUCATION SYSTEM

Governance and Decision Making

A reorganization of education administration in 1989 saw the functions of the former Department of Education distributed among several newly created agencies. A small Ministry of Education provides policy advice to the government, ensures the implementation of government policy, and oversees the use of resources allocated to education.

One of the agencies that carries out central functions is the Education Review Office. Its officers review school operations to ensure that schools are accountable for funds spent and are meeting the educational objectives set out in their charter. The New Zealand Qualifications Agency oversees the system of qualifications, certificates, and awards. In addition, it coordinates secondary school, academic, professional, and trade qualifications. The New Zealand Qualifications Agency prescribes criteria to be met for awards at senior secondary level, a fact that influences implemented curricula.

School administration is carried out by boards of trustees, the majority of whom are elected parent representatives. The role of the board of trustees includes developing a school charter that outlines goals and objectives and stipulates adherence to the national curriculum.

The intended curriculum is determined centrally by the Ministry of Education. However, schools do include locally determined curriculum material that reflects the special needs and resources of the community. At the elementary level, students and teachers are issued a centrally developed mathematics textbook series. There is no provision for a science textbook, but a large number of topic booklets are available. Secondary level textbooks are selected by heads of subject departments in collaboration with classroom teachers.

Teachers determine the pedagogical methods used in the classroom. They are guided by a department head with responsibility for providing leadership in the subject.

Structure of the System and Participation Rates

Structure of Education

The structure of the education system is shown in Figure 1, along with approximate participation rates for each level. Most students in urban centers begin primary education in Contributing Schools, which serve students from five years of age to Grade 5. Students progress to Intermediate Schools for Grades 6 and 7. Most rural schools are Full Primary Schools, enrolling students from entry through Grade 7. Virtually all primary schools are coeducational.

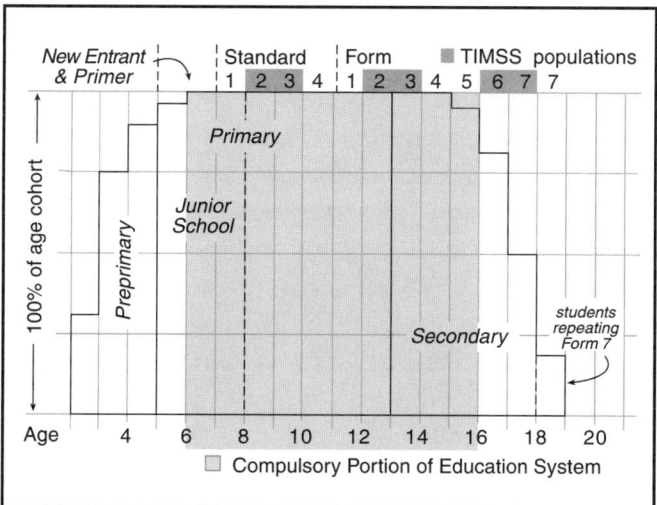

Figure 1: Structure of the Education System

Secondary education is offered in coeducational, comprehensive schools that serve Grades 8 to 12. Approximately 30 percent of students attend single-sex secondary schools. A small number of schools located in rural areas offer education at both primary and secondary levels within the same facility.

Special Education

Approximately 0.3 percent of students attend special schools for those who have visual, hearing, or other impairments that preclude their needs being met in a regular classroom. A further 0.4 percent are in special classes within regular schools.

Enrollment Rates

Children in New Zealand must attend school from the ages of 6 to 16, but almost all begin school on their fifth birthday. Virtually 100 percent of the cohort is enrolled in formal schooling during the compulsory years. In 1993, 52 percent of students in the age cohort were still in school midway through their Grade 12 year. Of the core tertiary age group, aged 18 to 24, more than 27 percent were enrolled in a formal tertiary education program, and others were participating in industry skills training.

Public and Private Systems

Most schools in New Zealand are part of the state system. These schools are funded by the central government according to nationally determined, publicly known formulae. However, there are a number of registered private schools that are run by religious or philosophical organizations, or by private individuals. Registered private schools receive reduced funding from the government and make up the balance through fees. Such schools must satisfy standards of accommodation and safety, and provide a satisfactory curriculum. In 1993, 3 percent of primary students and 5 percent of secondary students were educated in private schools. Most private schools are managed by boards of governors or trustees.

Mathematics and Science Programs

Mathematics, science, English, and social studies form a core of subjects that is compulsory for all students up to the end of Grade 9. From Grade 10 onward, students select from a wide range of subjects. Enrollment figures for mathematics drop in Grades 11 and 12; 71

percent of boys and 45 percent of girls are enrolled in mathematics. Science is taught as an integrated subject until the end of Grade 10, but physics, chemistry, and biology are also offered as separate subjects in Grade 10. A minority of students enrol in one or more of these separate subjects. Physics and chemistry enroll more boys in Grades 10, 11, and 12, while biology enrolls more girls in Grades 11 and 12. In recent years there has been strong encouragement given to girls to take mathematics and science in greater numbers, and teachers are reminded in the current mathematics and science curriculum guides to continue this encouragement. An increasingly high proportion of girls choose to study mathematics in the upper grades, and in chemistry the proportion of girls and boys is about the same. Substantial gender imbalances persist in physics and biology, with more boys favouring physics and more girls favoring biology.

Schools in the System

The School Year

Primary schools must be open for a minimum of 197 days, and secondary schools a minimum of 190 days each year. The school year begins the first week of February and ends the first or second week in December. The school year is divided into three terms of 12 to 14 weeks, with two- to three-week vacations between terms. The summer vacation between school years is of six to seven weeks' duration.

Instructional Times

Most primary and intermediate school students attend six-hour school days, five days per week, from Monday to Friday. The school day includes approximately one-and-a-half to two hours for lunch, room changes, assemblies, and other activities.

Class Size

The average class size in the primary system in 1994 was 33 students. Secondary mathematics and science classes at the Population 2 level average 29 students, but are smaller at senior levels.

Streaming and Tracking in Mathematics and Science

There is no official policy on within-school tracking, and schools are not tracked. Primary school classes generally include a wide ability range of students. In Grades 8 and 9 all students take mathematics and science; streaming takes the form of allocating students to three broad streams on the basis of achievement tests and student records from lower grades. Setting at these levels is not very common, but where it occurs it is usually for mathematics classes. Students in Grades 10, 11, and 12 choose from a wide range of options, and setting in mathematics and science sometimes occurs at these levels.

Teacher Certification

The Teacher Registration Board certifies successful candidates and administers regulations designed to ensure a quality teaching force. Teachers of primary and secondary schools must graduate from a college of education and complete two years of satisfactory teaching prior to being registered.

Training for primary teaching requires three years' full-time study at a College of Education, leading to a diploma in education. All colleges of education have affiliated college and university programs culminating in a university-awarded bachelor's degree for primary education trainees. Not all primary teachers have university degrees, but there is a strong trend towards this qualification. Training for secondary school teachers entails a minimum of four years of training, including university or polytechnic and a teacher training program. Most secondary school teachers complete at least a bachelor's degree in a discipline, followed by one year of full-time study in a college of education.

All schools offer a professional development program for teachers; these usually consist of ongoing guidance from principals, department heads, and subject specialists, training on "teacher only" days, and courses taken on a voluntary basis during vacations. Government-funded courses, many of which are designed to assist in the implementation of new curricula are also offered. Teachers are not required to upgrade or retrain unless they leave the teaching service for more than five years and wish to reenter the profession. Any teacher holding a diploma of education and wishing to upgrade may undertake study through the Advanced Study for Teachers Unit for the Higher and Advanced Diplomas of Teaching. Each level involves satisfactory completion of seven papers in a structure similar to that of a university degree. There is particular emphasis on professional knowledge and development, but specialist subjects may be included.

In addition to regular training programs, specialist training in particular areas of the teaching service is provided. Successful applicants are granted leave with pay, usually for one year, to complete additional qualifications in specialist areas. These include guidance, bilingual teaching, special education, workshop technology, home economics, and reading recovery. Specialist secondary teacher training in individual subject areas is provided from time to time when teacher shortages occur for particular subjects.

Teacher Profile

Teachers in New Zealand enjoy a relatively high socioeconomic status. On the Elley-Irving Socio-Economic Status Index (1985), secondary teachers are ranked in the first and primary teachers in the second of six categories of occupation. The Elley-Irving scale is in common use in New Zealand, and the categories are based on salary and educational levels.

The average age of regular primary school teachers in 1991 was 42 years, with women averaging two years younger in age than men. Forty-two was also the average age of regular secondary teachers, for both women and men. Most regular primary teachers are women (73 percent in 1991), while men are in the majority (55 percent) in secondary education. In general, primary teachers teach all subjects in the curriculum.

In secondary schools, teachers of mathematics are predominantly men. In 1981, two-thirds of mathematics teachers in the first secondary year and three-quarters in the final secondary year were men; it is believed that there has been an increase in the proportion of women teaching mathematics since this time. No comparable figures are available for secondary science, but while the proportion of female teachers of this subject appears to be higher than in mathematics, males are still in the majority. Physics graduates are more likely to be male, and biology graduates female. Proportions of primary and secondary teachers leaving the profession annually over the period 1992–93 to 1994–95 either permanently or for at least one year were 13 percent and 11 percent respectively.

3 TIMSS POPULATIONS

New Zealand, IEA, and TIMSS

New Zealand has been a member of IEA since 1968. Prior to joining TIMSS, New Zealand participated in the Six-Subject Survey, the Second International Mathematics Study, the Computer Education Study, the Written Composition Study, and the Reading Literacy Study.

New Zealand TIMSS Populations

Population 1 includes all students enrolled in Grades 3 and 4. These grades contain the largest proportion of students in the nine-year-old age cohort at the time of testing. At the midpoint of the testing year, 99 percent of nine-year-olds were in one of these grades.

Population 2 includes all students enrolled in Grades 7 and 8 in regular schools. These grades contain the largest proportion of students in the age thirteen cohort at the time of testing. This includes all students in Grades 7 and 8 in regular schools. At the midpoint of the testing year, 86 percent of thirteen-year-olds were in one of these grades. Less than 1 percent of nine-year-olds or thirteen-year-olds were excluded because they were in special schools, or receiving schooling at home.

Population 3 includes all students in their final year of secondary education. Since there are effectively two final years of secondary education in New Zealand, the sample was drawn from both Grade 11 and Grade 12. Mathematics and science literacy were the only areas assessed. An estimated 80 percent of the age cohort remained in Grade 11 and 53 percent of the age cohort remained in Grade 12 at the time of testing.

4 MATHEMATICS CURRICULUM AND PEDAGOGY

Goals for the Mathematics Curriculum

The intended curriculum in mathematics for New Zealand schools is described in Ministry of Education

(1992). This publication advises teachers on aims, content, teaching approaches, and methods of assessment. The curriculum statement was prepared by committees that included teachers from schools and universities, and curriculum experts from the Ministry of Education, and widely reviewed before adoption. The process of introduction of this new curriculum had not been completed at the time of TIMSS data collection, but its content was well known to mathematics teachers and was beginning to have an effect on classroom practice.

Content to be taught is incorporated in achievement objectives, and these are accompanied by suggested learning experiences designed to achieve the objectives and suggested methods of assessment. This document is not prescriptive, and teachers are encouraged to select from the suggestions offered, to incorporate material from a wide range of resources, and to adapt content and methods to the characteristics of particular classes or individuals. Nevertheless, the learning outcomes, teaching methodology, and modes of assessment officially favoured are obvious.

The mathematics curriculum is organized into six integrated learning "strands" and teachers are encouraged to make connections between these strands. Each strand is divided into eight "levels of achievement," defined by learning objectives.

The curriculum guide states that the general aims of mathematics education are to:

- help students to develop a belief in the value of mathematics and its usefulness to them, to nurture confidence in their own mathematical ability, to foster a sense of personal achievement, and to encourage a continuing and creative interest in mathematics;
- develop in students the skills, concepts, understanding, and attitudes that will enable them to cope confidently with the mathematics of everyday life;
- help students to develop a variety of approaches to solving problems involving mathematics, and to develop the ability to think and reason logically;
- help students to achieve the mathematical and statistical literacy needed in a society that is technologically oriented and information rich;
- provide students with the mathematical tools, skills, understanding, and attitudes they will require in the world of work;
- provide a foundation for those students who may continue studies in mathematics or other learning areas where mathematical concepts are central;
- help to foster and develop mathematical talent.

Major Changes in the Mathematics Curriculum

The new official curriculum statement resulted from a broad review and redesign of school curricula initiated by the Minister of Education in 1991. The design gave priority to improving achievement in mathematics and science. The statements replaced official syllabi for different grade levels, some of which had been in use for over a decade. Changes in emphasis in the curriculum have come about through a desire to make mathematics accessible and interesting for all students, and the influence of research into teaching and learning. There is less emphasis on specific topic content to be taught and more on learning outcomes to be achieved by students.

Increasingly, teachers are making use of new technologies for instruction in mathematics. Use of calculators and computers is encouraged and is widespread. Social considerations now play an important part in the curriculum, but it is intended that this complement and enhance academic achievement, not replace it.

Special attention is paid in the new curriculum to inclusion of aspects of the culture of the indigenous Maori people, and to making it gender inclusive. Cooperative learning is advocated, as is utilization of the experiences and interests of girls and minority groups in planning programs and lessons.

The curriculum statement encourages integration within and between topics. Primary school programs have traditionally incorporated a greater degree of between-subject integration than have secondary programs. Secondary teachers are subject specialists and usually teach a given class for only one subject; successful integration depends on a high level of cooperation between teachers.

Current Issues in the Mathematics Curriculum

Changes in the intended curriculum have more to do with approaches to teaching and learning than with changes in content. Concurrently, new methods of assessment have been encouraged. Teachers at all levels report a major increase in workload as a result.

One of the most problematic issues for classroom teachers of mathematics in the senior grades of secondary schools is the difficulty in reconciling the aims of the intended curriculum with the assessment requirements for student accreditation. The new curriculum emphasizes mathematical processes and recommends assessment methods to accommodate this. Accreditation requires students to demonstrate competence in an extensive series of subtopic areas; the most efficient assessment takes little account of process.

More attention than in the past is now being paid to finding methods of meeting the needs of gifted mathematics students.

Mathematics Textbooks

The intended curriculum recommends the use of textbooks as a resource rather than as the major determinant of what and how students should learn. In primary and intermediate schools the main resource is a standard textbook for each grade level, but this is frequently supplemented by the use of topic booklets and teacher-produced materials. The issued texts are produced by a central resource-producing agency under contract to the Ministry of Education. At secondary school level there is an increasing tendency for teachers to choose from a number of textbooks for specific topics, allowing them to utilize the approach most suitable for their classes.

Textbooks, which are issued free to secondary school students, are selected by heads of subject departments in each school. Of those in common use in mathematics, content ranges from reliance on practice exercises with minimal explanation and worked examples, to extensive explanations intended by the authors to be read and understood by students, with less emphasis on practice. In mathematics and science textbooks there is now more emphasis on activities and applications than previously.

Newer textbooks also tend to be more attractive in appearance. Use of color, illustrative photographs and sketches, better layout, and less dense print make them more interesting to students.

Expectations about how textbooks will be used vary from teacher to teacher and from class to class. Choice of textbook for a class depends on whether the teacher wishes it to be used largely as a source of exercises or activities, or whether a major part of what students learn

is to result from their studying the textbook and using it as a reference. At senior levels, the latter is more likely.

Pedagogy

The curriculum guide advises teachers that "A balanced mathematical program includes concept learning, developing and maintaining skills, and learning to tackle applications." The use of real-life problems with a problem-solving approach is advocated to assist students in learning mathematical thinking through the application of concepts and skills in meaningful contexts. Learning to communicate about and through mathematics, orally as well as in writing, is seen as an important part of becoming a mathematical problem solver.

Classroom teaching is inductive at primary and lower secondary grades, with deductive work forming an increasing component of learning in the senior grades. The use of apparatus to facilitate understanding of mathematical concepts has been a feature of mathematics teaching in New Zealand for many years, but the most recent curriculum statement also emphasizes the desirability of using apparatus to model problem situations. Teachers are urged to have students carry out investigations, and attention is drawn to the open-ended nature of many problems.

The new curriculum statements strongly encourage integration of topics within and between topics. Primary school programs have traditionally incorporated a greater degree of between-subject integration than have secondary programs. Secondary teachers are subject specialists and usually teach a given class for only one subject, so successful integration depends on a high level of cooperation between teachers.

5 SCIENCE CURRICULUM AND PEDAGOGY

Goals for the Science Curriculum

The curriculum guide *Science in the New Zealand Curriculum* (1993) lists twelve general aims of science education in the New Zealand curriculum. These include:
- helping students to develop knowledge and a coherent understanding of the living, physical, mate-

rial, and technological components of their environment;

- encouraging students to develop skills for investigating . . . in scientific ways;
- providing opportunities for students to develop the attitudes on which scientific investigation depends;
- promoting science as an activity that is carried out by all people as part of their everyday lives;
- portraying science as both a process and a set of ideas that have been constructed by people to explain everyday and unfamiliar phenomena;
- encouraging students to consider the ways in which people have used scientific knowledge and methods to meet particular needs;
- developing students' understanding of the evolving nature of science and technology;
- assisting students in using scientific knowledge and skills to make decisions about the usefulness and worth of ideas;
- helping students to explore issues and to make responsible and considered decisions about the use of science and technology in the environment;
- developing students' understanding of the different ways people influence, or are influenced by, science and technology;
- nurturing scientific talent to ensure a future scientific community;
- developing students' interest in and understanding of the knowledge and processes of science that form the basis of many of their future careers.

Major Changes in the Science Curriculum

Redevelopment of the science curriculum occurred alongside that of the mathematics curriculum as part of the broad overhaul of primary and secondary school curricula. The new curriculum statement replaces three syllabi in science for specific levels of the system. The syllabus for Grades 7 and 8 had been in use for three decades.

Features of the curriculum include provision for continuity of learning in science throughout schooling, the use of varied assessment techniques to support learning, and the setting of content to be learned in real life or familiar contexts. There is more emphasis on technology than in the past, both for its own sake and as a vehicle for scientific understanding.

Considerable thought has gone into the production of a curriculum guide that will lead to an inclusive curriculum in science. The guide states that " . . . the curriculum in science should recognize, respect, and respond to the educational needs, experiences, achievements, and perspectives of all students: both female and male; of all races and ethnic groups; and of differing abilities and disabilities."

Current Issues in the Science Curriculum

Science education is seen as essential for all students, whether as a preparation for high-level studies in science, for career purposes, or as a necessary grounding for effective living and citizenship. The curriculum statement recognizes the need for a sound knowledge and understanding of concepts, but also reflects the fact that New Zealand's trading networks consist of countries with rapidly increasing scientific and technological capacities. There is a strong desire to raise achievement in science and to broaden the base of expertise in technology.

Comments in Section 4 on current issues in mathematics curricula also apply to science curricula. Cooperative learning, extensive use of open-ended investigations, real-life contextualization of content, and greater emphasis on oral and written communication are all highlighted in the new curriculum.

Curriculum changes in science, still in the process of being introduced when TIMSS achievement data were being collected, have not been easy for teachers. While teachers generally support the new approaches, the combination of adapting to a restructured curriculum, making the curriculum inclusive, and beginning to use less familiar assessment methods has required a great deal of effort.

Science Textbooks

Textbooks are more often used to support the classroom program than as a basis for the program. The use of multiple print resources, including topic books and articles, in addition to one or more set textbooks is the mode.

Modern resources tend to be more appropriate to the developmental levels and learning needs of students. They are more readable, colourful, pictorial and have less dense text. Links between science and technology are more explicit.

Secondary school science students are issued with free textbooks chosen by department heads, usually in consultation with other science teachers. A number of science textbooks are produced in New Zealand, with most written to conform to the intended curriculum. Many textbooks produced in other countries are also on the market. Schools are free to choose the texts they deem most suitable for their students and teachers. It is unusual for primary schools to issue science texts, especially to Population 1 students, and teacher-produced materials and topic-specific booklets are common.

Pedagogy

Science in the *New Zealand Curriculum* (1993) states "Science is both a process of inquiry and a body of knowledge; it is an integrated discipline. The development of scientific skills and attitudes is inextricably linked to the development of ideas in science." The curriculum thus seeks a balance in teaching approach among knowledge and skill acquisition, development of conceptual understanding, and development of scientific attitudes. Integration of topics is seen as important and the curriculum is structured to facilitate this.

Teachers are advised to consider different rates and styles of learning among students. Cooperative learning is advocated both for its intrinsic value and to take advantage of the particular skills and preferences of girls and some ethnic minorities. Students are expected to communicate about science methods, results, and concepts to a much greater extent than in the past. They are to have the opportunity to do this in relation to open-ended projects and investigations, a prominent feature of the new curriculum. Laboratory work continues to underpin the science curriculum.

6 EVALUATION POLICIES AND PRACTICES

National Examinations

New Zealand sets two major national external examinations. The School Certificate Examination is administered at the end of Grade 10, which is close to the

minimum leaving age of 16 years for most students. The Bursary and Scholarship Examination is administered at the end of Grade 12. The examinations are optional, and are designed to gauge student attainment in each subject. Most students attempt five subjects. Results from the examinations may be used for several purposes: as a prerequisite for entry to higher level classes in those subjects, or for entry into certain occupations, or, in the case of the examination at Grade 12, for supplementary awards for study at a university.

The School Certificate examination incorporates an internal assessment component. Examination questions include free-response (extended and short answer) and multiple-choice questions. The instruments sample all that is taught in each subject, including practical work in the case of some subjects such as art and workshop technology. In subjects for which practical work forms an essential component, the school principal must declare that a student has completed the required practical work in order for the student to be permitted to write the examination.

The examinations undoubtedly affect the curriculum at lower levels. However, because those responsible for curriculum policy and those responsible for assessment policy have worked cooperatively, negative effects have been minimized and positive effects (such as accelerating curriculum implementation) maximized.

A National Qualifications Framework was introduced in 1995. The framework is structured in eight levels and encompasses all qualifications, including degrees. It identifies the units of learning necessary for particular qualifications, and establishes assessment standards for nationally recognized qualifications. One such qualification will be the National Certificate; attainments at Levels 1, 2, and 3 will be based on the Grades 10, 11, and 12 curricula respectively.

By 1998, schools will make annual reports to the community on improvements in curriculum delivery. Periodic national achievement testing of samples of 8- and 12-year-olds will be used to monitor the effectiveness of the system.

School District Examinations

There is no administrative level in New Zealand that corresponds with a school district. However, there are regional associations of subject-matter teachers that administer assessments at Grade 10 level to lower-achieving students in their subject. Assessments in mathematics consist largely of computations and simple applications of mathematics to real-life problems. Students receive a certificate that is informally recognized for entry to some occupations.

Assessment at the Classroom Level

Formative assessment in the classroom fulfills a variety of purposes: to provide feedback to students, to diagnose weaknesses in the teaching program, and to aid in the planning of future lessons. Summative assessment is used to report student progress to parents and to monitor long-term progress. Classroom assessment is seldom excessive or obtrusive and, in general, fulfills its intended purposes.

As a result of new approaches in the mathematics and science curricula, a wide variety of assessment procedures are now used at the classroom level. Teachers are becoming more conscious of the need to use multiple assessment techniques, including oral, demonstration, and group activities to complement more traditional written tests. The instruments used are almost exclusively teacher-made tests, supplemented by ongoing observation of student performance.

7 REFERENCES AND SOURCES FOR FURTHER READING

Common Sources

1   Husén T, N T Postlethwaite 1994 *International Encyclopedia of Education*. Pergamon Press, Oxford
2   Organisation for Economic Co-operation and Development 1995 *Education at a Glance: OECD Indicators*. Organisation for Economic Co-operation and Development, Paris
3   United Nations Educational, Scientific and Cultural Organization 1995 *Statistical Yearbook*. United Nations Educational, Scientific and Cultural Organization, Paris
4   World Bank 1996 *World Development Report*. Oxford University Press, New York

Other Sources

5   Clark M 1994 *Curriculum Development and Achievement Standards in Mathematics Education in the Asia-Pacific Region*. APEC Human Resources Development Working Group, Wellington, New Zealand
6   Ministry of Education 1994 *Better Beginnings: Early Childhood Education in New Zealand*. Learning Media Ltd., Wellington, New Zealand
7   Ministry of Education 1993 *Three Years On: The New Zealand Education Reforms 1989 to 1992*. Learning Media Ltd., Wellington, New Zealand
8   Wagemaker H (ed.) 1993 *Achievement in Reading Literacy: New Zealand's Performance in a National and International Context*. Ph.D. Research Section, Ministry of Education, Wellington, New Zealand
9   Ministry of Education 1994 *Education Statistics of New Zealand, 1994* Wellington, New Zealand
10 Robitaille D F, R Garden (eds.) 1989 *The IEA Study of Mathematics II: Contexts and Outcomes of School Mathematics*. Pergamon Press, Oxford
11 Evans J (ed.) 1994 *New Zealand Official Yearbook 1994*. Statistics New Zealand, Wellington, New Zealand
12 *Demographic Trends 1994*. Population and Demography Division, Statistics New Zealand, Wellington, New Zealand
13 *Mathematics in New Zealand Curriculum*. 1992 Learning Media Ltd., Wellington, New Zealand
14 *Science in New Zealand Curriculum*. 1993 Learning Media Ltd., Wellington, New Zealand

Statistical References

Section	Statistic	Reference	Page	Table	Year of Statistic
Country Profile	13%, 4%	5	96	4.15	1991
Country Profile	3.5 million	6	116–117	5.7	1994
Country Profile	19% to 20%	5	132	5.3	1991
Country Profile	3% to 7%	5	140	5.34	1991
Country Profile	>10,000	7	3	–	1994
Country Profile	1% and 2%	8	89	5.7	1994
Country Profile	high income GNP, GDP, <1%	4	239	–	1991
Country Profile	15%	4	181	10	1993
Country Profile	16%, 6%	9	14	4	1994
Country Profile	95%	4	163	–	1991
Structure of the System	0.3%	9	43	23	1993
Structure of the System	0.4%	9	44	24	1993
Structure of the System	100%	9	10	1	1993
Structure of the System	52%	9	32	–	1993
Structure of the System	27%	9	64	–	1993
Structure of the System	3%, 5%	10	1	–	1993
Structure of the System	math/science enrollment	9	54	31	1993
Teacher Profile	Elley-Irving rankings	13	25–26	–	1976
Teacher Profile	42 years	9	60	36	1983
Teacher Profile	two-thirds	14	42	Fig. 3.2	1981
Teacher Profile	13%, 11%	15	28	1.1, 1.2	1995
TIMSS Populations	<1%	9	43	23	1993
TIMSS Populations	80%, 53%, estimates based on July 1 stats	9	39	20	1983

Norway

Gard Brekke, Telemark Educational Research

Marit Kjærnsli, University of Oslo

Svein Lie, University of Oslo

1 COUNTRY PROFILE

Norway is situated in northern Europe and shares borders with Sweden, Finland, and Russia. Norway's total land area is 323,895 km^2, stretching 2540 km from north to south. Several thousand islands are scattered along its coast, which is deeply indented by fjords, while mountains and plateaus cover most of the interior.

The country has 4.3 million inhabitants, with a population density of 13 persons/km^2. Nearly half the population lives in the southeastern part of the country. The average annual growth of the population was 0.4 percent from 1983 to 1990. Forty-five percent of the population lives in urban centers of more than 10,000.

In addition to the Norwegian-speaking people, there are approximately 30,000 Sami, who constitute an ethnic minority with their own culture and language. In recent years there has been a considerable increase in the number of immigrants, including 27,000 regular immigrants as well as 13,000 refugees in 1992.

The Norwegian language has two official forms, *bokmål* and *nynorsk*, which differ somewhat in grammar and spelling. Children learn both forms in school. The Sami minority may choose whether to be instructed in Sami or Norwegian.

Norway was united around 900 AD. From 1380 to 1814, the country was under Danish rule, after which it entered a union with Sweden that continued until 1905. Norway is a constitutional monarchy governed by a parliamentary democracy, with a constitution dating from 1814. Executive power is formally vested in the king, and is implemented by the government. The assembly, with 165 seats, is elected every four years.

Norway's economy is ranked as high income by the World Bank, with a per capita gross national product of US$25,970 in 1993. The average annual growth rate of the GNP was slightly over 2 percent from 1980 to 1991, which is the average for high-income countries. The average annual rate of inflation was 5 percent for this period. More than 9 percent of total governmental expenditure was allocated to education in 1991. According to UNESCO, Norway's literacy rate is greater than 95 percent.

2 THE EDUCATION SYSTEM

Governance and Decision Making

The Norwegian school system consists of nine years of compulsory schooling, known as *Grunnskole,* and a one-, two- or three-year upper secondary school known as *Videregående skole.* Both stages are currently undergoing reform, the nature of which is explained later in this section and in Section 5.

The compulsory public school system is governed by municipalities according to national laws and regulations established by the Ministry of Education. These regulations include curriculum guidelines, teacher education regulations, and the number of lessons per subject per year. Salaries for teachers are determined by centralized negotiations with teacher organizations. In addition to the public system, there are some private schools, but fewer than 2 percent of school-aged children are enrolled in them.

Upper secondary schools, by contrast, are owned and administered by Norway's 19 counties and are governed according to the Upper Secondary Education Act. A substantial proportion of these schools, about 18 percent, are privately owned and enroll approximately 5 percent of the age cohort.

Although curriculum guidelines are centralized for both levels, schools have a substantial degree of freedom within these guidelines. Particularly in the compulsory grades, schools are able to set local content goals for each grade level. Classroom instruction is decided by teachers within the general framework given in curriculum guidelines.

The selection of textbooks is performed by the school's governing board, but teachers have a great deal of influence on the decision. For compulsory schooling, textbooks are provided to students by the schools.

Structure of the System and Participation Rates

Structure of Education

Compulsory schooling in Norway covers Grades 1 to 9 for ages 7 to 16. Parliament has recently decided to lower the entrance age to 6 years, which will extend the compulsory range to 10 years from 9. The current nine-year compulsory school is divided in a primary stage for Grades 1 to 6 called *Barnetrinnet*, and a lower secondary stage for Grades 7 to 9 called *Ungdomstrinnet*. The two stages are usually in separate schools, but approximately one-fifth of schools are combined. The current structure of the education system is shown in Figure 1, along with approximate enrollment rates for each level.

There is no streaming or tracking according to gender or ability during compulsory schooling, and almost no pupils repeat grades. Furthermore, there is a high degree of integration of pupils with physical and mental disabilities into ordinary classrooms.

Participation Rates

Enrollment at the compulsory stage is virtually 100 percent, and 95 percent of students move directly into one of the many areas of study in upper secondary school. On average, 87 percent of the 16- to 18-year-old age cohort attends upper secondary school. This figure varies between 79 and 91 percent for the 19 Norwegian counties: the lowest figures are for Finnmark in the far north and for the city of Oslo, due to the high number of students for whom Norwegian is a second language.

The upper secondary school system has up to now been complicated with a varied set of offerings ranging from general schooling to vocational areas of study with special one-, two-, and three-year programs for more than 200 vocational areas. Beginning in 1994, a simple, comprehensive system for upper secondary school was introduced. The number of programs was reduced considerably and all offerings were organized as three-year courses. At the same time, the government passed a law entitling young people to three years of free, upper secondary education.

Beginning in 1997, the school entrance age will be lowered to 6, increasing the years of primary schooling from 6 to 7. Norway will within a few years, therefore, offer 13 years of free schooling from primary at age 6 to the end of upper secondary at age 19. Studies are presently being conducted to examine the effects of the late school entrance age. One of Norway's reasons for participating in Population 1 is to examine the interplay between age and years of schooling for these grades. The arguments for earlier school entrance thus far seem to concentrate on parents' convenience, rather than on children's development.

Mathematics and Science Programs

Mathematics and science are compulsory for all students up to the end of Grade 9. Table 1 shows the enrollment rates for students beyond the compulsory years during the school year 1993. Due to the large number of optional courses, the figures are estimates only. In particular, the choice of which courses to include in "science" was made arbitrarily. However, in spite of these estimates, it is clear that for each year

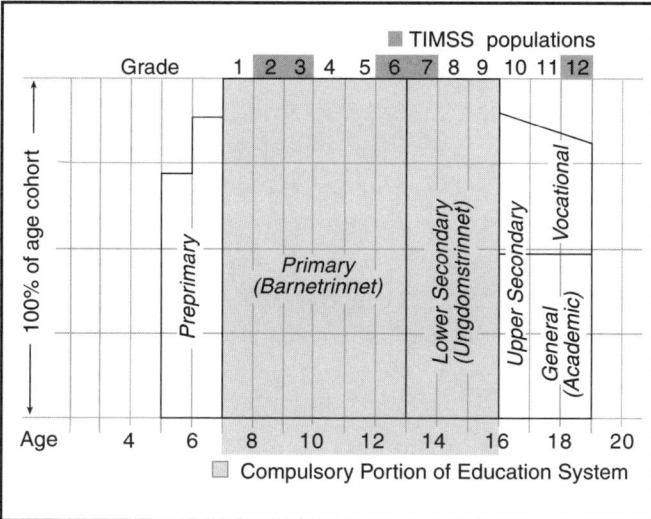

Figure 1: Structure of the Education System

Table 1: Postcompulsory Mathematics and Science Enrollment

Mathematics and Science Enrollment by Grade and Gender
(Percent of Grade Cohort)

Grade	Modal Age	Mathematics		Science	
		M	F	M	F
10	16	90	80	80	75
11	17	40	25	40	30
12	18	25	15	35	20

removed from the compulsory courses, there are significantly fewer students taking mathematics and science. Moreover, there are fewer females than males taking noncompulsory mathematics and science courses at all levels.

Science and mathematics enrollment in upper secondary schools is expected to increase due to the 1994 reform. Overall enrollment is expected to increase as all programs are changed in three-year courses. Science and mathematics are expected to play a more important role in all programs. As a result of the 1997 reform, more time will be devoted to science instruction beginning in the earlier grades and continuing throughout schooling. More generally, the interplay between national scores and instructional time (per year and total) is of great interest.

Schools in the System

The School Year

The school year runs from mid-August to mid-June and includes 190 instructional days. Most schools have holidays of a week or more in late autumn, at Christmas, in February, and at Easter. The school week is Monday to Friday, and each school day is approximately six hours long. The number of lessons per week starts at about 20 in the first year, increasing to 30 in secondary school, with each lesson being approximately 45 minutes long.

Class Time Allocated to Mathematics and Science

The proportion of class time per week devoted to mathematics decreases from 16 percent at the primary level

to 12 percent at lower secondary. The corresponding percentages for science are 7 percent, estimated by counting half of the time devoted to *O-fag* covering natural and social sciences, and 9 percent.

Class Size

Class size in compulsory schools can be as low as 19 students, particularly in small schools in the sparsely populated countryside. The maximum allowed class size is 29 at this level. At the upper secondary stage the maximum class size varies with area of study, from 30 students for the academic branch down to 15 and lower in some technical branches.

Streaming and Tracking

There is no tracking or streaming of pupils in the Norwegian compulsory school, Grades 1 to 9. There is a strong policy favoring integration of pupils with all types of disabilities into ordinary classes. This is a primary reason for the low pupil-teacher ratio of six to one in Norwegian schools. There are virtually no enhanced or accelerated programs for talented pupils in mathematics or science.

Certification of Teachers of Mathematics and Science

Teachers are certified by one of two routes: by a four-year (until 1992, a three-year) education from a college of education, or by a subject-matter bachelor's or master's degree with an additional year (until 1994, six months) of practical and pedagogical preparation. The first route leads to certification for teaching Grades 1 to 9 and some vocational fields of upper secondary school. The second route leads to certification for teaching Grades 7 to 9 and upper secondary school. Science and mathematics teachers at the lower secondary stage, Grades 7 to 9, are about evenly divided between the two routes.

In the compulsory school system there is no formal requirement in subject-matter background. Teachers at the primary level normally teach all subjects, and there are no science or mathematics specialist teachers at this level. Lower secondary school teachers are educated by either of the two routes described above. Those educated at a college of education may focus on mathematics or science by selecting either a half or full year of courses at the college. Teachers educated at a university may specialize in science or mathematics by taking a degree in one of these subjects.

Teachers at the upper secondary level are usually specialists. Mathematics and science teaching in the vocational division has up to now been done in a practical context, relating to a particular profession. The current reform will, however, introduce more general science and mathematics, thus creating a need for teachers with more knowledge of general and theoretical content.

Most teachers in the academic division, including teachers of science and mathematics, are highly specialized. Most of these teachers are designated *lektor*, meaning they have a master's degree and have taken practical pedagogical preparation. A majority of those teaching the physics specialist courses have a master's degree in physics, and usually also teach mathematics or chemistry.

There are no formal in-service or upgrading requirements for teachers at any level. However, school-based development work and innovation is usually included in the study and planning days, which constitute a total of one week of work per year. In addition, there is a national institution, called *Statens Lærerkurs,* that organizes and finances in-service courses taken on a voluntary basis during vacation periods, or, depending on permission to leave school, within the school year.

As a consequence of the extensive reforms at all levels of schooling, reforms to the teacher education process are also scheduled to take place within a few years. Due to the increased emphasis on science and information technology in schools, it is expected that this will also be reflected in teacher education programs.

Teacher Profile

The economic status of teachers in Norway is moderate, with annual salaries (before tax) ranging from US$26,000 for a new teacher from a college of education up to US$38,000 for a highly experienced *lektor* educated at a university. Compared to other professions, particularly in the private sector, salaries have decreased steadily over the past 30 years. The teaching profession, however, is still attractive to young people, offering a secure job in a time of relatively high unemployment.

The average age of full-time teachers is about 45 years. About two-thirds of all teachers in Grades 1 to 9

are female. However, due to the high proportion of part-time teachers among females, both genders are about equally represented among full-time teachers. Primary school teaching, including science and mathematics, is a predominantly female activity, particularly at the lower grades: 80 percent of Grade 2 teachers are women compared to 60 percent for Grade 6. At the lower secondary level, women represent 53 percent of mathematics teachers and 30 percent of science teachers.

At upper secondary level, 60 percent of all teachers (and close to 70 percent of full-time teachers) are men. Male dominance among mathematics and science teachers is more pronounced, particularly in physics and technology-oriented science courses. Ninety percent of specialist physics teachers are men.

3 TIMSS POPULATIONS

Norway, IEA, and TIMSS

Norway has been member of IEA since 1983. Prior to joining TIMSS, Norway participated in the Second International Science Study and the Reading Literacy Study. Norway has also joined the Foreign Language Study.

Norwegian TIMSS Populations

Norway is participating in TIMSS at all population levels. In Norway, Population 1 includes Grades 2 and 3. Population 2 includes Grades 6, which is the last year of primary school, and Grade 7 which is the first year of lower secondary school. At these grade levels, there is no streaming or tracking, and classes are kept together for all subjects.

At the Population 3 level, Norway is participating in the Mathematics and Science Literacy Study, with Grade 12 students from nine areas of study. Norway is also taking part in the study of physics specialists; all participants are enrolled in the academic branch of secondary school.

4 MATHEMATICS CURRICULUM AND PEDAGOGY

Goals for the Mathematics Curriculum

The most recent curriculum guide for compulsory schooling in Norway was issued in 1987 and is known as the *Ministry of Education and Research 1987*, or M87. It lays down a binding framework for the program of study and presents the subject matter that all pupils are required to study at their own ability level. For the years of compulsory education, municipal education committees, together with schools and teachers, may strongly influence the content to be taught within the given framework. The curriculum guide takes for granted that detailed content and specific applications of mathematics will be decided locally, through the preparation of local curricula and work plans.

The mathematics curriculum guides for upper secondary schools date from 1985. Secondary school teachers are required to follow the curriculum plans issued by the government, while compulsory school teachers have some flexibility in curriculum planning.

The official goals for mathematics teaching in compulsory school are to:
- teach students the fundamental topics and methods in mathematics, according to their abilities;
- develop students' knowledge and skills, enabling them to regard mathematics as a useful tool for solving problems in everyday life and at work;
- train students to think logically and to work systematically and accurately;
- train students to work through and evaluate data for themselves;
- enable students to make responsible decisions;
- preserve and develop students' imaginations and pleasure in creativity;
- stimulate students to help and respect each other, and to cooperate in solving problems.

The general aims for mathematics in the academic branch of upper secondary school state that students should develop:
- the necessary knowledge and skills both for their chosen route of education and to function in society;
- an understanding of concepts in many areas of mathematics;
- an understanding of mathematical inquiry and mathematical methods;
- an understanding of the role mathematics plays in the development of science and technology;
- the ability to carry on independent study in mathematics.

Curriculum statements are prepared by committees that include teachers at different levels of the school system, teacher educators, and representatives from the government. The committee's proposals are reviewed in a nation-wide process before being given government approval.

Major Changes in the Mathematics Curriculum

In 1987 a revised curriculum guideline, known as M87, replaced the earlier M74. Beginning in 1997 a new guideline, known as L97, will replace M87. The curriculum committee presented its proposal on the new curriculum in the fall of 1995, and a review process on this proposal has been ongoing since that time.

The following features were new in the 1987 curriculum:
- subject matter was presented for three-year periods, and year plans were developed as guides;
- subject matter was divided into ten main areas of study, thus structuring the content to be taught;
- problem solving is one of the most important areas of study, and should be incorporated in all instruction;
- personal and national economy is a main area of study, reflecting a growing emphasis on applications in society;
- computer science is its own area of study;
- more emphasis is placed on integration between topics.

In 1994, the Ministry of Education changed the structure of mathematics in upper secondary school. The mathematics courses for Grade 10 now consist of modules 1, 2A, and 2B. Module 1 is identical to the three-period course offered in vocational areas of study, while 2A and 2B are extensions of module 1. Grade 10 students must study module 1, and may choose either

2A or 2B. 2A is a practical extension, aiming directly at the application of mathematics in society and daily life. 2B is a more theoretical course that gives attention to the theoretical foundation for advanced study. Both alternatives lead to general qualifications for admission to tertiary education, but the advanced mathematics courses in the general academic branch are based on the 2B module.

Current Issues in the Mathematics Curriculum

The new mathematics curriculum, L97, adopts a constructivist view of learning, emphasizing understanding and the processes of learning. Communication and reflection are specifically mentioned as important components of learning, and there is some focus on the role of mathematics in everyday life. Creativity in mathematics is highlighted, and the history of mathematics is for the first time mentioned explicitly.

The proposal calls for calculators to be introduced in the new Grade 2, at the age of seven. Today, calculators may be used in primary school and should be used in lower secondary schools. Graphic calculators are used in upper secondary schools. Computers are available in most schools, and the development of mathematics software has made it possible to integrate computers into the traditional curriculum. The content of the curriculum, on the other hand, has not changed as a result of the availability of such software.

Mathematics Textbooks

Mathematics textbooks must be approved by the government to guarantee that they reflect the content of the curriculum. As a result of curricular changes, textbook content has been altered with, for example, more emphasis on problem solving. Overall, however, textbooks have not changed a great deal in the last 10 years. They follow a typical pattern: introduction of concepts and algorithms, explanation, applications, and exercises. Textbooks are used at all levels in school, and in the lower grades often serve as workbooks as well. In the higher grades, textbooks contain a combination of text that pupils are expected to read and learn from, and text summing up results of activities.

Textbooks are written by experienced teachers or employees of colleges of education, and published by professional companies. Each school, on the advice of its teachers, decides which books to use. Each student has his or her own textbook. The local municipality pays for the textbooks in compulsory school, while students pay for their own books in upper secondary school.

Pedagogy

Mathematics instruction is rather traditional, and textbooks guide the progression of topics. Classroom activities are teacher dominated, but a substantial amount of time is devoted to individual work while the teacher walks around providing help. Integration of mathematics with other subjects and the use of open-ended projects are not common but, in the 1997 curricular reform, both will be strongly encouraged.

5 SCIENCE CURRICULUM AND PEDAGOGY

Goals for the Science Curriculum

The Ministry of Education and Research established a detailed subject-matter framework for the science curriculum in compulsory schooling. Municipal education committees, schools, and individual teachers can strongly influence the content to be taught, often through the preparation of local curricula as specified in the national framework.

At the primary level, science is taught as part of *O-fag*, or civics, a combination of natural and social sciences. The natural science-related objectives for *O-fag* stated in M87 are to:

- provide students with all-round knowledge about themselves, the home and local environment, the natural resource base, and geography;
- teach students the interconnections between nature, society, and culture, and teach them to protect life, health, and the environment;
- stimulate students' imaginations and creative abilities, and encourage them to use their senses and explore their surroundings.

According to M87, the subject-oriented objectives for natural science at the lower secondary level are to:

- help students develop respect for and feel part of nature, life, and the local environment;
- help students develop the attitudes and acquire the knowledge which will lead them to accept responsibility for the use of natural resources and technology;
- communicate the knowledge, skills, and attitudes that will enable students to look after their own and others' health and lives;
- show students that natural science and technology are important elements of our culture, with both positive and negative sides;
- communicate scientific and technological information of importance for everyday life in the community;
- enable students to practice and understand scientific methods, and train them to use experimental equipment;
- maintain and develop the interest and curiosity of both girls and boys with regard to natural phenomena, natural science, and technology.

At the upper secondary level, the goals for the many science-related courses are given in the national curriculum guides for each area of study. The main goals for these courses cover such areas as training in vocational skills as well as general objectives such as the ones stated above. In addition, there are goals pertaining to preparation for tertiary education in the specialized sciences.

Major Changes in the Science Curriculum

In 1987, the old national curriculum guides for compulsory school were replaced by the current M87. Among the features of M87 are a new freedom for schools and teachers to teach topics of local interest. In addition, the integrated *O-fag*, or civics course, was introduced at the primary level. The principal change in content consisted of a stronger emphasis on environmental issues.

Beginning in 1997, new curriculum guidelines called L97 will replace M87. Two important changes will be implemented: first, requirements will be given for each year, instead of for three-year periods, and second, *O-fag* will be split into two separate subjects beginning in Grade 1, natural and social sciences. Natural science will be taught as a separate subject, with

stricter control of the content taught. The preliminary version of the guidelines puts particularly strong emphasis on technological issues and hands-on activities in science. There will also be an increase in the number of science lessons per week.

At the upper secondary level, the extensive structural reforms of 1994 combined the many areas of vocational studies into several well-defined three-year programs. These programs now focus more on theoretical and general aspects of the sciences and less on technical skills.

Current Issues in the Science Curriculum

Technological and experimental issues are coming more and more into focus in science courses, particularly at the primary level. There is a call for a departure from theoretical and textbook-based instruction and the inclusion of more hands-on activities.

The introduction of information technology is also an issue in science. The new guidelines require computers to be used at all grade levels to a much greater degree than has previously been the case. However, there are few computer programs available that are widely regarded as suitable, and teachers usually have a low competence level in computer handling.

As a result of the forthcoming reforms, science will be given more emphasis in the curriculum for all grade levels and for all areas of study. This raises the issue of what in science is both important and attainable for all students, an issue that is strongly related to the general goals for schooling as stated in the Core Curriculum (The Ministry of Education, Research, and Church Affairs 1994b).

Science Textbooks

Science textbooks must be approved by the Ministry of Education to ensure that they reflect the curriculum and pedagogical requirements of the curriculum. Textbooks are written by experienced teachers or employees of colleges of education, and published by professional publishing companies. In general, they seem to appeal to teachers and students at all levels. They all contain colored pictures and many graphics. Each school, on the advice of its teachers, decides which book to use among those approved by the Ministry of Education. Each student from Grade 1 has his or her own *O-fag* or science textbook. The local municipality pays for the textbooks in compulsory school, while students pay for their own books in upper secondary school.

Textbooks for *O-fag*, or Grades 1 to 6, usually contain separate chapters on science and social science topics such as history or geography. Science topics are deliberately spread throughout the books in a way that prevents the systematic presentation of science as such. The organizing principles are typically themes like seasons or daily life. From Grade 7 onward, science textbooks present science as an integrated subject with topics on physics, chemistry, biology, and some earth sciences, each chapter focusing on one of these areas.

Most science textbook series offer, in addition to the textbooks, separate workbooks containing laboratory tasks, problems, and exercises. There are also separate problem books, which include the previous year's examination items for advanced science in upper secondary school.

Pedagogy

Science instruction is so varied across science courses at different stages and areas of study that it is difficult to find a dominant theme. Some trends may be defined, however.

In Grades 1 to 6, teachers typically have little content knowledge in science, in particular physics and chemistry. Accordingly, they rely heavily on textbooks, and student activities are usually limited. Furthermore, many teachers tend to avoid science topics (particularly physics and chemistry) whenever they have the option to do so, focusing instead on the social sciences. At the secondary stage the teachers are more confident, but classroom activities tend to be teacher- and textbook-dominated. Curriculum guides encourage projects and explorations of students' ideas, but instruction is often confined to a traditional pattern, aiming at getting students to recall what is written in the textbook.

Instruction in the advanced science courses, particularly in chemistry and physics, puts strong emphasis on standard methods of handling quantitative problems. Calculating the correct answers is to a large extent the main goal of instruction. During recent years, however, qualitative and everyday-life problems have gradually come into focus.

In vocational areas of study, the sciences have traditionally been taught within a context of professional training. In the current reform, however, vocational courses will focus to a much greater extent on the academic side of the sciences. This more abstract approach already appears to be putting unrealistic demands on less academic students in these areas of study.

6 EVALUATION POLICIES AND PRACTICES

During primary school, no grades are given to students, and assessment is performed only in order to guide students' work and development. Grades are given beginning in lower secondary school. Consultations between parents and teachers take place twice a year.

At the end of compulsory school, there is a national written examination covering the three main subjects: mathematics, Norwegian, and English. The Ministry of Education is responsible for determining the sample of pupils to take each subject examination; each student is tested in one subject only. In mathematics, the national examination consists of both items that require a specific answer and items that require an explanation of the methods used.

In science there is no written examination, but students may be selected for a school-based, oral examination. A new feature introduced in recent years is the use of a practical activity as the starting point of the oral examination.

There are no other obligatory public assessments in regular use in primary and lower secondary school. Teachers give students a grade reflecting their overall achievement in a subject during the year. A separate grade for the examination is given for mathematics, Norwegian, and English. The average grade of all subjects is the basis for entrance to upper secondary school.

In upper secondary school, assessment practices vary between the different branches. In Grade 12, as a part of Norway's academic tradition, most students who take advanced courses in mathematics, physics, or chemistry must take a five-hour, written, national examination in these subjects. Students are graded in all subjects, with one grade given by the teacher for overall achievement during the year. In addition, there are examinations in the main subjects, for which students receive a separate grade.

In general, there have been few changes in assessment in Norway over the past 10 years. Assessment at the classroom level, both oral and written, fulfills a variety of purposes, both formative and summative: to provide feedback to pupils, to aid in future lessons, and to note pupils' progress. Such assessment is usually teacher-based and not very systematic. The Ministry of Education has initiated projects to develop instruments for systematic formative assessment.

There is a growing concern that we should have more knowledge about what is really going on in Norwegian schools. The absence of any kind of national testing has been a matter of concern. One can expect a growing debate about how assessment should reflect the more general aims given in the new plans. There seems to be some interest from the Ministry of Education in establishing a more permanent, national assessment system. TIMSS is regarded as an important way of creating expertise in the area of large-scale assessment and testing. For this purpose, TIMSS in Norway has been conducted as a project with heavy focus on research.

7 REFERENCES AND SOURCES FOR FURTHER READING

Common Sources

1 Husén T, N T Postlethwaite 1994 *International Encyclopedia of Education.* Pergamon Press, Oxford

2 Organisation for Economic Co-operation and Development 1995 *Education at a Glance: OECD Indicators.* Organisation for Economic Co-operation and Development, Paris

3 United Nations Educational, Scientific and Cultural Organization 1995 *Statistical Yearbook.* United Nations Educational, Scientific and Cultural Organization, Paris

4 World Bank 1995 *World Development Report 1995.* Oxford University Press, New York

Additional Sources

5 The Ministry of Education and Research 1987 *Curriculum Guidelines for Compulsory Education in Norway.* The Ministry of Education and Research, Oslo

6 Ministry of Education, Research and Church Affairs 1994 *The Development of Education 1992–1994, National Report, Norway, to The International Conference on Education, 44th Session, Geneva, October 1994.* Oslo

7 Ministry of Education, Research and Church Affairs 1994 *Education in Norway.* Oslo

8 The Ministry of Education, Research and Church Affairs 1994a *Upper Secondary Education in Norway: Structure and Excerpts of Foundation Course Curricula.* The Ministry of Education, Research and Church Affairs, Oslo

9 The Ministry of Education, Research and Church Affairs 1994b *Core Curriculum for Primary, Secondary and Adult Education in Norway.* The Ministry of Education, Research and Church Affairs, Oslo

10 OECD 1988 *Reviews of National Policies for Education, Norway.* Paris

11 Sjøberg, S 1995 *Science in School and the Future of Scientific Culture in Europe. National Report—Norway.* To be published by the EC Commission

12 Statistics Norway 1994a *Education Statistics, Primary and Lower Secondary Schools 1 September 1993,* Statistics Norway, Oslo-Kongsvinger

13 Statistics Norway 1994b *Statistical Yearbook 1992.* Statistics Norway, Oslo-Kongsvinger

14 Statistics Norway 1995 *Education Statistics, Upper Secondary Schools 1 October 1993.* Statistics Norway, Oslo-Kongsvinger

Statistical References

Section	Statistic	Reference	Page	Table	Year of Statistic
Country Profile	4.3 million, 13	3	1-6	1.1	1993
Country Profile	0.4%	4	289	26	1991
Country Profile	45%, 10,000	13	38	16	1992
Country Profile	30,000, 27,000, 13,000	13	66	51	1992
Country Profile	US$25,970, 2%, 5%, 95%	4	163	1	1993
Country Profile	9%	4	259	11	1991
Governance	<2%	12	25	11	1993
Governance	18%	14	63	28	1993
Governance	5%	14	77	38	1993
Structure of the System	95%	14	84	41	1993
Structure of the System	87%, 79% and 91%	14	18		1993
Schools in the System	19	12	18.4	3.22	1993
Schools in the System	6:1	4	295	29	1993
Teacher Profile	45 years	14	69	32	1993
Teacher Profile	45 years	12	45	25	1993
Teacher Profile	2/3	12	44	24	1993
Teacher Profile	80%, 60%, 53%, 30%	TIMSS data			1995
Teacher Profile	60% and 70%	14	68	31	1993

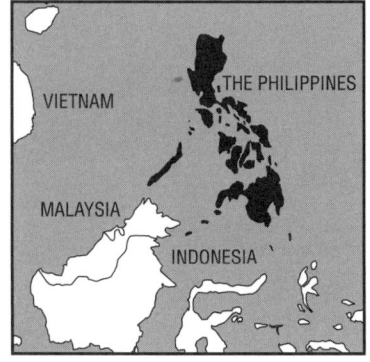

The Philippines

Milagros D. Ibe, University of the Philippines

Amelia E. Punzalan, University of the Philippines

1 COUNTRY PROFILE

The Philippines, a group of more than 7,000 islands, lies 966 km off the southern coast of Asia. From north to south, its greatest length is 1,851 km and from east to west, its greatest breadth is 1,107 km. It has a total land area of 300,000 km². Only 2,773 islands are named, and the 11 largest make up 92 percent of the total land and contain most of the population.

As of 1993, the country's total population was 65 million, with an average density of 216 persons/km². The density, however, varies across the 15 regions into which the country is divided, from 63 persons/km² in the Cordilleras to 12,500 persons/km² in the national capital region. Almost half the population lives in urban and semiurban areas. The average annual population growth rate during the 1980s was over 2 percent.

Filipinos, as the people of the Philippines are called, are derived from many ethnic groups, but the vast majority share an Indonesian-Malayan ethnicity. The Philippine languages belong to the Malayo-Polynesian family of languages. Eight major languages are spoken by close to 90 percent of the population, and an additional 120 minor languages have been identified. Filipino is the national language, but English is widely used in education, government, and commerce.

The Philippines is a constitutional democracy headed by a president elected every six years. The legislative branch of the government comprises a House of Senate with 24 senators and a House of Representatives with 220 members of congress. The country is divided into twelve geographical and administrative regions, two autonomous regions, and a national capital region. Each region has an average of five provinces or cities, for a total of seventy-six provinces and sixty-four cities. Each province or city is headed by an elected governor or mayor.

By United Nations classification, the Philippines is among the lowest of middle-income countries, with a per capita GNP income of US$850 in 1993. From 1980 to 1993, the average annual growth rate was negative, with an inflation rate of almost 14 percent. In June 1993 the inflation rate was almost 7 percent. The gross domestic product in 1993 was US$13 billion. The budget for education in 1993 was 16 percent of total central government expenditure, which was 19 percent of the total GNP. The budget of the Department of Education, Culture, and Sports that year was about 13 percent of the national government budget. The literacy rate is greater than 93 percent.

2 THE EDUCATION SYSTEM

Governance and Decision Making

The Department of Education, Culture, and Sports oversees education in the Philippines. The department is headed by a secretary who is assisted by three undersecretaries. At the central government level, primary, secondary, and nonformal education are administered by separate bureaus, each headed by a director. Tertiary education is administered by the Commission of Higher Education, headed by a chairperson and four commissioners.

The Bureau of Elementary Education and the Bureau of Secondary Education are responsible for the formulation and development of educational policies and programs nationwide. They also conduct studies, formulate educational standards, assess and evaluate the aims and objectives of education at these levels, and provide technical assistance to the Department of Education. Moreover, they are charged with undertaking curricular design including preparing materials; formulating guidelines for the improvement of school buildings, facilities, and equipment; and drawing up programs for upgrading staff education.

The three bureaus and the Commission of Higher Education have counterparts in each of the 15 regional administrative offices. Each regional office has a director and several chiefs of divisions corresponding to the bureau directors in the National Central Office. Regional directors are given autonomy to run basic education in a decentralized structure. Decision making on such matters as curriculum instruction is done at the regional level for local issues. For larger issues, decisions are initiated at the regional level and submitted to central administration for approval.

Structure of the System and Participation Rates

Structure of Education

Figure 1 shows the structure of the education system and approximate enrollment rates for each level. Nursery school and kindergarten are not part of the public education system, but a large number of five- to six-year-olds attend the many private institutions that are funded through tuition fees. Compulsory education consists of six years of primary school followed by four years of secondary school. Until 1994, students began

Grade 1 at the age of seven; they now begin at six and attend compulsory schooling until the age of sixteen. Pupils who leave before the age of sixteen, however, are not compelled to return because there are not enough places available for all school-age pupils.

Postsecondary schools include vocational and technical schools, which offer one-, two-, and three-year programs. There are also colleges and universities that offer degree programs leading to bachelor of arts or science degrees in selected fields of specialization.

Participation Rates

Participation rates for primary students are greater than 100 percent because some students begin school later than age seven, repeat grades, or drop out and return to school later. In 1995, participation rates at the secondary level averaged 75 percent, and at the tertiary level, 27 percent. Female participation rates are the same as those for the total population.

There are both public and private schools at all three levels. Approximately 5 percent of primary students, 31 percent of secondary students, and 77 percent of tertiary students attend private schools.

Schools in the System

The School Year

The school year for all three educational levels begins the first week of June and ends the last week of March or the first week of April. There is a two-week Christmas break in December. For both primary and secondary schools, the school year includes at least 200 scheduled instructional days. Graduating students, that is Grade 6 and Grade 10 students, normally complete the school year one or two weeks ahead of other students.

Primary Schools

Students attend school Monday to Friday. Most primary pupils take seven or eight subjects: language, health and science, reading, mathematics, grammar, civics, home economics and practical arts, and physical education. Each subject is allotted 30 or 40 minutes, and elementary students have between four and five hours of class time daily. Grades 1 to 3 average four to five hours and Grades 4 to 6 five to six hours of class time daily. Extracurricular activities are scheduled Saturdays or after class hours on weekdays.

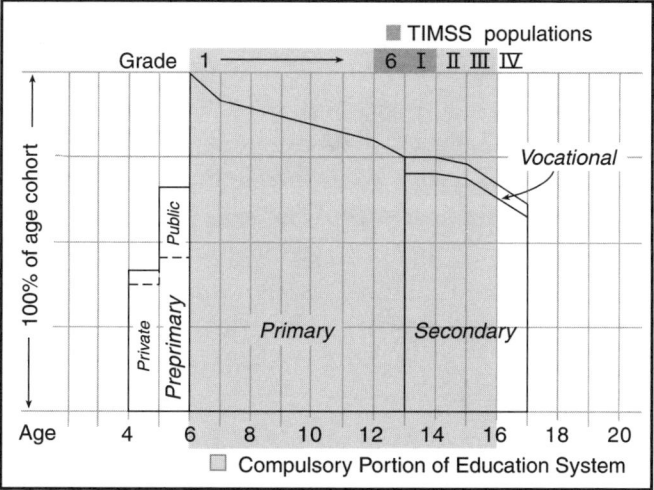

Figure 1: Structure of the Education System

Secondary Schools

Secondary schools fall into one of three categories. General schools offer the regular curriculum; comprehensive schools, both general and vocational curricula; vocational schools offer education in technical, agricultural, or fisheries-related subjects. In secondary school, the students take eight regular subjects. Classes meet 40 minutes daily or 60 minutes every other day. Biology, chemistry, physics, and practical arts are allotted double periods to allow time for practical exercises, field work, and experiments. Secondary school students spend an average of 6.5 hours daily in class, and students in special school science classes spend at least 1 extra hour each day on this subject.

In urban areas, schools run three overlapping five- or six-hour shifts to accommodate all students. In rural areas, however, schools have full-day programs, and the school day runs from 7:30 a.m. to 4:30 p.m. to give students some free time between subjects. Most students, therefore, spend between twenty-four to thirty-five hours a week in contact with their teachers. Of this time, 2 percent is devoted to science and another 12 percent to mathematics.

Class Size

The average class in elementary schools contains 31 students; in secondary schools, 38. However, class size varies widely in different areas of the country. There are classes in densely populated areas with as many as 60 or 65 students; in some rural areas, classes with fewer than 25 students may be found.

Mathematics and Science Programs

Mathematics and science are required subjects in Grades 1 to 10. There is at least one mathematics or science subject at each grade level. Generally, the same grouping schemes are used in science and mathematics as in other subjects in the curriculum. In large secondary schools, the first two or three sections are homogeneous and made up of high-ability students. In the remaining sections, medium- and low-ability students are grouped heterogeneously. Only in the 125 special science secondary schools and university laboratory schools—just over 2 percent of all secondary schools—are students streamed because of ability and interest in mathematics and science. Students in special science shools take entrance examinations and are chosen based on their interest and ability in mathematics and science. University laboratory schools are for the children of university faculty and staff, as well as for students not connected with the university. All students wishing to enter university laboratory schools must pass a qualifying entrance examination.

Public and Private Systems

There are both sectarian private schools, run by religious orders or groups, and nonsectarian ones administered by private individuals, families, or business corporations. Private schools depend largely on tuition fees, supplemented by donations and subsidies from foundations or religious orders, local and foreign.

Certification of Mathematics and Science Teachers

Primary school teachers must complete ten years of basic education and a four-year bachelor of elementary education degree from a university or teacher education college. The teacher education program includes courses in general education and pedagogy, and graduates of this program may teach any subject or grade. Some universities require six courses in an area of concentration like English or mathematics to ensure adequate specialization for teaching Grades 5 or 6.

Candidates for teaching science or mathematics in a secondary school must take a four-year bachelor of secondary education degree program at a university or teacher education college. Secondary school teachers specialize in one or two subjects, taking eight to eleven courses in the discipline(s) they choose.

Both primary and secondary school teachers must pass a professional board examination before they can be tenured. Beginning in 1996, the professional board examination will be replaced by a licensing examination. Specialization in mathematics or science requires completion of at least 10 courses in the field. The pedagogy component consists of about 10 professional education courses, including educational foundations, teaching methods, testing, research, and evaluation, and one or two semesters of teaching practice.

Noncredit, in-service courses for teacher upgrading are given by universities and the Department of Education. Universities also offer formal courses towards master's degrees and content courses for nonmajors who entered the profession with a bachelor of science rather than an education degree.

Teacher Profile

The 1991 Civil Service Commission survey showed that females dominate in the teaching profession; 84 percent of survey respondents were female. Seventy-six percent of teachers belong to the 26- to 50- year-old age bracket. Public school teachers are generally older than secondary teachers, with a median age of 43 years.

Thirty-four percent of all teachers have 10 or fewer years of teaching experience, while another 32 percent have been in the profession between 11 and 20 years. Teachers with between 21 and 30 years of experience account for 28 percent of the total population of teachers in the field, and teachers with 31 or more years, just over 6 percent.

The Civil Service Commission survey of 1991 found the monthly income of primary and secondary school teachers to range from US$124 to US$208; the average income was US$128. Additional benefits such as cost-of-living allowances and other cash benefits amounted to approximately US$60. These cash benefits included a bonus of a month's pay, overtime pay, special hardship allowance, and hazard pay for teachers serving in areas experiencing peace and order problems. Tenured teachers, whether in public or private schools, receive regular pay for twelve months, including the two-month summer vacation and the two-week Christmas break. They also receive a clothing and uniform allowance. Nontenured teachers are considered contractual employees, with short-term contracts that do not include full benefits. Teachers are paid overtime if they spend more than six hours a day in the classroom. Teachers, like most other government employees, usually receive a cash gift of US$40 at Christmas.

Generally, private school teachers receive lower salaries than their public sector counterparts, but they do receive benefits such as holiday pay, vacation pay, generous uniform and clothing allowances, and bonuses. Those with tenured appointments also receive full salary during school vacations.

The average monthly wage of teachers does not compare favorably to other occupations. According to the 1990 Current Labor Statistics, teachers earn lower average wages than messengers, bookkeepers and accounting clerks, miners and quarrymen, and electricians. The upgraded public school teacher's monthly salary, as of 1990, was still less than the average monthly compensation of national government employees.

3 TIMSS POPULATIONS

Philippines, IEA, and TIMSS

The Philippines has been a member of IEA through the Science Education Institute of the Department of Science and Technology since 1994. Prior to joining TIMSS, the Philippines participated in the Second International Science Study.

Philippine TIMSS Populations

The Philippines is participating in the study only at the Population 2 level. The sample of thirteen-year-old students was drawn from Grade 6 classes in 192 primary schools and first-year classes in 195 secondary schools, to represent both the lower and upper grades in which thirteen-year-olds are enrolled, in both public and private schools.

Participation in Population 1 was ruled out because the pupils are not formally introduced to English until Grade 3. A TIMSS Population 1 study would have included Grades 2 and 3. Translating tests to the national language would not have solved the language problem because in Grades 1 and 2, pupils are taught in local dialects. English is used as the language of instruction in Grades 1 and 2 only in single-sex schools, known as exclusive schools, and some private schools.

Population 3 is undefined in the Philippine context. Students graduate from secondary schools at the age of 16 or 17. Thereafter, they either stop schooling or proceed to university, where they take only the basic mathematics and science subjects at the college level.

4 MATHEMATICS CURRICULUM AND PEDAGOGY

Goals for the Mathematics Curriculum

The intended curriculum is determined by the Department of Education with the participation of regional officials and advice from subject specialists in teacher education institutions and curriculum development centres. The Department of Education works closely with the Institute for Science and Mathematics Education Development to develop curriculum.

The emphasis in mathematics and science follows that of "Education for All," a program developed by the National Committee on Education for All in 1991. During the past 10 years there has been an increased emphasis on science, mathematics, and English because national assessments have shown consistently poor performance in these subjects.

The mathematics curriculum is organized into 15 strands: whole numbers and operations; fractions, ratio and proportion; decimals; percentages; geometry; linear measurement including perimeter; area measurement; volume measurement; mass measurement; time; temperature; meter reading; maps and scales; and data handling. These strands carry over from primary into secondary mathematics classes. Algebra is introduced in the first year of secondary school.

The content of the courses is presented to educators as achievement objectives accompanied by suggested activities. The curriculum guide lists the following general objectives for mathematics education. Students are expected:

- to cultivate basic skills in numeracy;
- to develop logical and creative thinking skills;
- to develop problem-solving skills in daily life situations and related fields;
- to reinforce general skills in communication, manipulation of tools and equipment, social interaction, and valuing.

Major Changes in the Mathematics Curriculum

The secondary school curriculum for mathematics is now integrated and follows a spiral approach. In all four years of secondary education, general mathematics, algebra, geometry, and statistics and probability are taught at an increasing level of complexity. Until 1989, mathematics domains were assigned by school year: general mathematics in first year, algebra in the second, geometry in the third, and trigonometry in the final year. Subjects that were electives at that time (for example, consumer mathematics, statistics, and trigonometry) are now integrated into the curriculum.

Calculators are used in mathematics classes, but they are meant to be used to verify certain principles, not to perform basic operations. Calculators are also used for learning concepts and arriving at generalizations. Mathematics and science learning is shifting toward the use of activities that can make certain rules and principles concrete. Participatory learning methods are encouraged.

Methods that generate active participation and enjoyment are stressed in preservice and in-service teacher education to offset negative attitudes toward mathematics. Teacher education is also shifting towards an emphasis on concept formation, concept mapping, and hypothesis formulation skills.

Current Issues in the Mathematics Curriculum

The most pressing issue is the lack of trained mathematics teachers. Many teachers of mathematics are not specialists in this subject. In spite of the many innovative teaching methods being introduced, teachers commonly teach as they themselves were taught, that is, through lectures and other expository methods. In addition, few education students choose to specialize in mathematics teaching. The perception that mathematics is difficult is common among teachers, and this belief tends to filter down to their students. These students then tend to avoid mathematics in university and as a career.

In-service teachers need upgrading training to enable them to teach the integrated curriculum using a spiral approach. Many teachers feel incapable of teaching general mathematics, geometry, algebra, and statistics and probability simultaneously.

Recent research shows that at most 75 percent of textbook content is covered during a school year. Lack of teacher competence and familiarity with the subject matter are largely to blame for this problem; teachers tend to spend too much time on topics they know and hurry through topics and concepts they are unfamiliar with.

Mathematics achievement in national assessments is almost consistently the lowest among the subjects tested. Skill with fractions, in particular, receives among the lowest scores. College graduates appear to retain very little of the basic mathematics they learned in secondary school beyond the four fundamental operations with whole numbers.

Currently, the view is commonly held that mathematics textbooks cover too many concepts, and that the number of topics needs to be reduced to ensure that core topics are thoroughly learned. Many teachers also believe that 40 minutes of teaching daily is not enough time to cover all topics. Teachers are usually not able to supplement the textbooks with exercises, problems, and examples that involve students' real-life experiences.

Mathematics Textbooks

About 10 years ago, the textbook-to-pupil ratio was one to seven. Now, it is one to one, that is, one complete, free set of textbooks exists for each public school pupil. Private schools buy textbooks to be used by their students. Teachers are supplied with textbooks and the corresponding teacher's manuals. Textbooks are now present in all classrooms and are used daily. This is largely due to multilateral assistance from the World Bank, the Asian Development Bank, and the Canadian International Development Agency.

New mathematics textbooks take a more concrete approach than the textbooks of the past, with teaching strategies built into the textbook itself. More responsibility for learning is assigned to the student. For example, the student is asked to do some tasks independently and to check the results with a partner.

These textbooks were commissioned by the Department of Education and developed by the Institute for Science and Mathematics Education Development. Textbook development was guided by a list of expected learning outcomes formulated by the Bureau of Secondary Education. Mathematics textbooks developed by private corporations also adhere to the desired learning competencies. The competencies are itemized in the following list:

- to systematically perform counting, estimating, calculating, and approximating functions;
- to communicate and translate ideas from verbal to symbolic form and vice versa;
- to read and interpret tables, graphs, and other diagrams;
- to handle and manipulate measuring and calculating devices;
- to demonstrate skill in performing geometric constructions and graphing;
- to recognize patterns and relationships for making

predictions and inferences;
- to demonstrate logical, inductive, deductive, and statistical reasoning;
- to demonstrate decision-making skills in recognizing plausible and reasonable mathematics data;
- to apply mathematics to other fields, such as environmental science, population education, economics, statistics, and arts and culture;
- to apply problem-solving skills to daily life situations;
- to use numerical and reasoning skills in obtaining employment and for further studies;
- to appreciate the widening applicability of mathematics in the activities of humanity;
- to demonstrate an appreciation of the usefulness of mathematics as a language;
- to show interest and recognize values in mathematics as a discipline;
- to manifest the values and positive attitudes acquired from the study of mathematics.

Pedagogy

At present, classroom teaching is largely teacher-directed, with explaining and questioning being performed in a whole-group setting. Group work is done superficially, and discussion, when carried out among students, is seldom sustained or productive. Many teachers see group activities as taking too much time, and are therefore reluctant to encourage them. Slowly, teachers are introducing the use of devices like spinners and geoboards.

Assessment of student learning is still mainly of the paper-and-pencil type. Although calculators have become a part of daily use in shops and markets, their use in classrooms is limited since the majority of students cannot afford to buy them.

5 SCIENCE CURRICULUM AND PEDAGOGY

Goals for the Science Curriculum

The goals for science and technology education at the elementary level are formulated by the Department of Education. Students are required:

- to gain a functional understanding of science concepts and principles;
- to acquire the scientific skills, attitudes, and values needed to solve everyday problems pertaining to health and sanitation, nutrition, food production, and the environment and its conservation.

The science and technology program aims to develop competence, skills, and values relating to science. It is envisioned that, with these tools, students will become scientifically literate, concerned about the environment, self-reliant, productive, and active citizens engaged in nation-building.

Specific objectives are described by the Department of Education as desired learning competencies; these are listed as expected learning outcomes in different curricular areas of elementary and secondary education. These outcomes are stated in terms of cognitive, psychomotor, and affective behaviors. The desired learning competencies constitute the basis for developing and evaluating instructional materials, planning the teaching-learning process, defining the in-service and preservice needs of teachers, and assessing student performance.

Major Changes in the Science Curriculum

Before 1993, elementary school science teaching began in Grade 3 with a subject called science and health. The new elementary curriculum includes this subject for Grades 1 and 2 as well. The desired competencies develop in complexity within a conceptual theme of human beings and the environment. Instructional materials include a basic textbook for each level, teachers' guides, and the learning competencies.

At the secondary level, the science program is divided into the subdisciplines. In Grade 7, students have an integrated science program that includes material from earth science, biology, chemistry, physics, and technology. In Grade 8, the science course focuses on biology and technology; in Grade 9, on chemistry and technology; and in Grade 10, on physics and technology. The technology component in Grades 8 to 10 consists of application of concepts in everyday life.

At both levels, the science curriculum is student-centred and community-oriented. Scientific, cultural, social, human, and spiritual values are integrated into the program, and there is some emphasis on the practical application of scientific facts and concepts. Suggested teaching strategies are discovery, investigative approaches, and the science, technology, and society approach. The development of critical thinking to promote creativity and productivity is emphasized.

At both the elementary and secondary levels, additional time is allocated to science than was the case until 1993. Science and health are now taught for 40 minutes daily, while science and technology are taught for 80 minutes daily. Overall, science teaching occupies 13 percent of total classroom time.

Current Issues in the Science Curriculum

Studies of student achievement show poor performance in mathematics and science compared to other school subjects. A steady decline in achievement from Grade 1 to Grade 6 is also apparent. This implies that students have difficulty sustaining learning in science and mathematics from the beginning to the end of the elementary school program. That there was little change in the mean percentage scores of pupils between 1975 and 1991 suggests that there was little improvement in science teaching during this period.

A lack of equipment and other resources contributes to the poor performance of students in science. In addition, surveys have shown that most physics, chemistry, biology, and general science teachers are not qualified to teach these subjects, having neither a major nor minor specialization in these areas. In 1992, for example, only 8 percent of physics teachers were qualified to teach this subject. Many science supervisors do not have science as their area of specialization, although they should be able to provide adequate technical assistance to teachers.

Science Textbooks

As in the case of mathematics, textbooks in science are now provided free by the Department of Education to all public elementary and secondary students in a one-to-one ratio. Teachers are also provided with a textbook and a corresponding teacher's guide. Textbooks are the most important learning materials used in the classroom.

The Department of Education commissioned the University of the Philippines Institute for Science and

Mathematics Education to develop the secondary science curriculum following the desired learning competencies. The elementary science textbooks were written by independent writers, and evaluated and published by the Department of Education. These also followed the desired learning competencies.

Private corporations have also developed science textbooks that follow the desired learning competencies. Private schools usually buy from private publishers, since Department of Education-published textbooks are not sold commercially.

The elementary science and health curriculum is built around the theme of human beings and the environment. The knowledge, competencies, and attitudes are developed in a spiral pattern from the basic and simple in Grade 1 to the more detailed and complex in succeeding grades. Science and health are studied as one learning area.

The Grade 7 science course, Science and Technology I, is developed around basic principles and concepts from chemistry, physics, biology, and the earth sciences. It is intended to serve as a link between the elementary and secondary science programs, and as an introductory course for biology, chemistry, and physics. As at the other levels, attention is given to the development of scientific thinking, to the application of skills and knowledge in relevant technology, and to the acquisition of positive attitudes and values. Activities, laboratory experiments, and exercises or mastery tests are incorporated into the textbooks.

Pedagogy

Teachers are being trained in the different features of the new science curriculum such as the increased emphasis on technology, the organization of the subject, and the setting of learning competencies that must be mastered by teachers and students. The additional training is necessitated by the low numbers and inadequate preparation of teachers in some regions. New strategies, however, are needed to teach values education and to promote critical, creative, and analytical thinking.

The teaching strategies suggested for the teaching of science are discovery and investigative approaches and the science, technology, and society approach. Creative activities to encourage imaginative and scientific experimentation and discoveries are also suggested.

6 EVALUATION POLICIES AND PRACTICES

Evaluation is performed at the regional, provincial, city, and school levels. Tests are developed at each of these levels for the purpose of assessing achievement. Tests are administered by schools at the end of each grading period, or about every 10 weeks.

National assessment tests are given at the end of each grade in primary school and Grade 10 of secondary school. These are known respectively as the National Elementary Achievement Tests and the National Secondary Achievement Tests. A certain proportion of the grade average of students leaving school at these levels is based on test scores. Results of nationally-administered tests have been used as a basis for curricular reform and decisions on education development programs for the last 15 years.

Between 1973 and 1993, there was a National College Entrance Examination to screen for admission to university. This was abolished in 1994 because it was viewed as depriving most secondary school graduates of the opportunity to acquire a college diploma. Many colleges and universities now administer their own entrance tests to screen prospective candidates. Secondary school seniors may apply to take two or more such tests to raise the probability of admission to university.

Scholarship tests are also administered nationally. One such test is the Department of Science and Technology Science Education Institute Undergraduate Scholarship examination, which is used to screen students for the basic and applied sciences as well as for three-year technological programs.

The Department of Science and Technology Science Education Institute also administers tests in science, mathematics, English proficiency, and abstract reasoning to screen applicants to the special science secondary schools. In addition, these tests assess the achievement of students in these schools compared to those in general secondary schools.

In all these tests, multiple-choice items are used because of the large numbers of students writing them. Scannable answer sheets are used and the data are electronically processed. There are concerns about the validity and reliability of multiple-choice tests for measuring student capability, since these tests are the sole basis for important decisions such as admission to college and scholarship grants. Because of the sheer numbers of test papers to be scored, there appears to be little choice in the test format and approach.

7 REFERENCES AND SOURCES FOR FURTHER READING

Common Sources

1 Husén T, N T Postlethwaite 1994 *International Encyclopedia of Education*. Pergamon Press, Oxford
2 Organisation for Economic Co-operation and Development 1995 *Education at a Glance: OECD Indicators*. Organisation for Economic Co-operation and Development, Paris
3 United Nations Educational, Scientific and Cultural Organization 1995 *Statistical Yearbook*. United Nations Educational, Scientific and Cultural Organization, Paris
4 World Bank 1995 *World Development Report 1995*. Oxford University Press, New York

Other Sources

5 Yambot E M (ed) 1994 *Factbook Philippines Vol. 1*. Active Research Center Factbook. Philippines, Quezon City
6 Kolehiyo ng Sosyal Sayans at Pilosopiya, University of the Philippines 1995 *Mga Wika at Dayaleks ng Pilipinas*. University of the Philippines, Quezon City
7 Department of Budget and Management 1996 *Directory of Government Officials and Diary Planning*. Department of Budget and Management, Manila
8 National Statistical Coordinating Board 1994 *Philippine Statistical Yearbook*. National Statistical Coordinating Board, Makati City
9 Research and Statistics Division of the Department of Education, Culture, and Sports *Statistical Data Based on 1995–96 Enrollment*. Unpublished
10 Philippine Congressional Commission on Education (EDCOM) 1993 *Making Education Work—Book 1, Areas of Concern in Philippine Education, Vol. 2 EDCOM Report*. Congress of the Republic of the Philippines, Manila
11 Pascua L 1993 Secondary mathematics education in the Philippines Today. In: Bell G (ed.) 1993 *Asian Perspectives in Mathematics Education*. NSW pp. 160–172
12 Gloria R T 1994 Education Philippines 2000. In: *Fookien Times Philippines Yearbook*. The Fookien Times Yearbook Publishing Co. Inc., Manila, pp. 216–228
13 Ibe M D 1993 *Screening of Special Science Classes Students Through the Scholastic Achievement Test*. Science Education Institute, Department of Science and Technology, Manila
14 Curriculum Development Division, Bureau of Elementary Education, Department of Education, Culture, and Sports 1989 *New Elementary School Curriculum Final Reports*. Department of Education, Culture, and Sports, Manila
15 Asia and the Pacific Programme of Educational Training for Development (APEID) 1991 *Teacher Training for Science and Technology Education Reform*. UNESCO, Bangkok
16 APEID 1991 *Values and Ethics and the Science and Technology Curriculum*. UNESCO, Bangkok
17 APEID 1991 *Science for All and the Quality of Life*. UNESCO, Bangkok
18 APEID 1991 *Science Curriculum for Meeting Real-life Needs of Young Learners*. UNESCO, Bangkok
19 APEID 1991 *Future Content in Science and Technology Education at Secondary Level*. UNESCO, Bangkok
20 Roxas P S 1991 *In-Depth Study on Needs and Requirements in Science and Mathematics Education*. SEAMEO-RECSAM, Penang, Malaysia
21 Educational Development Projects Implementation Task Force and the Bureau of Secondary Education 1992 The Responsiveness of the Bachelor of Secondary Education Curriculum. *Synergia No. 2 Vol. 1*. Institute for Development Education, Pasig, Metro Manila
22 Science Education Institute 1993 *Science and Technology Education Plan*. Department of Science and Technology, Manila
23 Bureau of Secondary Education 1994 *Brief on SEDP*. Department of Education, Culture, and Sports, Manila
24 SEAMEO-RECSAM 1995 *Regional Workshop on Strategic Planning in Science and Mathematics Education in Southeast Asia for the Twenty-First Century*. SEAMEO-RECSAM, Penang, Malaysia

Statistical References

Section	Statistic	Reference	Page	Table	Year of Statistic
Country Profile	966km, 1851 km, 1107 km, 7100. 2773, 11, 92%, 300,000 km^2	5	1	Physiography	1994
Country Profile	65 million, 216/km^2	3	1–5	1.1	1993
Country Profile	63-125,000, almost half (48.8%)	5	12	1	1993
Country Profile	>2% (2.4%)	5	12	1	1980–90
Country Profile	8, 120	6	–	map	1995
Country Profile	24, 220, 12, 5, 76, 64	7	89–94, 119–120	–	1996
Country Profile	US$850	4	162	1	1993
Country Profile	negative (-0.6%), 15% (14.6%)	4	238	1	1980–93
Country Profile	7% (6.53%)	5	2	1	1992
Country Profile	US$13	5	3	2	1993
Country Profile	16%, 19%	4	180	11	1993
Country Profile	13%	8	10–16	10.8	1994
Country Profile	93%	8	252	–	1994
Structure of the System	100%, 75%, 27%, 5, 31%, 77%	9	–	–	1995–96
Schools in the System	200	10	86	–	1993
Schools in the System	7 or 8, 30 or 40, 8, 40 or 60	10	57	–	1993
Schools in the System	11%	11	166	–	1993
Schools in the System	125, 2.2%	13	1	–	1993
Schools in the System	40	12	226	–	1992–93
Country Profile	84%, 76%, 43	10	5	–	1991
Country Profile	34%, 32%, 6.2%	10	5	–	1991
Country Profile	US$128	10	6	–	1993
Country Profile	US$124, US$208, US$60, US$40	10	36	–	1989
Mathematics Textbooks	1:7, 1:1	12	220	–	1980–94

Romania

Gabriela Nausica Noveanu, Institute for Educational Sciences

Viorica Livia Pop, Institute for Educational Sciences

Marilia Ludu, Institute for Educational Sciences

Dragos Noveanu, Institute for Educational Sciences

1 COUNTRY PROFILE

Romania is a republic on the lower Danube River in southeastern central Europe. It borders Moldova and Ukraine to the north and northeast, Hungary and Yugoslavia to the west, Bulgaria to the south, and the Black Sea to the east.

The last census, in 1994, established Romania's population at almost 23 million. Fifty-five percent live in urban areas. The birth rate has been in decline since 1992, reaching -0.8 percent in 1994. Emigration has decreased since 1990 as well, and in 1994 more than 3000 citizens were repatriated. Eighty-nine percent of Romanian citizens are ethnic Romanians. A further 7 percent are ethnic Hungarians. Other minority groups include Germans, Turks, Tartars, Armenians, Slovaks, Czechs, Ukrainians, Poles, Romany, and Russians.

Romania is a republic governed by a parliament as the representative body and sole legislative authority. The Chamber of Deputies and the Senate, which make up Parliament, are elected by universal, direct suffrage. The president, who is also elected by universal suffrage, designates a prime minister and appoints a cabinet.

Romania's per capita GNP, calculated on the basis of purchasing power parity, was US$2910 in 1993. More than 3 percent of the GNP was spent on education in 1992; since 1994 the figure has been more than 4 percent. Expenditure on education in 1994 represented almost 14 percent of total government expenditure.

In 1991, government expenditure per student ranged from US$97 for preschool education to US$864 for higher education for music and arts students. The average per student cost was US$118 in 1993–94. The average cost per primary student in 1994–95 was US$157 and per secondary student cost was US$263. Romania's adult literacy rate is 97 percent.

2 THE EDUCATION SYSTEM

Governance and Decision Making

The Ministry of Education controls all aspects of the national, preuniversity education system. Its functions include: organizing public education; approving study programs, curricula, and textbooks; and establishing national competitions for textbooks and financing their printing costs. In addition, the Ministry is responsible for initial and continuing education of teaching staff, and for the appointment, transfer, dismissal, and posting of teaching and support staff.

At the regional level or *judet*, school inspectorates monitor the organization and operation of the public school system. The inspectorates are specialist bodies subordinate to the Ministry of Education. Among their responsibilities are admission competitions, graduation examinations, and school inspections within their jurisdiction. In addition, the inspectorates report to the Ministry of Education on the state of the local school systems under their jurisdiction. Education units are managed by directors appointed by the minister of education or by the general school inspector.

Structure of the System and Participation Rates

Structure of Education

The structure of the education system is shown in Figure 1, along with approximate enrollment rates for each level. Schooling is compulsory for Grades 1 to 8. Preschool education is provided in kindergartens for children aged three to six or seven. Most kindergartens

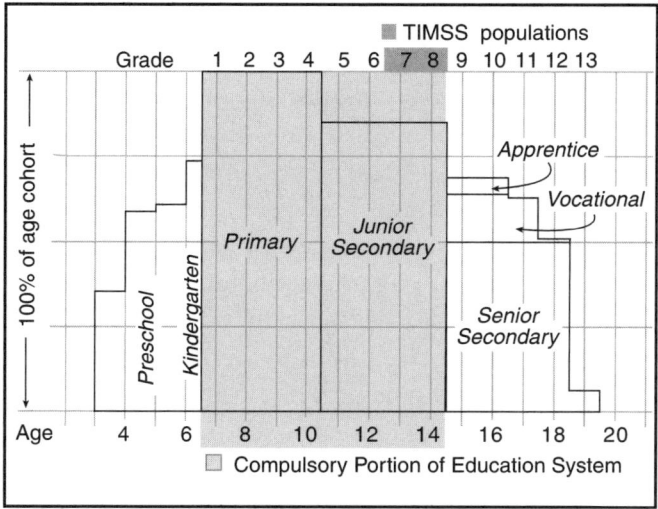

Figure 1: Structure of the Education System

offer weekly and daily full or part-time programs. Special kindergartens are provided for children with disabilities.

Primary education consists of Grades 1 to 4. Most children enrolled in Grade 1 are seven years old, although there are some six-year-olds; 99 percent of the age cohort is enrolled in primary school. One-room schools for Grades 1 to 4 exist in isolated settlements or areas with low numbers of children. These schools are maintained as long as there are at least four pupils of the appropriate age for the respective grade.

Junior secondary education consists of Grades 5 to 8, and 85 percent of the cohort is enrolled at this level. Sixty-nine percent of the age cohort is enrolled in secondary or vocational education. The vocational or secondary level of education, Grades 9 to 12 or 13, has a flexible structure designed to provide alternatives for students seeking specific academic or career paths. Secondary schools may specialize in academic, industrial, or agricultural subjects. There are secondary schools for forestry, economic and administrative studies, cybernetics, meteorology, fine arts, sports, and the military. In addition, there are normal schools which provide initial education for preservice teachers, and theological seminaries.

Vocational and Special Education

Vocational education is carried out in specialized vocational schools. Most programs are three to four years in length, according to the requirements of the trade. In addition to education for a specific trade, vocational schools also provide academic instruction equivalent

to Grades 9 and 10 in regular secondary schools.

Children with specific aptitudes in various fields such as music, fine arts, dance, or sports may attend Grades 1 to 8 in regular schools that provide integrated or intensive curricula. Intensive foreign language studies and alternative education forms such as Freinet, Montessori, or Waldorf also take place in junior secondary schools. The Ministry of Education has authority over these forms, establishing the number of pupils, syllabi and curricula, and mode of operation. The Ministry also provides education for children in orphanages and in institutions for those with sensory or intellectual disabilities.

Public and Private Systems

Most schools in Romania are part of the state system and are funded from central and community budgets. Accredited private education is an alternative or complement to public education. Private education institutions have organizational and operational autonomy, provided they meet national standards and use syllabi and curricula approved by the Ministry of Education. In 1994, 0.1 percent of primary and junior secondary students, and 0.4 percent of secondary students attended private schools. Private schools receive some funding from the central budget, but most of them supplement this through tuition fees, investment income, donations, internal and external sources.

Schools in the System

The School Year

Primary and secondary schools must be open for a minimum of 170 days each year. The school year begins September 15 and ends June 15. The school year is divided into three terms of 8 to 14 weeks, with a holiday period between terms. The winter holidays last from December 21 until January 10, the spring holidays from March 22 until April 7, and the summer holidays from June 15 until September 15.

Instructional Times

Primary students attend school from Monady to Friday for four to five hours per day. Junior secondary students attend five to six hours, and secondary students six to seven hours per day from Monday to Friday. The amount of class time devoted to mathematics varies from 20 to 22 percent in primary to 15 percent in junior secondary school. In senior secondary school, students in the humanities spend 6 to 9 percent of class time on mathematics, while mathematics and physics specialists may spend up to 17 percent. Similarly, the amount of time spent on science varies between 8 percent in primary school and 21 to 25 percent in junior secondary school. Humanities students in senior secondary school spend between 16 and 21 percent of class time on science, while mathematics and physics specialists spend between 30 and 33 percent.

Class Size

Education law stipulates that primary classes must have between 10 and 25 students, and most have approximately 20. In junior secondary school, the average is 25 students with an allowed range of 10 to 30. Classes for vocational, secondary, and postsecondary education also have an average of 25 students, with a range of 15 to 30.

Grouping in Mathematics and Science

The Ministry of Education allows tracking in schools, and has the authority to approve the creation of educational units or classes of highly capable students. Beginning in Grade 9, tracking takes the form of allocating students to different broad streams, including the sciences, mathematics and physics, chemistry and physics, biology and chemistry, and others on the basis of an entrance examination.

Students study a compulsory program of mathematics and science from Grade 1 to Grade 12. From Grade 8 to Grade 12, students may also take optional courses in mathematics or science.

Teacher Certification

Teacher education is carried out within several types of institutions. Normal schools provide a five-year secondary school level program for kindergarten and primary teachers. Students may enter normal school directly from Grade 8, after passing an entrance examination in Romanian language and literature, mathematics, and general aptitude. In regions where teaching is performed in the language of one of the ethnic minorities, the candidate must write an examination in the minority language and literature as well. Technical schools offer a one-and-a-half- to three-year program

for teachers of practical subjects in vocational schools. Institutions of higher education, such as universities, academies, and institutes, provide two to six-year programs for junior and senior secondary teachers. Preservice teachers enter these programs following completion of secondary school.

Mathematics and science teachers are required to study their subject at a university for four or five years before they can be licensed. The program includes a subject-specific curriculum, pedagogical instruction, and a unit of psychological and pedagogical training. Upon completion of these courses, the students graduate and are allowed to teach.

In-Service And Upgrading Requirements

Continuing education courses for teachers are, for the most part, carried out within initial training institutions. Specific subject and pedagogical upgrading are required every five years.

Didactic degrees such as a master's, second, and first degrees are granted by higher education institutions. A master's degree is acquired through an examination written after two or three years of teaching and an inspection of classroom work. The second degree is a qualification granted after an oral examination and classroom inspection four to five years after obtaining a master's degree. The first degree is granted after passing an entrance colloquium, a special class-teaching inspection, and an original thesis. The entrance colloquium consists of an oral presentation on a specific topic to a panel of experts, exhibiting scientific and methodological knowledge in the subject.

Teacher Profile

In 1995, the average teacher income before tax was US$126 per month, while the minimum cross-country income for federal government employees was US$56 per month. Self-employed professionals such as lawyers, accountants, and provincial and municipal government employees earned considerably more.

The average age of the teachers is 43. Women constitute the majority of teachers in Romania at all levels. Fifty-six percent of mathematics teachers and 72 percent of science teachers were women in the 1994–95 school year.

3 TIMSS POPULATIONS

Romania, IEA, and TIMSS

Romania has been associated with IEA since 1991. In the same year, Romania became involved in TIMSS.

Romanian TIMSS Populations

Romania is participating in TIMSS at the Population 2 level only, and all students enrolled in Grades 7 and 8 were part of this population. These grades contain the largest proportion of students in the thirteen-year-old age cohort at the time of testing.

4 MATHEMATICS CURRICULUM AND PEDAGOGY

Goals for the Mathematics Curriculum

The intended curriculum in mathematics is determined by the National Curriculum Commission for Mathematics under the authority of the Ministry of Education. A new national curriculum is under development. Since 1990, a transitional curriculum has been in place, with content and goals shifting from an informative to a formative teaching and learning mode in mathematics.

Under the authority of the Ministry of Education curriculum guides are prepared by national committees, which include both teachers from schools and universities. The guides are prescriptive in nature, and advise teachers on content, general and specific objectives, teaching approaches, and methods of assessment, although teaching methods may be adapted to particular classes or individuals. There are separate curriculum guides for primary, lower secondary, and upper secondary education.

The mathematics curriculum guide states the general aims of mathematics education are to enable students to:

- understand the specific concepts and symbols connected to numbers, operations, operation properties, and geometry;
- develop computational skills;
- develop logical and mathematical reasoning.

From Grades 1 to 5, the mathematics curriculum focuses on number properties, measurement, arithmetic, graphing, logic, and some simple concepts such as perimeter and area in geometry. Algebra and geometry are introduced as separate courses in Grade 6, and their basic concepts are studied until Grade 8. High-achieving students in Grades 9 to 12 may take advanced algebra, geometry, trigonometry, mathematical analysis, probability and statistics, and informatics.

Major Changes in the Mathematics Curriculum

The National Curriculum Commission for Mathematics is developing a new intended curriculum with goals and content corresponding to the standard of Romanian education reform. The standard focuses attention on a constructivist approach to teaching and learning, stressing that mathematics should be relevant to students' lives outside school. In addition, goals and learning opportunities for knowledge assimilation have changed, with less emphasis on memorization of rules, formulae, and procedures and more on the ability to operate them in order to solve problems. For this reason, the proposed goals for students are to value mathematics, to become confident in the ability to use mathematics for solving problems, to communicate mathematically, and to reason mathematically. The proposed changes will decrease the amount of content and increase attention to formative contents at all levels.

The new mathematics curriculum should recognize, respect, and respond to the educational needs, achievements, and perspectives of all students, both male and female, of all races and ethnic groups and of different abilities and disabilities. Its elaboration is a priority for preuniversity education reform. According to the Reform Education Project, which is co-financed by the World Bank, all intended curricula for preuniversity education will be changed before the end of 1997. Major changes include the following:

- reconsideration of curriculum guides from the perspective of a pedagogic conception centered on objectives;

- structuring content from a constructivist, didactic point of view, and moving the accent from information to the formation of mental skills;
- addition of new topics in applying mathematics, such as data handling and probability;
- promotion of aspects connected to communication and personal qualities in learning mathematics; promoting discipline correlation during the gymnasium and interdisciplinary integration during secondary school;
- changing from strictly delimited training to a single professional requirement, and finally to flexible training centered on students' capacity for solving problems, on basic cognitive abilities, and on the ability to readjust in order to meet the requirements of the changing social environment.
- improvement of the selection, evaluation, and examination systems, basing them on well-established criteria; on the transition from normative evaluation toward formative evaluation; on the use of standardized tests; and on the elaboration of minimum performance norms and item banks.

These changes were implemented in 1995–96 for Grades 1 to 5. Curricular change for the other grades will be implemented over the next five years.

Current Issues in the Mathematics Curriculum

The curriculum has given increased attention to social and cultural issues since 1989. Mathematics education has a role in primary, junior secondary, and senior secondary syllabi, whether as preparation for higher-level studies in science, for career purposes or as a necessary grounding for effective living and citizenship.

New technology has become a reality in Romanian schools. Computers allow new and more efficient ways of collecting, recording, and analyzing data. There is more emphasis on using calculators and computers than in the past, although there is regional variation in their availability and use. Computers are more often used for learning informatics than as a tool for learning other subjects.

Other changes in the curriculum focus on making mathematics accessible and interesting for all students. Research in teaching and learning indicates that new curricular contents and goals require new teaching methods in mathematics.

Changes in the intended curriculum have more to do with approaches to teaching and learning than with changes in content. Concurrently, new methods of assessment must be encouraged. Thus, the most significant issue in the mathematics curriculum is the shift in emphasis from transmitting information to developing enhanced operations and abilities to manipulate information.

Mathematics Textbooks

The textbooks used in the recent years are revised editions of those written before 1989. They were issued under the authority of the Ministry of Education, published by the Didactic Publishing House, and paid for by the state. In primary and secondary schools, the principal teaching resource is a standard textbook for each grade. Teachers use textbooks and supplementary collections of problems to support the classroom program. With the old textbook programs, there were no teacher's guides, and teachers relied on some general didactic works on mathematics. There were different textbooks for specific subjects such as algebra and geometry, but only one for each grade and subject. The number of topics ranged between four and nine (topics are identified as chapters in books) and most books were between 130 and 280 pages long. These textbooks were highly theoretical in nature as well as being densely written, making them difficult for students to learn from. The material contains few illustrations, and graphics and geometric figures are used only when strictly connected to the mathematical content.

Beginning in 1995, one of the most important education reform activities has been the development of alternative textbooks written to conform to the reformed mathematics curricula. The books are approved on the basis of intereditorial competitions organized by the National Textbook Assessment Committee. The publishers of the chosen textbooks also develop teacher's guides and supplementary learning materials. The National Textbook Assessment Committee selects several books for each grade, and schools are free to choose the text they think most suitable for their students and teachers. The text of these textbooks is more readable and not as dense. They are also more colorful and contain more pictures.

In 1994–95, the new, alternative mathematics textbooks were used in Grades 1, 2, and 5. These books were designed to conform with the goals and contents of the reformed intended curriculum. Their text is not as dense and contains less information, and focuses on independent activity for students. They include a variety of approaches such as pictorial representations of problems showing the usefulness of mathematics in everyday life. By the end of the decade, new textbooks will be used at all grade levels.

Pedagogy

Curricular change must be accompanied by changes in pedagogic goals, which in turn will necessitate special teacher education programs. Provisions for all of these have been included in the reform project. The new curriculum will encourage a shift from informational to operational aims in terms of student performance. It will be important for teachers at all grades to increase the extent to which mathematics instruction is integrated, both across topics and with other subjects. Additionally, inductive approaches will be used more frequently in mathematics instruction, and deductive approaches will be used less.

Manipulative learning materials, calculators, and computers are an important part of mathematics instruction. All mathematics teaching methods should be appropriate for investigations of real-world problems. Mathematics instruction should always involve students in communication activities such as debates and presentations, and cooperative-learning groups should be used whenever possible.

The present education system, generated by the principles and needs of the communist society, must gradually be replaced and adjusted to the requirements of the new educational model. The reform implies the restructuring of the whole organizational framework of education: aims, legal framework, institutions, curricula, financing, administration, evaluation, and training of teaching staff.

The major changes in the mathematics curriculum are integrated in a global vision of teaching the scientific disciplines, as outlined below.

Pedagogy:
- orienting teaching on objectives, not on "chapters" of information;
- promoting aspects connected to communication and developing cooperative learning.

Methodology:

- developing the continuity of topics at the level of each school year;
- using induction instead of deduction when the level of abstraction is up to the mental age of the children;
- building connections between concepts.

Content:
- permitting choice in the order of topics, leaving the option of extending some;
- updating the content by adding new topics: sequences, functional relationships, data analysis, data representation, estimating, and calculating probabilities.

Practical activities:
- using manipulative materials for introducing new concepts;
- using experiments to explore properties of numbers, shapes, and probabilities;
- using calculators to verify computational techniques or to explore algebraic properties.

5 SCIENCE CURRICULUM AND PEDAGOGY

Goals for the Science Curriculum

There is a curriculum guide for each science subject: physics, chemistry, biology, and geography; and schools are required to implement curricula as given. Curriculum guides advise teachers on content and in some cases on general and specific teaching approaches and assessment methods. The guides are prepared under the authority of the Ministry of Education by national committees that include teachers and university professors. These documents are prescriptive and implementation of the content is compulsory, although methods may be adapted to the characteristics of particular classes or students.

The chemistry curriculum guide states that the general aims of chemistry education are to develop:
- a scientific conception of life,
- a capacity for theoretical knowledge,
- concise ways of thinking,
- skills in scientific investigation,
- a specific scientific language,

- transfer capacity.

The geography curriculum guide states that the general aims of geography education are to:

- develop patriotic feelings about the geopolitical position of the country;
- educate students about human rights and equality between ethnic groups;
- develop environmental knowledge;
- develop geographical knowledge,
- develop intellectual skills such as memorization, geographical thinking, and scientific investigation.

The biology curriculum guide states that the general aims of biology education are to develop:

- scientific knowledge,
- the capacity for knowledge,
- specific biology skills,
- problem-solving abilities.

The physics curriculum guide states that the general aims of physics education are to:

- know the basic concepts such as physical measurement, models, theorems, and physical laws needed for comprehension of scientific explanations of phenomena;
- apply physical laws and theorems to the solving of specific problems;
- perform traditional laboratories and check hypotheses;
- construct hypotheses for explaining and interpreting phenomena and processes;
- discuss hypotheses, models, and theorems in a mathematical way.

Major Changes in the Science Curriculum

All changes that have occurred have been in the framework of curricular reform. (See Section 4 for further, general comments on curricular reform.) Educational objectives, processes, methods, and means have all been modified in the new curriculum. The implementation of the new curricula, therefore, is one of the priorities of preuniversity education. According to the schedule of the project, which is co-financed by the World Bank, all preuniversity curricula will be changed by the year 1997.

Major changes in science curriculum include:

- Updating content and its organization according to cognitive science and curriculum theory. Revised

curriculum guides articulate pedagogic conceptions centered on objectives.

- Developing general knowledge, promoting discipline correlation during lower secondary school and interdisciplinary integration during secondary school.
- Stimulating understanding of the contemporary world and its problems.
- Shifting the focus from teaching only professional requirements to a flexible training system centered on a capacity to solve problems, on basic cognitive abilities, and on a capacity for readjustment to meet the needs of a changing social environment.
- Improving the selection, evaluation, and examination system; basing it on well-established criteria; and changing it from normative evaluation into formative evaluation through the use of standardized tests, the elaboration of minimum performance norms, and the creation of item banks.

The curriculum in science recognizes, respects, and responds to the educational needs, experiences, achievements, and perspectives of all students, both female and male, of all races and ethnic groups, and of different abilities and disabilities. There is more emphasis on computer use than in the past, and science, technology, and society topics have been added.

Current Issues in the Science Curriculum

All students in primary, junior, and senior secondary school study science either as preparation for high-level studies in science, for career purposes, or as a necessary grounding for life and citizenship. Changes in the intended curriculum have more to do with approaches to teaching and learning than with changes in content. Currently, new methods of assessment are being encouraged. The most significant issue in the science curriculum is modifying the emphasis from transmitting information to enhancing operations and manipulating information.

Science Textbooks

All textbooks are approved by the Ministry of Education and published by the Didactic Publishing House. One of the most important reform activities began in 1995, with the development by competing publishers of alternative textbooks written to conform with the reformed science curricula. The old pattern of a unique textbook for each subject, subsidized and protected by the state, has been abandoned. There will be three different textbooks for each subjects and schools may choose which one they wish to use.

The number of topics in old textbooks ranges between five and seven in each textbook and the number of pages between 80 and 250. The greater part of the material included was nonpictorial; only about 10 percent was devoted to pictures, graphs, and other illustrations. Students use textbooks to read and to solve exercises. All students are issued with a free textbook. The new textbooks are more readable and colorful than the old ones.

Pedagogy

The transition from dictatorship to democracy has meant a fundamental change in the educational model. The present system, created by the principles and needs of a communist society, must be gradually replaced and adjusted to new requirements. The reform entails a restructuring of the whole organizational framework of educational aims, including the legal framework, institutions, curricula, financing, administration, evaluation, further education, and training of teaching staff.

Education is the specialized institution that has to train human resources at the quality level required by the future society. The way in which new ideas in pedagogy will be used, therefore, will determine the success of education reform as an important factor in social change.

The major changes in science curriculum are integrated into a global vision of science teaching. This vision includes a new approach to scientific disciplines that includes:

- integrating scientific disciplines into primary education;
- using modern technologies to understand phenomena;
- helping students learn to communicate about and through science orally as well as in writing;
- helping students learn concepts promoting direct contact with phenomena.
- moving the accent from deductive to inductive approaches in teaching science.

6 EVALUATION POLICIES AND PRACTICES

School-based Examinations

In Romania, assessment practices at the classroom level are similar in all types of schools, and rating in the national education system is based on a scale from 10 to 1. Ten indicates the highest mark, excellent; 5, pass; and under 5, failure. Continual assessment is the rule throughout the system and is performed by the teacher of each subject. At least two marks, oral and written, are given in each trimester of the school year to each student. The final mark is the average of all marks in each subject. In some subjects, there is a final written examination in each trimester, as determined by the Minister of Education. In middle school, for example, students in Grades 5 to 8 write examinations in Mathematics, and Romanian Language and Literature.

The Minister of Education periodically evaluates the professional and methodological level of teaching staff. The Minister of Education and the School Inspectorates evaluate the schools on the basis of rules approved by an order of the Minister of Education.

National Examinations

In Romania there are three national-level examinations: a minimum-competency examination, a competition for admission to secondary school, and a baccalaureate at the end of secondary studies. Education law stipulates that students leaving middle school must write a minimum-competency examination, a model of which is drawn up by the Ministry of Education, in Romanian language and literature, mathematics, Romanian history, and Romanian geography. Students who pass the examination are issued a minimum-competency certificate.

Admission to secondary school is based on a competition organized in conformity with a methodology developed by the Ministry of Education. The secondary school entrance examination includes the following tests: Romanian Language and Literature, native language and literature (written; taken by candidates for classes where tuition is offered in the language of a linguistic minority), mathematics (written), and a pass-fail aptitude test that precedes the entrance examination and that is specific to certain types of secondary school such as art or sports. Middle school graduates holding the minimum-competency certificate must write the examination, which consists of both free-response items and problem-solving questions, in order to be admitted to secondary school.

Secondary studies end with baccalaureate examinations, which differ according to secondary school type, class, and student options. The examinations use both free-response items and problem solving-questions. Graduates who passes the baccalaureate examination are issued a baccalaureate diploma and may seek admission to university.

7 REFERENCES AND SOURCES FOR FURTHER READING

Common Sources

1 Husén T, N T Postlethwaite 1994 *International Encyclopedia of Education*. Pergamon Press, Oxford

2 Organisation for Economic Co-operation and Development 1995 *Education at a Glance: OECD Indicators*. Organisation for Economic Co-operation and Development, Paris

3 United Nations Educational, Scientific and Cultural Organization 1995 *Statistical Yearbook*. United Nations Educational, Scientific and Cultural Organization, Paris

4 World Bank 1995 *World Development Report 1995*. Oxford University Press, New York

Other Sources

5 National Commission for Statistics 1996 *Romanian Statistical Yearbook 1995*. Editura si Atelierele Tipografice METROPOL, Bucharest

6 Ministry of Education Institute for Educational Sciences 1993 *The Reform of Education in Romania: Conditions and Prospects*. Ministry of Education, Bucharest

7 Institute for Educational Sciences 1995 *Carte Blanche of the Reform of Education in Romania*. Ministry of Education, Bucharest

8 Ministry of Education 1994 *The System of Education in Romania*. Bucharest

9 Ministry of Education 1995 *The Law of Education*. Bucharest

10 CNCP-PHARE 1995 *Child Protection, Component 2*. Bucharest

11 1995 *The Constitution of Romania 1991*. The Self-Managed Public Company, Bucharest

12 Ministry of Education 1996 *Educational Database* (not published)

13 1996 *The Official Guardian* No. 140

Statistical References

Section	Statistic	Reference	Page	Table	Year of Statistic
Country Profile	23 million	5	9	–	1994
Country Profile	55%	5	83	2.1.2	1994
Country Profile	-0.8	5	101	2.2.1	1994
Country Profile	<3000	5	138	2.3.5	1994
Country Profile	89%, 7%	5	81	–	1992
Country Profile	US$2910	5	914	18.11	1993
Country Profile	>3%	6	122	–	1992
Country Profile	4%	9	72	–	1995
Country Profile	14%	5	961	18.30	1994
Country Profile	US$97, US$864	6	122–123	–	1991
Country Profile	US$118	10	–	III	1993–94
Country Profile	US$157, US$263	12	–	–	1995
Country Profile	97%	10	–	III	1993–94
Structure of the System	99%	5	267	7.1.4	1994–95
Structure of the System	85%	5	267	7.1.4	1994–95
Structure of the System	69%	5	267	–	1994–95
Structure of the System	0.1%	12	–	–	1994–95
Structure of the System	0.4%	12	–	–	1994–95
Schools in the System	20–22%	8	14	–	1994
Schools in the System	15–16%	8	16	–	1994
Schools in the System	6%–9%	8	18	–	1994
Schools in the System	17%	8	20	–	1994
Schools in the System	8%	8	14	–	1994
Schools in the System	21–25%	8	16	–	1994
Schools in the System	16–21%	8	20	–	1994
Schools in the System	30–33%	8	20	–	1994
Schools in the System	20, 25	9	65	–	1995
Teacher Profile	US$126	12	–	–	1995
Teacher Profile	US$56	13	0	–	1994
Teacher Profile	43	12	–	–	1995
Teacher Profile	56%, 72%	12	–	–	1995

Russian Federation

Galina S. Kovalyova, Russian Academy of Education

George V. Dorofeev, Russian Academy of Education

Klara A. Krasnianskaia, Russian Academy of Education

1 COUNTRY PROFILE

The Russian Federation is the largest country in the world, occupying 17 million km^2 or one-seventh of the earth's surface, and is located in eastern Europe and northern Asia. The population is approximately 148 million, 81 percent of whom are ethnic Russians. One-hundred-twenty ethnic groups, most possessing their own languages, make up the other 19 percent of the population. The population density is 9 persons/km^2.

Russia is a parliamentary republic ruled by a president, a two-house federal parliament consisting of the Council of Federation and the *Duma*, and the courts of the Russian Federation. Legislative powers are exercised by the *Duma*. The country's political and economic situation is somewhat complicated. The transition of the national economy to a market basis has altered the social fabric of the country and significantly discriminated against certain sectors of the population. In 1994 one-fourth of the population had an income less than the living wage. The per capita GNP was US$2,260 in 1993.

The breakdown of the economic system, the decline in industrial production, and an acute budget deficit caused a reduction in the financing of all social programs, including education. In 1994, spending on education totaled only 0.9 percent of the GNP.

2 THE EDUCATION SYSTEM

Governance and Decision Making

The social, economic, and political reforms begun in 1991 have profoundly affected the education system in Russia. A highly politicized school system has given way to a democratic one, in which nonstate institutions are allowed and state schools no longer have to be identical.

According to the 1992 education law, the state guarantees citizens of the Russian Federation free general education and, on a competitive basis, free vocational education at state and municipal educational institutions. Compared with the highly centralized system of public education in the former Soviet Union, the Russian education system, according to the new law, is quite decentralized in funding and decision making.

Federal bodies are responsible for shaping and implementing policy in education. They develop the legislative basis for the functioning of the education system and establish the federal component of state educational standards. In addition, they elaborate model curricula as well as model programs of study for different school subjects on the basis of state educational standards, and they organize the publication of textbooks and supplementary literature.

The organization of educational processes at all institutions is regulated by the curriculum, an annual study schedule, and a timetable of classes developed by the institutions themselves and approved by the school council or school educational board. State educational and local government bodies may not change the curriculum or study schedule of an educational institution after it has been approved.

Structure of the System and Participation Rates

Structure of Education

The state system of education in the Russian Federation includes preschool education, elementary and secondary education, secondary vocational training, higher education, and postgraduate education. The structure of the education system is shown in Figure 1, along with approximate enrollment rates for each level. General education, including both elementary and secondary Grades 1 to 9, enrolled 21 million children in 1993. Annually, between two and two-and-a-half million students enter school. Seventy percent of these entered school at the age of seven for a 10-year program and 30 percent entered at the age of six for an 11-year program. The structure may follow either a 4-5-2 or a 3-5-2 model, depending on whether the student is enrolled in the 10- or 11-year program. Most students leave school at the end of Grade 11, and only a small number of schools offer Grade 12. After basic education, about 40 percent of graduates enter vocational institutions, for two- to four-year programs.

Virtually all students receive primary and junior secondary education within a general secondary school. In 1993–94, 60 percent of students finishing junior secondary school progressed to upper secondary school, 38 percent entered vocational institutions, and 2 percent entered the work force. Of the final 4 percent, some enrolled in evening and correspondence programs while in the work force. After completing secondary education, 42 percent of graduates entered institutions of higher education, 39 percent entered vocational training institutes, and 19 percent entered the work force.

General education is the core of the education system in Russia. It includes general schools, schools specializing in specific disciplines, gymnasia, lycea, evening schools, boarding schools, schools for children with special needs, and institutions for additional education. Some of these schools, such as the gymnasia and lycea, are old Russian school types that have recently been reintroduced. At present, only a small proportion of students attend these schools.

The general goals for secondary education are to provide favorable conditions for the intellectual, moral, physical, and emotional development of the individual; to develop in students a scientific world outlook; to impart systematic knowledge about nature, society, and humanity; and to teach the skills necessary for independent activity.

State and Nonstate Systems

A nonstate system of education is being created with the support of the Ministry of Education. At present, the system consists of preschool and general secondary institutions of different types, but the final goal is to develop a system paralleling the state one. In 1993, there were 8500 nonstate educational institutions.

Financing of educational institutions is determined by their organizational or legal form: state, including municipal and departmental; or nonstate, including private, public, and religious. Approximately 98 percent of general and secondary schools in Russia are state or municipal, meaning that the municipal budget is their principal source of financing. Nonstate schools are financed through tuition, fees, and occasionally private funds from foundations.

Native Language Education

All citizens have the right to be educated in their native language, and this right has been secured by the creation of national schools that teach in languages other than Russian. These include schools with native languages as separate or optional courses, as well as schools with no Russian-language teaching at all. During recent years, the number of national schools has increased significantly. The total number of students studying native languages increased from 708,000 in

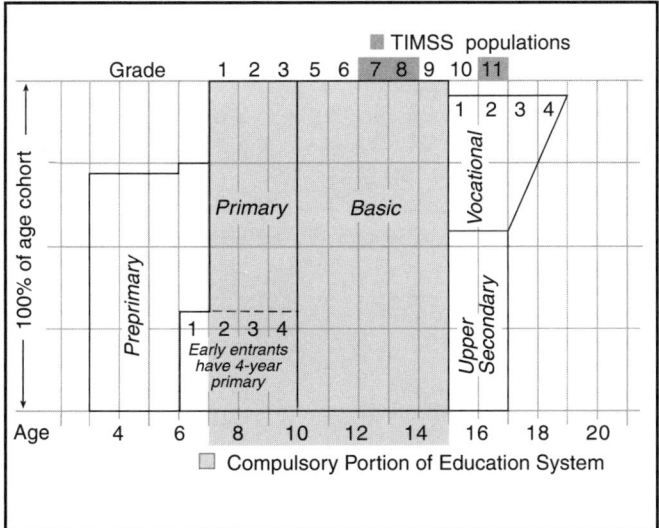

Figure 1: Structure of the Education System

1989 to 1.7 million in 1993. The total number of native languages taught at schools was 74 in 1993.

The Curriculum

The curriculum consists of two parts: invariant (core) and variant. The two components are presented through three major types of studies: compulsory studies comprising the core of general education, choice-based compulsory studies, and optional studies. The division of subjects is established through federal, ethnic-regional, and institutional components. The federal component ensures educational unity throughout the country and outlines the educational content that provides the introduction of world and national values into programs of study. These are Russian, mathematics, informatics, physics and astronomy, and chemistry. The ethnic-regional component meets the specific interests and needs of ethnic groups or regions. It includes the study of native languages and literature, regional history and geography, and similar subjects. Some subject-matter domains, such as history and social studies, arts, biology, physical education, and crafts, are included in both federal and ethnic-regional components. The school-based or institutional component, choice-based compulsory studies, and optional studies emphasize the specific features of the educational institution and promote active development within the school. Table 1 shows a model of core curriculum time distribution for general educational institutions in Russia. The document is the basis for development of all regional core curricula.

Schools in the System

In 1993 the system of secondary education in Russia included 17,200 primary schools, 14,100 basic schools, and 34,800 secondary schools. Primary schools provide primary education, basic schools provide primary

Table 1: Model of Basic Core Curriculum Time Distribution

	Number of hours per week in grades													
	Primary							Basic					Upper Secondary	
	3-year primary			4-year primary										
Subject-matter domains	1	2	3	1	2	3	4	5	6	7	8	9	10	11
Russian as a State Language	4	4	4	3	3	3	3	3	3	3	3	3	–	–
Language and Literature	4	4	4	4	4	4	4	8	8	6	5	5	4	4
Arts	2	2	2	2	2	2	2	2	2	2	2	–	–	–
Social Sciences	–	1	2	2	2	2	2	2	2	2	3	4	4	4
Natural Sciences								2	3	6	8	8	4	4
Mathematics	5	5	5	4	4	4	4	5	5	5	4	5	3	3
Informatics	–	–	–	–	–	–	–	–	–	2	–	–	–	–
Physical Culture	2	2	2	2	2	2	2	2	2	2	2	2	3	3
Design and Technology	2	2	2	2	2	2	2	2	2	2	3	3	2	2
TOTAL:	19	20	21	19	19	19	19	26	27	28	30	30	20	20
Compulsory, Choice-based Studies	5	4	3	1	3	5	5	3	3	4	2	3	12	12
Compulsory Students Academic Workload	24	24	24	20	22	24	24	29	30	32	32	33	32	32
Optional Individual and Team Studies	2	3	3	2	3	3	3	3	3	3	3	3	6	6
TOTAL:	26	27	27	22	25	27	27	32	33	35	35	36	38	38

and basic education, while secondary schools provide primary, basic (junior secondary), and upper secondary education.

The School Year

The school year usually starts at the beginning of September and finishes at the end of May. Schools operating on a five-day week have 170 instructional days, while schools on a six-day week have 210. Six-day schools are open from Monday to Saturday, and five-day schools from Monday to Friday. Six-day schools usually have 45-minute class periods, while five-day schools have 40-minute periods, both in six- to seven-hour days. Students have three holiday periods during the school year: one week at the beginning of November, two weeks from the end of December to the middle of January, and one week at the end of March.

School Size

School size varies significantly across the country, from 10 students in small villages to 3000 students in boarding schools. In 1994 the student-teacher ratio was 14 to 1, compared with 18 to 1 in 1985.

Streaming and Tracking

Students are usually not grouped by ability at schools, but those interested in some subject areas may, on the suggestion of a teacher, take optional courses or advanced studies beginning in Grade 8. There are also schools specializing in teaching specific disciplines. Thirteen percent of the country's students attend these general secondary schools, as well as gymnasium and lyceum, which provide advanced instruction in some areas.

Teacher Certification

Prospective secondary school teachers must study for four to five years at a teacher education institution. They normally specialize in one or two subjects and practice teaching them for a specified period before being fully certified. A second route to certification is through a regular degree program at a university, supplemented by some pedagogical courses. Teacher education is now being transformed into a two-level system that will correspond to bachelor's and master's degrees in education.

In the past, the qualification level and salary of teachers depended not on professional skill, but on length of service, political views, and evaluations by the school administration. Teachers were also required to take in-service education courses every five years. In 1993, the teacher certification system was changed. Teachers are now evaluated by specialists in their area, and assigned to the appropriate qualification level based on their professional level and experience. With additional qualification and in-service study, teachers may move to higher qualification levels and their salaries will increase accordingly.

Based on the new goals of education, in-service teacher education is changing its orientation from subject-specific content to student development. In-service courses promote active learning strategies such as investigations, discussions, and cooperative learning to develop critical thinking and higher-order thinking skills in students.

Teacher Profile

According to Ministry of Education statistics, among the 1.7 million teachers in Russia, 74 percent have a degree while the remainder have a diploma from a pedagogical college or no pedagogical education at all. Virtually all (97 percent) of mathematics and science teachers have a higher pedagogical education diploma. The average salary of teachers has always been less than the national average. In January 1995, the average monthly salary of teachers was US$60, while the national average was US$100, and a living wage was US$45.

3 TIMSS POPULATIONS

Russia, IEA, and TIMSS

Russia became associated IEA in 1991 and was accepted as a member in 1994. TIMSS is the first study in which Russia has participated. It is also participating in the Language Education Study and the Civic Education Study.

Russian TIMSS Populations

Russia is participating in TIMSS at the Populations 2 and 3 levels. Population 2 aims at 13-year-old students and includes all students enrolled in Grades 7 and 8 of general schools in Russia. Population 2 is a grade-defined population and all schools with Grades 7 and 8 formed the population. Only Grade 8 teachers were given teacher questionnaires.

The generalist division of Population 3 includes all students enrolled in Grade 11 as their final year at general secondary school. Students in vocational schools and colleges, about 10 percent of all students, and Grade 11 students in schools where Grades 7 and 8 are not offered, about two percent of all students, were excluded.

The specialist division of Population 3 includes all students enrolled in Grade 11 who are studying advances mathematics or physics. Specialists comprise about 5 percent of Population 3.

4 MATHEMATICS CURRICULUM AND PEDAGOGY

Goals for the Mathematics Curriculum

At the federal level there are two official documents regulating mathematics education: the state core curriculum and the national mathematics curriculum recommended by the Ministry of Education. The first establishes that mathematics is a required course for Grades 1 to 11, and sets out the minimum number of hours per week for mathematics teaching. The second document contains the goals for mathematics education, the required student outcomes, and a list of compulsory topics. The mathematics curriculum is written by researchers and experts in mathematics instruction from the Russian Academy of Education, in cooperation with professional mathematicians, experienced teachers from the Association of Mathematics Teachers, and specialists from the Ministry of Education.

Goals for mathematics instruction include:

- mastering mathematics necessary for real life activity, for the study of other courses, and for higher education;
- developing notions about mathematical ideas and methods and their role in the exploration of the world;

- developing the intellectual skills and habits of mind that are necessary both to the study of mathematics and to functioning in society;
- learning to value mathematics.

Major Changes in Mathematics Curriculum

Changes in mathematics education have been determined by the fundamental social transformations that have taken place in the country. An entirely new approach to mathematics has been adopted, "not a student for mathematics, but mathematics for a student." A humanitarian component to the teaching of mathematics, "mathematics for all," is strongly emphasized, and the general intellectual and cultural development of students is one of its principal objectives. In this paradigm, the pragmatic aspect of mathematical knowledge is also considered an important feature of mathematics instruction. In other words, a new balance has been established between academic, human, and social vectors in mathematics teaching.

The principal changes in content concern the teaching of stochastics and geometry. The former, including tables and diagrams, combinations, and probability and statistics, has been implemented in the Grades 5 and 6 curricula for the first time. This content is now considered an essential part of mathematics education, one that takes into account the important intellectual and pragmatic roles that mathematics plays.

The changes in geometry teaching are essential from a didactic point of view. The subject is not now treated as a vehicle for training students in deductive skills nor for the invention of exquisite heuristics. Instead, visual spatial notions are of the utmost importance.

Mathematics teachers have been slow to include computer technology in mathematics teaching, and it is expected that the national mathematics curriculum will require further development in this area. At the same time these technologies are now being considered an effective means of evaluating student achievement.

Current Issues in the Mathematics Curriculum

A draft document written by the Institute for General Secondary School (Russian Academy of Education) is recommended by the Ministry as a temporary mathematics standard while new national education standards are under development. The aim of the draft is to state realistic and accessible objectives that will provide the necessary mathematics education for all students, as well as stimulate the more capable students. The document describes two-level requirements for student attainment at the end of primary, junior, and senior secondary school. The first level characterizes the minimum that should be reached by all students, while the second describes possible goals (advanced or additional) that may be reached by students as a result of following the course. The implementation of the standards gives students real rights and opportunities in mastering courses at a level that corresponds to their interests and learning abilities.

Mathematics Textbooks

Textbooks are one of the principal teaching and study tools of the teaching-learning process. The most important change in recent years is the transition from the use of a compulsory textbook for each grade to the use of alternative textbooks chosen by schools. Schools and teachers may choose any published or experimental textbook that corresponds to the national curriculum.

Each year, the Ministry of Education issues a list of federally recommended textbooks. The federal government must provide regions with financial assistance for textbooks, and the regional authorities in turn pay textbook publishers with this money.

The number and nature of topics included in the textbooks for a given grade must meet national curriculum requirements regarding content and student outcomes. The textbooks currently in use include extensive explanations and practice exercises, and are designed for students to use after the teacher has explained the material. In some books, however, there is some material that is designed for self-education. Theory textbooks, for example, include rules, worked examples, exercises for compulsory and advanced levels, questions for self-testing, practical and laboratory activities, and subject indices. Special aids accompany the books, including worksheets and a set of quizzes and essay tests for student assessment. Textbooks for Grades 1 to 6 now contain more tables, pictures, sketches, and colors than in the past.

Pedagogy

The most characteristic changes of mathematics pedagogy are those produced by the demarcation between "mathematics for all" and traditional teaching, which was oriented to advanced students. Now each school determines, within stated guidelines, the number of mathematics classes taught per week. Mathematics teachers are allowed to choose a textbook that corresponds to the national curriculum, and to decide whether to intensify or expand the volume of content taught. In all these matters, teachers must take into consideration the interests and learning abilities of their students. At the same time, teachers must help each student to attain the minimum curriculum outcomes and ensure the development of students who wish to know more about mathematics.

Mathematics is designed to be an independent subject throughout schooling. The national curriculum suggests general mathematics courses for Grades 1 to 6. For the most part, the courses contain traditional arithmetic material, as well as some geometry and algebra topics. In Grades 7 to 11, two separate courses, algebra and geometry, are suggested. Streaming in mathematics education begins in Grade 8 (general and advanced courses) and diversifies in Grade 10 (general, advanced, and profiled courses).

In general education, the intellectual development of students has priority over the acquisition of concrete knowledge and skills. As a consequence, oral and written communication in mathematics have become more significant, and the formal aspect of computational skills in mathematics less so, particularly in "mathematics for all." Thus, the use of calculators and computers within the classroom will increase.

The inductive approach, when firmly grounded in concrete mathematical content, is considered of equal importance to the deductive one. Thus experimental methods, the discovery of mathematical laws and facts, and investigations of open-ended didactic and real-world problems are proposed as important activities. At the same time, deductive proof remains an essential tool.

5 SCIENCE CURRICULUM AND PEDAGOGY

Goals for the Science Curriculum

In Russian schools, two variants in the structure of science education have recently come into effect: the separate study of all science subjects in Grades 5 to 9, and the study of an integrated science course in Grades 5 to 7 followed by differentiated subjects in Grades 8 and 9. In upper secondary school, differentiated subjects are offered, and students may choose which subjects they will study in addition to the core. Different advanced courses in science subjects are available for students who choose physics-mathematics, science, or technical profiles.

The goals for science education may be summarized as follows:

- to master knowledge of experimental facts, concepts, laws, theories, and methods of science;
- to acquire a modern scientific picture of the world;
- to understand the wide possibilities of the applications of laws in technology and everyday life;
- to develop intellectual and learning skills, to acquire skills in applying knowledge, to observe and explain phenomena, to compare, to establish cause-and-effect relations, to generalize, to connect material studied with evidence observed, and to apply, gain, and systematize knowledge;
- to develop a cognitive interest in science and technology;
- to prepare for higher education or a profession;
- to foster morality, humanism, and a caring attitude toward nature.

The decentralization of curriculum development has meant that alternative curricula may be designed by professional organizations. Among the organizations involved are the Russian Academy of Education, the Academy of Science of the Russian Federation, universities, research institutes, and teacher associations. The Ministry of Education reviews all alternative curricula and prepares a special advisory document on the adopted curricula.

Schools have the right to make changes to the advisory science curriculum regarding number of hours in a year or amount of content, or to create a curriculum according to their individual goals, needs, and traditions. In practice most schools use the curriculum provided with the textbooks and, as a rule, do not make substantial changes. Thus, the textbooks included in the federal set of textbooks play a significant role in the educational process at school.

Major Changes in the Science Curriculum

Up to 1988–89, all general schools in Russia used the same science curricula and textbooks at every stage of schooling. The greater part of both the curricula and textbooks dated from the end of the 1960s, a period of reform aimed at increasing the scientific level of science education at school. School science courses became simplified images of the basic sciences.

The changes to the science curriculum that took place in the 1980s did not involve the logic of the curriculum structure. They were connected to a new emphasis on the learning process, a movement away from mastering knowledge and skills to mastering learning skills. The curricula reduced the quantity of material studied at the same time as overall instructional time was decreased. Increased attention was paid to the environmental aspects of science education.

The recent social changes in the country have produced dramatic changes in the system of public education. The democratic nature of the education law allows differentiated schools and gives a level of creative freedom to teachers' work. It allows students to choose the level of study in certain subjects in accordance with interests and abilities. Teachers are able to utilize methodological concepts and principles in teaching methods, organization, and course structure. School science is becoming more flexible, aiming to satisfy personal and social needs.

The effect of the rapid expansion of information and technical knowledge became apparent in the integration of the content of the separate courses on the basis of the main ideas and problems. In biology, for example, ideas of evolution may be shown to be related to the organization of living things; in physics, fundamental physical theories and the cycle of scientific cognition; in chemistry, the structure of substances and the dependence of substances' properties on their structure.

In all science courses, more attention than before is

given to the impact of science and technology on society. Whole topics in curricula and textbooks are devoted to this theme. In elementary science, for example, the use and protection of nature by humans is a topic, as is the protection of water from pollution in Grade 7 chemistry, heat engines and the protection of nature in Grade 10 physics, and the use of natural resources and nature protection in Grade 8 geography.

Current Issues in the Science Curriculum

The most pressing issues in science education are the introduction of standards, and the provision of a curriculum, textbooks, and instructional materials necessary for mastering the standards of science education. The comments in Section 4 on the introduction of education standards in Russia also apply to science education. The education standards outline a system of knowledge and skills in content areas that are appropriate to students' ages and psychological and physiological features; the level of knowledge and skills is gradually widened and deepened as students progress through elementary and secondary school. The standards describe a basic level of science education that curriculum developers, textbook authors, and classroom teachers must take into account.

According to science standards, the minimum level that must be mastered does not relate solely to learned knowledge, but incorporates elements of creative work into cognitive activity. Skills in observing and comparing phenomena and objects, identifying and classifying objects, carrying out experiments and drawing conclusions about their results, using theories to explain phenomena, and substantiating theory with facts are important intellectual activities in relation to science teaching.

Science Textbooks

The system of science textbook development and distribution is similar for all school subjects, as described in Section 4. Because the domain of science is part of both the federal (chemistry, physics, and astronomy) and ethnic-regional (biology) components of the core curriculum, schools use science textbooks developed at both the federal and regional levels.

Traditional school science textbooks are characterized by strict logic, exact and accurate statements and definitions, long descriptions and explanations, considerations of various aspects of science phenomena, and relatively advanced mathematics. These books principally contain explanations and exercises, and historical material is not often included. Applications of science are represented in different ways among the science subjects, and seem to be better treated in chemistry textbooks.

New textbooks are oriented toward student needs, and seek to develop motivation and interest in learning science. The inclusion of historical material, applications of science, environmental issues, and careers in science is geared toward this end. The character of exercises is changing, with more problems, practical applications, and real-life situations being used.

Pedagogy

The direction of modern science education is very similar to mathematics education and may be expressed as finding compromises between traditional science education in Russia and "science for all" with its democratic and humanistic aspects. The emphasis of traditional pedagogy is on working out the optimal system for organizing the teaching-learning process, while yielding a high level of student achievement.

During recent years, pedagogy covered various fields: an intensification of the teaching-learning process by means of rational organization and with the help of technology; the use of active methods and forms of teaching and learning; ways of raising the interest of students; development of creative abilities; enforcing interdisciplinary linkages; emphasizing major social, ecological, ethical, cultural, and historical problems in science education; enforcing a practical orientation in science courses; cooperative learning; and optimizing homework. Special attention was paid to methodological problems in science education. However, the academician's approach, with its emphasis on science content, made it difficult to represent modern educational technology and establish close linkages between education and real life. It is becoming more and more accepted among educators that only the balance between content and child-oriented teaching could be the basis of "good science for all."

6 EVALUATION POLICIES AND PRACTICES

The evaluation and assessment system in mathematics and science is in the process of change. Under the new system, both formative and summative assessment will be oriented toward mastering education standards. Students will have the opportunity to choose the compulsory or advanced level of material for both study and assessment purposes. The system, some elements of which were introduced in 1996, is designed to provide equal education in all regions of the country and promote the democratization of the teaching-learning process.

Assessment and marking criteria are also being changed. Earlier marking procedures were constructed based on the highest required level. Students achieving this level were given a mark of five, or excellent, while students not reaching this level received a lower mark. A new system of marking, which has its basis in the system of compulsory and advanced levels, is being introduced. Mastery of compulsory knowledge and skills is marked as satisfactory. Students who have achieved advanced knowledge and skills receive higher marks, based on the quantity and complexity of material learned.

A flexible system of school examinations has replaced the system of compulsory examinations. In order to receive a basic school leaving certificate, students must pass a compulsory national examination in mathematics and Russian, as well as a compulsory regional examination in a subject selected locally. For the secondary school leaving certificate, students must pass five compulsory examinations: national examinations in mathematics and Russian language and literature, a regional examination, and two examinations in subjects selected by the student. Both junior and senior secondary schools may decide to set examinations in additional subjects. Examinations are administered in oral or written form, using short-answer, essay, and occasionally multiple-choice questions according to the subject. Science examinations may include performance tasks. Some schools work with colleges or universities to administer combined school leaving and university entrance examinations. University specialists take part in preparing and administering such examinations.

Formative and summative assessment is conducted to ensure compliance with curriculum requirements and to monitor student progress. Results are occasionally used for teacher or school accreditation. The form and timing of assessment are chosen by schools, but summative assessment in all subjects usually takes place at the end of the school year. Forms of assessment include oral response; short-answer, extended-response, or essay questions; multiple-choice tests; *zachert* (a criterion-referenced, unscored pass-examination); and oral defense of an essay or examination. Schools use teacher-made tests, locally developed tests, or tests that have been centrally developed and published in special supplementary materials.

There is no longer a national assessment program, and schools are responsible for all assessment of student achievement. From time to time, local and large-scale assessments may be carried out by research institutions, or by educational bodies at the local or Ministry level. Assessment results are used to make decisions on curricular change and pedagogical improvement.

CONCLUDING REMARKS

In the Russian Federation, the TIMSS testing occurred during a complicated period that involved transition and great changes in all spheres of life, including education. For this reason, the TIMSS results for Russia are regarded as a cut-off point for measures of mathematics and science education in the former Soviet Union and Russian Federation, as well as the starting point in measuring the outcomes of the reformed system of education. The results of TIMSS will be used to find a balance between reforming and maintaining the system, as well as to identify the factors that had the most influence on Russia's performance in the testing.

7 REFERENCES AND SOURCES FOR FURTHER READING

Common Sources

1 Husén T, N T Postlethwaite 1994 *International Encyclopedia of Education*. Pergamon Press, Oxford

2 Organisation for Economic Co-operation and Development 1995 *Education at a Glance: OECD Indi-*

cators. Organisation for Economic Co-operation and Development, Paris

3 United Nations Educational, Scientific and Cultural Organization 1995 *Statistical Yearbook*. United Nations Educational, Scientific and Cultural Organization, Paris

4 World Bank 1995 *World Development Report 1995*. Oxford University Press, New York

Other Sources

5 Ministry of Education of Russian Federation 1995 *Basic Directions to Reformation of Educational Economics in Conditions of Passing to Market Relations*. Internal Ministry document, Moscow

6 Ministry of Education of Russian Federation 1995 CONCEPTS. *General Components of the State Educational Standard for Primary, Secondary and High Comprehensive School. General Provisions*. Internal Ministry document, Moscow

7 Ministry of Education of Russian Federation 1994 *Curricula for Educational Establishments. Biology*. Prosvetchenie, Moscow (in Russian)

8 Ministry of Education of Russian Federation 1994 *Curricula for Educational Establishments. Chemistry*. Prosvetchenie, Moscow (in Russian)

9 Ministry of Education of Russian Federation 1994 *Curricula for Educational Establishments. Mathematics*. Prosvetchenie, Moscow (in Russian)

10 Ministry of Education of Russian Federation 1994 *Curricula for Educational Establishments. Physics*. Prosvetchenie, Moscow (in Russian)

11 Ministry of Education of Russian Federation 1994 *Curricula for Educational Establishments. Primary grades (1–3)*. Prosvetchenie, Moscow (in Russian)

12 Ministry of Education of Russian Federation, Russian Academy of Education 1993 *Temporary State Educational Standard. General Secondary Education. Biology*. Draft Second edition Moscow (in Russian)

13 Ministry of Education of Russian Federation, Russian Academy of Education 1993 *Temporary State Educational Standard. General Secondary Education. Chemistry*. Draft second edition Moscow (in Russian).

14 Ministry of Education of Russian Federation, Russian Academy of Education 1993 *Temporary State Educational Standard. General Secondary Education. Concepts of Assessment of Pupils' Achievements of Educational Standards Requirements*. Draft. Moscow (in Russian)

15 Ministry of Education of Russian Federation, Russian Academy of Education 1993 *Temporary State Educational Standard. General secondary education. Mathematics*. Draft second edition Moscow (in Russian)

16 Ministry of Education of Russian Federation, Russian Academy of Education 1993 *Temporary State Educational Standard. General Secondary education. Physics*. Draft second edition Moscow (in Russian)

17 Ministry of Education of Russian Federation 1996 *Expenses in Education*. Internal Ministry document, Moscow (in Russian)

18 Ministry of Education of Russian Federation 1996 *Russian Education: Transition to the Second Stage of the Federal Program of Development of Education*. Moscow (in Russian)

19 Ministry of Education of Russian Federation 1994 *The System of Education of the Russian Federation*. Internal Ministry document, Moscow

20 Russian-European Center for Economic Policy. 1995 *Review of Russian Economy. No. 5 1995*. TACIS, Moscow (in Russian)

21 State Committee of Russian Federation on Statistics 1996 *Education in Russian Federation*. Goskomstat Rossii, Moscow (in Russian)

22 State Committee of Russian Federation on Statistics 1995 *Russian Federation in Numbers*. Goskomstat Rossii, Moscow (in Russian)

23 State Committee of Russian Federation on Statistics 1994 *Russian Statistical Yearbook*. Goskomstat Rossii, Moscow (in Russian)

24 State Committee of Russian Federation on Statistics 1995 *Socioeconomic State of Russia 1995 N7, January–June 1995*. Russian Federation in Numbers. Goskomstat Rossii, Moscow (in Russian)

25 State Committee of Russian Federation on Statistics 1995 *Social Sphere of Russia*. Goskomstat Rossii, Moscow (in Russian)

26 *State Law of Russian Federation on Education*. 1992 Moscow (in Russian)

27 TIMSS sampling data files, 1994, Moscow

Statistical References

Section	Statistic	Source	Page	Table	Year of Statistic
Country Profile	17 million	22	5	–	1994
Country Profile	148 million	21	15	–	1995
Country Profile	81%, 19%	18	5	–	1993
Country Profile	9	22	5, 17	–	1994
Country Profile	1/4	21	71	–	1994
Country Profile	US$2,260	4	–	–	1993
Country Profile	0.9	20	2	5	1994
Structure of the System	21 million	22	83	–	1994–95
Structure of the System	2–2.5 million	18	42	–	1993–95
Structure of the System	70%, 30%, 60%, 38%, 2%	18	58	–	1995
Structure of the System	42%, 39%, 19%	19	4	–	1993
Structure of the System	8500,98%	18	40	–	1993
Structure of the System	708000, 1.7, 74	19	5	–	1993
Schools in the System	17,142, 13,944, 35,222	21	57	–	1994–95
Schools in the System	10, 3000	27	–	–	1994
Schools in the System	14, 18	18	52	–	1995, 1985
Teacher Profile	1.7	21	68	–	1994
Teacher Profile	74%, 96–97%	21	113, 114	–	1994
Teacher Profile	US$60, US$100	25	40	–	1995
Teacher Profile	US$45	24	6	–	1995

Scotland

Brian Semple, Scottish Office Education and Industry Department

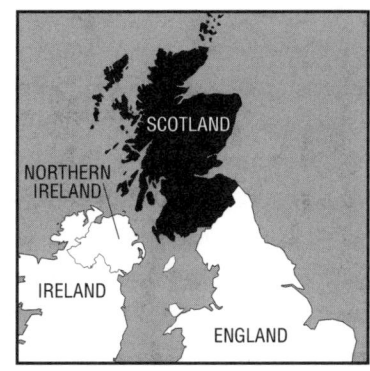

1 COUNTRY PROFILE

Scotland is the northern part of the United Kingdom, which lies off the northwestern coast of Europe. On the east it is bounded by the North Sea and on the west by the Atlantic Ocean. Scotland covers an area of nearly 79,000 km^2 and has a population of 66 persons/km^2.

Scotland's population is 5.1 million, just over 1 percent of which consists of ethnic minorities. English is the language of Scotland, but Gaelic is spoken by nearly 66,000 people and is the medium of instruction in 50 primary school units. Much of the population is concentrated in the central belt of the country where the two main cities, Edinburgh and Glasgow, are located. Emigration from Scotland slightly exceeds immigration by a marginal amount.

As part of the United Kingdom, Scotland is governed by the Parliament at Westminster, where 72 members are elected from Scottish constituencies. The Scottish Grand Committee, consisting of all Scottish members of Parliament, consider proposed legislation concerned exclusively with Scotland, and debate the financial estimates for Scottish Departments and wider aspects of Scottish affairs. Administration of the country is undertaken by The Scottish Office, which serves the ministers and Secretary of State for Scotland. Most powers are delegated to The Scottish Office, the major exceptions being defense, foreign policy, and social security. Local government in Scotland has been reorganized and there are 39 all-purpose authorities replacing the previous two-tier system.

The gross domestic product per capita in Scotland was US$5494 in 1992. Expenditure on education by government and local authorities was more than 13 percent of gross domestic product in 1992.

2 THE EDUCATION SYSTEM

Governance and Decision Making

Education is one of the functions of The Scottish Office. The Scottish Office Education and Industry Department defines education policy and administers further and higher education through the Further Education Funding Council and the Scottish Higher Education Funding Council. Most schools and nursery schools are administered by local authorities, but 4 percent of schools were independent in 1991. Public sector schools may opt for grant-maintained status, with direct funding from The Scottish Office Education and Industry Department. Quality control is achieved through inspections of educational institutions carried out by Her Majesty's Inspectors of Schools.

The school curriculum from age 5 to 14 is defined by The Scottish Office Education and Industry Department in a set of national guidelines that have been issued to schools. Beyond the age of 14, the school curriculum is determined by a system of certified qualifications—Standard Grade, Higher Grade, Sixth-Year Studies, and the National Certificate awarded by the Scottish Vocational Education Council. Standard Grade, Higher Grade, and Sixth-Year Studies are subject to external examination by the Scottish Examination Board.

Advice on teaching is issued by Her Majesty's Inspectors of Schools. Textbooks and other teaching and learning materials are selected by local authorities and schools. Most local authorities have advisers who assist schools in defining their needs for teaching materials.

Structure of the System and Participation Rates

Structure of Education

Education in Scotland is provided at preprimary, primary, secondary, further education, and higher education levels. The structure of the system and the approximate enrollment rates for each level are shown in Figure 1. There are a variety of preprimary institutions, including play groups and nursery schools. Nursery schools were attended by 37 percent of children aged three and four in 1993 and play-group places numbered 49,504 in 1992. In 1992 there were places for nearly 57 percent of three- and four-year-olds in nursery schools, day nurseries, children's centers, family centers, play groups, and registered child-minders.

Compulsory schooling starts at age 5 and ends at age 16, with seven years spent in primary school and the remainder in secondary school. Virtually all pupils attend school during those years. Seventy-six percent of the age cohort persisted to the fifth year of secondary school and 43 percent to the sixth year in 1993. All public sector secondary schools are comprehensive.

There has been a trend to educate pupils with special educational needs in mainstream schools. In 1993, however, more than 1 percent of pupils who suffered sensory impairment, physical or mental disorders, social or emotional difficulties, or learning difficulties attended 196 special schools.

Public and Private Systems

Over 4 percent of students attend private schools. Of these students, 3058 received assistance with fees from the government's Assisted Places Scheme in 1993–94. Public sector schools are funded by local authorities, with the exception of grant-maintained ones which are funded by The Scottish Office Education and Industry Department. Local authority funds are obtained partly from aggregate external funding from central government, and partly from council tax levies on domestic and commercial properties. Private sector schools are supported by fees paid by parents.

Enrollment in Mathematics and Science

Mathematics is taken throughout primary school and the first four years of secondary school. The proportion of the cohort taking a course in mathematics in Grades 12 and 13 fell to 71 percent in 1993. Science is taught in primary schools as a component of environmental studies, while in Grades 8 and 9 general science and a few subject-specific courses are taught. From Grade 10 onwards, there are gender differences among enrollments in specific science subjects, girls being overrepresented in biology courses and boys in physics ones. Table 1 shows the enrollment in mathematics and science courses, broken down by grade and gender.

In 1992–93, a total of 207,055 students were registered in vocational further education courses in Scotland, most of which lead to recognized qualifications. In the same year, 167,183 students attended institutes offering higher education courses, including the Open University. Over 48 percent of the full-time students among this group were female.

Schools in the System

The School Year

The academic year covers three terms starting in the middle of August and finishing at the end of June. There are 190 school days within 38 weeks, but teachers spend an additional week on in-service training as part of their contract. In addition to the break of six weeks from the end of June to the middle of August, there are breaks of one to two weeks in October, December, and April, the length of each break being defined by the local authority. Some authorities also have a short break in mid-

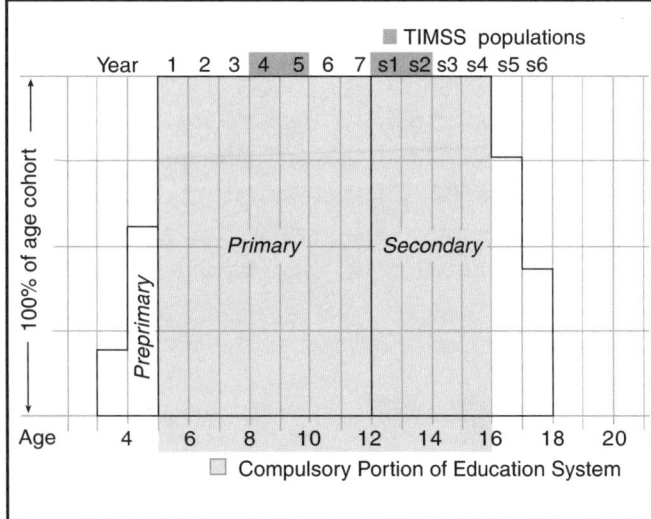

Figure 1: Structure of the Education System

Table 1: Enrollment in Mathematics and Science by Grade and Gender, September 1993

Subject	Grade 8		Grade 9		Grade 10		Grade 11		Grade 12		Grade 13	
	M	F	M	F	M	F	M	F	M	F	M	F
Mathematics	100	99	99	100	99	100	100	99	87	81	58	42
General Science	98	97	88	89	31	26	31	25	2	2	0.1	0.2
Biology	1	1	9	9	19	49	18	47	17	37	18	34
Chemistry	1	1	9	9	37	38	38	37	26	23	24	20
Physics	1	2	9	9	44	21	46	21	58	14	30	10

February and a small number of single-day holidays determined locally. Independent school tend to have a slightly shorter year, and some follow traditional English dates for holidays.

Class Time Allocated to Mathematics and Science

Pupils attend school for 27 hours from Monday to Friday, but there are no fixed daily or weekly timetables applying to all schools. There are variations in school opening and closing times, but in primary schools a common arrangement is for mornings of three hours and afternoons of two-and-a-half hours. The secondary school week is divided into periods, and some schools have an eight-period per day timetable, usually with 40 minutes per period. Others have five or six-period days, with periods of 55 minutes or an hour in length. In primary schools, at least 5 percent of pupils' time should be spent on science and 15 percent on mathematics. In 1993 in secondary schools, Grades 8 and 9 averaged 10 percent of pupils' time on science and 13 percent on mathematics, Grades 10 and 11 averaged 14 percent on science and 14 percent on mathematics, and Grades 12 and 13 averaged 16 percent on science and 17 percent on mathematics.

Class Size

In 1993 the average class size in primary schools was 25 pupils. In secondary school the average size in Grades 8 and 9 was 22 pupils; in Grades 10 and 11, 19; and in Grades 12 and 13, 14.

Streaming and Tracking

Teaching in Scottish primary schools in done in mixed-ability classes with pupils grouped by ability for many mathematics and English lessons. In secondary schools, mixed-ability classes operated in 97 percent of classes in 1993 in Grades 8 and 9, 67 percent in Grades 10 and 11, and 71 percent in Grades 12 and 13. Most of the remaining classes from Grades 10 to 13 were broadbanded, or composed according to academic ability as defined by the school. Setting is the most common mode of ability-grouping in mathematics. Set classes are organized within a subject department timetable to match levels of progress in that subject.

Certification of Mathematics and Science Teachers

All teachers must earn a teaching qualification and be registered with the General Teaching Council for Scotland before being employed by an education authority. Primary teachers and secondary teachers of technology, physical education, and music take a four-year course leading to a bachelor of education degree at one of five teacher education institutions. Primary and secondary teachers who already possess a university degree may take a one-year course at one of the five institutions. Entry qualifications for a four-year bachelor of education course are very similar to general entry qualifications to university, and include a pass in mathematics at Higher Grade or Standard Grade. Postgraduate certificate courses have a similar requirement.

For all preservice teachers there are three major elements in the course: professional studies, curriculum studies, and school experience. The curriculum studies element ensures that teachers have a reasonable level of competence and confidence in all areas of the primary school curriculum or specialist areas of the secondary school curriculum. The Scottish Office Education and Industry Department sets out the general and specific conditions for all teacher education courses.

In-service training is provided on both a compulsory and a voluntary basis to enable teachers to develop their own potential and to ensure they can cope with educational changes and innovations.

Teacher Profile

The average age of primary and secondary teachers in Scotland was 42 in 1992. Teachers in nurseries and special schools were marginally older at 43 and 44 respectively. Females comprised 92 percent of primary teachers and 48 percent of secondary teachers in 1992. In secondary schools, 43 percent of mathematics teachers were female, as were 4 percent of general science teachers, 4 percent of biology teachers, 25 percent of chemistry teachers, and 16 percent of physics teachers.

3 | TIMSS POPULATIONS

Scotland, IEA, and TIMSS

For many years Scotland was a member of IEA, with the Scottish Council for Research in Education holding actual membership. Membership lapsed during the 1980s, but Scotland rejoined in 1992 with membership vested in the Research and Intelligence Unit of The Scottish Office Education and Industry Department.

Scottish TIMSS Populations

Scotland joined IEA after TIMSS had started, but is participating in the study at the Populations 1 and 2 level. At Population 1 the grades involved are our Grades 4 and 5, which span ages 8 to 10. At Population 2, our Grades 8 and 9 are involved, spanning ages 12 to 14.

4 | MATHEMATICS CURRICULUM AND PEDAGOGY

Goals for the Mathematics Curriculum

The guidelines for the age 5 to 14 mathematics curriculum were published in 1991 and have been imple-

mented in most primary and secondary schools. The guidelines set out attainment targets at five levels within three outcomes: information handling; numbers, money, and measurement; and shape, position, and movement. There is also a further outcome, problem solving and inquiry, for which there are no defined attainment targets. The aims of the 5 to 14 curriculum are to enable students:

- to understand the nature and purposes of mathematics;
- to acquire skills in mathematical thinking with a supporting network of concepts, facts, and techniques;
- to develop confidence in using mathematics and to enjoy its challenges and aesthetic satisfaction.

The Standard Grade Curriculum covering the Grades 10 and 11 levels has three elements: knowledge and understanding, reasoning and applications, and investigating. It is based on the concept that problem solving lies at the heart of mathematics. Higher Grade and Certificate of Sixth-Year Studies courses are based on similar principles and cover similar work at a higher level, with courses designed to prepare pupils for the demands of adult life, employment, and further study, and to develop awareness of the importance of mathematics in society and in the development of technology.

Major Changes in the Mathematics Curriculum

The 5 to 14 guidelines are the first national curriculum guidelines to be produced in Scotland. They are based on deliberations by expert groups convened by the Scottish Consultative Council on the Curriculum, which included teachers in its membership. The guidelines set out a series of attainment targets at five levels, which are designed to provide for the progression of pupils through these stages and to give continuity between primary and secondary schools.

Beginning in 1993, Standard Grade was introduced by schools in Grades 10 and 11, providing mathematics courses for all pupils to follow up to the age of 16. Standard Grade candidates take external examinations at one of three levels: credit, general, and foundation. These examinations are set by the Scottish Examination Board and cover knowledge, understanding, reasoning, and applications. Longer investigative tasks assessed by schools also form a component of the examinations.

Higher Grade and Certificate of Sixth-Year Studies have been revised based on a similar analysis of the mathematics curriculum. As an alternative, many pupils opt to take National Certificate modules in mathematics, which are designed to be completed in 40 hours.

Current Issues in the Mathematics Curriculum

One of the current issues concerns teaching approaches. Teachers are encouraged to use a range of approaches, including exposition, discussion, teaching through activity, and teaching through inquiry, to enable pupils to experience mathematics in a range of contexts. Teachers have been asked to review their teaching practices with a view to making more use of problem-solving in mathematics.

Advances in educational technology are influencing what happens in classrooms. Teachers are making full use of graphics calculators and microcomputers in their teaching. At the same time, they must ensure that pupils have the ability to perform basic mathematical skills including mental calculations.

Continuity and progression when students transfer from primary to secondary schools is still an area of some difficulty. The introduction of the 5 to 14 guidelines has resulted in improved communication between schools, but there is a need to ensure the satisfactory transfer of information on attainment, both between and within schools.

Mathematics Textbooks

Textbooks have remained an important teaching resource in Scottish schools over the past 10 years. The textbooks currently in use have kept pace with curricular reforms in 5 to 14, Standard Grade, and revised Higher Grade. There is now more emphasis on problem solving and setting mathematics in meaningful contexts. The presentation of textbooks has been improved by greater use of color, illustrations, and graphs. In primary schools, much teaching consists of presentation to groups followed by group work using teaching schemes, or graded teaching programs for individual pupils. In Grades 8 and 9 in many secondary schools,

learning is based on individualized schemes that involve students working at their own pace through graded materials, sometimes supplemented by group investigative work. Individualized work is usually supplemented by other course components that allow for class and group teaching and for discussion. In Grades 10 to 13, teacher exposition followed by working from textbooks is the norm, with some individual work on modules.

Pedagogy

Classes in primary schools are mixed ability, and much of the teaching is done in ability groups. There is widespread use of individualized schemes, and calculators and microcomputers are used regularly, particularly in the upper stages. In Grade 8, classes are usually mixed ability. This is also common in Grade 9, although some classes are then set by ability. Grade 10 and 11 classes are normally formed by ability, with pupils taking Standard Grade courses at credit, general, or foundation level. Microcomputers and graphics calculators are increasing in use at this level.

Pupils do not need to study mathematics to Higher Grade and Certificate of Sixth-Year Studies in Grades 12 and 13. About 20 percent of the cohort take a Higher Grade course in Grade 12, and less than 5 percent a Certificate of Sixth-Year Studies course in Grade 13. In addition, many less-able pupils take a National Certificate module that caters to a wide range of mathematical abilities. Some of these modules, when taken in the fifth year of secondary education, can lead to Higher Mathematics in the sixth year.

5 SCIENCE CURRICULUM AND PEDAGOGY

Goals for the Science Curriculum

The 5 to 14 national curriculum guidelines on environmental studies, which includes science, were published in 1993. These adopt a broad definition of the environment, encompassing social, physical, and cultural aspects. The guidelines require pupils to:

- achieve knowledge and understanding of the environment;

- develop skills to enable them to interact effectively with the environment;
- progressively recognize the knowledge, understanding, and skills associated with science and technology;
- develop informed attitudes and values relating to the care and conservation of the environment.

The 5 to 14 curriculum has three science attainment outcomes: understanding living things and the processes of life, understanding energy and forces, and understanding earth and space. Learning is described in six strands: knowledge and understanding, planning, collecting evidence, recording and presenting, interpreting and evaluating, and developing informed attitudes.

The goals of Standard Grade, Higher Grade, and Certificate of Sixth-Year Studies courses are essentially the same and include mention of the need to develop knowledge and understanding, problem-solving skills, practical abilities, and positive attitudes. Most courses also draw attention to the social, economic, and industrial applications of science and to the close relationship between science and technology.

Major Changes in the Science Curriculum

Until publication of the 5 to 14 curriculum guidelines in environmental studies, primary schools had considerable leeway in deciding how much and what kind of science they would teach. As a result, there is still great variation across primary schools in the emphasis given to science. This situation will gradually improve over the next few years as the guidelines are implemented.

In the first two years of secondary education, most schools teach integrated science courses, which include equal amounts of biology, chemistry, and physics, as well as smaller amounts of other sciences. In the last 10 years there has been a shift in emphasis from scientific knowledge and understanding towards scientific skills and processes.

Over the last 10 years, Standard Grade courses have been introduced in the third and fourth years of secondary education. All pupils now study at least one science subject up to age 16. Standard Grade candidates take external examinations set by the Scottish Examination Board, which defines the syllabi. Practical abilities, including investigative skills, are assessed by the school and contribute to the pupil's overall grade.

New courses were introduced in 1991 at the Higher Grade and Certificate of Sixth-Year Studies levels in the fifth and sixth year of secondary education. There is increased emphasis on the everyday applications of science. In chemistry, for example, a new topic on petrochemicals was introduced to take account of the North Sea oil industry. There are also 40- or 20-hour modular courses covering a wide range of science topics. These are assessed by schools and national certificates are awarded by the Scottish Vocational Education Council. A major initiative called "Higher Still" is currently underway to unite the academic and vocational systems, and to better meet the needs of the larger number of pupils remaining in secondary school for Grades 12 and 13.

Current Issues in the Science Curriculum

With the introduction of the 5 to 14 curriculum, one of the emerging issues is the training and qualification of the primary teachers who are expected to implement it. Many teachers with little or no background in science have limited confidence in their ability to teach it, and many hold misconceptions about some major scientific ideas. Support materials and in-service training are being introduced to address this situation.

Another issue is the differentiation of teaching and learning geared toward pupil ability in mixed-ability classes. A broad range of strategies going beyond worksheets and similar resources is being promoted. The role of the teacher in supporting pupils' learning is becoming more widely recognized.

The transfer of pupils from primary to secondary schools has presented problems in ensuring the continuity and progression of their learning. In science, the lack of a consistent curriculum in primary schools has meant that pupils must start their learning over again at secondary school. Greater integration of primary and secondary courses and, therefore, improved continuity and progression should follow implementation of the 5 to 14 guidelines.

The use of computers in science has been increasing in recent years but there is a need for more sophistication in systems and usage in order to reap the benefits of their use. The development of educational software and the connection of schools to internet resources may have a significant effect on science education.

In order to meet the needs of the increasing number of pupils staying in school beyond the compulsory age, a major review of the upper school curriculum is being undertaken. This will address the needs of the full ability-range and will unite academic and vocational provisions in schools and in further education.

Science Textbooks

Pupils are expected to play a greater part in their own learning, for example, by following written instructions and accessing a variety of resources. Most science courses use work cards or work-sheets, either commercially or school produced, rather than a single textbook. Many of these refer to textbooks, requiring pupils to read for information, answer questions, extract data, draw diagrams, or carry out experiments. Pupils may also be encouraged to read books at home in order to answer homework questions. In general, textbooks have become more colorful with less text and more space for photographs, diagrams, and sometimes cartoons. They often contain problem-solving or practical activities, subject applications, and assessment materials. They are often tailored to the needs of particular courses, and contain little content beyond the course. In general, schools choose the textbooks they wish to use, sometimes with advice from their education authorities.

Pedagogy

In primary schools, science is typically taught within a general topic or within a more science-oriented one, but there is considerable variation across schools and even within schools. Pupils generally work in mixed-ability groups for science, and there is often a practical component to the classes.

In the first two years of secondary education, mixed-ability classes are taught an integrated science course that includes biology, chemistry, and physics. In these courses a balance is struck between teaching knowledge, and helping pupils develop scientific skills. A great deal of science teaching involves resource-based learning, in which pupils assume responsibility for their own learning by accessing and using a variety of resources independently. This is usually balanced by some direct teaching, including practical demonstrations. Laboratory work is an integral part of these courses.

The introduction of Standard Grade courses in the third and fourth years of secondary education means that all pupils study biology, chemistry, physics, or general science, the latter being taken by pupils who are less academically able. The science course is presented at three levels of difficulty: credit, general, and foundation, while the specific subject courses are only presented at credit and general level. These courses provide some coverage of the social, economic, and cultural effects of science as well as skills such as problem solving.

6 EVALUATION POLICIES AND PRACTICES

In primary school and the first two years of secondary school there is informal assessment in classrooms to monitor pupils' progress and report to parents. There are also national tests at five levels in mathematics, reading, and writing, corresponding to the targets in the 5 to 14 guidelines. Pupils take these tests, which mainly consist of short-response items, when their teachers think they are ready to move from one level to another.

At the end of Grade 11 all pupils take Standard Grade examinations, which are based on grade-related criteria at three levels. In mathematics, the examinations are mainly short-response questions, while in science they contain a mixture of multiple-choice, short-answer, and data-handling tasks. The assessments include evaluation by teachers of pupils' performance on more extended investigations or practical techniques, which are externally moderated. Standard Grade replaced the former Ordinary Grade examinations, which were intended for academic pupils in order to provide criterion-related assessment for all students.

In Grades 12 and 13, students take Higher Grade examinations and Certificate of Sixth-Year Studies examinations based on Scottish Education Board syllabi, and they may also take national certificate modules. In the case of the science examinations, there are extended investigations that are assessed by teachers and moderated by an external examiner. The modules are all assessed by schools. A development program is currently in progress to bring these examinations and modules into a unified curriculum and assessment system. The new system is aimed at higher levels of attainment, recognized qualifications for all, an even gradient of progression, good breadth attainment, and competence in core skills.

There is also a national monitoring system in Scotland called the Assessment of Achievement Programme, in which pupils' attainment is assessed in annual surveys. The subject areas assessed are mathematics, science, and English, using written and practical tasks. Surveys in each subject area are carried out at three-year intervals with samples of Grades 4, 7, and 9 pupils.

7 REFERENCES AND SOURCES FOR FURTHER READING

Common Sources

1 Husén T, N T Postlethwaite 1994 *International Encyclopedia of Education*. Pergamon Press, Oxford

2 Organisation for Economic Co-operation and Development 1995 *Education at a Glance: OECD Indicators*. Organisation for Economic Co-operation and Development, Paris

3 United Nations Educational, Scientific and Cultural Organization 1995 *Statistical Yearbook*. United Nations Educational, Scientific and Cultural Organization, Paris

4 World Bank 1995 *World Development Report 1995*. Oxford University Press, New York

Other Sources

5 *Scottish Abstract of Statistics*. 1993 Edition No 22 The Scottish Office

6 General Register Office for Scotland 1994–*1991 Census*. Gaelic Language HMSO, Edinburgh

7 The Scottish Education Department 1994 *Provision for Gaelic Education in Scotland*. The Scottish Office, Edinburgh

8 The Scottish Office Statistical Bulletin Education Series 1993 *Schools, Pupils, and Teachers in Scotland*. Government Statistical Service, Edinburgh

9 The Scottish Office Statistical Bulletin Education Series 1994 *Provision for Preschool Children*. Government Statistical Service, Edinburgh

10 The Scottish Office Statistical Bulletin Education Series 1994 *Scottish school leavers and their qualifications 1982–83 to 1992–93*. Government Statistical Services, Edinburgh

11 The Scottish Office Statistical Bulletin Education Series 1995 *Provision for Education of Pupils with Special Educational Needs*. Government Statistical Services, Edinburgh

12 The Scottish Office Statistical Bulletin Education Series 1995 *The Assisted Places Scheme 1983–84 to 1993–94*. Government Statistical Services, Edinburgh

13 The Scottish Office Statistical Bulletin Education Series 1995 *The curriculum in education authority secondary schools in Scotland 1989–93*. Government Statistical Services, Edinburgh

14 The Scottish Office Statistical Bulletin Education Series 1995 *Students Registered in Vocational Further Education in Scotland, 1985–86 to 1992–93*. Government Statistical Services, Edinburgh

15 The Scottish Office Statistical Bulletin Education Series 1995 *Scottish Higher Education Statistics 1992–93*. Government Statistical Services, Edinburgh

16 The Scottish Office Education Department 1994 *Education and Training in Scotland. A National Dossier*. Eurydice, Edinburgh

17 The Scottish Office Education Department 1993 *Curriculum and Assessment in Scotland. National Guidelines. The Structure and Balance of the Curriculum 5–14*. Edinburgh

18 The Scottish Office Statistical Bulletin Education Series 1994 *Pupils and Teachers in Education Authority Primary and Secondary Schools*. Government Statistical Service, Edinburgh

19 The Scottish Office Statistical Bulletin Education Series 1995 *Teachers in Scottish Schools: September 1992*. Government Statistical Services, Edinburgh

Statistical References

Section	Statistic	Reference	Page	Table	Year of Statistic
Country Profile	79,000 km^2	5	195	16.1	1985
Country Profile	66/km^2	5	5	1.5	1992
Country Profile	5.1 million	5	3	1.3	1992
Country Profile	1 %	5	7	1.7	1991
Country Profile	66,000	6	24	1	1991
Country Profile	50	7	61	4	1993
Country Profile	US$5494	5	144	11.1	1992
Country Profile	13%	5	188,190	15.5a, 15.7, 15.8	1992
Governance	4%	6	6,7,8,9	5,6,8,10,11	1991
Structure of the System	37%	9	1	–	1993
Structure of the System	49,504, 57%	9	10	13	1991–92
Structure of the System	76%, 43%	10	5	4	1993
Structure of the System	1%	11	2	1	1993
Structure of the System	196	11	14	5	1993
Structure of the System	3058	12	2	1	1993–94
Structure of the System	71%	13	4	2.1	1993
Structure of the System	207,055	14	3	1	1992–93
Structure of the System	167,183	15	3	1	1992–93
Structure of the System	48%	15	4	3	1992–93
Schools in the System	190, 27	16	47	–	1994
Schools in the System	5, 15	17	16	Fig. 1	1993
Schools in the System	10%, 13%, 14%, 14%, 16%, 17%	13	7	3	1993
Schools in the System	25, 22, 19, 14	18	4	2	1993
Teacher Profile	97%, 67%, 71%	13	12	6	1993
Teacher Profile	42, 43, 44	19	4	2	1992
Teacher Profile	92%	19	3	–	1992
Teacher Profile	48%	19	5	–	1992
Teacher Profile	43%, 48%, 46%, 25%, 16%	19	7	7	1992

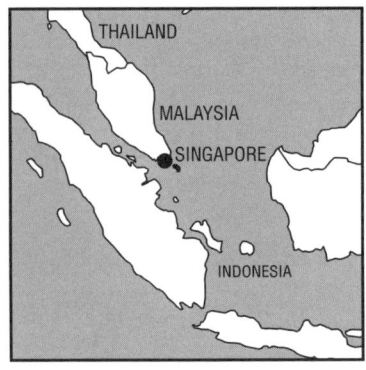

Singapore

Chan Siew Eng, Ministry of Education

Chang Swee Tong, Ministry of Education

Mary Toh, Ministry of Education

INTRODUCTION

TIMSS not only measures student achievement in mathematics and science worldwide, but also investigates differences in curriculum and instruction. Singapore's participation in TIMSS provides an opportunity to compare pupils' performance in mathematics and science to an internationally recognized standard. The study will also enable us to correlate achievement in mathematics with achievement in science.

In view of the improvements made to Singapore's education system in 1991, our participation in TIMSS is timely. The baseline data yielded by the study will be a useful yardstick for measurement of achievement in mathematics and science, and future standards will be used to chart progress against that baseline.

1 COUNTRY PROFILE

Singapore is situated approximately 137 km north of the equator. It is about 42 km in length, 23 km in breadth, and 646 km² in area. The resident population, including citizens and permanent residents, was nearly three million in 1994. The population density rose from 3893 persons/km² in 1983 to 4481 persons/km² in 1993, with Chinese residents forming 78 percent of the population, Malays 14 percent, Indians 7 percent, and other ethnic groups 1 percent. Residents below 15 years of age formed 23 percent of the population, while those above 60 years comprised nearly 10 percent. The median age of the resident population in 1993 shifted upwards to 31 years from 26 years in 1983.

Rising standards of living, health, and hygiene have reduced the infant mortality rate to under 5 per 1000 live births in 1993, compared with 12 in 1980, 21 in 1970 and 35 in 1960. The life expectancy at birth for residents also increased from 71 in 1983 to 74 years in 1993 for males and from 76 years to 78 years for females for the same period.

Singapore is a republic with a parliamentary system of government headed by a president as head of state. Authority to govern is vested in the cabinet, which is headed by the prime minister. The prime minister and the other cabinet members are appointed by the president from among members of parliament.

Singaporeans enjoy a high standard of living. The gross national product per person was US$20,414 in 1994. About 87 percent of Singapore's population live in Housing and Development Board flats, which are apartments built by the national housing authority. Other indicators of the standard of living in 1993 were 296 persons per public bus, 693 persons per doctor, 275 persons per hospital bed and 4 persons per residential telephone line. The general literacy rate, defined as the number of literate persons for every 100 residents aged 10 years and older, was estimated to be about 92 percent in 1993.

Total government expenditure, including operating and development expenditure, stood at US$9.2 billion in 1994. Of this, 22 percent was spent on education.

2 THE EDUCATION SYSTEM

Governance and Decision Making

The Ministry of Education consists of a Ministerial Committee and 10 divisions. The Ministry's role is to develop national education goals and a coordinated education program for the whole country. The Ministry aims to provide a quality education to every child in school and to ensure that each pupil reaches his or her full potential. Responsibility for curriculum development, textbook selection, instruction, and examination standards is centralized in the Ministry.

Structure of the System and Participation Rates

Structure of Education

Every child in Singapore receives at least 10 years of general education, including 6 years of primary education and 4 years of secondary education. While there is no compulsory period of education, enrollment is virtually 100 percent in the first 10 years. Secondary school leavers may proceed to technical-vocational, preuniversity, or tertiary courses. The structure of the system and the participation rates for each level are shown in Figure 1.

At the primary level, pupils study a foundation stage from Grades 1 to 4 and an orientation stage from Grades 5 to 6. The foundation stage emphasizes basic literacy and numeracy skills. All pupils at this stage follow a common curriculum that includes English, mathematics, and their heritage language, either Chinese, Malay, or Tamil as core subjects.

At the end of Grade 4, pupils are streamed according to their abilities. There are three streams at the orientation stage, EM1, EM2, and EM3. Pupils in the EM1 and EM2 streams take English, their heritage language, mathematics, and science. EM1 pupils study their heritage language at a higher level. Pupils in the EM3 stream study English, their heritage language, and

mathematics. All pupils, regardless of the stream they are in, also take other nonexamination subjects such as health education and social studies.

All students write a national placement examination at the end of Grade 6. The examination assesses their suitability for secondary education, and pupils are placed in a secondary school course appropriate to their learning ability.

At the secondary level, Grades 7 and above, the majority of students take four-year special and express courses while the rest enter a four- to five-year normal course. Pupils in the special and express courses study essentially the same curriculum, with the exception that the former study their heritage language at a higher level. Both these courses prepare students for the Singapore-Cambridge General Certificate Ordinary Level Examination in four years. Within the normal course, pupils are divided into academic and technical streams, both of which lead to the Singapore-Cambridge General Certificate of Education Normal Level Examination at the end of the fourth year. Those who do well go on to write the Ordinary Level Examination at the end of the fifth year.

Students completing secondary education may enter junior college for a two-year preuniversity course, or a centralized institute for a three-year preuniversity course. Admission is based on a point system computed from the pupil's General Certificate of Education Ordinary Level aggregate. At the end of the preuniversity course, students write the Singapore-Cambridge General Certificate of Education Advanced Level Examination, the results of which determine eligibility for tertiary education.

Students interested in technical and commercial studies may join the polytechnic stream instead of taking a preuniversity course. They may also pursue degree courses at university following graduation. Alternatively, secondary school leavers may pursue a wide range of higher-level vocational courses at the Institutes of Technical Education.

School Types

There are 192 primary schools, 144 secondary schools, five combined primary and secondary schools, 14 junior colleges, and 4 centralized institutes in Singapore. There are government schools, which are fully funded by the government; government-aided schools which were set up by religious missions and other organiza-

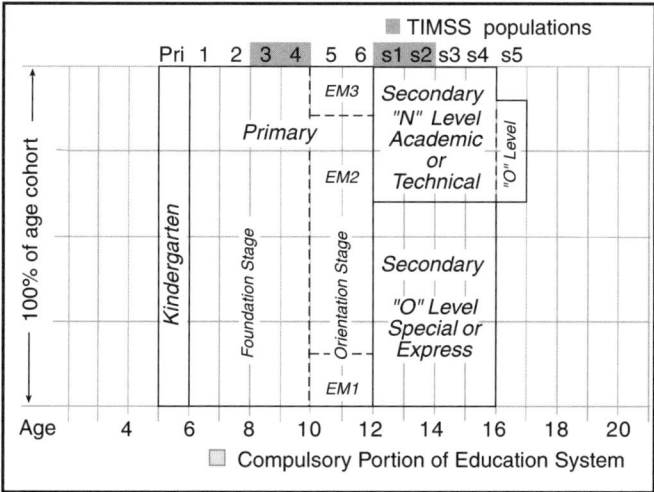

Figure 1: Structure of the Education System

tions, and whose running costs are subsidized by the government; and independent schools which receive substantial funding from the government but which are run by Boards of Governors that decide on personnel and policy matters within the institution. Some government and government-aided schools are given greater autonomy and more funds to provide a wider range of innovative enrichment programs to enhance the quality of education provided to pupils. These are known as autonomous schools. In addition, there are also the Special Assistance Plan schools where both the special and express courses are offered. These schools were established to maintain high standards in both English and Chinese while preserving the traditional ethos existing in the schools.

At the primary level, 73 percent of students are in government schools and the remaining 27 percent in aided schools. At the secondary level, 66 percent of pupils are in government schools, 19 percent in aided schools, 9 percent in autonomous schools, and 6 percent in independent schools.

Streaming and Tracking in Mathematics

All students study mathematics from Grades 1 to 10. At the secondary level, all pupils learn elementary mathematics, and students in the Grades 9 and 10 express and special course may also opt to do additional mathematics. At the end of Grade 10, special- and express-course pupils write the General Certificate of Education Ordinary Level mathematics examination, while normal-course pupils write the General Certificate of Education Normal Level mathematics examination. About 75 percent of these normal-course pupils will proceed to the next year and write the General Certificate of Education Ordinary Level mathematics examination at the end of the school year.

At the preuniversity level, most pupils study mathematics (Syllabus C) and write the General Certificate of Education Advanced Level mathematics examination. A smaller group of pupils, who are more able in mathematics, will also study for and write the General Certificate of Education Advanced Level further mathematics examination. Enrollment of females in mathematics at this level is somewhat lower than enrollment of males; 90 percent of females and 94 percent of males are enrolled in one or more mathematics subject at this level.

Streaming and Tracking in Science

Primary school students study science from Grades 3 to 6, and all students in Grades 7 and 8 learn general science. In Grades 9 and 10, students may choose various science discipline combinations, including courses in physics, chemistry, biology, combined science, and human and social biology. Students may choose one to three courses based on school recommendations and parental input.

At the preuniversity level, pupils in the science stream may specialize in physics, chemistry, biology, or physical science. Students may choose to study one to three of these subjects, based on school recommendations and parental input. Enrollment of females in science at this level is much lower than enrollment of males: 41 percent of females and 74 percent of males are enrolled in one or more science subject at this level.

Schools in the System

The School Year

The school year consists of four terms of 10 weeks each, beginning on January 2. There is a 1-week vacation between the first and second terms and another between the third and fourth terms. A 4-week break follows midyear school examinations and a long vacation of 6 weeks occurs at the end of the year.

Most schools are double-session, with pupils attending either the morning session from 7:30 a.m. to 1:00 p.m. or the afternoon session from 1:00 p.m. to 6:30 p.m. Pupils attend school from Monday to Friday and participate in extracurricular activities either before or after school hours or on Saturday.

Class Time Allocated to Mathematics and Science

In Grades 1 to 4, 20 percent of curricular time is spent on mathematics and 4 percent on science. In Grades 5 and 6, pupils spend about 20 to 27 percent of curricular time on mathematics and 6 to 10 percent on science. In Grades 7 and 8, the figures are 13 to 20 percent on mathematics and 10 to 15 percent on science; and in Grades 9 and 10, the range is between 13 and 25 percent for mathematics and 13 and 38 percent for science. At the junior college level, students spend between 0 and 37 percent of class time on mathematics, and from 0 to 51 percent on science.

Class Size

The average class size at the primary level is 37 and at the secondary level, 35. For junior colleges, the size is smaller at 22.

Certification of Teachers

Professional education of teachers is conducted by the National Institute of Education, which offers the following courses:

- A four-year undergraduate course leading to a bachelor's degree; graduates may teach in primary or secondary school.
- A one-year postgraduate diploma program in education, which trains university graduates to become teachers in primary or secondary schools. In addition, this program allows a specialization in physical education, a two-year course.
- A two-year diploma in education program, which qualifies general certificate of education advanced level holders and polytechnic diploma holders to become generalist primary school teachers. There is also provision for the education of specialist teachers in heritage languages, art, music, home economics, and physical education.
- In-service courses are also offered at the Institute for practicing teachers. These include the further professional diploma in education program, which educates teachers to be department heads, and the diploma in educational administration program, which trains vice-principals to be principals.

Teacher Profile

In 1995, there were more than 20,000 teachers in primary and secondary schools and junior colleges. About 70 percent of these teachers were female. Slightly more than one-third were university graduates. Thirty-three percent of teachers have less than 10 years of teaching experience; 19 percent have teaching years varying from 10 to 19, and the remaining 48 percent have taught for at least 20 years in Singapore. Teachers in primary and secondary schools were, on average, 41 years old while teachers in junior colleges were slightly younger at 37.

3 TIMSS POPULATIONS

Singapore, IEA, and TIMSS

Singapore became associated with IEA in 1982, participating in the Second International Science Study, the International Item Banking Project, and the Reading Literacy Study. Singapore is presently involved in the Third International Mathematics and Science Study.

TIMSS Populations

Singapore participated in TIMSS at the Populations 1 and 2 levels. Population 1 consisted of more than 14,000 Grade 3 and 4 students, the two grades containing the largest population of 9-year-olds at the time of testing. Population 2 consisted of more than 8,000 Grade 7 and 8 students, the two grades containing the largest population of 13-year-olds at the time of testing. All primary and secondary schools in Singapore took part in the main survey in October and November of 1994. All school principals and mathematics and science teachers of the selected classes were also involved in the study.

4 MATHEMATICS CURRICULUM AND PEDAGOGY

Goals for the Mathematics Curriculum

Mathematics syllabi for primary and secondary schools are developed by the Curriculum Planning Division of the Ministry of Education. They emphasize the development of mathematical concepts and skills as well as underlying mathematical processes.

The goals of the mathematics curriculum for primary and secondary schools are to enable students to:

- acquire the necessary mathematical knowledge and skills, develop thinking processes, and apply these in the mathematical situations of everyday life;
- use mathematics as a means of communication;
- develop positive attitudes towards mathematics;

- appreciate the importance and power of mathematics in the world around them.

Major Changes in the Mathematics Curriculum

The mathematics syllabi were revised in 1990, and the new documents place greater emphasis on the development of mathematical concepts and the ability to apply them to solve mathematical problems. The focus is on adopting effective teaching strategies that will bring out the emphasis of the revised syllabus. These strategies are to:
- develop concepts through meaningful activities;
- develop competence in basic skills;
- use mathematical communication, investigative, work and problem solving;
- use mathematical thinking;
- use technologies such as computers in teaching and learning mathematics.

Current Issues in the Mathematics Curriculum

The main challenge in the implementation of the revised syllabi is to ensure that pupils are actively engaged in problem-solving activities through the use of teaching strategies that reflect the emphasis of the syllabi. In-service courses are being conducted to familiarize teachers with the revised syllabi and to equip them with various teaching strategies to further improve their instruction. The provision of appropriate instructional materials, assessment programs, and in-service courses and possibly the availability of technology, will help to speed the pace at which teachers adapt to the revised syllabi.

Mathematics Textbooks

The Ministry of Education provides a list of approved textbooks and instructional materials to assist principals, department heads, senior subject teachers, and subject coordinators in selecting suitable texts for their pupils. The list includes mathematics textbooks that are commercially produced as well as those published by the Curriculum Development Institute of Singapore, which is part of the Ministry of Education.

Textbooks normally follow the intended syllabi very closely. Features include explanations, exercises, and enrichment activities, and, where necessary, illustrations, graphs, and pictures. While teachers are encouraged to use textbooks as instructional aids, they are not expected to follow them rigidly.

All students have their own mathematics textbooks, and the majority of students pay for them. Students are encouraged to read the textbooks but they do not generally learn from the textbooks on their own. They are used mainly for teaching, as well as for review and assignments.

Pedagogy

Teachers are encouraged to adopt the teaching strategies that best suit the abilities, needs, and interests of their pupils. In learning mathematics, pupils are allowed to progress from the concrete and pictorial to the abstract. They are actively involved in learning through a variety of materials and media, including computers. The teacher's role is to provide appropriate learning experiences by careful selection of resource materials, and to facilitate the learning process.

To enable pupils to develop a better understanding of mathematical concepts and problem-solving skills, the following elements are an integral part of mathematics teaching and learning:
- practical and investigative work, which provides opportunities for pupils to explore, experiment, and discuss mathematical ideas;
- mathematical communication, which enables pupils to illustrate, to interpret, to explain, and to discuss mathematical ideas and experiences in doing mathematics;
- problem solving which includes solving familiar and unfamiliar problems where a repertoire of heuristics can be applied;
- mental calculation, when the computations involved are within the pupils' mental capability.

5 SCIENCE CURRICULUM AND PEDAGOGY

Goals for the Science Curriculum

The goals of the primary science program are to:

- provide students with first-hand experience of real objects and situations in order to extend their natural interest in their environment;
- provide students with knowledge and understanding of scientific concepts and facts to help them understand themselves and the world around them;
- enable students to develop the skills and attitudes necessary for scientific inquiry and problem solving;
- equip students with the understanding necessary for responsible decision making;
- enable students to develop skills for lifelong learning.

The goals of the secondary science program are to enable students to:

- acquire knowledge and understanding;
- develop the ability to inquire and to problem-solve;
- develop positive attitudes and values towards scientific investigation and exploration, including curiosity and open-mindedness;
- promote a critical awareness of the interaction between science, technology, and society;
- develop an awareness and understanding of the environment, creating positive attitudes and behaviors toward its use.

Major Changes in the Science Curriculum

In primary schools, the emphasis is on learning content, acquiring process skills, and developing positive attitudes. At the secondary and preuniversity levels, understanding and applications are emphasized, along with process skills such as handling and interpreting data and problem solving. The assessment of skills and attitude through project work is encouraged. Topics relevant to a technological society, such as electronics, human health, food technology, and the impact of humanity on the environment, have been introduced. With the accessibility of computers, some teachers are beginning to use computer-based learning in instruction, remedial work, and enrichment.

Current Issues in the Science Curriculum

One of the concerns in the science curriculum is encouraging more pupils to embark on science and engineering courses at the tertiary level. With a greater pool of scientists and engineers, the country can move to the next phase of its development, including research and development, innovation, and design in high value-added and high-tech industries. Programs have been mounted to attract more young talents to these fields.

Science Textbooks

Primary and lower secondary levels use textbooks written by the Ministry of Education, while upper secondary levels use a variety of textbooks approved by the Ministry. At the lower levels, textbooks are used extensively by students and teachers for all activities. At the upper secondary level, a wider variety of materials is used by students. Students buy their own textbooks at all levels.

The most recent textbooks include more features such as summaries, assignment questions, suggestions for projects, and further reading. Over the last 10 years, the presentation of information has improved, and better layouts, color photographs, and more illustrations are now used.

Pedagogy

The principal role of the teacher is to provide opportunities for using the tools required to understand and internalize knowledge. These tools include both laboratory equipment and scientific inquiry skills.

Teachers use a variety of methods in science teaching to accommodate different learning styles. These methods include lectures, demonstrations, discussions, role-play, instructional films, field trips, case studies, projects and practical work, debates, and field work. When choosing a teaching method, teachers must consider factors such as student ability, the concepts, skills, and attitudes to be developed, and the availability of time and resources.

Within a given curriculum time, a variety of teaching methods are often used to capture and sustain student interest. A period of teacher-centered activity, for example, is often followed by a period of pupil-centered activity. The choice of teaching method depends largely on the topic or skill to be learned, since certain topics lend themselves to didactic teaching and others to more pupil-centered strategies.

6 EVALUATION POLICIES AND PRACTICES

Assessment of student achievement in mathematics is an integral part of the teaching and learning processes. It is meant to:

- ascertain whether learning has taken place;
- assess readiness to learn new topics;
- provide feedback to students on their progress;
- give teachers feedback on the effectiveness of their teaching;
- assist teachers in planning follow-up action;
- grade student achievement.

Assessment is used for both formative and summative purposes. Formative assessment enables teachers to identify learning weaknesses and to monitor progress during instruction. Summative assessment enables teachers to determine to what extent pupils have achieved the overall course aims, and to decide what grades to award. It is usually formal, broad in coverage, and tests a representative sample of what is taught. Assessments may be conducted informally through observing and questioning pupils as they work, or formally through examinations and standardized tests. Mathematics is tested at all grades from primary to preuniversity.

Evaluation is an integral part of the teaching-learning process in science. It involves gathering information through various assessment techniques as a basis for making value judgments and decisions, and providing information about student achievement in relation to learning objectives. With this information, teachers make informed decisions about what should be done to enhance learning or to improve teaching methods.

Assessment in science is carried out regularly and through the use of different techniques such as oral questioning, observation, assignments, and practical and written tests. Test instruments include multiple-choice, short response, essay questions, project work, and practical items. Students are tested in science at all levels from Grade 3 to preuniversity.

7 REFERENCES AND SOURCES FOR FURTHER READING

Common Sources

1 Husén T, N T Postlethwaite 1994 *International Encyclopedia of Education.* Pergamon Press, Oxford
2 Organisation for Economic Co-operation and Development 1995 *Education at a Glance: OECD Indicators.* Organisation for Economic Co-operation and Development, Paris
3 United Nations Educational, Scientific and Cultural Organization 1995 *Statistical Yearbook.* United Nations Educational, Scientific and Cultural Organization, Paris
4 World Bank 1995 *World Development Report 1995.* Oxford University Press, New York

Other Sources

5 Chan Y T 1992 *Science Curriculum Expert Questionnaire.* Ministry of Education, Singapore
6 Chang S T, K W Lee 1993 *Mathematics Curriculum Expert Questionnaire.* Ministry of Education, Singapore
7 Curriculum Planning Division, Ministry of Education, Singapore 1990 *Mathematics Syllabus Grade 1 to 4, Grade 5 & 6 (EM1 & EM2 Streams).* Ministry of Education, Singapore
8 Curriculum Planning Division, Ministry of Education, Singapore 1994 *Mathematics Syllabus Grade 5 & 6 (EM3 Stream).* Ministry of Education, Singapore
9 Curriculum Planning Division, Ministry of Education, Singapore 1990 *Mathematics Syllabus (Lower Secondary).* Ministry of Education, Singapore
10 Curriculum Planning Division, Ministry of Education, Singapore 1992 *Mathematics Syllabus for Normal (Technical) Course.* Ministry of Education, Singapore
11 Curriculum Planning Division, Ministry of Education, Singapore 1990 *Science Syllabus (Primary).* Ministry of Education, Singapore
12 Curriculum Planning Division, Ministry of Education, Singapore 1992 *Science Syllabus (Lower Secondary).* Ministry of Education, Singapore
13 Curriculum Planning Division, Ministry of Education, Singapore 1993 *Science Syllabus, Lower Sec-*

ondary Normal (Technical) Course. Ministry of Education, Singapore

14 Department of Statistics, Singapore 1995 *Monthly Digest of Statistics May 1995*. Department of Statistics, Singapore

15 Ministry of Education, Singapore 1995 *Education Statistics Digest*. Ministry of Education, Singapore

16 Ministry of Education, Singapore 1995 *Directory of Schools and Educational Institutions 1995*. Ministry of Education, Singapore

17 Ministry of Education, Singapore 1995 *Principals' Handbook*. Ministry of Education, Singapore

18 Ministry of Information and the Arts, Singapore 1994 *Singapore 1994*. Ministry of Information and the Arts, Singapore

19 Ministry of Information and the Arts, Singapore 1994 *Singapore Facts and Pictures 1994*. Ministry of Information and the Arts, Singapore

20 Schools Division, Ministry of Education, Singapore 1993 *The Implementation Guidelines for Secondary Schools*. Ministry of Education, Singapore

21 Nanyang Technological University / National Institute of Education, Singapore 1995 *Bachelor of Arts/ Science with Diploma in Education / Bachelor of Arts/ Science with Diploma in Education (Physical Educa-

tion) 1995–96*. Nanyang Technological University / National Institute of Education, Singapore

22 Nanyang Technological University / National Institute of Education, Singapore 1995 *Diploma in Education / Diploma in Physical Education 1995–96*. Nanyang Technological University / National Institute of Education, Singapore

23 Nanyang Technological University / National Institute of Education, Singapore 1994 *General Information 1994–95*. Nanyang Technological University / National Institute of Education, Singapore

24 Nanyang Technological University / National Institute of Education, Singapore 1994 *Postgraduate Diploma in Education 1994–95*. Nanyang Technological University / National Institute of Education, Singapore

25 Public Affairs Branch, Ministry of Education, Singapore, 1994 *Education in Singapore*. Ministry of Education, Singapore

26 Research and Statistics Department, Ministry of Labor, Singapore *Labor Statistics in Brief 1994*. Ministry of Labor, Research and Statistics Publication, Singapore

27 The Sunday Times (6 August 1995), Singapore *Sunday Review*. The Straits Times, Singapore

Statistical References

Section	Statistic	Reference	Page	Table	Year of Statistic
Country Profile	137 km, 42 km, 23 km	19	1	–	1994
Country Profile	646 km^2	26	Pamphlet	–	1994
Country Profile	nearly 3 million	26	Pamphlet	–	1994
Country Profile	3893/km^2, 4481/km^2, 78 %, 14%, 7%, 1% 23%, nearly 10%, 31, 26 under 5, 12, 21, 35, 71, 74, 76, 78	18	29	–	1993
Country Profile	US$20,414	27	1	–	1994
Country Profile	87%, 296, 693, 275, 4, 92%	18	30	–	1993
Country Profile	US$9.2 billion, 22%	14	104 & 105	15.2 & 15.3	1994
Schools in the System	192, 144, 5, 14, 4	16	Table of Contents	1995	
Schools in the System	73%, 27%, 66%, 19%, 9%, 6%	15	33	21	1995

Continued

Statistical References continued

Section	Statistic	Reference	Page	Table	Year of Statistic
Schools in the System	75%	6	5	–	1993
Schools in the System	90%, 94% 41%, 74%	From Examination Files: GCE ordinary GCE normal GCE advanced	–	–	1993 & 94 1994 1994
Schools in the System	20%, 4%, 20% to 27%, 6% to 10%	17	C11-12	–	1995
Schools in the System	13% to 20%, 10% to 15%, 13% to 25%, 13% to 38%	17	C21-28	–	1995
Schools in the System	0% to 37%, 0% to 51%	17	C41-50	–	1995
Schools in the System	37, 35, 22	15	14	6	1995
Teacher Profile	more than 20,000, 70%	15	7	2	1995
Teacher Profile	slightly more than a third	15	9	3	1995
Teacher Profile	33%, 19%, 48%	15	20	11	1995
Teacher Profile	41, 37	15	19	10	1995
Timss Populations	more than 14,000, more than 8000	Statistics from Timss Main Data Collection			1994

Slovak Republic

Silvia Matúsová, National Institute for Education

Mária Berová, National Institute for Education

Renata Tothova, National Institute for Education

1 COUNTRY PROFILE

The Slovak Republic covers an area of 49,039 km² and is situated in central eastern Europe. It has a population of 5.3 million, consisting of 86 percent Slovaks, 11 percent Hungarians, and almost 4 percent Romany, Czechs, Ruthenians, Ukrainians, and Germans. The average population density is 106 persons/km². The official language is Slovak.

The Slovak Republic has been an independent nation since January 1, 1993, the date of the dissolution of the former Czechoslovakia. On January 19, 1993, the Slovak Republic became a member of the United Nations and on June 30, 1993, a member of the Council of Europe. The Slovak Republic is governed by a democratically elected parliament, the sole legislative body. It has 150 deputies who are elected for a four-year term in direct elections. The president is elected for a five-year term by a three-fifths majority in parliament. The president has the power to appoint and dismiss the prime minister and the members of government.

The Slovak Republic has had a declining birth rate since 1980. In 1995, 23 percent of the population was under the age of 14. This figure is expected to drop to 20 percent by 2015. The proportion of the population over 65 rose from 9 in 1970 to 11 percent in 1995, and is expected to reach 12 percent in 2015.

Registered unemployment, after peaking at 15 percent in 1994, is in decline and will reach 13 percent by 1996. There were 3.2 million people in the workforce in 1995. Youth unemployment was registered as 340,000 people in May of 1995, with 12 percent of these being school leavers.

Like other postcommunist countries, Slovakia's economy is in transition. Stringent policies have yielded a convertible currency and the lowest inflation rates among the postcommunist countries. The growth of the gross domestic product has gone from minus 4 percent in 1993 to almost 5 percent in 1994 and to more than 6 percent in 1995. The per capita GDP was US$2,505 in l994.

Inflation, reported at 23 percent in 1993, declined to 11 percent in 1995. Spending on education totaled 6 percent of net government expenditure in 1994. Per-pupil expenditure in 1995 was US$305 in gymnasia, US$367 in vocational schools, and US$600 in special education.

2 THE EDUCATION SYSTEM

Governance and Decision Making

Slovakia's bill of fundamental rights and freedoms includes the right to education and establishes a legislative framework for education. In the former Czechoslovakia, responsibility for education was divided between Slovak and Czech national ministries of education, with no unified federal body. The transition to independence was therefore relatively smooth as it related to education.

The language of instruction is Slovak, although there is some teaching in minority languages in regions where a linguistic minority such as Hungarian or Romany exists. All minority-language students, however, must study Slovak. There is a proposal to offer courses in Slovak as a second language in order to assist in the integration of minority groups into higher education and the labor force.

School administration is carried out by head teachers, local or regional educational authorities, and the

Ministry of Education. Head teachers report to school boards and carry out administrative duties such as curriculum implementation, maintenance of educational standards, and financial management of schools. Local education authorities in 42 districts administer the first and second stage primary schools, and four regional educational authorities administer secondary schools. Local and regional educational authorities allocate funds for salaries, textbooks, and the maintenance of school facilities.

The Ministry of Education is the main policy-making body and has budgetary responsibility and control. In addition, it determines the intended curricula for primary and secondary schools through the national institute for education. Curriculum specialists at the institute, together with committees of teachers, inspectors, and teacher educators write mandatory curricular documents, which they submit to the Ministry of Education for approval. The most recent complete curriculum revision took place between 1976 and 1984, and some changes and amendments occurred after 1989.

Education is provided by the state, but parents pay for supplementary teaching and consumable materials. Schools choose books from a Ministry-approved list, and textbooks are provided free of change at all preuniversity levels. In general, the Ministry list contains at least one Slovak textbook per subject as well as a limited number of minority-language books.

Structure of the System and Participation Rates

Structure of Education

The school system consists of three levels: preschool and primary education, secondary education, and higher education. The structure of the system, along with approximate enrollment rates for each level, is shown in Figure 1.

Children up to the age of three may attend crèches managed by the Ministry of Health. Children aged three to six may attend kindergarten, although preference is given to children five years and older, and two-year-olds are admitted if there is space. Over 90 percent of the age cohort attends kindergarten.

Compulsory education lasts nine years. It covers four years of first-stage and four to five years of sec-ond-stage primary school. Grade 9 is taken either at primary school or within secondary education. First and second-stage primary schools provide pupils with an education in general subjects such as Slovak and mathematics, as well as in health, environmental, physical, and religious education. Many primary schools are small: 40 percent enroll fewer than 100 pupils.

The transition rate into the upper levels of schooling is high, with 98 percent of second-stage primary school students entering secondary school. Just over 1 percent of primary school leavers entered the labor market or were registered as unemployed in 1992–93.

School Types

There are three types of upper secondary schools: gymnasia or academic schools, vocational schools, and apprentice schools. Approximately 21 percent of the cohort is enrolled in gymnasia, and of these 7 percent attend denominational or private gymnasia. The principal goal of gymnasia is to prepare students for further study at institutions of higher education. They are subdivided into four-year standard, five-year bilingual, and eight-year specialized programs. The standard program begins in Grade 9, while the eight-year program begins in Grade 5. Bilingual gymnasia, first introduced in 1990, enable graduates to continue their studies at institutions of higher education abroad. In 1995 there were 14 bilingual gymnasia offering instruction in English, French, German, Italian, and Spanish. The eight-year gymnasium offers an internally differentiated academic program for higher-ability students after completing Grade 4 and passing an entrance examination.

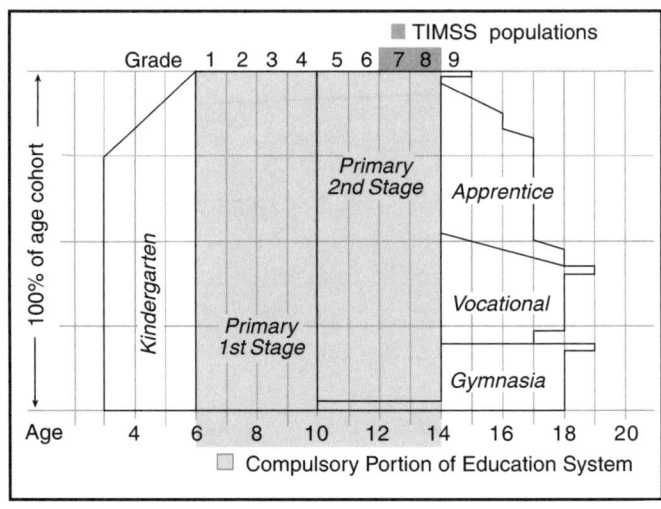

Figure 1: Structure of the Education System

In 1995, 101 schools were registered in the network as providing the eight-year program.

Vocational schools, sometimes referred to as secondary specialized schools, offer four- and five-year programs. The transition to a market economy has profoundly affected vocational education and training. Many new areas of study have been introduced, notably technical and information services, commerce, banking and finance, hotel services and catering, manufacturing industries, engineering, chemistry, civil construction, food processing, health care, and social services.

Other school types include conservatories, schools of forestry, agriculture, viticulture, fine arts, library science, education for kindergarten teachers, home economics, and specialized schools for girls. All of these prepare students for occupations in technology, economics, education, health care, art, social work, culture, and administration, and are attended by a total of 32 percent of the age cohort.

Secondary apprentice schools prepare students for skilled occupations. These schools are attended by approximately 46 percent of the cohort, but enrollment has been declining in recent years. The majority of training courses last three years, about 10 percent last two years, and 10 percent four years. General academic and vocational education is organized within schools, while practical training takes place in the work place on two days per week.

Public and Private Systems

Private schools have existed at all school levels since 1990, although the number of nonstate preschool and primary schools is relatively low. In 1994–95, 86 out of 2,480 schools were denominational. Nonstate schools receive subsidies from the state budget, but these are lower than those received by state schools. Most private schools make up the balance through fees, or donations from churches, the community, or private donors.

Schools in the System

The School Year

The school year begins the first week of September and ends the last week in June. The school year is divided into two terms of 18 to 20 weeks, with a Christmas vacation of 2 weeks, a spring vacation of 1 week, and several midterm or other breaks of three days in length. The summer vacation between school years lasts for 9 weeks. There are between 190 and 200 instructional days per year, as determined by the Ministry of Education.

Students in Grades 1 to 4 attend five-hour school days, while students in Grades 5 to 8 attend six-hour days, five days per week from Monday to Friday. The school day includes approximately one to one-and-a-half hours for breaks, room changes, and lunch. Instructional time ranges between 21 and 25 lessons per week in Grades 1 to 4, and between 28 and 30 lessons per week from Grades 5 to 8. Gymnasia have between 30 and 31 lessons per week, while all other secondary school types have 30 to 34. Most lessons are 45 minutes long.

Class Size

The average number of students in primary classes is 24, although the maximum allowed is 34. The average class size in secondary schools is 32, at secondary specialized schools 31 and at secondary apprentice schools 24. All secondary schools have a maximum class size of 38 students.

Streaming and Tracking

Extended instruction in foreign languages, mathematics, science, and physical education and sport have been established at various levels. Extended instruction in mathematics begins in Grade 5; extended instruction in physics begins in Grade 6, and extended instruction in chemistry does not start until Grade 7. In 1993–94, just over 1 percent of all students attended extension classes in mathematics and science. Students in extension programs generally receive two additional lessons per week.

Teacher Certification

The Ministry of Education establishes policies concerning teacher education and certification. Currently, primary and secondary school teachers must graduate from a faculty of education or a subject faculty within a university. There is also a four-year program to qualify for teaching both stages of primary school, and a five-year program to qualify for teaching both second-stage primary and secondary schools. Students usually choose two subjects as majors, such as mathematics and physics, and generally take a supplementary pedagogical study program. The minimum period of preservice teacher education is four years, including pedagogical studies and practice. Within the school, beginning teachers are supervised by experienced teachers for a one-year probationary period.

In-service training is available at the National Institute for Education, at four regional teacher in-service training centers, and the teacher training departments of the local education authorities. Teacher in-service programs are not meeting the needs of practicing teachers. Adequate training in curriculum planning, for example, is not available. Individualized teaching and learning and the teaching of higher-level thinking skills and problem solving also need to be addressed. Training in understanding attainment standards as well as in process planning and curriculum development should be stressed in initial teacher education. A change from subject-oriented to student-oriented instruction should be promoted in order to achieve better educational results in schools and in teacher training.

Teacher Profile

Basic data on teachers in Slovakia appears in Table 1. With the exception of higher education, the majority of teachers are female at all levels. Over 40 percent of primary teachers are more than 50 years old. This situation is considered critical since there are not enough young teachers to replace this age group when it retires. Preservice teachers represent only 3 percent of primary school teachers. Many preservice teaching students do not plan to apply for a teaching post. Developments in the private sector are facilitating the continuation of these trends; people tend to prefer profes-

sions other than teaching, since at present teachers are underpaid. Numerous highly qualified specialists are leaving schools because of their low salaries. The relatively low economic status of teachers and lack of incentive on the part of teachers to participate in change are major barriers to the improvement of the education system. Teachers tend not to participate in school-level decision making, to learn new curricula and teaching methods, or to enroll in in-service education. There is a critical shortage of teachers of foreign languages and computer science, principally because of the availability of lucrative jobs outside education.

3 TIMSS POPULATIONS

Slovak Republic, IEA, and TIMSS

The Slovak Republic has been a member of IEA since 1991, the same year it joined TIMSS. Until 1994, both the Slovak and Czech Republics participated together in the study. The National Institute for Education is the national center for IEA studies in the Slovak Republic. This Institute is directly governed by the Ministry of Education. Before joining TIMSS, the Slovak Republic was an observer country in the Computer Education Study and the Reading Literacy Study.

Slovakian TIMSS Populations

Slovakia is participating in TIMSS at the Population 2 level, which includes Grade 7 and 8 students. These grades included students aged 13 to 15, with 50 percent of students aged 14 at the time of testing. The

Table 1: Basic Data on Teachers in Slovakia

Type of School	Full-time Teachers	Women (%)	Part-time Teachers	Qualified Teachers (%)
Nursery Schools	17,218	100	–	n.a.
First- and Second-Stage Primary Schools	39,867	82	–	82
Gymnasia	4,659	67	719	95
Secondary Vocational (Technical) Schools	7,812	60	3,037	91
Secondary Apprentice Schools	6,112	61	1,007	91
Higher Education	8,103	34	1,248	100

average age of pupils in Grade 7 was 13.29 years old and in Grade 8, 14.26 years old. There was equal representation of boys and girls in the study. The sample included students from regular primary schools, eight-year gymnasia, and schools with extended programs in mathematics, physics, foreign language, or physical education. Special schools for children with mental or physical disabilities and minority language schools were excluded from the sampling.

4 MATHEMATICS CURRICULUM AND PEDAGOGY

Goals for the Mathematics Curriculum

The Ministry of Education issues curricula for all schools types, according to suggestions made by subject committees made up of education experts, school principals, and methodology specialists in the particular subject. The mathematics curriculum includes a core curriculum for specific school types. Introductory comments define the tasks and aims of mathematics education and explain the basic concept of education for a given type of school. The content is classified into particular study levels and topics. For each grade the number of lessons per week is suggested, as well as a total number for the school year. There are several goals in mathematics education: to reason through logical deduction; to learn how to think creatively and precisely; to learn how to use mathematical knowledge for problem solving; and to learn the skills to create and interpret mathematical models.

Major Changes in the Mathematics Curriculum

The current mathematics curriculum was approved by the Ministry of Education in 1991. Changes include the following:
- The overall number of compulsory classes decreased from 18 classes a week to 14 in the four grades of gymnasium.
- Lessons have been divided into two groups, basic and extra. Basic lessons are concerned with the pre-

scribed new material, while extra lessons are for content reinforcement or the teaching of additional topics, as determined by the teacher.
- Teachers are to take increased responsibility for the quality of mathematics teaching. They now decide the order in which topics are taught, their difficulty level, and their distribution among grades.
- Principals may alter the recommended plan by increasing or decreasing the number of lessons per week.
- Originally, programming and algorithms were included in the curriculum. These topics were later excluded, although informatics and computing technology are now taught at secondary schools.

Current Issues in the Mathematics Curriculum

Recently, subject standards that define fundamental content were introduced for both the first and second stages of primary schools. Based on these standards, evaluations of students and schools will be performed in the near future. A definition of the standards required for graduation will make it possible to discontinue using entrance examinations for secondary schools. As of the end of 1995, the standards were still being drafted for secondary schools.

New educational issues, such as the focus on the acquisition of facts and rational thinking, are closely connected to changes in Slovakia's political system. The core of contemporary education is designed around items deemed important for today and the future: communication; information processing; problem solving; and understanding of self, others and nature. The abilities specific to mathematics education that should be cultivated include: the utilization of empirical life experiences; development and maintenance of a natural hunger for understanding; evaluation of information; and critical reasoning.

Mathematics Textbooks

Until 1990, schools used uniform textbooks approved by the Ministry of Education. Textbook publishing houses now prepare sample books in accordance with the curriculum and submit them in a competition. A board of 15 teachers reviews and rates the samples and submits them to the Ministry of Education for approval.

The Ministry chooses one textbook per subject, and this is used throughout the country.

The competition for board approval means new textbooks tend to have better, more interesting layouts and employ more color. Most mathematics textbooks include text providing definitions, rules, propositions, and evidence. However, they also incorporate questions and stimuli for reflection, including pictures, illustrations, schemata, tables, and graphs. Some books also devote space to motivational elements such as the practical use of mathematical knowledge in various areas of technology, economic, and social practice, and the historical development of mathematics and its applications. Attached to the textbooks is usually an exercise book that includes problems, solutions, problem analysis, and construction of mathematical models of real situations. Nontechnical language is used in the most recent textbooks in order that students may use the books for self-study.

Pedagogy

Due to the rigidity of the approved curriculum and the heavy load of material to be learned, the learning process is strongly teacher based. Students are able to acquire knowledge and memorize facts, but seem to have problems with higher-level thinking such as interpreting data and problem solving. The next curriculum revision should promote these abilities.

Mathematics teaching tends to be spiral: the same topics are covered at a deeper level from year to year according to the students' maturity. In successive years, efforts are made to provide a context within other subjects for mathematical concepts. From a methodological point of view, students are becoming active participants in mathematical activity and are learning about mathematics as a medium.

A one-year, compulsory computer subject, informatics, was introduced in secondary schools in 1986 with the goal of making all students computer literate. Groups of up to 10 students attend classes in the computer laboratories which are a feature of all secondary schools. Equipment quality varies markedly between schools, with some urban secondary schools having modern multimedia laboratories with connections to the internet, while others have equipment of lower quality. The lack of educational software is an additional problem.

The didactic system of mathematics was designed to give pupils in Grades 5 to 8 many opportunities for using algorithms. The content is based on knowledge gained in informatics and computer classes, and the effectiveness of these classes is increased by the use of pocket calculators. The education process is directed at acquiring not only knowledge but also skills in different forms of counting, including counting with a calculator.

The algorithmic character and its application to practical task solving is stressed in all curricular documents, including curriculum guides and textbooks. The use of calculators for accuracy and speed is emphasized in the following topics: decimal numbers, percentages, area, volume and surface of solids, power, goniometrics. Calculators are not considered an educational aid, that is, they are not provided free of charge by the school. Recently, however, the calculator has become an essential part of numeric counting in mathematics and natural science subjects, and all students own their own calculators.

5 SCIENCE CURRICULUM AND PEDAGOGY

Goals for the Science Curriculum

The Ministry of Education issues curricula for all school types according to suggestions made by a subject committee of education experts, school principals, and methodology specialists in the particular subject. In Slovakia, science is not taught as an integrated subject; physics, chemistry, biology, and geography are taught separately. The curriculum emphasizes understanding the subject and its relationship to other natural science subjects, at the appropriate level. It also defines a time frame and lessons for practical exercises. In primary schools, 3 percent of class time is spent on physics, and 8 percent on chemistry and biology, while a number of field trips and excursions are recommended for geography. In gymnasium, 23 percent of class time is devoted to physics, 14 percent to chemistry, and 25 percent to biology.

The curriculum has been in place in primary schools since 1988, and in secondary schools since 1990. The geography curriculum was updated in 1994. The goals of science education are:

- In physics: the acquisition of knowledge and skills that will enable students to understand and explain phenomena, processes, and laws of the real world.
- In chemistry: the acquisition of an overview of fundamental notions of chemistry, an awareness of a range of inorganic and organic substances, an understanding of the fact that features of chemical substances influence their composition, and an understanding of activities relating to experiments.
- In biology: the acquisition of current biological and geological terminology, skills and habits in observing and recognizing natural materials, the fundamentals of environmental thinking, and an understanding of sanitary principles relating to health.
- In geography: knowledge of phenomena and processes taking place in a landscape, and an understanding of the laws of its development in order to cultivate feelings of personal responsibility for the state and the landscape.

Major Changes in the Science Curriculum

Teachers now have greater independence from the syllabus, broadening the scope of their work considerably. As a result of social changes there has been more contact with foreign countries and thus the possibility of incorporating new trends into science teaching. A general reduction in the quantity of material to be taught has taken place, especially in physics and chemistry. At the same time the number of teaching hours for these subjects has been reduced. In biology the number of teaching hours has not changed, while those for geography have increased.

New topics such as physiology, molecular biology, genetics, ecology, and ethnology have been introduced into the biology curriculum. Since 1994, animal dissections and laboratory exercises with animals have not been performed.

In 1992, the Schools for Health project was launched with the participation of kindergartens, elementary, and secondary schools. The fundamental aims of this project concentrate on nutrition, physical education, stress, marriage and parenthood, as well as the purposeful spending of leisure time, philanthropy, and good human relations.

An integrated science teaching project is being tested in selected eight-year schools. It is possible that the results of this project will influence the formation of a new syllabus.

Current Issues in the Science Curriculum

Recently, there has been a great effort to orient teaching goals to methods of cognition. A process of basic content definition is now taking place, with additional definition for individual school types. This process is complete for primary schools and the curriculum comprises newly drafted subject goals, including a demand for increased environmental and health education in each science subject. Beginning in 1996, the content for secondary schools will be restructured and reduced in the same manner. It is expected that in physics and chemistry the number of practical activities will be increased, while in biology activities requiring direct sensual perception and observation will be included.

Science Textbooks

University professors, secondary school teachers, and education researchers write science textbooks following the official curriculum. The publishers pay the authors and submit the books in a competition. The Ministry of Education selects the books to be used, but teachers may use complementary literature that they have chosen themselves.

Textbooks are designed for individual study and serve as a starting point for assignments and daily lessons. The majority of textbooks used in Slovakia have an identical format. They contain explanatory readings, questions, exercises to test the comprehension of the readings, and ideas for reflection. In physics, chemistry, and biology there are also laboratory exercises.

Textbook use depends on the individual teacher. Students do not use textbooks at all in certain cases; in others, individual study is encouraged. In elementary school, pupils tend work from textbooks with the help of teachers, while in secondary school students are encouraged to work on their own.

Elementary school textbooks are full color, with pictures and photographs taking up two-thirds of the book. These elements are seen as motivating students to read and learn. Secondary school textbooks, on the other hand are largely black and white, with text taking up most of the space, few diagrams, and no photographs. Some books do, however, contain short historical comments, portraits of noted scientists, and anecdotes. In general, textbooks are only suitable for individual study for intellectually capable students, since they contain a great deal of professional terminology and tend to use an advanced writing style.

Pedagogy

Science subjects are taught by specialists, either graduates of the faculty of pedagogy or of natural sciences. Teachers are free to choose their preferred teaching method. The explanatory method is still widely employed, because the amount and complexity of subject matter is often too extensive to be dealt with in more progressive but time-consuming ways. Contemporary trends, however, lean toward the active participation of students in the education process. Problem solving and group learning are promoted as well.

Conceptual change as a teaching method is currently being tested in a sample of schools. The basis of this method is the teacher's active work with the student's primary concept of a phenomenon. This method aims at profound comprehension of the phenomena rather than shallow memorization.

Great attention is paid to the role of experiments, through which students gain valuable and lasting knowledge. A well-equipped network of school laboratories for physics, chemistry, and biology has been created to meet this requirement. The curriculum emphasizes the need to form intersubject references; this is being realized through the introduction of common science concepts into individual subjects.

Computer technology is used to an increasing degree in science teaching since schools now have a wide choice of educational software. However, the varying standard of computer equipment in schools remains a problem.

In recent years an effort to change schools into humanistic, creative institutions has been growing in Slovakia. Schools should not only give information and prepare students for their future occupations, but also develop the individual, the noncognitive elements of the personality, and value systems and moral standards.

6 EVALUATION POLICIES AND PRACTICES

Evaluation is considered part of the educational process, and includes informative, corrective, and motivating functions. In addition, it has a didactic function, that of ensuring students' knowledge, and a guiding function in relation to the adoption of knowledge and skills. In the educational process, students are regularly evaluated and have a right to know the result of each evaluation. Assessment is carried out by means of grading and verbal evaluation according to directives issued by the Ministry of Education.

There are two types of evaluation: progressive and summative. The progressive type has a motivational function, and considers personality and age differences among students, as well as mental and physical abilities. Summative evaluations are carried out at the end of each semester, and evaluate the level and quality of knowledge and skills within a given subject. The school year lasts 10 months, beginning on September 1 and finishing on June 30. This period is divided into four parts. During each reporting period students usually undergo several oral examinations, as well as a minimum of two written examinations. Twice during the school year a written assessment of students' knowledge is provided. Four written mathematical assignments, prepared by teachers and based on topics from the curriculum, are completed during the school year.

Rules of evaluation in elementary and secondary school students are established by the Ministry of Education. Students in their final year of elementary school write entrance examinations to determine acceptance to secondary school. These examinations test the abilities required for success at the chosen type of school. At the end of secondary school all students write a school leaving examination or *Maturita,* consisting of a minimum of four subjects: Slovakian, mathematics or a second language, and two optional subjects. Students may write a fifth subject if they wish. Students must pass this examination in order to write university entrance examinations. The aim of the school leaving examination is to test students' ability to perform in their future occupation, or their ability to pursue further studies at university.

Practical examinations are an important part of the evaluation of specialized secondary students, as well as of elementary students who are examined in science subjects. These practical examinations are to show the skills gained in projects, computer programs, or experiments to verify laws of nature. In general, teachers create their own forms of evaluation, tests being only one of the many common forms. Tests are usually created and used only within individual schools.

7 REFERENCES AND SOURCES FOR FURTHER READING

Common Sources

1 Husén T, N T Postlethwaite 1994 *International Encyclopedia of Education*. Pergamon Press, Oxford
2 Organisation for Economic Co-operation and Development 1995 *Education at a Glance: OECD Indicators*. Organisation for Economic Co-operation and Development, Paris
3 United Nations Educational, Scientific and Cultural Organization 1995 *Statistical Yearbook*. United Nations Educational, Scientific and Cultural Organization, Paris
4 World Bank 1995 *World Development Report 1995*. Oxford University Press, New York

Other Sources

5 Beníčková M, L Horáková 1994 *Základné skoly v SR [Primary Schools in the Slovak Republic]* Ustav informácií a prognóz SMT, Bratislava
6 Gábor O, O Kopanev, K Krisalkovie 1989 *Teória vyuèovania matematiky 1*. Slovenské pedagogické nakladatestvo, Bratislava
7 Hejny M 1989 *Teória vyuèovania matematiky 2*. Slovenské pedagogické nakladatesstvo, Bratislava
8 Institute of Information and Prognoses of Education, Youth and Sports 1993 *Education in Figures — the Slovak Republic*. Institute of Information and Prognoses of Education, Youth and Sports, Bratislava
9 Institute of Information and Prognoses of Education, Youth and Sports 1994 *Statistická rocenka skolstva SR 1994. [Statistical Yearbook of Education SR.]* Institute of Information and Prognoses of Education, Youth and Sports, Bratislava
10 International Conference on Education, 44th session, Geneva 1994 Development of Education. *National Report of Slovakia 1992–1994*. Institute of Information and Prognoses of Education, Youth and Sports, Bratislava
11 Kristín J 1995 *Standards, Curricula, Modules, Teaching Materials and Tests in Vocational Education and Training*. National Institute for Education, Bratislava, SPU
12 Kristín J 1992 *Tradition and Transitions: Slovak Vocational Education and Training on the Crossroad*. CEDEFOP, Berlin
13 Ministry of Education SR 1995 *Curriculum Guide for Chemistry for Grades 7–8 of Primary School*. Príroda a.s., Bratislava
14 Ministry of Education SR 1995 *Curriculum Guide for Geography for Grades 5–8 of Primary School*. Príroda a.s., Bratislava
15 Ministry of Education SR 1995 *Curriculum Guide for Gymnasia (four-year study program)*. Príroda a.s., Bratislava
16 Ministry of Education SR 1995 *Curriculum Guide for Mathematics for Grades 5–8 of Primary School*. Príroda a.s., Bratislava
17 Ministry of Education SR 1995 *Curriculum Guide for Natural Study for Grades 5–8 of Primary School*. Príroda a.s., Bratislava
18 Ministry of Education SR 1995 *Curriculum Guide for Physics for Grades 6–8 of Primary School*. Príroda a.s., Bratislava
19 The Economist Newspaper Limited 1995 *Business Central Europe, The Annual*. December 1995, ISSN No: 1350–1240
20 The Economist Newspaper Limited 1995/96 *Business Central Europe*. December 1995/January 1996, Volume 3, Number 27, ISSN No: 1350–1240
21 Vantuch J, M Kostelníková 1994 *Problems of Teachers and Teacher Education in Slovakia*. Yearbook of Teacher Education. Brussels
22 Written Communication between Ministry of Education SR and National Institute for Education 1996 *Conception of Secondary Schools*. No: 9/96–1521

Statistical References

Section	Statistic	Reference	Page	Table	Year of Statistic
Country Profile	49,039	1	101	–	1995
Country Profile	5.3 million	19	32	–	1994
Country Profile	86%, 11%, 4%	1	101	–	1995
Country Profile	106	1	101	–	1995
Country Profile	23%, 20%	1	101	–	1995
Country Profile	9%, 11%, 12%	1	101	–	1970, 1995
Country Profile	15%,13%	19	32	–	1994, 1996
Country Profile	3.2 million	19	32	–	1995
Country Profile	340,000, 12%	1	101	–	1995
Country Profile	-4%, 5%, 6%	19	32	–	1993, 1994, 1995
Country Profile	US$2,505	1	101	–	1994
Country Profile	23%, 11%	19	32	–	1993, 1995
Country Profile	6%	1	101	–	1994
Country Profile	US$305, US$367, US$600	1	101	–	1995
Structure of the System	90%	10	13	–	1994
Structure of the System	40%	1	102	–	1994
Structure of the System	98%, 1%	1	103	–	1994
Structure of the System	21%, 7%	1	103	–	1994
Structure of the System	14	1	103	–	1995
Structure of the System	101	22	21	–	1995
Structure of the System	32%	12	30	–	1994
Structure of the System	46%	12	32	–	1994
Structure of the System	10%, 10%	12	32	–	1994
Structure of the System	86	9	52	–	1994
Structure of the System	2,480	9	17	–	1994
Schools in the System	1%	5	18	–	1994
Teacher Profile	Table 1	21	298	Table 1	1994
Teacher Profile	40%	21	298	–	1994
Teacher Profile	3%	21	298	–	1994

Slovenia

Marjan Setinc, Educational Research Institute

1 COUNTRY PROFILE

Slovenia is located in central Europe and shares borders with Italy, Austria, Hungary, and Croatia. Slovenia has only two cities with more than 100,000 inhabitants, 13 towns with between 10,000 and 50,000, and 40 towns between 5000 and 10,000. Approximately, 75 percent of the population of two million live in cities and towns, while the balance live in the more than 5000 hamlets or villages.

Slovenia has a total area of 20,000 km^2. Half of the territory historically populated by Slovenes is now part of Austria and Italy; Slovenes living there have at times been subjected to rigorous assimilation policies. Many Italian and Austrian Slovenes emigrated overseas, and many have since returned to Slovenia. After the Second World War, Slovenia accepted large numbers of immigrant workers from Bosnia, Serbia, Croatia, and Macedonia. Today these groups constitute almost 9 percent of Slovenia's population. Slovenian is the official spoken and written language, but Hungarian and Italian are also recognized in the territories where these minority languages predominant.

Some years before it gained independence in 1992, Slovenia adopted a multiparty parliamentary system of government. Since that time, three branches of power have been established: legislative, executive, and judicial. Elections are based on the proportional system with regional representation adjustment.

Slovenia has one of the lowest birthrates in the world. In 1994, the ratio of births to deaths was 19,463 to 19,359. This ratio is quite low compared with those of the 1960s and 1970s, when on average 30,000 children were born and 18,000 people died. From 1980 onward, the birth rate declined by an average of 3 percent per year.

Slovenia's per capita gross national product in 1995 was US$9,210, placing it among the upper third of upper-middle-income countries. After the dissolution of Yugoslavia, economic growth slowed considerably, reaching a low of minus 8 percent in 1991. This figure has increased steadily since then, reaching 5 percent in 1994 and was estimated at 6 percent for 1995.

Slovenia's spending on education constitutes 18 percent of total government expenditure. The total sum spent on education is somewhere between 5 and 6 percent of the gross national product; this includes central government expenditure plus the expenditures of self-governing municipalities and local communities. There is no illiteracy in Slovenia.

2 THE EDUCATION SYSTEM

Governance and Decision Making

Public education in Slovenia is funded through the state budget, and schools are financed directly by the Ministry of Education and Sport. During the past 40 years, education in Slovenia was strongly centralized and allowed very little diversity. The goal was to provide an equal education, in both academic and social terms, to all students. During the 1980s, individual schools and teachers began to use different teaching approaches, and are now encouraged to do so.

The education system from preschool to tertiary education is still centrally administered by the Ministry of Education and Sport, but universities are fully autonomous. The council for preuniversity education is a permanent advisory body elected by parliament. Its 20 members represent various institutions involved in education, such as research institutes, universities, and counseling centers. The council's role is to give advice to the minister on curriculum and on measures to be carried out at the school level, including the number of periods per school year, the maximum number of students per class, and other matters. The council also approves all textbooks, although textbook selection is done by teachers if more than one is available. Instructional practices in public schools vary, but tend to be determined by teacher's educational background; teachers do, however, have considerable freedom in instructional approaches. Schools are managed by a head or principal who is responsible for pedagogical process as well as for all financial management except salaries.

The quality of school facilities is quite evenly distributed across the country, since general requirements are nationally defined and vary only slightly based on the wealth of the municipality. If a large discrepancy exists among municipalities, national authorities provide the means to bring the schools up to standard. Every school has a library.

Structure of the System and Participation Rates

Structure of Education

The structure of the education system and the approximate enrollment rates for each level are shown in Figure 1.

Approximately 52 percent of children between 2 and 7 are enrolled in preschool or day-care institutions.

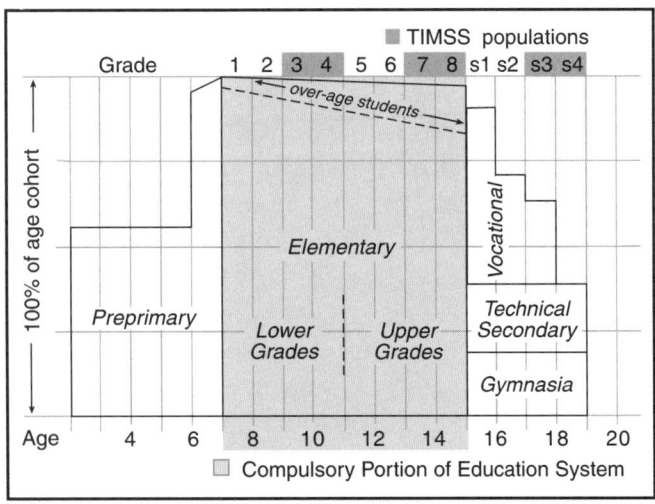

Figure 1: Structure of the Education System

Children enter school at the age of seven and must complete four lower and four upper grades of elementary school. At present, the minimum leaving age is 15, which corresponds to the end of elementary school. Enrollment at the elementary level is virtually 100 percent. According to the new education act, children will enter school at the age of six and remain in elementary school for nine years.

Forty-nine percent of all students are female. Fifty-seven percent of children up to three years of age and 53 percent of those between 3 and 5 are in kindergartens. Ninety percent of children aged 6 are enrolled in kindergarten preschool preparation classes; the 40 to 50 percent who are not in kindergarten daycare are in preparation classes for at least two days per week. Ten percent of students do not continue any education after elementary school at age 15.

All elementary schools are comprehensive, but in the upper grades, some schools have adopted a practice of limited streaming. Students are given the choice of learning Latin, additional mathematics, or additional science, and separate classes are formed on this basis. At the end of elementary school, virtually all students write an externally assessed examination in two subjects, Slovene language and mathematics. Students are free to take this examination, but those who apply to secondary schools with more applicants than places are required to take the examination for placement purposes. Successful students may proceed to secondary school.

Ninety percent of elementary school leavers continue their education in one of the three types of secondary education: four-year gymnasium, four-year technical and professional school, and two- or three-year vocational school. Students may write an entrance examination to enter tertiary education after completing any four-year upper secondary school. A regulation enacted in 1995 established a five-subject externally assessed baccalaureate examination as the new requirement to enter university. Prior to 1995, universities were the only form of tertiary education; since that time, two, three, and four-year professional schools have been introduced.

Gymnasia are in principle comprehensive, but some offer a science-heavy curriculum while others emphasize humanities and languages. All students must study mathematics, physics, chemistry, biology, two foreign languages, and a social sciences program of psychology, sociology, and philosophy. The five-subject baccalaureate consists of three compulsory subjects: Slovenian, mathematics, and a foreign language, as well as two subjects chosen by the student. The technical and professional baccalaureate features the same required subjects, but students choose from economics, electronics, engineering, or a similar subject for the final two sections. Vocational schools offer programs varying in length from two to four years, and usually involve practical work experience as well as classroom time. All vocational schools end with a final examination that may differ from school to school. There are five private, upper secondary Catholic schools, but 90 percent of their financing comes from the state.

Schools in the System

The School Year

The school year consists of 190 teaching days for all grades of elementary school, except Grade 8, which has a two-week period for the school-leaving examination. School begins at the beginning of September and ends June 25. Upper secondary schools have a four-week examination period in June. Students have four one-week holiday periods around November 1, between Christmas and the New Year, in February, and the beginning of May.

Class Time Allocated to Mathematics and Science

Students attend school from Monday to Friday. Lessons are 45 minutes long at all levels of education. In 1993–94, the average class size was 23 students in elementary schools, 31 in gymnasia, and 32 in all other secondary schools. In elementary school, students spend between 13 and 24 percent of class time studying mathematics, and between 8 and 19 percent on science. At the secondary level, class time devoted to mathematics and science varies considerably, depending on the program students are enrolled in. Table 1 shows all figures for elementary and secondary schools.

In Grades 1 to 3, science teaching includes both natural and social sciences. In Grades 4 and 5, science normally includes geography, which would account for 7 percent of class time; the figures in the table exclude geography. Beginning in Grade 6, geography is taught as a separate subject. At the upper secondary level, some schools also offer geology, which is also not included in these figures.

Certification of Mathematics and Science Teachers

Elementary school teachers must obtain a four-year undergraduate degree from one of the country's two universities. Faculties of education within the universities educate all preschool and primary school teachers. Approximately 50 percent of the program consists of psychology and pedagogical subjects, such as didactics, education theory, teaching methods, and educational psychology. Lower secondary school teachers are educated either at faculties of education or the faculty of their specialty subject. As a rule, students at faculties of education major in two subject areas, such as mathematics and physics, or history and geography, while students within faculties become one-subject specialists. These specialists usually become upper secondary teachers although they are required to take some pedagogical courses before they are certified.

Teachers must pass a written and practical licensing examination after one year of teaching. Candidates who do not pass the examination within two years are not allowed to continue. A seniority system was recently established, in which teachers who perform such activities as attending in-service training, marking external examinations, or performing research are promoted and receive one of three titles that carry additional pay.

Teacher Profile

While the economic status of teachers is not on par with some of the country's highest paid professions,

Table 1: Percent of Class Time Spent on Mathematics and Science

Grade	1	2	3	4	5	6	7	8	9 to 12
Mathematics	24	24	23	21	16	14	13	13	13 to 18
Science	14	14	14	8	8	14	21	19	18 to 25

like law or medicine, it is a job which holds a fair amount of esteem in Slovene society. The average age of teachers is approximately 45. Eighty-five percent of elementary school teachers are women, as are 59 percent in upper secondary schools.

3 TIMSS POPULATIONS

Slovenia, IEA and TIMSS

Slovenia joined IEA in 1988 and participated in all studies conducted since then.

Slovenian TIMSS Populations

Slovenia is participating in TIMSS at all three population levels. Since students in Slovenia begin school at the age of 7, one year later than many other countries, Slovene students are on average older than students from other countries. Slovenia's Population 1 included students in Grades 3 and 4 who were 9 or 10 years old at the time of testing. Population 2 included students in Grades 7 and 8 who were 13 or 14 years old at the time of testing. The generalists at the Population 3 level included students in Grades 11 and 12. Mathematics and physics specialists were also sampled at this level.

4 MATHEMATICS CURRICULUM AND PEDAGOGY

Goals for the Mathematics Curriculum

The mathematics curriculum for elementary and secondary schools is determined centrally and schools are required to implement it as it is given. Achievement standards for the elementary school leaving examination, the baccalaureate examination, and final examinations of other secondary schools are also established centrally. Examination standards in both mathematics and science are described in baccalaureate catalogues prepared by university professors and secondary school teachers in consultation with international experts. The catalogues are particularly useful to teachers as they

provide a concrete definition of the content that must be taught during the final years of secondary education. The overall goals are to meet the standards achieved elsewhere in the world and to enable students to pursue higher study. The National Examination Center issues official curriculum guides that meet these criteria for the more than 30 baccalaureate subjects; similar guides will be issued when the new education act establishes official educational standards for lower school grades. Although the mathematics taught in elementary schools is uniform, high-ability students may attend supplementary classes, taught by their mathematics teacher, in which they work at problems of greater difficulty. Upper secondary school mathematics is taught at two levels. Students attempting the baccalaureate receive 700 hours of mathematics instruction over four years, while students at the basic level receive 560. At technical and professional schools students take an internal final examination after 490 hours of mathematics instruction over four years.

Major Changes in the Mathematics Curriculum

At the elementary level, mathematics is now introduced through set theory. Probability and statistics are being introduced into the elementary school curriculum, covering graphs, bar graphs, charts, and interpretation of data. Students in elementary schools are now permitted to use calculators in mathematics classes.

In upper secondary schools, several changes have been incorporated into the curriculum. Computer science was introduced as a separate subject in the gymnasium. In technical and professional schools, numerical methods were incorporated into the mathematics curriculum for nautical engineering, surveying, and other courses. Statistics and probability were recently established as separate units. More emphasis is placed on logic. Very little time is spent using logarithmic or square root tables as a result of increased calculator use.

Current Issues in the Mathematics Curriculum

A weakness in primary education is the late introduction of decimal numbers. It is expected that decimal numbers and decimal place value will be introduced into the curriculum at the Grade 3 level. Basic measurement is weakly covered in the curriculum, and an improved understanding of this will be possible when students are able to use decimals and the metric system to describe the world around them. At the upper secondary level, links between mathematics and other subjects are poorly demonstrated. Applied mathematics, geometry, statistics, probability, and combinations should be given more emphasis.

Mathematics Textbooks

The same compulsory textbooks are used throughout the country, but workbooks and specific topic booklets are to an extent replacing traditional textbooks in elementary school. Textbooks rely too heavily on text, have few or no historical highlights, and rarely any diagrams or figures even in topics such as geometry. Changes to textbooks and materials have been very slow in coming.

At the upper secondary level the number of textbook types offered has been reduced to two that must suit the mathematics curricula of all school types. There are no specific mathematics textbooks, for example, for economists, surveyors, or electrical technicians.

There is a shortage of qualified mathematics teachers because of the movement of mathematics teachers to other professions such as those within the computer industry. Students therefore rely heavily on the explanations provided in textbooks, and often seek instruction outside school.

Textbook authors are normally sought through open tender, but in many cases university professors or teachers within the school system initiate the process. Draft textbooks must pass an independent review process and be approved by the council for preuniversity education. Textbooks likely to achieve a fairly high circulation, such as mathematics textbooks, are sold at market value. Textbooks with a low circulation, such as those for subjects offered at only a few schools, are subsidized. The price of textbooks is monitored by the Ministry of Education. Textbooks are not provided by schools, but are purchased by students. Students from low income families may receive subsidized textbooks.

Mathematics Pedagogy

Teachers are free to choose their instructional techniques, and consequently there is a great deal of variation from classroom to classroom. However, some general trends may be noted. Frequently, emphasis is placed on the development of computational techniques with too little time devoted to their practice. Rules and concepts are mastered through rote learning. Both inductive and deductive teaching approaches are used. There is little attempt at the integration of topics across subjects. The nature of the curriculum is to describe the content in disparate topics, and teachers follow this model in teaching and assessment; the entire process does not encourage integration.

The use of calculators is permitted in the higher grades of elementary school and in upper secondary education. Computers are used very little for instructional purposes since there is almost no software available in Slovenian. Extended projects are not often required, since mathematics is taught as an academic and not an investigative subject. Oral communication on the part of students is not encouraged.

5 SCIENCE CURRICULUM AND PEDAGOGY

Goals for the Science Curriculum

Science as an integrated subject exists only in Grades 1 to 5. From Grade 6 onward it is divided into biology, chemistry, physics, technical education, and geography. In elementary school the same science content is studied by all students. At the upper secondary level, particularly in the gymnasium, students have the option of taking additional courses at a more demanding level. Many choose an in-depth study of science subjects, specially if they plan to write the science baccalaureate or final examination.

The science curriculum is determined at the national level and must be followed by all schools. It is prepared and constantly renewed by a group of university experts and secondary and elementary teachers. The curriculum must be approved by the education council before it is issued.

Major Changes in the Science Curriculum

The science subjects are now much more integrated than in the past, when competition between them precluded an interdisciplinary approach. The science curriculum has been gradually enriched over the past 10 years by a new emphasis on environmental issues and renewable resources. The implementation of these topics is the teachers' responsibility, and they have incorporated these topics into existing science teaching. New technologies have had an impact on the way science is taught, but not directly on the curriculum. Real-life problems are tackled in group work, as a part of student practical and research work done at school, or in collaboration with research institutions or universities.

Current Issues in the Science Curriculum

The science curriculum is still divided into at least three subjects: chemistry, biology, and physics. This is a product of a tradition in which each discipline tried to convince the educational community it needed more attention, school hours, and funding. This struggle led to the science subjects at the upper grades of elementary school occupying a disproportionately large number of hours.

Science Textbooks

Teachers have become more independent in selecting among available textbooks and supplementary materials, despite various problems linked to the use of the alternative sources. Many teachers prefer to use old textbooks, which cover the whole curriculum, rather than the new textbooks, which are structured by topic.

Students are not expected to learn from science textbooks, which, on their own, tend to be complex. Teachers provide explanations and supplementary material to assist students. Most science textbooks, however, require highly qualified teachers to provide such explanations, and they are not always available. Less capable students are particularly disadvantaged by having to use textbooks that were not meant for independent use.

Textbook authors are sometimes sought through open tender, but in many cases university professors or teachers within the school system initiate the process. Draft textbooks must pass an independent review process and be approved by the council for preuniversity education. Textbooks that are likely to achieve a fairly high circulation, such as science textbooks, are sold at market value. Textbooks with a low circulation, such as those for subjects offered at only a few schools, are subsidized. The price of textbooks is monitored by the Ministry of Education. Textbooks are not provided by schools, but are purchased by students. Students from low-income families may receive subsidized textbooks.

Pedagogy

Inductive and deductive teaching approaches are used at all levels. Students are required to memorize rules and concepts, although the use of reasoning in mastering concepts is highly valued. Cooperative learning is not used systematically in science learning; it is used most frequently in elementary school during project work. There is a tendency towards increased integration of topics in science teaching despite the separation of subjects at the secondary level.

Science teaching is in large part based on experimental and laboratory work, and all elementary and secondary schools have laboratories for chemistry, physics, and technical education. New technologies, including electronic media for presentations and illustrations, are present and used in virtually all elementary and secondary schools. Oral communication is not highly valued, but it is used frequently in laboratory and group work. Project work constitutes a considerable part of the learning process in science.

6 EVALUATION POLICIES AND PRACTICES

Students in all grades are examined and given marks every semester. Marks are averaged over the semester, and are based on formal and informal assessment.

Assessment of student progress in all subjects is done in the classroom and is determined solely by the teacher. The most widely used approach in evaluating knowledge is oral questioning, which is based on the assumption that oral examining improves communication skills. In practice, however, many students find that being questioned in front of the whole classroom is stressful and limits communication. Teachers are now advised to prepare written exercises, open-ended questions, multiple-choice tests, and other approaches to evaluate students.

At the end of Grade 8, there is an external assessment in two subjects, Slovene language and mathematics. Tests are a combination of multiple-choice and open-ended questions. This external evaluation was introduced in the early 1990s to provide a uniform framework for secondary schools to qualify candidates without having to administer individual entrance examinations.

Students who have successfully completed Grade 12 and wish to proceed to university must write a five-subject baccalaureate. Topics on the examination include three compulsory subjects, Slovenian, mathematics, and a foreign language, as well as two subjects chosen by the student. Questions on the examination are a combination of open-ended and multiple-choice. The examinations are externally evaluated by secondary school teachers and university professors. As with the external examination at the end of Grade 8, the baccalaureate was introduced so that universities would not have to administer entrance examinations. Both external assessments have a significant impact on teaching.

7 REFERENCES AND SOURCES FOR FURTHER READING

Common Sources

1 Husén T, N T Postlethwaite 1994 *International Encyclopedia of Education*. Pergamon Press, Oxford
2 Organisation for Economic Co-operation and Development 1995 *Education at a Glance: OECD Indicators*. Organisation for Economic Co-operation and Development, Paris
3 United Nations Educational, Scientific and Cultural Organization 1995 *Statistical Yearbook*. United Nations Educational, Scientific and Cultural Organization, Paris
4 World Bank 1995 *World Development Report 1995*. Oxford University Press, New York

Other Sources

5 Statistical Office of the Republic of Slovenia 1995 *Statistical Yearbook of the Republic of Slovenia*. Statistical Office of the Republic of Slovenia, Ljubljana
6 Ministry of Education and Sports 1995 *Bela knjiga o vzgoji in izobrazevanju v RS. [White Book on Education in Slovenia]* Ministry of Education and Sports
7 National Examination Center 1993 *Predmetni izpitni katalog za maturo Matematika. [Baccalaureate Catalogue for Mathematics.]* Ljubljana
National Examination Center 1993 *Predmetni izpitni katalog za maturo Fizika. [Baccalaureate Catalogue for Physics.]* Ljubljana
National Examination Center 1993 *Predmetni izpitni katalog za maturo Kemija. [Baccalaureate Catalogue for Chemistry.]* Ljubljana
National Examination Center 1993 *Predmetni izpitni katalog za maturo Biologija. [Baccalaureate Catalogue for Biology.]* Ljubljana
8 Porocevalec Drzavnega zbora Republike Slovenije 1994 *The Official Parliamentary Gazette*. Drzavni zbor Republike Slovenije, Ljubljana

Statistical References

Section	Statistic	Reference	Page	Table	Year of Statistic
Country Profile	population figures	5	70	4.2	1991
Country Profile	19,463 births	5	80	4.14	1994
Country Profile	US$6490	4	163	1	1993
Country Profile	18%	8	31–40	33	1994
Country Profile	National 1 Public 60	3	7–16	7.1	1989, 1991
The Education System	52%	6	39	1	1993
The Education System	49%	3	3–101	3.4	1992
The Education System	57%, 53%, 90%, 100%	5	114	6.2	1994–95
The Education System	10%	5	117	6.9	1993–94
The Education System	curriculum	6	160	14	1991
The Education System	teaching days	6	77	5	1993
The Education System	school, class	6	79	6	1993
The Education System	mathematics, science	6	88	9	1984–94
The Education System	elementary teachers - 85% secondary teachers - 59%	5	114	6.3	1993–94
The Education System	population levels	6	75	4	1993
The Education System	mathematics and science curricula	7			1994
The Education System	hours of mathematics instruction	6	160	14	1991
Science Curriculum	science curriculum	6	88	9	1984–94
Evaluation	external assessment	6	88	9	1984–94

South Africa

Derek Gray, Human Sciences Research Council

INTRODUCTION

At the time of writing this chapter, January 1996, education in South Africa was in the process of metamorphosis. The Government of National Unity and all political parties have made forceful policy statements to the effect that education and its reform are high on their priority lists. A new dispensation for education is being prepared and implemented. In the context of these affirmative political statements concerning the importance of education, science and mathematics are given an added importance.

1 COUNTRY PROFILE

The Republic of South Africa occupies the southern portion of the African continent bordered by Namibia to the northwest, as well as Botswana, Zimbabwe, Mozambique, and Swaziland. The Kingdom of Lesotho is an enclave within South Africa's 1,220,088 km². The physical features of South Africa resemble an inverted saucer with a narrow coastal rim and elevated highland plateau in the center. The climate shows great variety, ranging from arid desert in the west to tropical mangrove forests on the eastern coastline. Inland, on the plateau, upwards of 2000 m in altitude for most of the country, the climate is cool in winter and hot in summer, which is also the rainy season for most of the country. Only the South Western Cape has a Mediterranean, or winter rainfall, climate.

The population of South Africa is about 39 million according to the 1991 census. As is characteristic of many developing countries, there is considerable population drift from rural areas into the cities. The heaviest densities are in Durban and the industrial heartland surrounding Johannesburg in the province of Gauteng. The population is characterized by considerable variance in the distribution of wealth and opportunities for advancement. Rapid population growth, both natural and immigrant, is another cause for concern.

South Africa is often referred to as the "rainbow nation." There are 11 principal languages and a number of minor ones. The dominant 11 are recognized by the constitution, mass media, and the state. Historically, however, the population has been classified as black, white, Indian, and colored. In today's population, these groups would correspond to 81 percent black, 14 percent white, broken down into 8 percent Afrikaans-speaking and 6 percent English-speaking, 3 percent Indian, and 4 percent colored, a group that includes those of mixed race and descendants of the Malay slaves. In the TIMSS testing, over 80 percent of sampled students wrote in a language other than their native one.

Government structure is a bicameral, centralized parliamentary system. At present, there are nine provinces, but there is also a vocal platform for an Afrikaner homeland. The Interim Constitution provides for education to be a provincial matter with centralized control of norms and standards.

Relative to much of Africa, South Africa is economically stable and, in World Bank terms, on the margins of the per capita income for a developing country. This is despite the fact that there is an estimated unemployment figure of 40 percent. The Northern Province, the Eastern Cape, and KwaZulu/Natal are particularly poor and overpopulated in the context of employment opportunities. There is, however, a rapidly expanding informal trading sector.

Spending on education is a matter of concern to the government, with some 24 percent of the national budget being allocated. In monetary terms this amounts to US$8,836 million. With growing amounts being demanded from the national exchequer for health, housing, civil service wages, and many other causes, all sectors of the budget are under intense scrutiny, competition, and pressure.

2 THE EDUCATION SYSTEM

Governance and Decision Making

In the interim constitution, education is defined as a provincial responsibility. There are nine provinces each with a Department of Education in the new South Africa, including Eastern Cape, Gauteng Province, KwaZulu/Natal, Mpumalanga Province, Northern Cape Province, Northern Province, North West Province, Free State Province, and Western Cape. All provinces are in the final stages of reforming their previous systems into unified provincial education administrations. In the Northern Province, for example, this involves the merging of no fewer than eight preexisting authorities into a single administrative unit.

Curricula that will better represent the needs of the new South Africa are being constructed by 41 national subject curriculum committees, consisting of government representatives, nongovernment organization representatives, recognized experts, and teachers. In the general climate of administrative uncertainty, the activities of these subject committees are somewhat hesitant. It should also be noted that the curricula of 1993 were revised to some extent as an interim measure. No new policies have emerged regarding the approval and selection of textbooks, but there has been considerable discussion about revising the old system of authoring, publishing, and approving textbooks. The present policy is that textbook approval is performed by provincial authorities; in the future, direction will come from the central Department of Education.

There is still some uncertainty as to the final form of the new constitution, scheduled for final presentation to Parliament later in 1996. The degree to which South Africa will become a unitary state and the extent of the devolution of responsibilities to federal structures are yet to be determined. At this time, decision-making with regard to education policy-making, core

curricula, and standards is essentially a centralized matter while the new Provincial Education authorities are in the process of reform.

Structure of the System and Participation Rates

Structure of Education

Within the constitution, provision has been made for education to be compulsory and free for Grades 1 to 9, leading to the General Education Certificate. Education will likely be voluntary and fee paying for Grades 10 to 12, or the Further Education Certificate level. There are no available data concerning participation rates.

Preprimary education and school readiness programs are marked for attention by the government, but are likely to prove too costly an exercise. South African schooling is currently divided into four phases: junior primary for Grades 1 to 3, senior primary for Grades 4 to 6, junior secondary for Grades 7 to 9; and senior secondary for Grades 10 to 12. This structure is shown in Figure 1, along with approximate enrollment rates for each level.

Grades 1 to 7 are normally grouped together in comprehensive, nonspecialist primary schools, as are the higher grades in secondary schools. However, with the large number of rural and what were termed farm schools, this division is not universal. In two of the provinces, schools are divided into three levels, primary for Grades 1 to 6, middle for Grades 7 to 9, and

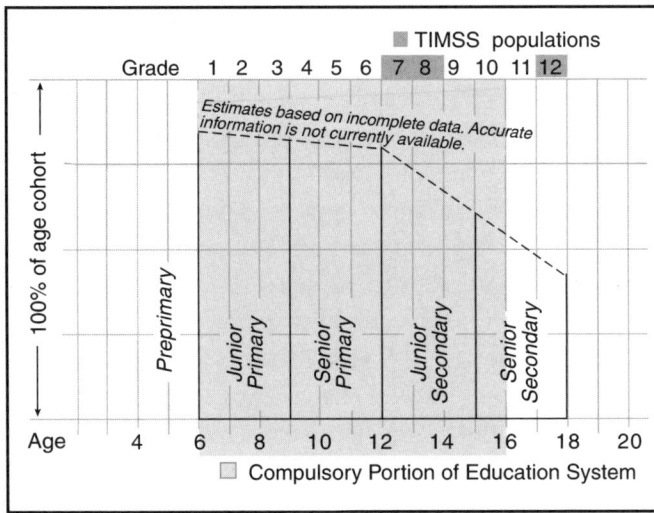

Figure 1: Structure of the Education System

senior secondary schools for Grades 10 to 12. Several other combinations of grades in schools also occur.

Because of the previous absence of compulsory schooling, there is a wide range of entry age in South African schools, a problem compounded by large numbers of students repeating classes and high drop-out rates. In some provinces it has been estimated that the number of students repeating years approximates the number of school-age children not in school. School attendance control is made more difficult by the existence of a highly mobile population as well as political and social violence in the poorer areas and in what are termed informal settlements. This latter factor leads to students attending schools in quieter areas, often far from home.

The majority of South African secondary schools are comprehensive. There are a limited number of schools that provide commercial or technical subjects and a few that provide art, music, and ballet as specialties. There are also a number of technical colleges offering a wide range of vocational instruction. Provisions for students with mental and physical disabilities are very limited.

Public and Private Systems

There are private schools in South Africa, some of which are denominational in nature, some commercial enterprises, and others operated by nongovernmental organizations. All are funded by means of fees, corporate donations, and, to a varying extent, by federal and provincial subsidies. In total, the private school system accounts for less than 1 percent of school enrollments. Public schools are funded by provincial administrations from a budget determined by the national Department of Education, together with considerable subsidies from the private sector.

Mathematics and Science

Students must study either general science, which includes elementary physics and biology, or environmental science and mathematics as compulsory subjects up to Grade 9. However, this situation is clouded by the lack of qualified teachers. Biology and physical science, including physics and chemistry, are offered as optional subjects beyond Grade 9. Mathematics becomes an optional subject at this stage also. A very small fraction of 1 percent of students take additional mathematics, and even fewer students take science or math-

ematics subjects in the thirteenth year. Only 18 percent of students write the physical sciences examination in the national examinations for the School Leaving Certificate. Gender distribution in mathematics and physical science classes is not available, but it is commonly accepted that there is a serious imbalance in enrollments with regard to girls.

Schools in the System

The School Year

The school year approximates 200 school days, although nonscheduled public holidays may decrease this number. The school year usually starts the first or second week in January and ends early in December. There are short vacations in April, July, and early October. Schools terms and holidays are a matter of provincial concern and determination.

The school week is normally Monday to Friday, with a large number of local features such as a half-day on Wednesdays. Hours of instruction are nominally from 7:30 a.m. to 1:30 or 2:00 p.m., depending on the free time allocated. More than 50 percent of schools do not function for more that four hours per day due to late starts, early closing, and class and teacher absences. The erratic functioning of South African schools is one of the major problems that have been identified but have yet to be addressed.

Class Time Allocated to Mathematics and Science

Class time allocated to science and mathematics varies from two to three-and-a-half hours per week each, but the effectiveness of this time allocation is quite varied. Expressed as a percent, this ranges from 7 to 15 percent for each subject.

Class Size

Class size and teacher-to-student ratios are currently issues for debate. Class size in primary schools may range from 24 to 60, with classes of 90 or more on record. Secondary school classes tend to be somewhat smaller, particularly mathematics and science classes for Grade 10 and beyond, since enrollments tend to be lower. There is a target of 42 students per class in primary and 35 students in secondary schools.

Streaming and Tracking

Streaming and tracking are practiced in only a small percentage of schools, and at present streaming is widely regarded as undesirable. In at least two provinces, there are the remains of a specialist science and mathematics school system, but these were never fully developed.

Certification of Mathematics and Science Teachers

Most teachers have a three- or four-year diploma or certificate issued by a provincial teacher education college. The 18 universities enroll science and mathematics teachers at the postgraduate level. Initial education and certification of teachers is currently a matter of intense political and professional debate. Major changes in the characteristics of teacher education are anticipated during 1996–97. A recent departmental publication, *Norms and Standards and Governance Structures for Teacher Education,* is still being debated. In these proposals for all junior primary and senior primary courses, there are requirements for courses in general science, environmental science, mathematics, and technology.

In-service education and upgrading are not required, except for courses organized by Department of Education officials. In two or three provinces there are upgrading units offering full-time programs. In other provinces, these courses are the responsibility of teachers' centers. There are discussions in process concerning a proposal that several days each year be set aside for professional development. However, this is unlikely to take effect until a compromise is reached regarding salaries based on qualifications and other criteria.

Teacher Profile

Compared to most of Africa, teaching is a well-paid profession. With the growth of the trade unions, however, productivity has been given less emphasis than demonstrations of power and influence. There is some evidence that this trend is reversing; with the most recent poor results from School Leaving Certificate examination, the teaching profession is showing signs of greater commitment to its pupils. At the same time, with the development of an industrial base, career opportunities away from the uncertainties of teaching have become an attractive possibility.

At present, salaries are based exclusively on qualifications and years of service. Relative to other professions, such as lawyers and medical doctors, teachers are not well paid although they do tend to earn more than nurses or social workers. Average salaries are shown in Table 1.

3 TIMSS POPULATIONS

South Africa, IEA, and TIMSS

South Africa is associated with IEA through the Human Sciences Research Council. South Africa joined TIMSS with the intention of establishing baseline data upon which postapartheid education developments might be gauged.

South African TIMSS Populations

South Africa is participating in TIMSS at the Populations 2 and 3 levels only. Because of the diversity of languages and the fact that instruction in English begins in Grade 5 for the majority of the population, it was considered unfair to enter South Africa into Population 1. South Africa has completed the Populations 2 and 3 (generalist) studies. The definition of Population 2 is the two adjacent grades containing the majority of 13-year-olds. South Africa tested Grades 7 and 8, although only 48 percent of this age group falls into these grades. Students tested in Population 3, which is targeted at students in their final year of schooling, were selected from Grade 12. Because of the paucity of reliable data concerning the entry status of students into the final School Leaving Certificate, South Africa decided to opt out of all but the generalist papers.

4 MATHEMATICS CURRICULUM AND PEDAGOGY

Goals of the Mathematics Curriculum

The Department of Education's *Core Curriculum for Mathematics* discusses the goals for mathematics edu-

Table 1: Teachers' Average Salaries by Category of Qualification

Category	Qualification	Salary	Comment
Category A2	Teachers with less than a Standard 10 School Leaving Certificate	US$5600 per year, fixed	Fewer than 3 percent of teachers fall into this category
Category C	Teachers with three years of education at a teacher education college	US$10,170 to US$15,460 per year	The majority of teachers fall into this category
Category D	Teachers with four years of university education	US$11,555 to US$18,030 per year	
Category G	Teachers with a doctorate-level qualification	US$13,125 to US$20,800 per year	Classroom teachers with these qualifications are remarkably rare

cation. The syllabus is aimed at fostering and developing the following specific aims :
- to enable pupils to gain mathematical knowledge and proficiency;
- to enable pupils to apply mathematics to other subjects and in daily life;
- to develop insight into spatial relationships and measurement;
- to enable pupils to discover mathematical concepts and patterns by experimentation;
- to develop a number sense and computational capabilities and to judge the reasonableness of results by estimation;
- to develop the ability to reason logically, to generalize, specialize, organize, draw analogies, and prove;
- to enable pupils to recognize a real-world situation as amenable to mathematical representation, formulate an appropriate mathematical model, select the mathematical solution and interpret the result back in the real-world situation;
- to develop the ability to understand, interpret, read, speak, and write mathematical language;
- to develop an inquisitive attitude towards mathematics,
- to develop an appreciation of the place of mathematics and its widespread applications in society;
- to provide basic mathematical preparation for future study and careers;
- to create an awareness of and an appreciation for the contributions of all peoples of the world to the development of mathematics.

The national Department of Education distributes interim core curricula for adaptation by provincial authorities. Provincial departments write and distribute curriculum guides. At the present time, each province issues the Grade 12 school leaving certificate based on its own curricula. By the end of 1996, it is expected that there will be common national school leaving certificate examinations, and this has been a matter of some debate.

Major Changes in the Mathematics Curriculum

There is currently a major debate concerning the shift toward a constructivist approach in mathematics curricula. In addition, there is a growing emphasis on individual problem-solving approaches.

There have been several initiatives by private, departmental, and university groups to introduce video and television-based education to underprivileged schools. These initiatives focus on mathematics, biology, and the physical sciences, among other subjects. With the majority of students being taught in English, language skills also receive considerable attention.

The national policy paper on technology education proposed that technology become a compulsory core subject in all schools from Grade 1 to Grade 9. This is a strongly debated issue; and, while several proposed schemes are in trial stages in various provinces, none has been formally implemented. When the subject comes into being, it will most certainly have considerable impact on mathematics and science, both on content and the way in which these subjects are taught.

Within the context of the new political dispensation, there is now considerable emphasis on equity and equalization of opportunity. Social equity is a very strong issue since it is linked directly to the national Reconstruction and Development Program. Gender issues in the enrollment of students in senior science and mathematics classes are being addressed at national and provincial levels.

Current Issues in the Mathematics Curriculum

At present, a serious examination of the implementation of constructivist approaches to mathematics education is being conducted. The approach has ardent proponents but it has not as yet been resolved on a national basis. Fairly extensive trials on this approach are being run in areas of the Western Cape. Equity in the face of different resources and teacher qualification resources are other major issues.

Mathematics Textbooks

All policies concerning authorship, approval, purchase, and distribution of textbooks are under discussion, including the possible institution of guidelines from the national Department of Education. There is a policy statement that all education beyond Grade 10 will be modular, but little movement has taken place in this direction. A study module will be a defined topic of content, skills, and outcomes of stated teaching duration and examinable at the end of the teaching period. If this policy is implemented, the textbook industry will undergo radical change.

Textbooks are usually written by mathematics subject advisers, academics, and teachers. Publishers tend to invite authors who are well known or have sufficient influence to ensure a market for their products. Textbooks are approved by the textbook committees of provincial departments of education. Purchasing of textbooks is done by the provinces.

In general terms, teachers rely heavily on textbooks for teaching. In situations where teachers are lacking in subject mastery or confidence, this reliance is even greater. One of the major problems in student use of textbooks is their short supply. Cases have been documented where there are either no books available in important subjects, or there are so few as to be of little use. The major criticism of South African mathematics textbooks is that they are more compilations of exercises than teaching and learning aides.

Pedagogy

When considering classroom practice it must be remembered that there is a substantial gap between intentions and what actually happens. A mathematics teacher faced with a class of perhaps as many as 70 students cannot operate on what are generally regarded as good classroom techniques. These problems are compounded by inadequate classrooms that are too small for the numbers they are expected to accommodate, insufficient desks and chairs, and often inadequate ventilation in hot climatic conditions. Some classes in the Northern Province, in fact, are held outside under trees.

Pedagogy

The use of approved hand-held calculators is encouraged in all classroom and testing situations. The use of computers, on the other hand, is limited by cost and accessibility to electricity in rural schools. Computers are expensive and often targets for theft, particularly from schools. In addition, suitable and curriculum-relevant software is also costly or unavailable.

Investigations and open-ended projects are performed in the context of the National Science Fair movement, of which mathematics categories are an integral part. Mathematics entries, particularly at the Grades 8 to 11 level, are reasonably popular. This type of work, however, is not really suitable for poorly qualified teachers who are a long way from professional support systems. Notwithstanding these problems, a large entry from rural schools is achieved each year. In a minority of schools, projects are a component of regular classroom activities. Cooperative learning has its proponents but has not yet made a great impact on mathematics education.

5 SCIENCE CURRICULUM AND PEDAGOGY

Goals of the Science Curriculum

The general aims of physical science education, as outlined in the Physical Science Core Curriculum, are to:
- provide pupils with the necessary subject knowledge and comprehension;

- develop the necessary skills, techniques, and methods of science, such as handling of certain apparatus and techniques of measuring;
- develop desirable scientific attitudes, such as interest in natural phenomena, desire for knowledge, and critical thinking;
- introduce pupils to the scientific explanation of phenomena;
- introduce pupils to the use of scientific language and terminology;
- introduce pupils to the applications of science in industry and in everyday life;
- help pupils obtain perspective in life, for example to develop an esteem for the wonders of the universe through contact with the subject matter.

Major Changes in the Science Curriculum

The process of curricular change has been halted until a new curriculum is launched. As with mathematics, social equity will be the driving force behind the changes, but at present there are no clear indications of the essential character of the curricula envisaged.

Within the context of the new political dispensation, there is now considerable emphasis placed on equity. Equalization of opportunity and social equity are strong issues since they are directly linked to the national Reconstruction and Development Program. There are identified gender issues in enrollment in the more senior science classes, and these are being addressed at the national and provincial levels.

Current Issues in the Science Curriculum

The focus of debate in this area lies in that of relevance. Science, technology, and society curricula materials are being developed for distribution to teachers by some universities. In the fields of both science and mathematics there are a substantial number of nongovernment organizations and private initiatives working in the field of curriculum development. These include organizations such as the Primary School Science Project, controlled by the Urban Foundation, the Science Education Project for high schools, the Shell Centre in Durban, and the Primary Schools Curriculum Project.

Science Textbooks

As in the case of mathematics, in situations where science teachers are underqualified, lacking in subject mastery or confidence, their reliance on textbooks is heavy. Textbooks are usually written by science subject advisers, academics, and teachers. Publishers invite well-known authors who have sufficient influence to ensure a market for their products. Science textbooks are approved by textbook committees of the provincial Departments of Education. One of the major problems is the short supply of textbooks. Cases are documented where there are either no textbooks available in important subjects or there are so few as to be of little use.

Textbooks are all quite similar in the number and nature of topics that are included. For a science textbook to be placed on the approved list, it must be structured to reflect the curriculum. The topics covered in science texts precisely reflect the topics of the core curriculum. Criticisms of South African science textbooks note that they are too academic and formal and not relevant in their presentation of material.

Pedagogy

When considering classroom practice, it must be remembered that there is a substantial gap between aims and ideals and what actually happens within classrooms. A science teacher faced with a large class, perhaps as many as 70 students, cannot operate with what are generally regarded as good classroom or laboratory techniques. These problems are compounded by inadequate and underequipped laboratories that are too small for the numbers they are expected to accommodate, insufficient desks and chairs, and often inadequate ventilation in hot climatic conditions. Many teachers simply transfer the text to the blackboard and students copy the writing down into their notebooks. Laboratory equipment is often either absent, unused, or insufficient. Replacement of consumed and broken items usually does not happen.

Calculator use is encouraged in all classroom and testing situations in science. As in the case of mathematics, the use of computers is limited by cost and accessibility to electricity in the rural schools. Computers are expensive and often targets for theft. Suitable, curriculum-relevant software is also costly or unavailable.

Investigations, laboratories, and open-ended projects are encouraged, often in the context of the National Science Fair movement. Science, engineering, and technology categories are main components of the entry category list. Entering the fair is particularly popular in Grades 8 to 11. This type of work, however, is not really suitable for poorly qualified teachers and teachers who are a long way from professional support systems. Notwithstanding these problems, a large entry from rural schools is achieved each year. In a minority of schools, projects are a component of classroom activities. There are also a few local, small-scale initiatives involving project work.

Approaches focusing on interactions of science, technology, and society will be largely dependent on the outcome of discussions about the place of technology as a core school subject. The current preamble to the grade curricula refers to the need to develop an awareness of the interrelationships and interactions of science, mathematics, and technology without specifying how and where.

6 EVALUATION POLICIES AND PRACTICES

Evaluation is generally regarded as a school and teacher matter. In reality, this means that with large classes and the lack of duplication equipment, evaluation receives little attention. Evaluation methods and techniques are given remarkably little time in teacher education courses. Regular testing is required, as are term and cycle tests, midyear, and end-of-year examinations.

The preamble to the official syllabus documents lays out the following policy on assessment:

- Assessment is an integral part of the learning process.
- Assessment is a means to an end. It is a process closely related to each aspect of the total program of pupil development and with the ultimate goal of improving the effectiveness of the mathematics program and the students' mathematical competence commensurate with their abilities, aptitudes, and needs.
- It is essential that the assessment program should reflect the broad classroom approaches to the teaching and learning of mathematics.

- Approved calculators may be used during written tests and examinations.
- Frequent short tests are advocated. This form of assessment is formative in that it provides information that teachers can use in deciding if any pupil needs help with any section.
- In the final examinations, multiple-choice questions will form one-third of the total marks allocated in the examination papers.

As in all matters to do with education, these policies are under review and are likely to change drastically in 1996.

The public examinations administered at the end of Grade 12 are extremely important to students, since entrance to tertiary education and many forms of employment depend on this certification. Both mathematics and physical science are examined at Higher and Standard Grade. Students decide at which level they will enter, usually late in the Grade 12 year. Many students enter Higher Grade in the hope of being awarded a condoned pass at Standard Grade. Many enrichment, upgrading, and equity programs are in fact preparation courses for these final examinations.

A considerable amount of time has been spent in setting up a national item bank that is expected to become one of the national Department of Education's tools for controlling norms and standards nationally. A new General Certificate of Education examination for Grade 9 students is planned as the first important exit point from the schooling system, and present indications are that this will be, to a large extent, internally managed by schools in conjunction with external examinations. The final Grade 12 examinations, however, are the most important examinations.

All mathematics and science examinations are written examinations, usually of two to three hours in length. There is a multiple choice section followed by questions requiring short or longer answers. A contributory year mark accumulated by the school is merged with public examination marks to give a final raw mark. Provincial results are consolidated and taken through a controversial normalization process before release to the students. The national item bank is intended to exercise control of norms and standards within these tests.

Traditionally, South Africa has been heavily examination-centered. Recent trials included optional topics in the curriculum; in many cases, these topics were not taught since they were not examined. End-of-year revision and review for examinations is a well-ingrained process in schools. Furthermore, there is a well-developed secondary industry surrounding the buildup to the examination program each year. To supplement the school provisions in mathematics and science, the National Television broadcasts educational material for approximately 15 hours per week. Towards the end of the academic year, many national newspapers contain mathematics and science supplements. Between 40 and 50 nongovernment organizations also participate in providing extra classes. In urban centers, universities provide venues for supplementary teaching. Many teachers supplement their income by teaching extra lessons and "cram" courses. Many of the private education faculties offer outreach programs as goodwill measures and as a means of raising additional income.

Issues in evaluation are surrounded by the political issues of equity and equalization of opportunity for traditionally disadvantaged students. The issue of equity, in particular, is receiving concentrated attention. The national Examination Board and the Independent Examination Board are involved in this process. The political and social issues of language of instruction and the overburden of language problems are also being examined.

7 REFERENCES AND SOURCES FOR FURTHER READING

Common Sources

1 Husén T, N T Postlethwaite 1994 *International Encyclopedia of Education*. Pergamon Press, Oxford
2 Organisation for Economic Co-operation and Development 1995 *Education at a Glance: OECD Indicators*. Organisation for Economic Co-operation and Development, Paris
3 United Nations Educational, Scientific and Cultural Organization 1995 *Statistical Yearbook*. United Nations Educational, Scientific and Cultural Organization, Paris
4 World Bank 1995 *World Development Report 1995*. Oxford University Press, New York

Other Sources

5 Department of Education 1995 *White Paper on Education*
6 ANC 1994 *Policy Statement on Education*
7 Department of Arts, Culture, Science, and Technology 1996 *Preparing for the 21st Century*
8 Research Institute for Education Planning, University of the Orange Free State 1996 *Education and Manpower Development*
9 The Education Foundation December 1995 *Edusource*
10 Department of Education 1995 *Teachers' Salary Scales*
11 Department of Education 1995 *Norms, Standards and Governance Structures for Teacher Education*
12 Department of Education 1995 *A Proposal for a National Qualifications Framework*

Statistical References:

The tremendous social and political changes that have taken place in South Africa, involving the merging 18 racially divided Education Departments into 9 new geographical ones, has made accessing reliable education data extremely difficult. The principal sources for this chapter are unpublished statistics from the National Central Statistical Office and from the Human Sciences Research Council Education Database. Other statistical information may be found in the Research Institute for Education Planning *Education and Manpower Development*, cited above, as well as in the several National Department of Education Publications also cited above.

Sweden

Kjell Gisselberg, Umeå University
Sigurd Johansson, Umeå University

1 COUNTRY PROFILE

Sweden, part of the Scandinavian peninsula in northern Europe, is bounded by Norway and the North Atlantic to the west, and by Finland and the Baltic Sea to the east. The distance from the north to the south end of the country is 1574 km, and the total land area is 450,000 km². Sweden has a population of close to 9 million with an average density of 20 persons/km². About 85 percent of the population lives in the southern half of the country, while the northern areas are more sparsely inhabited.

Sweden's population is homogenous. Ethnic minorities constitute only 0.5 percent of the population, including Sami in the Arctic region and Finnish-speaking people along the border with Finland. Immigrants and refugees account for approximately 9 percent of the population. Immigration has come from neighboring Scandinavian countries as well as southern and southeastern Europe. Most refugees to Sweden are from Iran, northwestern Africa, and the former Yugoslavia. New immigrants and refugee groups tend to concentrate in specific areas, particularly in the suburbs of large cities. Schools in these areas have special concerns in accommodating students for whom Swedish is a foreign language.

Sweden is a constitutional monarchy governed by a parliamentary democracy. The government is headed by the *Statsminister* or prime minister, who is the leader of the political party in power. The country is divided into 25 counties and 280 municipalities. Members of municipal councils are chosen through public elections.

Sweden is a member of the OECD, and in 1995 became a member of the European Union. Its per capita gross national product for 1993 was US$24,740 and its economy is ranked as high income by the World Bank.

The government allocated more than 7 percent of total public expenditure to education in 1993. The adult literacy rate exceeds 95 percent.

2 THE EDUCATION SYSTEM

Governance and Decision Making

Education goals and guidelines for public sector schooling are outlined by the parliament or *Riksdagen* in both the education act and the national curricula. Goals, guidelines, and curricula are compulsory for all schools, public as well as private.

The national agency for education, the *Skolverket*, was founded in 1991 when the centralized and highly bureaucratic system of school governance was changed to a decentralized, goal-oriented one. The *Skolverket* develops, evaluates, administers, and supervises all schools in the public sector. It is also responsible for drawing up syllabi and grade criteria, issuing educational directives, and operating the national training program for principals. The *Skolverket* gathers information and stimulates change through evaluation, development, research, and supervisory programs. Every three years it presents a comprehensive report on the state of education to the *Riksdagen*.

In each municipality, a municipal council appoints a political body known as the school board. The board's mandate is to ensure that activities and plans are carried out at the school level, and it issues plans containing guidelines and goals for all schools in the municipality. Each school devises a local school plan that describes how the goals are to be achieved. It is the responsibility of the school board to ensure that, despite the decentralization of the system, all schools maintain uniform standards. The municipality provides students with free transportation, text books, health care, and lunch throughout the education system.

Structure of the System and Participation Rates

Structure of Education

The structure of the education system is shown in Figure 1, along with approximate enrollment rates for each level.

Preschool in Sweden refers to all day-care activities for children aged 1 to 6. This reflects the opinion that learning is not limited to situations where adults try to teach children something, but that it takes place in almost all situations. The municipalities are responsible for providing day-care for all children aged 1 to 12, so that all schoolchildren are cared for after school. About 60 percent of children aged 1 to 6 are in different types of day-care programs.

Schools provide activities for six-year-olds with the intention of preparing the children for compulsory school. These activities last for three hours per day and almost all six-year-old children participate (except the ones in Grade 1, see below), even those who are not enrolled in a day-care program.

Traditionally, students in Sweden have always entered school at the age of seven, but beginning in 1991–92, six-year-old children were allowed to enter Grade 1. The new system allows a flexible introduction to school. Children who have started Grade 1 but are not ready for a full day of lessons may leave the class to join the preparatory program. Similarly, six-year-old

children in the preparatory program who show an interest in the more academic portions of that program may join a Grade 1 class for part of the day. This type of exchange occurs for a very small number of students.

Enrollment of six-year-old children in Grade 1 has increased in recent years. In 1991–92, approximately 2 percent of children in Grade 1 were six years old. By the following year the figure was over 3 percent, by 1993–94 over 5 percent, and by 1994–95 almost 7 percent.

Day-care programs for students in Grade 1 to 6 are organized by schools. The programs are provided in the afternoon after regular classes finish, and offer leisure activities that occasionally intersect with school activities.

Previously, compulsory school was organized in three levels: junior, comprising Grades 1 to 3; intermediate, comprising Grades 4 to 6; and senior comprising Grades 7 to 9. The different levels were in many cases housed in separate schools. Senior level was sometimes called lower secondary school, while the *Gymnasieskolan* was called upper secondary school. *Gymnasieskolan*, which offered three- or four-year academic tracks and two-year vocational tracks, was not compulsory, although it attracted almost all compulsory school leavers. At the time that the three-tiered system was in place, teacher education used the same pattern, producing junior, intermediate, and senior level teachers. Senior level teachers also served in upper secondary school.

A new, national curriculum came into operation in the autumn of 1995. It defines the underlying values, basic objectives, and guidelines of the school system. For each subject, there is a nationally defined syllabus in which the purpose, content, and objectives are described. Objectives are of two kinds: those that the school must pursue, and those that the school must give the chance of achieving to all pupils. The second type of objective is articulated for the end of Grades 5 and 9. The concept of levels within schools has disappeared.

Subject to the goals and frames defined by the *Riksdagen* and the government, each municipality is free to decide how its schools are to be run. One result of this freedom is that in many municipalities there is a tendency to organize schools so that students of all grades attend the same school. Teacher education has also changed to correspond to the new system; there

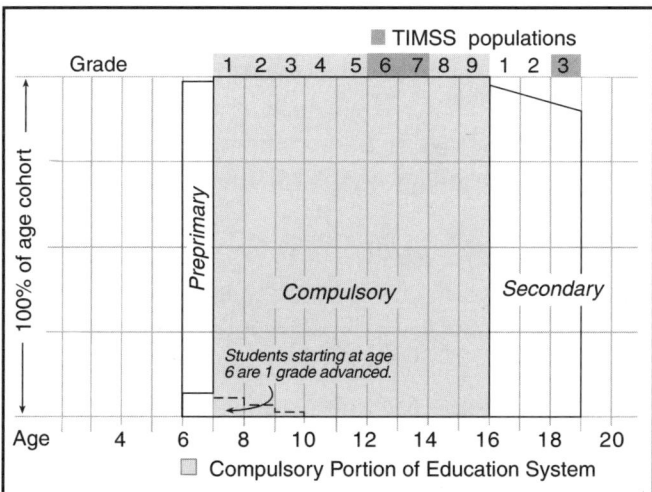

Figure 1: Structure of the Education System

are now teachers for Grades 1 to 7 and for Grades 4 to 9. Teachers of Grades 1 to 7 can choose either natural science and mathematics, or social science and Swedish as their major interest. Teachers of Grades 4 to 9 specialize in two to three different subjects. Because the change took place recently, Sweden has relatively few such teachers. Most teaching is done as in the old system, by junior-level teachers in Grades 1 to 3, intermediate-level teachers in Grades 4 to 6, and senior-level teachers in Grades 7 to 9.

Streaming and Tracking

Upper secondary school is divided into 16 national programs, all of which last for three years. There are academic programs like the Natural Science Program and the Social Science Program, and there are Vocational programs like the Construction Program, the Business and Administration Program, and the Vehicle Engineering program. Most of the programs are subdivided into different branches. Thus the Electrical Engineering Program is split into three branches: Automation, Electronics and Installation. In addition to the sixteen programs, there are individual programs, the duration and content of which are determined by the needs of the individual pupil. Individual programs are often used as a way for students to test different courses before making a final decision; the aim is for the student to commit to a national program at a later date. Apprenticeship training is another kind of individual program; the student combines school study and vocational training as an employee.

All programs must include the eight core subjects: Swedish, English, civics, religious education, mathematics, general science, physical and health education, and arts activities. The programs derive their different emphases from their various specialty subjects. Graduates from any of the 16 programs are eligible to enter university. The two academic programs, Natural Science and Social Science, are specifically intended to prepare students for university study.

Enrollment Rates

Virtually all children are enrolled in compulsory school. Of all the students who left Grade 9 in 1995, 98 percent went on to upper secondary school. More than 90 percent of students entering upper secondary school complete their education within four years.

With a few exceptions, all schools are part of the state or municipal system. Schools are funded by the municipalities who receive support from the government for their different activities. The council of each municipality, together with the local school board, determines levels of funding and distributes funds to local schools. Registered private schools, comprising about 2 percent of students, receive reduced funding. Those schools must satisfy specific standards and provide a satisfactory curriculum.

Most pupils in need of special support are taught in ordinary compulsory basic school and upper secondary school classes, but special teaching groups are also organized to some extent for pupils with functional impairments, as well as special groups for pupils with social and emotional problems. Special schools exist for deaf, hard-of-hearing, vision-impaired and speech-or language-impaired children with secondary disabilities. Most children with intellectual disabilities attend compulsory school for the intellectually disabled.

Municipalities are obliged to organize adult education up to the end of upper secondary school. These programs are free of charge. Typically, students in these programs are adults with little education or former school leavers who wish to enroll in higher education. Immigrants and refugees also study at these schools.

Schools in the System

The School Year

A typical school year in Sweden would include 170 instructional days divided between two terms. The first term runs from the second half of August to the third week in December. The second term begins the second week of January and ends the second week of June. Schools are closed for over two weeks during Christmas and New Year. There is a one-week holiday in March and another at Easter. In addition, there are a few national holidays on which schools are closed.

Instructional Time

The school week is from Monday to Friday. A typical school day in the lower grades of primary school begins at 8:00 a.m. and ends at 2:00 p.m., with a one-hour lunch break. As students move upward through school, the length of the school day is gradually prolonged. Most upper secondary schools end classes at 3:30 p.m.

Class Time Allocated to Mathematics and Science

In Grades 1 to 9, students spend a minimum of 6665 hours of study time on compulsory subjects. Of these, 900 hours are spent studying mathematics and 800 on science subjects, including biology, chemistry, physics, and technology. The remaining hours are divided between the social sciences, geography and earth science, history, religion, and civics. Each school decides how much time to allot to the different science subjects, and the same applies in the social sciences. The distribution of teaching time over grades is also decided by schools. In secondary school the minimum hours of instruction for each subject are specified in the curriculum guide. There are compulsory courses in mathematics and science at all levels of the system.

Class Size

It is commonly accepted that class sizes in Grades 1 to 6 should not exceed 25 students. In Grades 7 to 9 and in upper secondary school the number of students seldom exceeds 30. Class sizes vary to a great extent, depending on the number of children in the school's neighbourhood. In some elementary schools, students from different grades work together in a group. This often happens when the number of students in each participating grade is small, but in some cases it is done as a progressive pedagogical method.

Tracking in Mathematics and Science

Mathematics and science are compulsory subjects from Grades 1 to 9. At this level, each is taught as an integrated subject with its own curriculum. In secondary school, mathematics is organized into five courses known as A to E. Course A in mathematics (110 hours) is compulsory for all students. A+B (150 hours) is compulsory for the liberal arts branch of the social science program, as is A+B+C (180 hours) for the economics and social science branches. Students in the natural science program must take A to D (240 hours) in mathematics, and may choose to take the optional course E (60 hours) as well.

Science is organized in a similar way. A basic science course, 30 hours, is compulsory for all students. Integrated science (100 hours) is compulsory for the social science program. Biology (110 hours), chemistry (180 hours), and physics (220 hours) are compulsory for the natural science program.

Certification of Mathematics and Science Teachers

Beginning in 1988, the education of compulsory school teachers was consolidated into one general program, known as the compulsory school teaching program. Two overlapping subprograms allow teachers to work with a broader range of grades. One subprogram is a 140-point program for Grades 1 to 7, the other is a 180-point program for Grades 4 to 9. A normal study program comprises 40 points per year.

Teacher education for upper secondary school consists of a 180- to 220-point program, with 80 points allocated to the study of a major subject and 60 points to a second subject. Methodology and pedagogy courses are usually sandwiched between subject-matter courses. There is a 40-point program for those who have completed at least 140 points of university education with at least 80 of those points in a subject relevant to school teaching.

Municipalities are responsible for in-service development, and each school must devise a plan for developmental work within the school. During the school year there are thirteen days designated for developmental work, including in-service training if planned. Sometimes schools themselves invite lecturers to give seminars on different topics related to the development plan. Teachers may apply for funds to attend shorter courses in various subjects, organized and offered by teacher organizations and universities. There are courses that are subject specific—relating to either content or method—but there are also general courses aimed at teachers in all subjects. These might include computer training in word-processing, spreadsheet applications, or internet research.

Teacher Profile

Teachers in Sweden have a moderate socioeconomic status. The total number of teachers is 112,000, and 65 percent of these are women. Twelve percent of teachers are under 34 years old, while 20 percent are over 55. Sixty-eight percent are aged between 35 and 54.

Among mathematics and science teachers, the figures are quite different. Only 23 percent are female. More teachers of mathematics and science are young; 26 percent are under 34, while 52 percent are between 34 and 54.

3 TIMSS POPULATIONS

Sweden, IEA, and TIMSS

Sweden has been a member of IEA since the beginning of the organization. For many years, IEA headquarters were situated in Stockholm. Sweden has participated in almost all studies conducted by IEA.

Swedish TIMSS Populations

Sweden is participating in TIMSS at the Populations 2 and 3 levels. In Sweden, Population 2 included 13-year-olds in Grades 6 and 7. These grades were still following the old school system at the time of testing, and students in Grade 6 were in elementary school while students in Grade 7 were in secondary. Most Grade 6 students were in a class in which all subjects were taught by the same teacher, while most Grade 7 students had a specialist teacher for mathematics.

At the Population 3 level, there were similar changes in school structure. Many two-year programs were being changed to three-year programs, and there were many schools in which there was no final grade at the time of testing. The Population 3 sample was drawn from students in their final year of secondary schooling, and in the year of testing this population was somewhat smaller than usual.

4 MATHEMATICS CURRICULUM AND PEDAGOGY

Goals for the Mathematics Curriculum

The mathematics curriculum begins with a description of the relationship between the specific aims of mathematics and the aims of the national curriculum. It is followed by a list of goals for mathematics education. These are expressed in terms of students' development in knowledge, skill, understanding, and familiarity. The goals place no limit on the development of students.

According to the curriculum for compulsory school administered by the Ministry of Education in 1994, the goals for mathematics education are to enable students:

- to develop self-confidence in thinking as well as the ability to learn mathematics and use it in different situations;
- to realize the important role that mathematics has in different cultures and contexts, and to understand the historical contexts in which important mathematical concepts and methods have been developed and used;
- to understand and use basic mathematical concepts and methods;
- to use and realize the value of mathematical language, symbols, and forms of expression;
- to understand and be able to use logical reasoning, make conclusions and generalize, and through oral and written expressions explain and argue their own thinking;
- to understand and be able to formulate and solve problems with the help of mathematics, and to interpret and evaluate solutions to the original problem;
- to organize and use simple mathematical models; to evaluate them critically with respect to preconditions, limitations, and practical use;
- to explore the possibilities of calculators and computers with skill and judgment.

The goals are followed by a description of the characteristics of mathematics describing what the subject is about, as well as phenomena, aspects, and structures. The mathematics curriculum ends with explicit statements about the goals that must be achieved at the end of Grades 5 and 9.

In secondary school, the goals for mathematics education are quite similar to those in the compulsory school. However, there is a stronger focus on development of individual competencies such as self-confidence, curiosity, creativeness, openness, criticism, and persistence. Analytic and communicative skills are explicitly expressed.

Major Changes in the Mathematics Curriculum

Earlier curricula for mathematics consisted of lists of mathematics content, telling teachers what to teach in different grades or stages. The shift from rule-based directives toward decentralization and goal orientation has changed this. The new curriculum, issued in 1994, states only what schools and teachers should strive to accomplish. Decisions about content, methods, organization, and the ways and means of achieving the goals are made cooperatively by principals, teachers, and students.

The new curriculum also implies a new role for teachers. The lecturer role is gradually being replaced by a tutorial approach to teaching. The teacher's task is to listen, understand, work to develop the student's thinking, tutor toward curricular goals, discuss and ask questions, clarify relations, and create activities that will assist students in making discoveries.

The opinions held by teachers on calculator use in compulsory school vary widely. Some want to introduce calculators as early as possible and do not see any need to teach algorithms, while others want students to be skilled in algorithms first, introducing calculators in Grade 8 or 9. Until Grade 6 calculators are not used frequently, but by Grade 9 all students have calculators and about 80 percent use them in school. Half of students at this level are allowed to use calculators in tests. Computers are used by approximately 50 percent of mathematics teachers, mainly for training in mental arithmetic, but also for the theory in functions and statistics. Teachers are not satisfied with the availability of high-quality educational software, and computers appear to be used more frequently for word processing than for mathematical purposes. In 1995 there were 19 students per computer in compulsory school and 8 students per computer in upper secondary school. In many schools—even in Grades 4 to 6— there is a computer in each classroom.

In upper secondary school, calculators are used whenever needed. Most students have a simple calculator of their own, but normally the school provides calculators for use in mathematics classes. In advanced programs more sophisticated calculators are used, and for students who take the most advanced program in mathematics, programmable and graphic calculators have been introduced. This has of course affected teaching, but much more research must be done in this area to develop effective teaching methods and to educate teachers in their use. Computers are, of course, much more common at this level than in compulsory school. There is great variation between schools and between teachers in the extent to which computers are used and the way they are used. Enthusiastic teachers may use

them frequently and in a very advanced way while many teachers do not use them at all.

Current Issues in the Mathematics Curriculum

The changes in society and the increasing use of calculators and computers have profound implications for the teaching of mathematics. Algorithmic practice is decreasing in favor of teaching about number sense. Technical aids like calculators are used to facilitate the solving of problems, which are often set in an everyday context to emphasize the importance and usefulness of mathematics.

The current reform of the Swedish school system means that teachers are given great freedom in planning their teaching and in choosing their working methods and subject matter. Each school must develop a plan for the teaching in all subjects. Making up these plans has revitalized discussions about content. In the debate, mathematics as a way of describing situations and processes, of communicating, and of solving problems has received more emphasis. The need for qualitative skills and knowledge in mathematics has increased in both professional and everyday life. There is a sense, however, that traditional mathematics should not be left out of the curriculum, even though it is often perceived as not useful. Experiences from earlier reforms in curricula, for example in connection with the introduction of the new mathematics, have proven that it is necessary to introduce changes with care and respect for the complex history of the subject.

When the new criterion-referenced marking system was introduced, the criteria for "passed" were specified in the curriculum. Criteria for "passed with distinction" were distributed from the national agency for schools. Within each school teachers are now determining the criteria for "excellent."

Mathematics Textbooks

Textbooks now tend to be structured in a way that makes it easier to accommodate students of differing ability levels. The newer books also include tests that may be used as formative assessments. Students may use the tests on their own to find out the areas in which they need further practice.

There is a growing tendency to use other material

as a starting point for mathematics instruction, but textbooks still have a strong position. They are used as the sole resource by most teachers.

Textbooks are produced and sold by publishing companies. Publishers have been working to make new textbooks more appealing visually, and they now include color as well as many pictures and diagrams. Schools decide which textbooks to use, and provide them free of charge.

Pedagogy

There is a shift in all education from an atomistic to a holistic approach to learning. This shift is apparent in attempts to increase the level of integration between subjects, and to utilize mathematics in other subjects. Communication in mathematics is highly valued.

Constructivist approaches, in which experience and reflection play a role in understanding concepts, are strongly emphasized. The teacher's role is to listen and understand student thinking in order to develop their mathematical abilities. Teachers are to guide, discuss, ask questions, explain, and organize activities; throughout this process students can develop their own knowledge and understanding.

Even though these new ideas have been discussed in education literature for some years, they have only recently been included in the national curriculum. It will be some time before they will have a noticeable impact on the teaching of mathematics in the schools.

5 SCIENCE CURRICULUM AND PEDAGOGY

Goals for the Science Curriculum

The principal goal in science teaching is to create an interest in the natural sciences and to show that they are a part of Sweden's cultural heritage. The curriculum also notes the importance of science knowledge for every citizen, since knowledge of basic science concepts is necessary in the discussion of many of today's social questions.

The science curriculum describes two types of goals. Common goals for science as a whole describe the direction of teaching as aiming at developing the pupil's knowledge, insight, understanding, attitudes, and skills. Attainment goals express the minimum knowledge that all pupils are to have acquired by the end of Grades 5 and 9. There are common goals for biology, chemistry, and physics, as well as specific goals for each subject.

According to the curriculum for compulsory school administered by the Ministry of Education in 1994, common goals in biology, physics, and chemistry include:

- to experience the joy of discovery and experimentation and to develop the inclination and ability to ask questions about phenomena in nature;
- to develop knowledge of scientific concepts and models and an awareness that they are human constructs;
- to acquire an understanding of the scientific way of working and develop the ability to present observations, conclusions, and knowledge in written form;
- to become aware of how knowledge about nature develops, and how it is formed by, and forms, humanity's picture of the world;
- to develop respect for nature and responsibility both for the local and global environments;
- to develop knowledge about the evolution of the cosmos, earth, life, and humanity;
- to acquire insight into how matter is studied at various levels of organization;
- to develop knowledge about the flow of energy from the sun through various natural and technical systems on earth, and about the natural cycles of matter.

In secondary school there are courses in general science and in biology, chemistry, and physics. General science, an extension of the science taught in Grades 7 to 9, is a core subject and all students must take a minimum of 30 hours. In this course, the intention is that students will acquire the knowledge about environmental, energy, and resource issues that will make it possible for them to take a personal position on those issues.

In biology, chemistry, and physics, the goals in upper secondary school are academically oriented. Science processes and the scientific method are stressed, and there are specific periods set aside for practical laboratory work. During the lessons, particularly in physics, much of the time is spent solving problems.

Major Changes in the Science Curriculum

In the old national curriculum for compulsory school, science and social science content was described under the common heading "general subjects." The descriptions contained disparate content on biology, chemistry, physics, civics, history, religious knowledge, and geography. For Grades 1 to 6, the heading "general subjects" appeared in the timetable for students, meaning that teachers were able to determine which area of study received the most attention. In many cases biology, history and geography received the most class time, at the expense of physics, chemistry and religious knowledge. In Grades 7 to 9, "general subjects" was split into science and social science, where science comprised biology, chemistry and physics. In about half of schools, science was organized as an integrated subject, while in the other half as subject-specific courses in biology, chemistry, and physics.

In the new national curriculum for compulsory school, all subjects are specified and the schools must set aside a specific number of hours for each. One consequence of this arrangement is that it will force schools and teachers to introduce chemistry and physics at a much earlier age. It is hoped that this will increase interest those subjects and improve the conditions for learning at later stages. Science is now also compulsory for all students in secondary school. The minimum course is 30 hours in environmental studies.

Current Issues in the Science Curriculum

Even though separate subjects are emphasized in the curriculum, there is a strong tendency toward integration. Research in science education has shown that we introduce too many difficult science concepts in Grades 7 to 9. The tendency now is to concentrate on basic concepts like energy, the laws of conservation, and photosynthesis. These concepts should then be used to create an understanding of fundamental issues in the environment, such as pollution and energy production and use. This approach will show that science is interesting and useful to all, not solely to a handful of specialists. The orientation of science education has thus shifted from memorizing terminology and facts to gaining understanding and insight. Expressed in another way, science teaching has shifted from an emphasis on academic tradition to one which recognizes the individual's need to understand the world.

Science Textbooks

Textbooks are a subject of discussion at present, since some people consider them to be too subject oriented and to have little relation to everyday life. Research has shown that difficult concepts are introduced rapidly in textbooks, thereby making science difficult and less interesting for students. Another concern is that illustrations concentrate on boys and boys' activities, implying that science is not for girls.

Some textbooks for compulsory school are organized in terms of themes rather than the traditional science subareas. These books are not in wide usage, but there seems to be a certain interest in them. Many teachers are also choosing to work with material from different sources such as newspapers and company brochures, and are not relying solely on textbooks. Schools are rapidly becoming more computerized, and students are searching for information both via CD-ROMs and the internet. Networks between schools are being established.

Pedagogy

Direct instruction still plays an important role in the delivery of the science curriculum. The constructivist influence is perhaps stronger in science, however, than in other subjects. Recent trends include less whole-class instruction and increased individual and group work. Open-ended questions, applications, and real-world problems are treated extensively. The role of the teacher has shifted from informing to administering and tutoring. Furthermore, an experimental project that would integrate teaching in different subjects is under discussion. Such a project would entail an increased level of cooperation between teachers.

6 EVALUATION POLICIES AND PRACTICES

National Level Examinations

There are no national examinations in Sweden. Under the new system, there will be nationally distributed tests in mathematics, Swedish, and English for Grades 5 and 9. The purpose of these tests will be to assist teachers when judging the quality of their students' work and the standard of the school. In Grade 9 they will also serve as benchmarks for grading.

At the secondary level there will also be national tests for the various courses in mathematics, Swedish, and English. The dates for the tests are announced well in advance in order for schools to plan their courses. The course structure varies between schools and a test may well be taken in Grade 2 in some schools and in Grade 3 in others. Only the tests in Grade 9 are compulsory, and schools may choose not to participate if the corresponding course can not be timetabled to fit the testing date. Schools, however, can make participation compulsory for the students.

At the system level, the National Agency for Education evaluates schools every third year. This evaluation includes tests in different subjects, and serves as the basis for a report to the government about the status of the school system.

School Level

Marks are given for the first time in Grade 8, and thereafter issued each term. In all grades teachers are obliged to follow and analyze student work and inform students and parents about individual achievement and the ways it can be improved.

The five-mark relative scale has been changed to a criterion-related scale. In compulsory school there are three levels: passed, passed with distinction, and excellent. In secondary school there are four levels: not passed, passed, passed with distinction, and excellent. For each subject, the National Agency of Education issues criteria for "passed" and "passed with distinction," while teachers set the criteria for the remaining levels. The national curriculum states that when marking a student, teachers must use all available information about achievement related to curricular demands, and make an overall judgment. In this procedure, there is a tendency to move away from tests that measure simple factual knowledge and to emphasize students' skill in applying knowledge. As a consequence, national tests in mathematics contain many problems related to real-life situations.

CONCLUDING REMARKS

Being a highly industrialized nation, Sweden is dependent on its export of industrial products. To maintain the quality of these products and to develop new products of high quality it is necessary to have skilled, qualified workers at all levels. In a modern society it is also important that all citizens have a basic knowledge of mathematics and science, not only for handling everyday situations like using different types of advanced equipment but also in order to follow the public debate and to take a personal stand on different social issues. Thus mathematics and science are important subjects and the quality of the teaching as well as the recruitment of students into this area is of great concern.

7 REFERENCES AND SOURCES FOR FURTHER READING

Common Sources

1 Husén T, N T Postlethwaite 1994 *International Encyclopedia of Education*. Pergamon Press, Oxford
2 Organisation for Economic Co-operation and Development 1995 *Education at a Glance: OECD Indicators*. Organisation for Economic Co-operation and Development, Paris
3 United Nations Educational, Scientific and Cultural Organization 1995 *Statistical Yearbook*. United Nations Educational, Scientific and Cultural Organization, Paris
4 World Bank 1995 *World Development Report 1995*. Oxford University Press, New York

Other Sources

5 Skolverkets Rapport nr 74 *Skolan i siffror 1995: Del 1* Ljunglöfs Offset AB, Stockholm
6 Statistics Sweden *Statistical Yearbook of Sweden 95* Norstedts Tryckeri AB, Stockholm
7 Skolverket *Den Nationella Utvärderingen av Grundskolan våren 1992: Matematik Åk 9* Katarina Tryck, Stockholm
8 Skolverket *Bilden av Skolan 1996* AB C O Ekblad & Co, Västervik
9 Ministry of Education and Science 1993 *The Swed-*

ish Way Towards a Learning Society. Allmänna Förlaget AB Stockholm

10 Besmanoff A, Stenborg-Blom 1994:1 *Teacher Education in Sweden.* NARIC-Report National Agency for Higher Education Stockholm

11 *A New Curriculum for Upper Secondary School.* Ministry of Education Stockholm

12 *1994 Curriculum for Compulsory Schools* (Lpo 94) Ministry of Education Stockholm

13 *1994 Curriculum for Non-compulsory Schools* (Lpf 94) Ministry of Education Stockholm

14 The National Agency of Education (Skolverket, S-106 20 Stockholm) publish pamphlets in English where the various parts of the Swedish school system are described more in detail.

Statistical References

Section	Statistic	Reference	Page	Table	Year of Statistic
Country Profile	450,000 km^2	3	1–6	1.1	1993
Country Profile	9 million	4	163	26	1993
Country Profile	US$24,740.	4	163	1	1993
Country Profile	7 percent	4	181	11	1993
Country Profile	95%	4	163	1	1993
Structure of the System	60%	6	–	38,307,308 33,360,362	1993
Structure of the System	2%	5	–	1,5A	1995
Structure of the System	3%	5	–	1,5A	1995
Structure of the System	5%	5	–	1,5A	1995
Structure of the System	7%	5	–	1,5A	1995
Structure of the System	98%	8	16	diagram 8	1995
Structure of the System	90%	8	16	diagram 8	1995
Structure of the System	2%	5	1, 2	A–C	1995
Mathematics Curriculum	80%	7	35	–	1992
Mathematics Curriculum	50%	7	35	–	1992
Mathematics Curriculum	19	8	57	5	1995
Mathematics Curriculum	8	8	57	5	1995

Switzerland

Urs Moser, Office of Educational Research (Canton of Bern)

Armin Gretler, Swiss Coordination Center for Research in Education

Erich Ramseier, Office of Educational Research (Canton of Bern)

Peter Labudde, University of Bern

1 COUNTRY PROFILE

Switzerland is a small country in the heart of Europe, covering an area of roughly 41,300 km². Geographically, Switzerland is made up of three main parts: the Jura, the Central Plain, and the Alps. At the end of 1994, the population reached seven million inhabitants, 19 percent of whom were foreigners. The average density of the population is 170 persons/km², although the figures are higher in the plain than in the mountains. With an average growth rate of less than 1 percent between 1980 and 1990, the total population is increasing slowly. There have been, however, significant changes in the age structure of the population: between 1950 and 1994, the 0 to 19 age group decreased from 31 to 23 percent, whereas the over-65 age group increased from 10 to 15 percent. Due to its geographical location, bordering Germany, Austria, France, Italy, and Liechtenstein, and due to its political will, Switzerland is a multicultural and multilingual country. Some 65 percent of the population, including foreigners, speak German, 18 percent French, 10 percent Italian, and 1 percent Romansch, the remainder being foreigners of other native tongues.

Switzerland is a direct democracy, where citizens vote several times a year on a series of issues. Politically and administratively, the country is structured in three main levels: the Confederation, the 26 cantons, and the 3029 communes, each of which has its own legislative and executive power. The central government is formed by seven ministers, while legislative power at the federal level is vested in two houses, the National Council with 200 deputies representing 10 different political parties, and the Council of States with 46 deputies.

Economically, Switzerland has followed the general evolution of European countries from an agrarian to an industrial, and finally to a service-dominated economy. Along with this change in economic structure came a process of urbanization: by 1990, 60 percent of the population lived in urban and 40 percent in rural settings. The Swiss economy depends heavily on exports, mainly of high-quality products, and on a high proportion of foreign labor, 27 percent in 1990. Women account for 38 percent of the total labor force. The national income per inhabitant, US$33,453 in 1993, is one of the highest in the world.

In 1992, public expenditure on education reached US$17 billion, which is 19 percent of total public expenditure. Of this total, 10 percent was spent on compulsory schooling, 3 percent on vocational training, 3 percent on universities, and the remainder on other types of schools. Compared to other countries, Swiss public expenditure on education as a percent of total public expenditure is high, since the average among OECD countries is about 12 percent.

2 THE EDUCATION SYSTEM

Governance and Decision Making

In accordance with the federalist structure of the country, the Swiss education system is decentralized. Education lies within the jurisdiction of the cantons, and consequently there is no federal ministry of education. Each canton has its own Department of Education, school law, and system of education. In a number of cantons, the government is assisted by an elected, consultative body in educational matters, the *Erziehungsrat*. The next level of supervisory structure in many but not all cantons, is the district, or *Bezirksschulrat*. Local decisions are made by the local educational authority, the *Schulpflege* or *commission scolaire,* a body elected by and out of the communal population, in which local teacher representatives have a consultative voice. All teachers are supervised by an inspector. In some cantons, inspectors are part-time fellow teachers or lay persons and in other cantons full-time staff of the Department of Education. Curriculum and textbook decisions are normally made at the cantonal level, whereas the choice of instructional methods lies with the teacher.

As a result of this decentralized, multilevel structure, the bonds between the population and its education system are very close. However, since all important decisions are put to a popular vote, changes in the education system are often slow. An education system characterized by decentralization and by cantonal authority requires harmonization and coordination, and this is the function of the Swiss Conference of Cantonal Directors of Education. Its main instruments are the Concordat on School Coordination dating from 1970, a number of commissions, and a few institutions. The Concordat includes provisions aimed at common principles for the curricula as well as for common textbooks and manuals.

Structure of the System and Participation Rates

Structure of Education

There is no single, Swiss system of education, and therefore no definitive structure. Each canton determines its own system and structure. A generalized version of the different systems is given in Figure 1 along with approximate enrollment rates for each level.

Most Swiss children spend one or two years in kindergarten before going to primary school. Preschool attendance in 1990–91 was 25 percent for four-year-olds, 75 percent for five-year-olds, and 98 percent for six-year-olds. Primary education, with all pupils in the same type of school, lasts four years in four cantons, five years in four, and six years in the other 18 cantons.

Lower secondary education takes place in different types of school of varying difficulty. Some cantons organize lower secondary schools in a streamed, comprehensive format. In others, schools types are divided into those with basic and those with extended require-

ments. In 1991, 31 percent of students attended schools with basic requirements and 69 percent schools with extended requirements.

The duration of compulsory schooling, including primary and lower secondary education, is nine years. The minimum school leaving age is 16 in most cantons, and 15 in cantons where schooling starts at age 6.

School Types

Upper secondary education is divided into four major types: maturity schools designed for university entrance, general education schools, teacher education institutions, and vocational training institutions. Unlike other schools, maturity schools are governed by federal regulations. Until now, there have been five types of maturity programs, all of which have final examinations providing access to university: A, with emphasis on Greek and Latin; B, on Latin and modern languages; C, on mathematics and science; D, modern languages; and E, economics. A decision has been made but not yet fully implemented to introduce a unique type of maturity offering a number of options during the two last years of upper secondary education.

General education schools at the upper secondary level prepare students for nonuniversity professions, such as in the paramedical and social fields. There are two main types of teacher education institutions. The traditional one, known as the seminary, begins after compulsory schooling and educates candidates in a four-year program. Most students in this program are between 16 and 20 or 21 years old and this type of

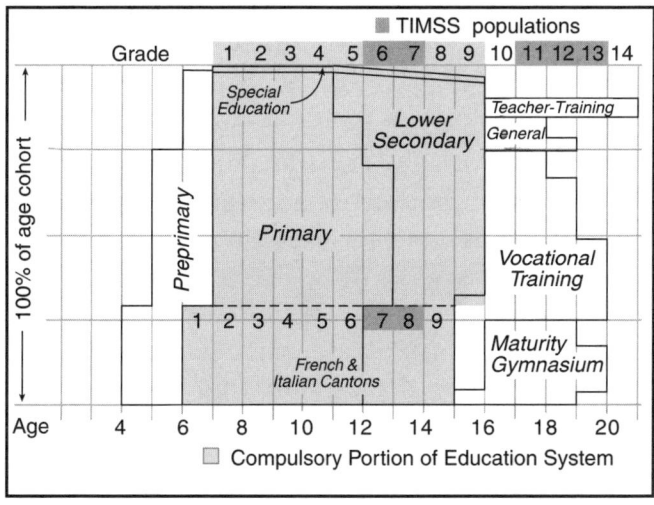

Figure 1: Structure of the Education System

teacher education is part of upper secondary education. The other type, starting after the maturity examination and lasting for two or three years, is part of tertiary education. Most students in this program are between 19 and 23 years old.

More than 70 percent of the 16- to 19-year-old age cohort enters vocational training after compulsory schooling. For most students, this takes the form of an apprenticeship program consisting of two basic elements: three to four days per week in practical training at a business, and one day per week in theoretical and general instruction in a vocational school. A third element, involving bringing together apprentices of the same craft or trade for basic courses at training centers, is becoming increasingly popular. Contrary to most other types of education, vocational training is regulated by federal law. There are recognized apprenticeships of two to four years' duration in approximately 280 vocations in the industrial, handicraft, and service sectors. An important reform was recently introduced in the field of basic vocational training: apprentices are able to obtain an additional 1200 lessons of general education, leading to a vocational maturity offering access to specialized university institutes in the corresponding vocational field.

The most common forms of tertiary education are the nine cantonal universities and two federal institutes of technology. There is also a growing and increasingly diversified extrauniversity tertiary sector, known as higher vocational training. Most higher technical and vocational schools will be converted into specialized university institutes, or *Fachhochschulen* by the end of the 1990s.

Public and Private Systems

Public education is funded through the Confederation (12 percent), the cantons (54 percent), and the communes (34 percent). Private schooling is quantitatively not significant; estimates indicate only a few percent of students attend private institutions.

Enrollment in Mathematics and Science

Mathematics and science are compulsory until the end of Grade 9. After this, enrollment in these courses depends on which senior secondary program the student chooses; some require mathematics and science and some do not.

Grouping of students in mathematics and science

takes place in lower secondary education. In a few cantons, streaming students by ability level is practiced. In the majority of cantons, however, students are tracked into different schooltypes based on their ability and interest. The decision determining which stream or track students will attend is normally based on some or all of the following elements: academic performance during the period preceding the decision, teacher's advice, parents' opinion, and results of entrance examinations.

Schools in the System

The School Year

Because the Swiss education system is decentralized, there are, in effect, 26 cantonal systems. Much information such as number of hours per school day and week, therefore, differs significantly from canton to canton. The length of the school year is stipulated by the Concordat on School Coordination at a minimum of 38 weeks; the actual average is 39 to 40 weeks. The total number of hours during the nine years of compulsory education varies between 7300 and 9000. The average class size in 1993–94 was 20 pupils in primary and 19 students in secondary education, with a variance of five to six between cantons with highest and lowest average.

Certification of Mathematics and Science Teachers

In primary schools, all teachers have the same educational background and teach virtually the whole range of subjects. As mentioned above, preservice teachers may attend one of two types of institution, seminaries or tertiary level institutions. Seminaries, or teacher education institutions as they are correctly known, offer a five-year program that begins after compulsory schooling as a part of upper secondary education. Tertiary teacher education is normally a two-year program at a teacher education institution. Both programs lead to certification for primary teaching.

In lower secondary education, future mathematics and science teachers undergo special preparation in their field of specialty. Content, methodology, and pedagogy requirements vary from canton to canton or region to region. Three or four-year programs at a uni-

Table 1: Socioeconomic Status of Teachers

	Cantons with lowest salary	Cantons with highest salary
Primary education		
Starting salary	US$48,000 (Vaud)	US$68,000 (Zurich)
Maximum salary*	US$76,000 (Ticino)	US$ 109,000 (Zurich)
Lower secondary education		
Starting salary	US$54,000 (Appenzell)	US$75,000 (Zurich)
Maximum salary*	US$77,000 (Neuchâtel)	US$116,000 (Zurich)

* normally reached after about 20 years of service

versity or teacher education institution lead to certification for lower secondary teaching. In cantons, however, with basic requirement and extended requirement lower secondary schools, teacher education for these two types may be different: shorter and more general study for the former, longer and more specialized studies for the latter. Full academic studies and additional courses in didactic methods are required in order to teach mathematics and science in upper secondary schools or Gymnasium.

Teacher Profile

As with other data, teacher salaries differ from canton to canton. Table 1, with information from 1994, gives an overview of the socioeconomic status of teachers, including starting and maximum salaries for the cantons with the lowest and the cantons with the highest salaries. These are given in US dollars at an exchange rate of 115.

Comparing these figures to the overall average salary, including all trades and professions, of US$ 58,000, it becomes obvious that the economic status of teachers in Switzerland is excellent. The percent of women teachers in primary education is 69, in lower secondary education 37.

Table 2: Age Distribution of Teachers in 1993–94

	Primary education (percent)	Lower secondary education (percent)
29 and younger	24	13
30–39	31	30
40–49	29	34
50–60	14	20
60 and older	2	3

3 TIMSS POPULATIONS

Switzerland, IEA, and TIMSS

Switzerland has been associated with IEA since 1987. In addition to TIMSS, Switzerland has participated in the following projects: Phase 1 of the Computer Education Study, the Reading Literacy Study, the Language Education Study, and the Civic Education Study.

Swiss TIMSS Populations

Switzerland is participating in TIMSS at the Populations 2 and 3 levels only. Population 2 is restricted to 22 of the 26 cantons, which include 86 percent of the national population. Since the age at school entrance varies, Population 2 includes all students enrolled in Grades 7 and 8 in the French and Italian-speaking regions and all students enrolled in Grades 6 and 7 in the German-speaking region. As an international option, Grade 8 has been added to the German part for analytical purposes. About 4 percent of thirteen-year-olds were excluded because they were in special schools for the disabled, or in very small schools with fewer then five students in both target grades. About 3 percent were excluded because they were nonnative language speakers.

Population 3 includes all students in their final year of general academic or vocational secondary education. The final grade varies from 11 to 13, depending on the type of educational program. About 90 percent of an age cohort finishes secondary education. Seventeen percent of Population 3 are regarded as specialists in mathematics and physics.

4 MATHEMATICS CURRICULUM AND PEDAGOGY

Goals for the Mathematics Curriculum

As the structure of the school system differs considerably from canton to canton, so do curricula. Jurisdiction over curricula depends on the school level. In primary and lower secondary schools, Grades 1 to 9, the intended curriculum in mathematics is determined by the 26 cantonal Ministries of Education. National recommendations are made in order to achieve a measure of comparability from canton to canton. The recommendations are not compulsory, but they have been consistently adopted by cantonal authorities.

Typical goals for mathematics education include:
- recognizing and analyzing mathematical properties and relations in daily and professional situations;
- solving problems in daily and professional life with mathematical methods;
- abstract and formal thinking and reasoning;
- developing the ability to express ideas precisely;
- developing spatial imagination;
- assessing deduced solutions;
- discriminating between significant and unimportant information;
- estimating orders of magnitude;
- inferring relevant information from tables and graphs.

In some regions there is a high degree of cooperation between cantons. The French-speaking part of Switzerland, in particular, has virtually one common mathematics curriculum. The elaboration of such a cantonal curriculum is an extensive process and involves an inquiry among persons concerned with the mathematics curriculum, including teacher organizations, administrators, researchers in mathematics education, textbook authors, and other opinion leaders in mathematics and mathematics education. The mathematical content of these curricula is compulsory, but teachers are generally free to use different instructional methods.

Postcompulsory education is more standardized, since national authorities coordinate cantonal school curricula in order to prepare students for the university entrance certificate. In Grades 10 to 13 at the higher education level, there is a national catalogue of basic knowledge in mathematics defined by the association of mathematics teachers. The intended curriculum is determined at the cantonal or even at the school level, in some cantons. These curricula are advisory rather than compulsory; a certain core is defined, but teachers have a great deal of freedom in choosing further topics.

At upper secondary schools for vocational education, Grades 10 to 14, the scope of the training, the subject taught, and the curricula are determined for each profession in the relevant vocational programs at the national level. There are special programs for different professions. Mathematics may or may not be an important component of the professional program. Programs with mathematics courses have a well-defined curriculum; governmental and professional organizations inspect and control curricula as well as examinations in mathematics education.

Major Changes in the Mathematics Curriculum

During the last 20 years, the basic concepts of mathematics instruction have been particularly affected by new developments in three sectors:
- the growing importance of mathematics in a highly technical and industrialized environment as a tool for problem solving in numerous fields and as a basis for scientific and technological progress;
- the significant advances made in the study of reasoning and learning processes;
- further research results in mathematics education.

Mathematics instruction has undergone a change from the teaching of arithmetic skills toward mathematical understanding and problem solving. The emphasis is now on integrated, contextual reasoning, problem solving, applications of mathematics in science, technology, and social and economic studies. Instruction has increasingly been oriented towards the process of learning, or how to learn mathematics. Thus, a shift has occurred away from mere knowledge toward learning in a wider context.

At some primary and lower secondary schools, new curricula have been discussed and in some cases implemented. These curricula are based on a more constructivist view of learning, and on the theories of Jean Piaget. The subject-matter structure in the new curriculum of the Canton of Bern, for example, includes arithmetic and algebra, contextual reasoning, geometry, and a new section, statistics and probability. In upper secondary schools, there is increasing use of computers as a tool in mathematics instruction, and in some cantonal curricula computer science is included.

Current Issues in the Mathematics Curriculum

Mathematics instruction today, in its view and contents, has become a guide to judicious learning and purposeful reasoning. These possibilities have, of course, always been inherent in the discipline and have always had some influence. But it is only their factual expression that will make them effective. Four points are now more salient than previously:
- abstract thinking;
- strategic and productive thinking;
- independent research and problem solving;
- logical reasoning, that is, reasoning by ordering, judging, classifying, and comparing.

Other contentious topics in the mathematics curriculum are formal reasoning, and the applications of mathematics in science, technology, social, and economic studies. A concentration and interconnection of subject-matter topics has occurred. Even more important than changes in content are changes in teaching approaches. There is a new focus on a broader repertoire of teaching methods in mathematics and other domains, particularly on individualized methods.

At the upper secondary level, vocational education provides courses in basic theoretical knowledge for the different trades as well as general subjects such as mathematics or science. The general role of vocational education has been more strongly emphasized in recent years, giving access to tertiary education in engineering and other fields. In the long run, Switzerland's two-stream system might become a three-stream system: higher education, vocational education, and the new combination of higher and vocational education.

Mathematics Textbooks

Teaching aids consist of a textbook and exercise books for students, and instructional materials such as didac-tic-methodological supports for teachers. The textbook is conceived as an aid to the student, but it often contains only exercises. Teaching aids are designed to assist in the achievement of the objectives of mathematics instruction. The subject matter is presented in a spiral rather than a linear way. Mathematics textbooks remain the principal resource of teachers and students in most educational systems in Switzerland.

At the primary and lower secondary levels teachers must use the official textbooks. Most cantons have their own textbooks written by a team of teachers and published by the Ministry of Education. Each student has his or her own textbook, usually purchased by the school. Textbooks, even the newest ones, contain very little material except exercises, but a growing tendency toward richer textbooks has been noted.

In upper secondary schools, textbooks include many exercises, explanations, and applications. The number of applications, relations to everyday life, historical remarks, humorous sketches, and mathematical games has increased within the last 10 years. Textbooks for this level are published by commercial publishers. Teachers at upper secondary schools are free to use any textbook or none at all, and many teachers develop their own mathematics curriculum as a result.

Pedagogy

Teachers are free to choose their own teaching methods. Desirable teaching approaches are often included in the curricula, but as suggestions only and written in general terms. In the new curriculum of the canton of Berne, some didactic comments are included, such as active-discovering learning, social learning, using more than one representation (pictures, concrete materials, or symbol sets), and presenting subject-matter in a spiral way. Above all there are two pedagogical approaches apparent in many schools:
- individualization and social learning;
- integration within a topic and between topics.

Individualization does not mean that students must learn on their own. In fact, social and cooperative learning must be part of this approach. The application of both approaches depends on the level of teaching. While teachers in primary schools are general educators who teach all subjects, those at the secondary schools are subject specialists. Secondary teachers are thus usually subject-matter oriented, and individualization is of less importance. Integration between topics requires more organizational activities. Specialists need to cooperate with other teachers to come to a successful between-subject integration.

At the upper secondary level classrooms are usually well equipped. Teachers have the opportunity to apply computer-based mathematics education, but computers have not had a marked influence on teaching methods. In most cantons, computer courses have been implemented into the curricula. Most students are able to use computers, and all students in Grades 7 to 13 use calculators, many of them programmable and graphic calculators.

5 SCIENCE CURRICULUM AND PEDAGOGY

Goals for the Science Curriculum

All general remarks pertaining to mathematics curriculum and pedagogy apply to science education as well, for example, that there are 26 cantonal science curricula for primary and lower secondary schools. There are, however, some general characteristics specific to science education.

At the primary and lower secondary levels, Grades 1 to 9, the intended science curricula are not compulsory and allow a great deal of freedom in the implementation of the science curriculum in the classroom. In most curricula, only a few core topics are defined, while other topics may be chosen by teachers and students. Particularly in Grades 1 to 4, teachers tend to develop their own science curriculum. At this level, science is not a subject on its own, but science topics are integrated in a more general subject. Science includes not only topics in natural science, but elements of the social sciences as well, such as history, anthropology, geography, and even ethics. Examples of these programs are *Natur-Mensch-Mitwelt* (nature, humanity, and the age we live in) or *Mitwelt-Umwelt* (home and environment). Science includes the exploration of objects and phenomena common to students' environment.

There are some official statements of overall goals for science instruction at the cantonal level. Typical goals include:

- understanding science, including the facts, principles, concepts, and methods pertaining to the discipline;
- giving order to observations of the natural and technical world;
- studying the fundamental laws of nature;
- mastering conceptual and practical skills as a result of involvement in scientific activity;
- analyzing scientific information critically, and recognizing the limitations of scientific knowledge;
- applying knowledge and skills in order to generate new knowledge.

At the upper secondary level for higher education, Grades 10 to 13, the intended curricula are determined at the cantonal or in some cases the school level. There are separate curricula for physics, chemistry, biology, and geography and earth science. Recommendations are prepared by associations of chemistry, physics, and biology teachers and disseminated on a national level to establish minimum standards in the subjects.

At the upper secondary level for vocational education, only certain vocational programs include science courses. In these cases there are official national directions which differ from program to program.

Major Changes in the Science Curriculum

There has been a continued emphasis on both science topics and science processes during the last 10 years. At all school levels, environmental concerns and resource issues have become important topics in science courses. Integrated science curricula that include biology, chemistry, and physics have been developed at the lower secondary level in many cantons. Other health-related topics, such as drug use and AIDS, have been incorporated as topics into the Grades 7 to 13 curricula.

The application of science in everyday life has become an important issue in the development of science curricula. At the primary level topics are adapted to a child's way of thinking, using everyday language rather than scientific or academic terms. This trend exists together with a broader repertoire of teaching methods in science. Specifically, teachers are using more individualized teaching methods. The focus on science processes such as observing, classifying, and communicating as well as collecting, recording, and analyzing data have become general goals at all school levels. In lower and upper secondary schools, computers are used in physics and chemistry, particularly for computer simulations. Typically, at that level an increased number of laboratory experiments are performed in science courses. Teachers, school administrators, and textbook authors have discussed treating technology as a subject in its own right, but this has not yet been realized.

Current Issues in the Science Curriculum

At all school levels the following tendencies can be observed:

- more emphasis on procedural knowledge instead of factual knowledge;
- a focus on scientific method;
- increased attention to applications of science in everyday life;
- more emphasis on independent research, problem solving, and logical reasoning, including ordering, judging, classifying, and comparing.

At the lower secondary level, the teaching of separate biology, chemistry, physics, and geography and earth science courses has been replaced by integrated science courses. These courses typically include topics like "energy matters," "a day at the farmhouse," and "a world in motion." Some courses include projects like reducing our energy consumption, building a model aircraft, and investigating our food. As in mathematics and other subjects, there is increased concern for a broader repertoire of teaching methods.

At the higher secondary level at the Gymnasium, the number of compulsory courses and lessons in biology, chemistry, geography, and physics will be reduced because of the new nationwide policy which sets the programs leading to the *Matura*, or university entrance certificate. All programs will include an extended essay in a subject, including science topics, chosen by the student. Many programs are developing a more integrated science course, but this is far from being realized in a regular manner. The new track in the Swiss educational system, combined higher and vocational education, includes physics in all programs that lead to engineering.

Science Textbooks

The general remarks concerning mathematics textbooks also apply to science textbooks. Teachers in Switzerland are free to use science textbooks or not, as they choose. In Grades 1 to 6, there is no official science textbook and teachers use many different textbooks to teach the integrated science curriculum. At this level, teaching aids consist of different books used as a basis for the production of teacher-made worksheets, textbooks, and other material. It is common for teachers to prepare material that better fits the child's way of thinking and uses language more appropriate to the age group than some academic textbooks. In Grades 7 to 13, teachers use common science textbooks for physics, biology, geography, and chemistry. There are some tendencies common to these textbooks:

- the number of topics and pages has increased;
- textbooks contain more pictures, diagrams, and schematics;
- there are more supplementary materials for teachers, including manuals, didactic information, exercises for students, slides, videos, and software;
- there is a shift in content from factual knowledge towards methodology, cognitive and experimental skills, and genetic-historical learning;
- textbooks are used by students for information, assignments, reference, and review, and not solely for exercises.

Pedagogy

Teaching methods are not prescribed by the curriculum. There are, however, some recommendations that are common:

- individualized learning; for further comments, see Section 4;
- social learning;
- improving the self-confidence of the individual.

In Grades 1 to 9, many teachers follow these recommendations, while teachers in Grades 10 to 13 are somewhat less likely to do so. Methods used include individualized programmed instruction, case studies, individual learning centers, projects, peer tutoring, the Keller plan technique, and reciprocal teaching.

Learning as an active, constructive process has become an important issue, at least in all intended curricula, and science lessons are meant to integrate students' preconceptions and interests. Until now this constructivist view of learning has not become a mainstream practice among teachers. Integration within a subject domain and also between different subject domains has increased in recent years, in particular at the primary and the lower secondary levels.

6 EVALUATION POLICIES AND PRACTICES

Switzerland does not have a national assessment program. Assessment performed within schools is used for the placement of students within the streamed system. Typically, mathematics and science are tested as well as other subjects. Grades in mathematics and science are combined with the grades of other subjects, and annual promotion is determined by an aggregate of these grades.

At the end of primary school, Grade 6 in most cantons, an assessment is used to place students in one of the two (in some cantons three or four) different tracks of lower secondary school, Grades 7 to 9. At the end of Grade 9, an assessment is used to place students in higher education programs. All students receive a final report, and a certain average grade is required for entrance to upper secondary schools like the Gymnasium. If students obtain good results in the ongoing tests within schools, they often do not need to write the entrance tests for the Gymnasium. In some cantons, however, allocation is done through a special placement test written by all students.

At the end of upper secondary school for higher education, all students must write the final examination, or *Matura*, which leads to a university entrance certificate. Passing this examination allows access to all universities and higher studies. There is a special mathematics and science program taken by about 4 percent of the national age group. These students write examinations in mathematics, physics, chemistry, and biology as well as in languages, history, and other subjects. The foci of the other programs are subjects other than mathematics and science, such as languages or economics; these are taken by about 13 percent of the age group. These students write an examination in mathematics and at least one science subject besides examinations in language, history, and other subjects. Examinations are designed by the schools themselves and supervised by experts from universities and authorities.

At the end of upper secondary school for vocational education, the final examination depends on the trade chosen. Students attend different numbers of mathematics and science courses depending on the requirements of the trade. Students studying mathematics or science must pass final examinations in these courses in Grade 13 or 14.

All tests vary from canton to canton, and often from school to school as well. Tests are teacher-constructed, not standardized, and most include short-response and multiple-choice questions, and in some cases an essay question. Until now, no comparison has been possible between schools or cantons because of the diverse nature of their assessments.

There has been some discussion in political circles concerning the development of new assessment practices and the evaluation of educational achievement. There is national interest in research about the effectiveness of educational systems. Evaluation seems to have become an instrument of quality control in Swiss schools.

7 REFERENCES AND SOURCES FOR FURTHER READING

Common Sources

1   Husén T, N T Postlethwaite 1994 *International Encyclopedia of Education*. Pergamon Press, Oxford
2   Organisation for Economic Co-operation and Development 1995 *Education at a Glance: OECD Indicators*. Organisation for Economic Co-operation and Development, Paris
3   United Nations Educational, Scientific and Cultural Organization 1995 *Statistical Yearbook*. United Nations Educational, Scientific and Cultural Organization, Paris
4   World Bank 1995 *World Development Report 1995*. Oxford University Press, New York

Other Sources

5   Federal Statistical Office [Bundesamt für Statistik] 1991 *The Swiss Educational Mosaic. A Study in Diversity*. Federal Statistical Office, Bern

6   Federal Statistical Office [Bundesamt für Statistik] 1995 *Swiss Educational Indicators. [Bildungsindikatoren Schweiz]*. Federal Statistical Office, Bern

7   Gretler, A Switzerland: System of education. In Husén T, NT Postlethwaite (eds.) 1995 *The International Encyclopedia of Education, Research and Studies* Vol. 8. Pergamon Press, Oxford

8   National Federation of Swiss Teachers [LCH Dachverband Schweizer Lehrerinnen und Lehrer] 1995 *Salary Statistics* [Besoldungsstatistik/ Statistique des salaires] Zürich

9   1995 *Statistical Yearbook of Switzerland* [Statistisches Jahrbuch der Schweiz]

Statistical References

Section	Statistic	Reference	Page	Table	Year
Teacher Profile	Table 1	8	9	–	1995
Teacher Profile	Table 2	6	96	–	1995
	all other statistics	9	–	–	1995

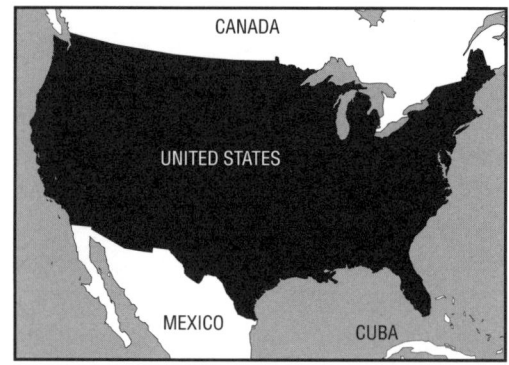

United States

Edward C. Robeck, Hastings College

1 COUNTRY PROFILE

The United States covers an area of close to 10 million km², the majority of which is found in the middle region of North America from the Pacific Ocean to the Gulf of Mexico and the Atlantic. Two noncontiguous states are Alaska in the far northwest of North America and the Hawaiian Islands in the Pacific Ocean. The United States also has jurisdiction over a number of territories in the Pacific and the Caribbean. The District of Columbia, the seat of the federal government, functions as a separate jurisdiction that is similar to a state in many respects.

The population of the United States is characterized most significantly by its diversity. Before European contact in the fifteenth century, there was a native population of approximately 80,000. Since then, the population of the United States has grown to approximately 260,651,000 through immigration and domestic births. The annual growth rate of the population has been approximately 1 percent per year since the early 1970s with immigration playing a significant role. While English is the official language of the United States, in 1990 more than 13 percent of school-aged children spoke a language other than English at home. The population is increasing.

For many purposes, demographic information in the United States is reported in terms of race, ethnicity, gender, and other background demographic factors. When considering such reports, it is important to note that categories, especially race, are defined by social convention. That is to say, differences among groups reflect social conditions that are contingent on historical antecedents, rather than having a biological basis. For example, that race is socially defined can be seen in the fact that sources differ on what is meant by the related terms. "Black" is conventionally used to refer to people of African-American descent, but can also include people from South Asia, the Middle East, and their descendants. "White" usually refers to people of European descent, but is sometimes relegated to Europeans excluding Spain and other Spanish-speaking individuals. Although "Hispanic" is often used in parallel with Black and White designations, it is not considered a race and individuals of any race can be considered Hispanic. Hispanic is used by the U.S. Bureau of the Census to refer to people from Mexico, Puerto Rico, Cuba, South and Central America, and Caribbean countries who are of Spanish or mixed Spanish and native ancestry, but may also refer to native populations that have adopted the Spanish language. In addition to these three groups, others are often recognized for specific purposes. For most of the statistics below, racial and ethnic categories are determined by self-report of the individuals responding to census questions, or other questionnaires. The purpose of using categories here is not to be exhaustively inclusive, but to demonstrate that in the United States "race" and ethnicity are afforded a high degree of social significance that has an impact in many facets of social life, including education, despite the ambiguities inherent in any definitions of the groupings. In 1994, the U.S. Bureau of the Census estimated that 83 percent of the population was White, 13 percent Black, and 10 percent Hispanic, demonstrating some overlap among categories.

The United States government consists of an executive branch headed by an elected president; a judiciary branch headed by an appointed supreme court; and a legislative branch encompassing a two-chamber Congress. Members of Congress are elected by the people; members of the House of Representatives are elected in numbers proportional to the state's population, while the Senate seats two senators for each state. Within each state is an elected governor and a representative body to govern internal affairs. Although the legislative branch sets laws, these are often interpreted within the judiciary branch. For example, in the mid to late 1970s, racial desegregation of schools that had been mandated by the legislative branch was tested and upheld through court proceedings in the judiciary. This led to many school districts having to transport students to schools away from their residences in order to comply with the court-ordered desegregation program.

In 1993, the per capita GNP of the United States was US$25,744. However, there are important income disparities associated with conventionally determined demographic categories. While 12 percent of families were living below what is considered poverty, only 9 percent of White families were living at that level, whereas the figure was 31 and 27 percent for Black and Hispanic families respectively. Such discrepancies persist even when considering the educational attainment of the householder. Nationally, only 2 percent of families with the householder having a bachelor's degree or more lived in poverty. However, the figure was just under 2 percent for Whites, over 5 percent for Blacks, and over 8 percent for Hispanics. The incidence of poverty among children shows even more pronounced patterns associated with race. In 1993, 23 percent of children 18 years old or younger lived in economic poverty. This included 18 percent of White children, 46 percent of Black children, and 41 percent of Hispanic children.

Most funding for public schools comes from state and local governments. Traditionally, the federal government has had a less significant role in educational finance than the states, always providing less than 10 percent of public school funds. In 1993 the federal budget included about US$31 billion in outlays for elementary and secondary education programs, including some programs outside the conventional school structure, such as those for disabled veterans. This was less than 5 percent of the federal budget. Nevertheless, an estimated total of approximately US$263 billion was spent on elementary and secondary schooling in the 1993, meaning that state and local governments were responsible for well over 90 percent of funding for public schools.

Statistics from 1995 indicate that a large proportion of the U.S. population is educated beyond the compulsory level. As with other factors, the educational attainment of the population also exhibits important differences associated with race. In the case of secondary school education, for example, 82 percent of the White population had completed four years of secondary school education or more in 1995, whereas the percentages were 73 percent for Blacks and 53 percent for Hispanics. In the case of higher education, differences were greater. Income levels also show considerable differences, even within groups with similar educational attainment. For example, women that held professional degrees in 1995 earned on average about 55 percent of the average wage for men with similar degrees. Racial differences were also notable, with Black and Hispanic professional degree holders earning about 60 and 40 percent of the average salary for Whites in this category respectively.

2 THE EDUCATION SYSTEM

Governance and Decision Making

In general, education is a responsibility of the individual state. However, all states are required to operate schools meeting federal standards. The United States Department of Education, with the participation of more than 30 federal agencies, is responsible for federal educational policy. State legislation establishes the regulations by which public schools operate and the criteria by which private schools are accredited. The day-to-day operation of schools is primarily a local matter, with state-level administration varying widely in its impact on local school boards.

All states have guidelines for school accreditation that specify such things as the credentials of teaching staff and what resources must be available for instruction. The majority of states also develop curriculum guidelines, at least in the major academic areas, and professional standards for teachers.

There are currently several standards-setting projects under way at the national level. In the near future, both mathematics and science curricula will be influenced in significant ways by an ongoing federal project initiated in response to what has been perceived as the unacceptably low international standing of U.S. students' achievement, and the concern this raises for the future global economic position of the United States. This project was initiated in a meeting of state governors in 1989, and was formally legislated by Congress in 1994 as the Educate America Act. One of the key goals of the act is to have American students rank first in international comparisons by the year 2000. Among the provisions of the act, and perhaps the most influential, is the setting of rigorous standards for academic subjects. The role and impact of these standards is currently being debated. For example, it is not clear whether students will be evaluated with direct reference to these standards and, if so, at what levels (e.g. as a requirement for promotion to the next level, or for graduation). One of the key points at issue is how the project will affect the governance of schools in terms of standards setting, standards implementation, and curriculum decisions. It is likely that such decisions will remain a matter of state-level and local concern, but perhaps with some level of federal participation in the setting of minimum competencies and programs of evaluation.

Structure of the System and Participation Rates

Public and Private Systems

Education is provided primarily by public schools, although there are a number of private schools. Of the approximately 64 million students attending school in the 1994–95 school year, 86 percent were enrolled in public schools. In kindergarten through Grade 8, 88 percent of all school children were enrolled in public schools. In Grades 9 through 12, 90 percent were enrolled in public schools.

Structure of Education

Public schooling in the United States generally begins with a kindergarten year prior to Grade 1. Most students enroll in kindergarten at age 5, and must enroll in most states by the age of 7. Most continue through the public system to graduation at the end of Grade 12, which is considered the final year of compulsory schooling, finishing at age 18. In most states, it is possible for students to choose to leave school before finishing Grade 12, but attendance is mandated by law until age 16 to 18 (age 16 in 33 states; age 17 in 8 states and the District of Columbia, age 18 in 9 states.)

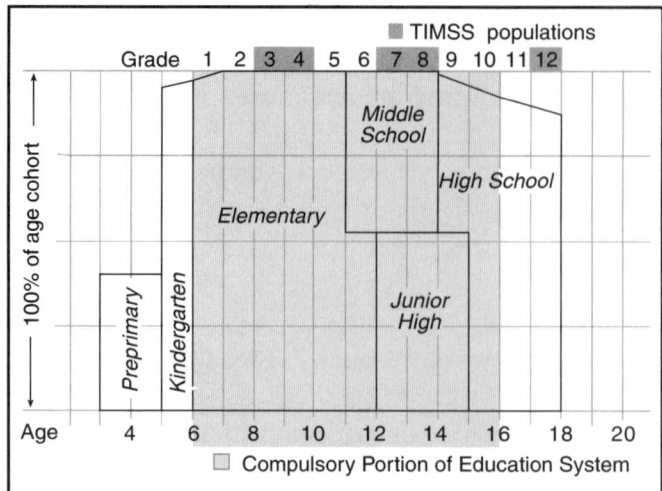

Figure 1: Structure of the Education System

Grades 1 to 12 are divided into primary (kindergarten to 2), intermediate (3 to 6), junior secondary (7 to 9), and senior secondary (10 to 12). Variations on these divisions include a middle school (6 to 8) model, the incorporation of Grade 6 into junior secondary, and the incorporation of Grade 9 into senior secondary. Schools are arranged with different grade ranges, but elementary (kindergarten to 5 or 6), junior secondary or middle (7 to 9 or 6 to 8), senior secondary (10 to 12) schools are common. In some locations there are public or private preschools or other programs serving three-year-old children. Up to Grade 6, students typically attend school in self-contained classrooms and are taught by one teacher, although specialist teachers may be available for physical education, music, art, and science. Beyond that level, students typically move among specialist teachers for some number of subjects, and for all subjects in the upper grades.

Enrollment Rates

In 1993, enrollment of three- and four-year-olds in some kind of education program, primarily preprimary programs, was 40 percent. Over 95 percent of five- and six-year-olds were enrolled in school. The enrollment rate for 16- and 17-year-olds was 94 percent.

Nationally, about 10 percent of students leave secondary school without graduating or obtaining an equivalent terminal credential. In 1994, just under 8 percent of White students between the ages of 16 and 24 were secondary-school dropouts; the figure was over 12 percent for Black students and 30 percent for Hispanic students. Some students who did not complete secondary school will earn a General Educational Development Credential, which for many purposes replaces a secondary school diploma. Of those who do complete secondary school or an equivalent credential, 62 percent enroll in college. In the general population, about 23 percent of adults 25 years or older have completed four or more years of college.

Schools in the System

School Types

Most students attend schools in the public system, although both secular and nonsecular private schools are strongly attended. There were 111,486 schools operating in the 1993–94 school year, of which 85,393 were public schools. Of those public schools, 60,052 were elementary schools, 20,059 secondary, and 2674 combined (2608 other). Student enrollment in public schools range from fewer than 100 to over 3000; public elementary-only schools averaged about 468 pupils per school, public secondary-only schools about 695, while combined elementary and secondary schools averaged 418.

The School Year

Most schools hold classes between the beginning of September and the end of May, and a typical calendar includes between 175 and 190 instructional days. The school week is from Monday to Friday, and each day is approximately six hours long. Schools have extended holidays from late December to early January, and often have a week-long holiday in late March or early April. Students generally have at least one or two holidays each month. In the 1993–94 school year, pupil-teacher ratios averaged 17 in elementary grades and 11 in secondary grades.

Streaming and Tracking

Grouping by ability occurs in many schools, although most do not support overt tracking by ability. Students with higher grades may specialize in mathematics and science early, increasing the number of optional, specialized mathematics and science courses as they reach higher levels. Twenty-five states require enhanced or accelerated programs for talented students and another 18 states provide such services through discretionary programs. Some of these offer specializations in mathematics or science. Nationally, approximately 10 percent of students aged 3 to 21 are served under federal or state-operated programs for students with disabilities or economic disadvantages.

Teacher Certification

Teachers may be certified by two routes: by earning a four-year degree in education, or by earning an education certificate through one or two years' study after a four-year degree in another area. In the former case, science and mathematics teachers may take as many as 15 courses in an area of specialty, and almost as many in education. Teachers certifying after a bachelor's degree often take fewer courses in both areas. In both cases, courses in specific mathematics and science teaching methods are usually required.

In the 1993–94 school year, 52 percent of working public school teachers held a bachelor's degree, and 42 percent held a master's degree. Almost all states require some kind of professional examination for initial teacher certification. These may include basic skills questions, content questions in specialized areas, or professional skills. Some states include in-class observations of teaching performance.

Professional development for teachers after they are certified may take several forms. In some districts, teachers are required to take courses every year in order to retain certification. These courses are sometimes offered by the district itself. School districts also provide in-service workshops on several days each fall and spring. In addition, some school districts require teachers to earn a master's degree within the first five to six years of teaching.

Teacher Profile

Most elementary and secondary teachers are female; the proportion of males is just under 30 percent. Demand for teachers is expected to increase over the next few years, particularly at the secondary school level and in learning-disabled, remedial, and preprimary education. Specialist teachers are most in demand, mathematics and science teachers particularly. Rural schools experience more problems in recruiting teachers than urban schools. Teacher salaries are set by individual school districts and vary considerably. The national average for a public teacher's salary in 1979–80 was US$15,970 while in 1993–94 the average had increased to US$36,933. When considered in constant 1994–95 dollars, this represents an increase of about 14 percent.

3 ## TIMSS POPULATIONS

United States, IEA, and TIMSS

The United States is a member of IEA. Membership is supported by the National Center for Educational Statistics and the National Science Foundation. The official member sent to IEA is selected by a National Academy of Science, Board on International Comparative Studies in Education. Currently, the member, Gordon Ambach, is an official of the Council of Chief State School Officers, which is an organization supporting state education systems, each of which is represented on the Council by its chief operating officer in the state's department of education. The Council conducts a variety of programs to support the provision and improvement of education at the state and national level. The Council's representative is the principal point of contact with IEA, and also reports to the Board on International Comparative Studies in Education.

United States TIMSS Populations

The population definitions for TIMSS in the United States follow those given for the study in general. Population 1, the two adjacent grades with the most nine-year-olds, is represented by students in Grades 3 and 4. Population 2, the two adjacent grades with the most thirteen-year-olds, is represented by students in Grades 7 and 8. Population 3 becomes more diverse, but includes students in the final year of compulsory education, Grade 12, which includes students who are typically 17 or 18 years of age.

Mathematics and science enrollments for Populations 1 and 2 are considered to be 100 percent, since virtually all schools require instruction to all students at those grades. At the Population 3 level, it is estimated that at least 40 percent of students (43 percent of males and 38 percent of females) are enrolled in mathematics and 27 percent (29 percent of males and 25 percent of females) are enrolled in science.

Goals for the Mathematics Curriculum

The intended curriculum in mathematics is determined at the school district level in accordance with state guidelines. Almost all of the states have curriculum documents or guidelines. The curriculum guidelines vary considerably in the specifications of content for mathematics education. However, overall, there has been a greater press in the state guides for achieving mathematical literacy for all students in the population. There has also been a greater emphasis on developing students' abilities to solve mathematical problems situated within the context of real-world applications. Rote memorization of content is being de-emphasized, with the influence of contemporary trends in educational psychology and pedagogy being seen in the increased emphasis on regarding students as constructors of meaning.

The ability to solve novel problems is considered an important general goal of mathematics instruction. In 1989 the National Council of Teachers of Mathematics, a professional organization with no official government status, published a set of standards that has had tremendous influence on subsequent state and local curriculum standards.

Major Changes in the Mathematics Curriculum

In recent years there has been increased attention to social and cultural issues in curriculum and instruction. Most districts have explicit mandates stating that learning materials will be nonracist and nonsexist. Textbooks are frequently reviewed not only for the racial and gender distribution of persons shown or discussed, such as historical figures, but also for the roles, cultural references, and gender-bias of examples and applications, such as examples of uses of mathematics that focus on traditional male interests. Ethnic and racial groups, for example, must be shown in contemporary activities, rather than in strictly historical representations. Much of this has been sparked by the recognition of explicit inequities in learning opportunities among groups of students that lead to differences in achievement test scores, and the apparent connection these have to stereotyping and other implicit forms of prejudice.

The National Council of Teachers of Mathematics standards make specific recommendations for changes in content. In the kindergarten to Grade 4 curriculum, increased attention is suggested in several areas, including the following: number sense, meaning of fractions and decimals, mental computation, geometric relationships, spatial sense, and pattern recognition. In Grades 5 to 8, the following are suggested: reasoning in spatial contexts, connecting mathematics to other subjects and to the world, identifying and using functional relationships, and developing an understanding of variables, expressions, and equations. In Grades 9 to 12, the following are suggested: the use of real-world problems to motivate and apply theory, integration across topics, deductive arguments expressed orally and in sentence or paragraph form, and the use of scientific calculators. In all grades, the standards recommend decreased attention be paid to the memorization of rules, formulae, and procedures, and to practicing tedious pen-and-paper computations.

Current Issues in the Mathematics Curriculum

Although there has been a great deal of support generated for instruction that focuses on students' conceptual understanding, enjoyment, and ability to see the relevance of mathematics, changes in actual teaching practice have been slow in coming. At the same time as attempts have been made to broaden the kinds of strategies being used, there has also been a "back to basics" movement that has a more conservative agenda. Instruction in the basics has been promoted as a way of improving the skills of students and, therefore, to promote students' ability to compete in the labor market, which in turn is expected to improve the individual's standard of living. In contrast to past generations, efforts are being made to encourage parents to take a direct role in school governance and a supportive role in their children's education.

Mathematics Textbooks

A generation ago mathematics textbooks were filled with exercises for students to solve, with some explanation of the procedures they were to use. Currently, textbooks show a variety of approaches, including pictorial representations of problems and photographs showing ways that mathematics is useful in everyday life. However, the degree to which the actual approach to mathematics has changed in textbooks is debatable.

Textbooks are produced by private corporations, with very few such corporations dominating the industry. A number of states recommend textbooks for local use after some process of review against state curriculum guidelines. In some cases, state education funds may only be used for the purchase of books approved at the state level. In all cases, specific textbook selection is a local decision, although this selection may be limited to those approved by the state.

Pedagogy

Textbooks are one of several media now used in mathematics instruction. Manipulative learning materials such as counters and blocks are an important part of instruction in the primary grades so that students develop a conceptual understanding of what numbers and symbols mean. There is a great deal of teaching that fosters rote learning, and repetitive practice exercises are still in wide use. One trend that is quickly taking hold in the United States is the use of hand-held calculators. Many school districts make calculators available to all students and allow them to be used at all times to avoid mundane calculations.

Along with movements toward more active learning in mathematics, the variety of instructional technologies available has become a significant factor in instruction. Recent instructional technologies available in schools are included in Table 1. Numbers represent percent of schools having the technology at various levels in 1995.

Television has been an important part of classroom instruction for the past twenty years and the variety of programs available through instructional television networks has increased dramatically in that time.

Microcomputers began entering schools in the early 1980s, and by 1993–94 there were computers in 99 percent of public schools, 98 percent of Catholic pri-

Table 1: Instructional Technology in Schools in 1995

Media	Elementary	Junior Secondary	Senior Secondary
videodisk players	24	34	34
modems	29	39	51
computer networks	22	32	48
CD-ROM players	33	45	54
satellite dishes	10	22	37
cable television	73	83	78

vate schools, and 86 percent of other private schools. Nationally, there was on average one computer for every 11 students, with this ratio moving to approximately one to nine in public secondary schools. In 1992, half of the school-based computers were in computer labs and another 37 percent were in classrooms. Overall they were used on average 20 hours per week for instructional purposes, with mathematics learning representing the most prevalent use (15 percent of use) and science instruction also being affected, but to a lesser extent (7 percent of use). In addition, almost 32 percent of students in Grades 1 to 8 and over 37 percent of students in Grades 9 to 12 had access to computers at home, although this varies dramatically by both income and racial categories with fewer than 15 percent of Black and Hispanic students but more than 43 percent of White students having such access at home.

5 SCIENCE CURRICULUM AND PEDAGOGY

Goals for the Science Curriculum

Science is a required subject of instruction at all levels. However, at the elementary school level, students may have limited exposure to science depending on the proficiency of the classroom teacher. By Grades 7 and 8 almost all students take a required science course. Some schools offer separate life, earth, and physical science courses that may be taken by students in Grades 7, 8, and 9 respectively, while others offer general science at that level. In senior secondary school most students take one or two years of science required for graduation; at this level these are typically taught as separate biology, chemistry, and physics courses. Some schools

are integrating technology courses into the curriculum, either as separate courses or in combination with physics. Secondary school students are often able to take several science courses concurrently, and many schools have specialized courses, including advanced chemistry and physics, zoology, astronomy and cosmology, ecology, and others.

Major Changes in the Science Curriculum

Since 1983, educational policies have emphasized scientific literacy for all citizens. This circumstance has led to increased emphasis on earlier and more science instruction in elementary school where, traditionally, little science had been taught. There has also been a growing concern with addressing environmental topics in school, and increased attention to the interactions between science, technology, and society. Some schools offer courses based on science-technology-society themes, although these are sometimes considered less academically rigorous than science discipline-based courses. Some schools incorporate an element of social awareness into their science-technology-society programs, sending students into the community to research and address social problems such as pollution and environmental degradation.

Supplemental curriculum materials are growing in their impact in science education, and also in mathematics. These materials enrich the program by adding to or replacing specific units of study and are not intended to provide a complete course. Several major projects focusing on environmental issues, resource-based industries, and integration of mathematics and science have been produced and made available commercially or at no cost to schools as a public service. Many of these programs are developed by public agencies such as utilities companies and parks boards, as facets of public awareness campaigns.

Current Issues in the Science Curriculum

Science education is often justified by the need for a science-literate work force, although it is recognized that the majority of students will not go into technical fields. In 1992, more than 5 percent of students in Grade 12 indicated that they expected to go into a technical field that directly related to science. Science learning can also be justified on the basis of the need to understand science and technology in order to make informed decisions, both personally and as a member of society. There is, therefore, a constant tension in curriculum debates regarding the extent to which students should learn science as part of an academic preparation, and to what extent science learning should be focused on more practical issues for the purposes of informed citizenship. Much of what is discussed with respect to United States science curriculum has to do with finding the proper balance among these concerns.

Schools in the United States have taken on a good deal of responsibility for students' social well-being, which is reflected in recent discussions regarding topics to be addressed in the science curriculum. Topics such as Acquired Immune Deficiency Syndrome, alcohol and drug use, adolescent sexuality, and other issues have all found a place in the debate regarding what should be included in science education, especially in regions where there is no separate health curriculum. When these topics are considered, as throughout curricula that address science-technology-society topics, there is a recognition of the role of personal and social values in curriculum topics. With this recognition have come questions and debates regarding the relationship between the values inherent in curriculum and the values that are promoted in the home and general community.

Two organizations have developed and published science standards within the past few years: the American Association for the Advancement of Science, and the National Research Council. Both were developed with extensive input from the public, educators, and the scientific community, and are considered voluntary. Neither is officially part of the Educate America Act, described above. However, both are expected to have a significant impact on ongoing reform movements, and are consistent with the emphases described above. As states develop their own standards in response to the Educate America Act, it is likely that these two projects will be quite influential.

Science Textbooks

Science textbooks often include a significant amount of material showing how science applies to everyday life, industry, health, new technologies, and other ar-

eas. Textbooks are present in most classrooms although they may not be used on a daily basis. Instead, students perform activities, experiments, laboratory exercises, or projects of some sort. The balance between textbook use and instructional strategies that involve students with objects and equipment varies greatly. Curriculum guides often assume 90 to 120 hours of instructional time will be used in a course, and topic coverage is planned accordingly. Many courses therefore cover a great deal of content, which may have the effect of limiting both student discussion and active teaching strategies.

The fact that both science and mathematics textbooks are developed by private corporations means that large commercial markets can influence what is included in textbooks. Both California and Texas have large student populations and purchase many textbooks; both states also have state textbook adoption processes allowing them substantial influence over textbook design. Since textbooks produced for California and Texas are also sold in other states, their guidelines have an broad impact on the national textbook market.

Pedagogy

Cognitive science and research into children's thinking about science have become important influences on the instructional methods that are promoted in teacher training. There has been a growing recognition that students often do not use their formal science knowledge in daily activities outside of school. In addition, it has been found that students who learn science as a set of facts or algorithmic processes are often not able to use that knowledge flexibly in novel situations. Virtually all reform efforts include some attention to increasing the degree of inquiry-based science learning. Many schools have specialized laboratory facilities, especially at the senior secondary level. At all levels, however, some schools are well supplied with resources for science education, while others are not.

Another growing field in both mathematics and science education has to do with the effect of social interactions on learning. Theories have been developed that indicate that students learn through a process of enculturation much like an apprenticeship model. While the effect of this research on actual classroom practice is questionable, it does signal a shift in emphasis from instruction based on the logical structure of the discipline to a basis in cognitive and social factors, from the content of science and mathematics to the students and their social context.

With this shift in science education from discipline-based strategies has been a greater movement to student-centered instruction. The rise in the use of cooperative learning strategies are an indication of this. In cooperative learning, students are grouped and encouraged to work toward a common goal, focus attention not only on the learning but on the social dynamics of the learning environment. One rationale for cooperative learning argues that if students learn to cooperate in schools, rather than to compete with one another, they will carry this pattern into society. This demonstrates an increasing recognition of the social role of schools. Many experts in this area also advocate the grouping of students with different abilities and other characteristics, to foster students' ability to work successfully with others who are different from themselves. Overall in American schools, issues surrounding diversity are at the forefront of many changes in curriculum and practice. Most teacher education programs, school district training opportunities, and reform initiatives include facets related to this area, such as those focused on increasing multicultural sensitivity and the active implementation of multicultural education.

Attention to diversity has also been brought about by discrepancies in achievement between girls and boys, and among members of different ethnic and racial groups. Pedagogical practices have come under increased scrutiny in an attempt to address the issue. It is now common for educators to attend to the distribution of teacher attention among students, the roles that different students take in classroom activities, and the potential influence of teachers' expectations for students of a particular gender or ethnic background. Many curricula now call for educators to endeavor to ameliorate patterns of unequal treatment that, through the effects they have on learning and grades, may help perpetuate unequal social conditions.

6 EVALUATION POLICIES AND PRACTICES

Assessment in the United States is primarily undertaken at state and local levels, and there are no national examinations. There are, however, periodic large-scale assessments undertaken by states on a voluntary basis. These assessments administer tests in a range of subject areas at several levels separated by two to three years. The assessment instruments are placed in normative categories by age and results are compared locally and nationally. The goal of these assessments is to identify any serious discrepancies in achievement.

Assessment systems are operated by individual states for various purposes. Thirty-four states use some kind of state-mandated system of assessment. About half of the states test students in intermediate and junior secondary grades, and about as many require some kind of test for graduation from secondary school. Students' scores have a bearing on school accreditation and very low scores can lead to increased state-level involvement in local school administration. These policy-level assessments occur and are important in some areas, while other regions have few such large-scale efforts. Other states have examinations used for ranking students or establishing proficiency.

A current trend in assessment is toward alternative and authentic assessments in all areas, and performance-based assessments are having a growing impact nationwide. Several of the states using some form of state-level testing now incorporate practical tasks into the examinations. Some states are using assessment techniques that were developed at the classroom level, such as portfolios, at the state level.

The most widespread use of assessments is for instructional purposes. Teachers develop their own forms of assessment; tests are only one of many forms in common use. Textbook publishers provide tests with virtually all programs, but there has been an expansion of available options with portfolio, performance, and project assessment becoming increasingly standard in commercial programs. Overall, however, classroom assessment is the prerogative of the teacher, and this is an area where there is much staff development undertaken. This reflects the emphasis on the local control of education so that community standards and priorities can take precedence.

7 REFERENCES AND SOURCES FOR FURTHER READING

Common Sources

1   Husén T, N T Postlethwaite 1994 *International Encyclopedia of Education*. Pergamon Press, Oxford
2   Organisation for Economic Co-operation and Development 1995 *Education at a Glance: OECD Indicators*. Organisation for Economic Co-operation and Development, Paris
3   United Nations Educational, Scientific and Cultural Organization 1995 *Statistical Yearbook*. United Nations Educational, Scientific and Cultural Organization, Paris
4   World Bank 1995 *World Development Report 1995*. Oxford University Press, New York

Other Sources

5   Altbach, P G (ed.) 1991 *Textbooks in American Society: Politics, Policy, and Pedagogy*. State University of New York Press, Albany
6   Blank, R K, D Gruebel 1995 *State Indicators of Science and Mathematics Education*. Council of Chief State School Officers, Washington DC
7   National Center for Education Statistics 1995 *Digest of Education Statistics: 1995*. US Department of Education, Office of Educational Research and Improvement, Washington DC
8   U.S. Bureau of the Census 1995 *Statistical Abstract of the United States: 1995* (115th edition) Washington, DC
9   U.S. Congress, Office of Technology Assessment 1995 *Risks to Students in School*. OTA-ENV-633, Washington, DC; U.S. Government Printing Office, September 1995

Statistical References

Section	Statistic	Reference	Page	Table	Year of Statistic
Country Profile	260,651,000	8	8	2	1994
Country Profile	1%	8	8	2	1970–1994
Country Profile	13%	8	53	57	1990
Country Profile	US$25,744	8	456	706	1994
Country Profile	12%, 9%, 31%, 27%, <1%, 5%, 8%	8	484	753	1994
Country Profile	23%, 18, 46, 41	8	481	747	1994
Country Profile	less than 10%	7	50	38	1993
Country Profile	31 billion (30,834,300) less than 5% (4.7)	8	154	232	1993
Country Profile	US$262,525,000 (constant 1990–91 dollars)	8	151	229	1993 estimated
Country Profile	82%, 73%, 53%	8	157	238	1994
Country Profile	55% (6312/month for men, 3530/month for women)	8	158	241	1993
Country Profile	60% 40% (average monthly salary for Whites with professional degree = US$5590; Blacks=US$3445; Hispanics=US$2317	8	158	241	1993

Continued

Statistical References continued

Section	Statistic	Reference	Page	Table	Year of Statistic
Structure of the System	63,939,000	7	11	2	1994 estimated
Structure of the System	86%, 88%, 90%	7	11	2	1994 estimated
Structure of the System	95%	7	68	55	1993
Structure of the System	10%, <8%, 12%, 30%	7	110	101	1994
Schools in the System	61.9	7	187	177	1994
Schools in the System	23%	7	17	8	March 1994
Schools in the System	111,486; 85,393; 60,052; 20,059; 2674; 2608	7	14	5	1993–94
Schools in the System	100 - > 3000	7	104	94	1993–94
Schools in the System	468; 695; 418; 16.8; 11.2	7	74	63	1994 preliminary data
Schools in the System	25; 18	7	66	52	1991–92
Schools in the System	approximately 10% of 4,586,000	7	66	53	1993
Schools in the System	52%, 42%	7	77	66	1993–94
Teacher Profile	just under 30% (1991) note 1986 - 31% males 1981 - 33.1% males	7	79	68	1991
Teacher Profile	US$15,970; US$36,933 14% (US$30,941 - US$36,933 in constant 1994–95 dollars)	7	84	76	1979–80; 1994–95
Mathematics Pedagogy	Table 1	8	171	261	1995
Mathematics Pedagogy	99%, (98.6); 98% (97.9); 86% (85.7)	8	169	258	1994
Mathematics Pedagogy	12, 11, 9	8	169	258	1993–94
Mathematics Pedagogy	approximately half	8	169	259	1992
Mathematics Pedagogy	20 hours	8	169	259	1992
Mathematics Pedagogy	15	8	169	259	1992
Mathematics Pedagogy	almost 32% (31.9)	7	451	414	1993
Mathematics Pedagogy	over 37% (37.2)	7	451	414	1993
Mathematics Pedagogy	fewer than 15%. more than 43%	7	451	414	1993

Appendix A

CHAPTER OUTLINE

1 Country Profile

Country location;
Population, including the impact of immigration and emigration and the principal ethnic and linguistic groups;
Government structure;
Socioeconomic conditions;
Importance of education.

2 The Education System

Governance and Decision Making

Administration of the education system;
Decision-making processes regarding curriculum, textbook selection, and instructional methods.

Structure of the System and Participation Rates

Structure of the system, including levels of education, age group that attends each level, compulsory attendance age, and minimum leaving age;
School types;
Public and private systems;
Sources of funding for the system;
Compulsory and noncompulsory mathematics and science programs, including participation rates.

Schools in the System

Length of the school year, holiday periods, length of school day;
Instructional time spent on mathematics and science;
Average class size in elementary and secondary schools;
Streaming, tracking, or grouping of students, particularly in mathematics and science.

Certification of Mathematics and Science Teachers

Routes to certification for mathematics and science specialists;

Content and pedagogy requirements for mathematics and science specialists;
In-service and upgrading requirements for practicing mathematics and science teachers.

Teacher Profile

Economic status of teachers;
Average age of teachers;
Gender split among elementary and secondary teachers, and among mathematics and science teachers.

3 TIMSS Populations

IEA and TIMSS

Country's relationship to IEA and to TIMSS.

TIMSS Populations

TIMSS populations the country participated in, noting ages and grades.

4 Mathematics Curriculum and Pedagogy

Goals for the Mathematics Curriculum

Major goals of the intended mathematics curriculum;
Development and status of official curriculum guides;
Extent to which curriculum guides are used by teachers.

Major Changes in the Mathematics Curriculum

Significant changes to the mathematics curriculum during the past 10 years, with reference to the impact of technology on the curriculum, changes in curricular emphases, and attempts to influence social change through the curriculum.

Current Issues in the Mathematics Curriculum

Current, major issues in the mathematics curriculum

Mathematics Textbooks

Major changes in the nature and use of mathematics textbooks in the past 10 years, with reference to:
The extent to which teachers at different levels use mathematics textbooks;
The number and nature of the topics included, and the way materials are presented;
Student use of mathematics textbooks;
Ownership of mathematics textbooks;
Writing and publishing of mathematics textbooks;
Selection process for mathematics textbooks.

Pedagogy

Major changes and current issues in mathematics pedagogy, with reference to the use of:
Concept development;
Teaching rules and computational techniques;
Cooperative learning;
Integration of topics within the mathematics curriculum;
Calculators and computers in the classroom;
Deductive and inductive approaches in teaching mathematics;
Manipulative materials;
Investigations and open-ended projects;
Activities focusing on oral and written communication.

5 Science Curriculum and Pedagogy

Goals for the Science Curriculum

Major goals of the intended science curriculum;
Development and status of official curriculum guides;
Extent to which curriculum guides are used by teachers.

Major Changes in the Science Curriculum

Significant changes to the science curriculum during the past 10 years, with reference to: the impact of technology on the curriculum, changes in curricular emphasis, and attempts to influence social change through the curriculum.

Current Issues in the Science Curriculum

Current, significant issues in the science curriculum.

Science Textbooks

Major changes in the nature and use of science textbooks in the past 10 years, with reference to:
The extent to which teachers at different levels use science textbooks;
The number and nature of the topics included, and the way materials are presented;
Student use of science textbooks;
Ownership of science textbooks;
Writing and publishing of science textbooks;
Selection process for science textbooks.

Pedagogy

Major changes and current issues in science pedagogy, with reference to the use of:
Concept development;
Cooperative learning;
Integration of topics within the science curriculum;
Computers, videodisks, and other electronic media in the classroom;
Deductive and inductive approaches in teaching science;
Investigations, laboratories, and open-ended projects;
Activities focusing on oral and written communication;
Approaches focusing on interactions of science, technology, and society.

6 Evaluation Policies and Practices

Evaluation policies as they affect mathematics and science, including national or system-level examinations, school district examinations, and classroom- or school-level examinations, with reference to the following:
Purposes and goals of testing;
Ages and grade levels at which students are tested;
Types of instruments used in testing;
Impact of assessment on curriculum and instruction;
Changes in assessment over the past 10 years and the reasons the changes were implemented;
Current issues in evaluation.

Appendix B

AUTHOR INFORMATION

The Importance of Context for International Comparisons

Dr. David F. Robitaille
Head
Department of Curriculum Studies
Faculty of Education
2125 Main Mall
University of British Columbia
V6T 1Z4
e-mail: dfr@unixg.ubc.ca

Edward C. Robeck
Hastings College
800 Turner Avenue
Hastings, Nebraska
68901
e-mail: erobeck@hastings.edu

Cross-National Similarities and Differences

Dr. David F. Robitaille
Head
Department of Curriculum Studies
Faculty of Education
2125 Main Mall
University of British Columbia
V6T 1Z4
e-mail: dfr@unixg.ubc.ca

Alan R. Taylor
Reseach Associate
Applied Research and Evaluation Services
Faculty of Education
2125 Main Mall
University of British Columbia
V6T 1Z4
Tel: 604-434-6315
Fax: 604-434-7830
e-mail: ataylor@ares.ubc.ca

Argentina

Carlos A. Mansilla
Scientific Researcher
Universidad del Chaco
Av. Italia
3500 Resistencia, Chaco, Argentina

Australia

John G. Ainley
Associate Director
Australian Council for Educational Research
19 Prospect Hill Road (Private Bag 55)
Camberwell, Vic 3124
Australia

Belgium

Christian Monseur
SPE
University of Liège
Bld Du Rectorat, 5 (B32)
4000 Liège
Belgium

Christiane Brusselmans-Dehairs
Rijksuniversiteit Ghent,
Seminarie en Laboratorium voor Didactiek
Henri Dunantlaan 2
B-9000 Ghent
Belgium

Bulgaria

Dr. Kiril G. Bankov
Foundation for Research, Communication,
Education, and Informatics (INCOBRA)
Tzarigradsko shausse 125, bl. 5
1113 Sofia
Bulgaria
e-mail: kbankov@bgearn.acad.bg

Canada

Alan R. Taylor
Reseach Associate
Applied Research and Evaluation Services
Faculty of Education
2125 Main Mall
University of British Columbia
V6T 1Z4
Tel: 604-434-6315
Fax: 604-434-7830
e-mail: ataylor@ares.ubc.ca

Colombia

Carlos Jairo Díaz
Dean
Faculty of Science
Universidad del Valle
C.U.V. Apartado 25360
Cali, Colombia

Efrain Solarte
Miltitaller de Materiales Didacticos
Universidad del Valle
C.U.V. Apartado Aereo 25360
Cali, Colombia

Jorge Arce
Miltitaller de Materiales Didacticos
Universidad del Valle
C.U.V. Apartado Aereo 25360
Cali, Colombia

Cyprus

Constantinos Papanastasiou
Associate Professor
Department of Education
University of Cyprus
Kallipoleos 75
PO Box 537
1678 Nicosia, Cyprus

Czech Republic

PhDr. Jana Svecová, Csc.
Education Policy Center
Faculty of Education, Charles University
M.D. Rettigové 4
116 39 Prague 1
Czech Republic

RNDr. Jana Straková
Institute for Imformation on Education
Senovázné nám. 26
111 21 Prague 1
Czech Republic
e-mail strakova@mbox.cesnet.cz

Denmark

Peter Weng
The Danish National Institute for Educational
Research
28 Hermodsgade
DK 2200
Copenhagen N
Denmark

England

Claudia J. Davis
Assistant Research Officer
The National Foundation for Educational Research
The Mere, Upton Park
Slough
Berkshire SL1 2DQ
England

France

Anne Servant
Ministère de l'Education Nationale, de
l'Enseignement Supérieur et de la Recherche
Direction de l'Evaluation et de la Prospective
142, rue du Bac
75007 Paris
France

Germany

Dr. Kurt Riquarts
Senior Researcher
IPN - Institute for Science Education
at the University of Kiel
Olshausenstrasse 62
24098 Kiel, Germany
e-mail kurt@ipn.uni-kiel.de

Greece

Georgia Kontogiannopoulou-Polydorides
Department of Education
University of Patras

Patras, 26500
Greece

Vasilis Koulaidis
Department of Education
University of Patras
Patras, 26500
Greece

George Stamelos
Department of Education
University of Patras
Patras, 26500
Greece

Joseph Solomon
Department of Education
University of Patras
Patras, 26500
Greece

Hong Kong
Frederick K. S. Leung
Department of Curriculum Studies
The University of Hong Kong
Pokfulam Road
Hong Kong
e-mail: hraslks@hkucc.hku.hk

Nancy W. Y. Law
Department of Curriculum Studies
The University of Hong Kong
Pokfulam Road
Hong Kong
e-mail: nlaw@hkucc.hku.hk

Hungary
Judit Krolopp
Researcher
National Institute of Public Education, Centre for
Evaluation Studies
Dorottya u. 8, Pf. 120
1364 Budapest
Hungary
krolopp@ppp.ces.hu

Péter Vári
Director
National Institute of Public Education, Centre for
Evaluation Studies

Dorottya u. 8, Pf. 120
1364 Budapest
Hungary
e-mail: vari@ppp.ces.hu

Iceland
Einar Gudmundsson
Institute for Educational Research
Department of Measurement and Testing
Sudurgata 39, 101 Reykjavik
Iceland

Anna Kristjansdóttir
Professor
University College of Education

Stefan Bergman
Associate Professor
University College of Education

Thorlakur Karlsson
Associate Professor
University of Iceland

Iran
Ali Reza Kiamanesh
Head, Center for Educational Research
Organization of Research and Educational Planning
Martyr Mossavi Bldg.
Iranshahr Shomali Ave.
Tehran, Iran

Fatemeh Faghihi
Center for Educational Research
Organization of Research and Educational Planning
Martyr Mossavi Bldg.
Iranshahr Shomali Ave.
Tehran, Iran

Israel
Pinchas Tamir
Professor of Education and Science Teaching
& Director of High School Biology Project
Science Teaching Center
The Hebrew University of Jerusalem
Jerusalem 91904
Israel
e-mail: portnyb@vms.huji.ac.il

Italy

Professor Anna Maria Caputo
Centro Europeo Dell'Educazione
Villa Falconieri
Via F. Borromini 5
I-00044 Frascati, Roma
Italy
e-mail: cede@vm.bdp.fi.it

Japan

Mr. Masao Miyake
Chief, Science Education Section
National Institute for Educational Research
6-5-22 Shimomeguro, Meguro-ku
Tokyo 153
Japan
Tel: 03-5721-5077
Fax: 03-3714-7073
e-mail: miyake@nier.go.jp

Mr. Eizo Nagasaki
Chief, Mathematics Education Section
Research Centre for Science Education
National Institute for Educational Research
6-5-22 Shimomeguro, Meguro-ku
Tokyo 153
Japan
Tel: 03-5721-5080
Fax: 033714-7073
e-mail: enagasaki@nier.go.jp

Republic of Korea

JinGyu Kim
Assistant Professor
National Board of Educational Evaluation
15-1, Chungdam 2-dong, Kangnam-gu
Seoul 135-102
Korea

Kuwait

Dr. Mansour G. Hussein
Assistant Under Secretary for Education
Development
Ministry of Education
P.O. Box 7
Safat, 13001
Kuwait

Latvia

Andrejs Geske
Faculty of Education and Psychology
University of Latvia
Jurmalas gatve 74/76
Riga, LV-1083
Latvia
e-mail: andrejs@eduinf.lu.lv

Lithuania

Dr. Algirdas Zabulionis
Faculty of Mathematics
University of Vilnius
Naugarduko 24
2600 Vilnius
Lithuania
Fax: +370-2-651684
e-mail: algiz@auste.elnet.lt

Netherlands

Wilmad Kuiper
University of Twente
Faculty of Educational Science and Technology
Department of Curriculum
P.O. Box 217
7500 AE Enschede
The Netherlands
e-mail: kuiper@edte.utwente.nl

Anja Knuver
University of Twente
Faculty of Educational Science and Technology
Centre for Applied Research in Education (OCTO)
P.O. Box 217
7500 AE Enschede
The Netherlands
e-mail: knuver@edte.utwente.nl

New Zealand

Robert A. Garden
Education Consultant
4 Tregear Place
Porirua
New Zealand
e-mail: 100251.3527@compuserve.com

Norway

Gard Brekke
Telemark Education Research
Laererskoleveien 35
3670 Notodden
Norway
e-mail: gard.brekke@not.hit.no

Marit Kjaernsli
Department of Teacher Education and School
Development
P.O. Box 1099
Blindern, N-0316
Oslo, Norway
e-mail: marit.kjarnsli@ils.uio.no

Svein Lie
Department of Teacher Education and School
Development
P.O. Box 1099
Blindern, N-0316
Oslo, Norway
e-mail: svein.lie@ils.uio.no

The Philippines

Dr. Milagros D. Ibe
Director
University of the Philippines Institute for Science
and Mathematics Education Development (UP ISMED)
UP, Diliman 1101
Quezon City, Philippines
Tel: 63-2-928-2622
Fax: 63-2-928-2625
e-mail: ismed@nicole.upd.edu.ph

Dr. Amelia E. Punzalan
Science Education Specialist
University of the Philippines Institute for Science
and Mathematics Education Development (UP ISMED)
UP, Diliman 1101
Quezon City, Philippines
Tel: 63-2-928-2622
Fax: 63-2-928-2625
e-mail: ismed@nicole.upd.edu.ph

Romania

Gabriela Nausica Noveanu
Scientific Researcher
Institute for Educational Sciences
37, Stirbei Voda St,
RO-70732 Bucharest
Romania

Viorica Livia Pop
Scientific Researcher
Institute for Educational Sciences
37, Stirbei Voda St,
RO-70732 Bucharest
Romania

Marilia Ludu
Scientific Researcher
Institute for Educational Sciences
37, Stirbei Voda St,
RO-70732 Bucharest
Romania

Dragos Noveanu
Scientific Researcher
Institute for Educational Sciences
37, Stirbei Voda St,
RO-70732 Bucharest
Romania

Russian Federation

Galina Kovalyova
Director, Centre for Evaluating the Quality of
Education
Institute of General Secondary Education
Russian Academy of Education
Pogodinskaya st. 8
119905 Moscow
Russia
e-mail: gkovalev@sovam.com

Klara Krasnianskaia
Leading Researcher, Center for Evaluating the
Quality of Education
Institute of General Secondary Education
Russian Academy of Education
Pogodinskaya st. 8
119905 Moscow
Russia
e-mail: gkovalev@sovam.com

George Dorofeev
Head, Mathematics Education Department
Institute of General Secondary Education
Russian Academy of Education
Pogodinskaya st. 8
119905 Moscow
Russia
e-mail: gkovalev@sovam.com

Scotland

Brian Semple
Principal Research Officer
Research and Intelligence Unit
The Scottish Office Education Department
New St. Andrew's House
Edinburgh, EH1 3SY
Scotland

Singapore

Chan Siew Eng
Research and Evaluation Officer
Ministry of Education
Kay Siang Road
Singapore 248922
e-mail: mse@moe.edu.sg

Chang Swee Tong
Assistant Director of Sciences
Ministry of Education
Kay Siang Road
Singapore 248922
e-mail: cst@moe.edu.sg

Mary Toh
Specialist Inspector of Science
Ministry of Education
Kay Siang Road
Singapore 248922
e-mail: mth@moe.edu.sg

Slovak Republic

PhDr. Silvia Matúsová
Director
The National Institute for Education
Pluhova 8
P.O. Box 26
830 000 Bratislava
Slovak Republic

RNDr. Mária Berová
The National Institute for Education
Pluhova 8
P.O. Box 26
830 000 Bratislava
Slovak Republic

Renata Tothova

Slovenia

Marjan Setinc
Senior Researcher
Center for IEA Studies
Educational Research Institute
(Pedagoski institut)
Gerbiceva 62, P.O. Box 2976
1111 Ljubljana
Slovenia
tel: +386-61-331-625
Fax: +386-61-331-637
e-mail: marjan.setinc@uni-lj.si

South Africa

Derek Gray
Education Consultant
Human Sciences Research Council
Private Bag X41
Pretoria 0001
South Africa
Tel: 12-202-2598
Fax: 12-202-2481
e-mail: djg@tutor.hsrc.ac.za

Sweden

Dr. Kjell Gisselberg
Department of Mathematics and Science
Umeå University
S-901 87 Umeå
Sweden
e-mail: kjell.gisselberg@educ.umu.se

Sigurd Johansson
Department of Education
Umeå University
S-901 87 Umeå
Sweden

Switzerland

Dr. Urs Moser
Office of Educational Research
Amt für Bildungsforschung
Sulgeneckstrasse 70
CH-3005 Bern
Switzerland
e-mail: umoser@ed.unibe.ch

Armin Gretler
Director of the Swiss Coordination Center for
Research in Education
Schwizerische Koordinationsstelle für
Bildungsforschung
Entfelderstrasse 61
CH-5000 Aarau
Switzerland
e-mail: skbf-csre@ping.ch

Erich Ramseier
Office of Educational Research
Amt für Bildungsforschung
Sulgeneckstrasse 70
CH-3005 Bern
Switzerland
e-mail: abf@ed.unibe.ch

Professor Peter Labudde
Universtät Bern
Hoeheres Lehramt
uesmattstrasse 27a
CH-3012 Bern
Switzerland
e-mail: labudde@kl.unibe.ch

United States

Edward C. Robeck
Hastings College
800 Turner Avenue
Hastings, Nebraska
68901
e-mail: erobeck@hastings.edu

Appendix C

THE MATHEMATICS FRAMEWORK

This section contains a breakdown of the TIMSS mathematics framework into its main categories within each of the three aspects or dimensions. This is followed by a more detailed breakdown of each category into its sub-categories. Appendix C contains amplified descriptions and a number of examples to illustrate the intended scope of each of the categories and sub-categories.

The Three Aspects

1. Content
2. Performance expectations
3. Perspectives

Major Categories of the Mathematics Framework

Content

1.1 Numbers
1.2 Measurement
1.3 Geometry: position, visualization, and shape
1.4 Geometry: symmetry, congruence, and similarity
1.5 Proportionality
1.6 Functions, relations, and equations
1.7 Data representation, probability, and statistics
1.8 Elementary analysis
1.9 Validation and structure
1.10 Other content

Performance Expectations

2.1 Knowing
2.2 Using routine procedures
2.3 Investigating and problem solving
2.4 Mathematical reasoning
2.5 Communicating

Perspectives

3.1 Attitudes towards science, mathematics, and technology
3.2 Careers involving science, mathematics, and technology
3.3 Participation in science and mathematics by underrepresented groups
3.4 Science, mathematics, and technology to increase interest
3.5 Scientific and mathematical habits of mind

SUB-CATEGORIES WITHIN THE MATHEMATICS FRAMEWORK

Content

The content categories are based primarily on functional considerations. The main goal was to select categories that were useful for coding.

1.1 **Numbers**
- 1.1.1 **Whole numbers**
  - 1.1.1.1 Meaning
  - 1.1.1.2 Operations
  - 1.1.1.3 Properties of operations
- 1.1.2 **Fractions and decimals**
  - 1.1.2.1 Common fractions
  - 1.1.2.2 Decimal fractions
  - 1.1.2.3 Relationships of common and decimal fractions
  - 1.1.2.4 Percentages
  - 1.1.2.5 Properties of common and decimal fractions
- 1.1.3 **Integer, rational, and real numbers**
  - 1.1.3.1 Negative numbers, integers, and their properties
  - 1.1.3.2 Rational numbers and their properties
  - 1.1.3.3 Real numbers, their subsets, and their properties
- 1.1.4 **Other numbers and number concepts**
  - 1.1.4.1 Binary arithmetic and/or other number bases
  - 1.1.4.2 Exponents, roots, and radicals
  - 1.1.4.3 Complex numbers and their properties
  - 1.1.4.4 Number theory
  - 1.1.4.5 Counting
- 1.1.5 **Estimation and number sense**
  - 1.1.5.1 Estimating quantity and size
  - 1.1.5.2 Rounding and significant figures
  - 1.1.5.3 Estimating computations
  - 1.1.5.4 Exponents and orders of magnitude

1.2 **Measurement**
- 1.2.1 Units
- 1.2.2 Perimeter, area, and volume
- 1.2.3 Estimation and error

1.3 **Geometry: position, visualization, and shape**
- 1.3.1 Two-dimensional geometry: coordinate geometry
- 1.3.2 Two-dimensional geometry: basics
- 1.3.3 Two-dimensional geometry: polygons and circles
- 1.3.4 Three-dimensional geometry
- 1.3.5 Vectors

1.4 **Geometry: symmetry, congruence, and similarity**
- 1.4.1 Transformations
- 1.4.2 Congruence and similarity
- 1.4.3 Constructions using straight-edge and compass

1.5 **Proportionality**
- 1.5.1 Proportionality concepts
- 1.5.2 Proportionality problems
- 1.5.3 Slope and trigonometry
- 1.5.4 Linear interpolation and extrapolation

1.6 **Functions, relations, and equations**
- 1.6.1 Patterns, relations, and functions
- 1.6.2 Equations and formulas

1.7 **Data representation, probability, and statistics**
- 1.7.1 Data representation and analysis
- 1.7.2 Uncertainty and probability

1.8 **Elementary analysis**
- 1.8.1 Infinite processes
- 1.8.2 Change

1.9 **Validation and structure**
- 1.9.1 Validation and justification
- 1.9.2 Structuring and abstracting

1.10 **Other content**
- 1.10.1 Informatics

Performance Expectations

2.1 Knowing
- 2.1.1 Representing
- 2.1.2 Recognizing equivalents
- 2.1.3 Recalling mathematical objects and properties

2.2 Using routine procedures
- 2.2.1 Using equipment
- 2.2.2 Performing routine procedures
- 2.2.3 Using more complex problems

2.3 Investigating and problem solving
- 2.3.1 Formulating and clarifying problems and situations
- 2.3.2 Developing strategy
- 2.3.3 Solving
- 2.3.4 Predicting
- 2.3.5 Verifying

2.4 Mathematical reasoning
- 2.4.1 Developing notation and vocabulary
- 2.4.2 Developing algorithms
- 2.4.3 Generalizing
- 2.4.4 Conjecturing
- 2.4.5 Justifying and proving
- 2.4.6 Axiomatizing

2.5 Communicating
- 2.5.1 Using vocabulary and notation
- 2.5.2 Relating representations
- 2.5.3 Describing/discussing
- 2.5.4 Critiquing

Perspectives

3.1 Attitudes towards science, mathematics, and technology

3.2 Careers involoving science, mathematics, and technology
- 3.2.1 Promoting careers in science, mathematics, and technology

3.2.2 Promoting the importance of science, mathematics, and technology in non-technical careers

3.3 Participation in science and mathematics by underrepresented groups

3.4 Science, mathematics, and technology to increase interest

3.5 Scientific and mathematical habits of mind

THE SCIENCE FRAMEWORK

This section contains a breakdown of the TIMSS science framework into its main categories within each of the three aspects or dimensions. This is followed by a more detailed breakdown of each category into its sub-categories. Appendix D contains amplified descriptions and a number of examples to illustrate the intended scope of each of the categories and sub-categories.

The Three Aspects

1. Content
2. Performance expectations
3. Perspectives

Major Categories of the Science Framework

Content

1.1 Earth sciences
1.2 Life sciences
1.3 Physical sciences
1.4 Science, technology, and mathematics
1.5 History of science and technology
1.6 Environmental and resource issues
1.7 Nature of science
1.8 Science and other disciplines

Performance Expectations

2.1 Understanding
2.2 Theorizing, analyzing, and solving problems
2.3 Using tools, routine procedures, and science processes
2.4 Investigating the natural world
2.5 Communicating

Perspectives

3.1 Attitudes towards science, mathematics, and technology
3.2 Careers in science, mathematics, and technology
3.3 Participation in science and mathematics by underrepresented groups
3.4 Science, mathematics, and technology to increase interest
3.5 Safety in science performance
3.6 Scientific habits of mind

SUB-CATEGORIES WITHIN THE SCIENCE FRAMEWORK

Content

The content categories are based primarily on functional considerations. The main goal was to select categories that were useful for coding.

1.1 Earth sciences
    1.1.1   **Earth features**
        1.1.1.1 Composition
        1.1.1.2 Landforms
        1.1.1.3 Bodies of water
        1.1.1.4 Atmosphere
        1.1.1.5 Rocks, soil
        1.1.1.6 Ice forms
    1.1.2   **Earth processes**
        1.1.2.1 Weather and climate
        1.1.2.2 Physical cycles
        1.1.2.3 Building and breaking
        1.1.2.4 Earth's history
    1.1.3   **Earth in the universe**
        1.1.3.1 Earth in the solar system
        1.1.3.2 Planets in the solar system
        1.1.3.3 Beyond the solar system
        1.1.3.4 Evolution of the universe

1.2 Life sciences
    1.2.1   **Diversity, organization, structure of living things**
        1.2.1.1 Plants, fungi
        1.2.1.2 Animals
        1.2.1.3 Other organisms
        1.2.1.4 Organs, tissues
        1.2.1.5 Cells
    1.2.2   **Life processes and systems enabling life functions**
        1.2.2.1 Energy handling
        1.2.2.2 Sensing and responding
        1.2.2.3 Biochemical processes in cells
    1.2.3   **Life spirals, genetic continuity, diversity**
        1.2.3.1 Life cycles
        1.2.3.2 Reproduction
        1.2.3.3 Variation and inheritance
        1.2.3.4 Evolution, speciation, diversity
        1.2.3.5 Biochemistry of genetics
    1.2.4   **Interactions of living things**
        1.2.4.1 Biomes and ecosystems
        1.2.4.2 Habitats and niches
        1.2.4.3 Interdependence of life
        1.2.4.4 Animal behaviour
    1.2.5   **Human biology and health**
        1.2.5.1 Nutrition
        1.2.5.2 Disease

1.3 Physical sciences
    1.3.1   **Matter**
        1.3.1.1 Classification of matter
        1.3.1.2 Physical properties
        1.3.1.3 Chemical properties
    1.3.2   **Structure of matter**
        1.3.2.1 Atoms, ions, molecules
        1.3.2.2 Macromolecules, crystals
        1.3.2.3 Subatomic particles
    1.3.3   **Energy and physical processes**
        1.3.3.1 Energy types, sources, conversions
        1.3.3.2 Heat and temperature
        1.3.3.3 Wave phenomena
        1.3.3.4 Sound and vibration
        1.3.3.5 Light
        1.3.3.6 Electricity
        1.3.3.7 Magnetism
    1.3.4   **Physical transformations**
        1.3.4.1 Physical changes
        1.3.4.2 Explanations of physical changes
        1.3.4.3 Kinetic theory
        1.3.4.4 Quantum theory and fundamental particles
    1.3.5   **Chemical transformations**
        1.3.5.1 Chemical changes
        1.3.5.2 Explanations of chemical changes
        1.3.5.3 Rate of change and equilibria
        1.3.5.4 Energy and chemical change
        1.3.5.5 Organic and biochemical changes
        1.3.5.6 Nuclear chemistry
        1.3.5.7 Electrochemistry

Performance Expectations

Perspectives

DETAILED MATHEMATICS FRAMEWORK CATEGORIES

Content

1.1 **Numbers**
    1.1.1   **Whole numbers**
        1.1.1.1 Meaning (the uses of numbers, place value and numeration, ordering and comparing numbers)
        1.1.1.2 Operations (addition, subtraction, multiplication, division, mixed operations)
        1.1.1.3 Properties of operations (commutative property, distributive property, etc.)
    1.1.2   **Fractions and decimals**
        1.1.2.1 Common fractions (meaning and representation of common fractions, computations with common fractions and mixed numbers)
        1.1.2.2 Decimal fractions (meaning and representation of decimals, computations with decimals)
        1.1.2.3 Relationships of common and decimal fractions (conversion to equivalent forms, ordering of fractions and decimals)
        1.1.2.4 Percentages (all work with percent computations and various types of percent problems)
        1.1.2.5 Properties of common and decimal fractions (commutative, distributive, etc.)
    1.1.3   **Integer, rational, and real numbers**
        1.1.3.1 Negative numbers, integers, and their properties
        1.1.3.2 Rational numbers and their properties (terminating and recurring decimals)
        1.1.3.3 Real numbers, their subsets, and their properties
    1.1.4   **Other numbers and number concepts**
        1.1.4.1 Binary arithmetic and/or other number bases
        1.1.4.2 Exponents, roots, and radicals (integer, rational, and real exponents)
        1.1.4.3 Complex numbers and their properties
        1.1.4.4 Number theory (primes and factorization, elementary number theory, etc.)
        1.1.4.5 Counting (permutations, combinations, etc.)
    1.1.5   **Estimation and number sense**
        1.1.5.1 Estimating quantity and size
        1.1.5.2 Rounding and significant figures
        1.1.5.3 Estimating computations (mental arithmetic and reasonableness of results)
        1.1.5.4 Exponents and orders of magnitude

1.2 **Measurement**
    1.2.1   **Units** (concept of measure and standard units [including metric system], use of appropriate instruments [precision and accuracy], common measures [length, area, volume, capacity, time and the calendar; money; temperature; mass and weighing; angles; quotients and products of units; km/h, m/s, etc.], dimensional analysis)
    1.2.2   **Perimeter, area, and volume** (concepts of perimeter, area, surface area, volume; formulae for perimeters, areas, surface areas, and volumes)
    1.2.3   **Estimation and errors** (estimation of measurements and errors of measurement, precision and accuracy of measurements)

1.3　Geometry: position, visualization, and shape

　　1.3.1　**Two-dimensional geometry: coordinate geometry** (line and coordinate graphs, equation of line in the plane, conic sections and their equations)

　　1.3.2　**Two-dimensional geometry: basics** (points, lines, segments, rays, angles; parallelism and perpendicularity)

　　1.3.3　**Two-dimensional geometry: polygons and circles** (triangles; quadrilaterals: their classification and properties; Pythagorean Theorem and applications; other polygons, circles, and their properties)

　　1.3.4　**Three-dimensional geometry** (three-dimensional shapes and surfaces and their properties; planes and lines in space; spatial perception and visualization; coordinate systems in three dimensions; equations of lines, planes, and surfaces in space)

　　1.3.5　**Vectors**

1.4　Geometry: symmetry, congruence and similarity

　　1.4.1　**Transformations** (patterns, tessellations, friezes, stencils, etc.; symmetry [line and rotational symmetry, symmetry in three dimensions, symmetry in algebra and number patterns]; transformations: symmetries and congruence, enlargements [dilations], combinations of geometric transformations, group structure of transformations, matrix representation of transformations)

　　1.4.2　**Congruence and similarity** (congruences [congruent triangles and their properties; SSS, SAS], congruent quadrilaterals and polygons and their properties, similarities [similar triangles and their properties])

　　1.4.3　**Constructions using straight-edge and compass**

1.5　Proportionality

　　1.5.1　**Proportionality concepts** (meaning of ratio and proportion, direct and inverse proportion)

　　1.5.2　**Proportionality problems** (solving proportional equations, solving practical problems with proportionality, scales [maps and plans], proportions based on similarity)

　　1.5.3　**Slope and trigonometry** (slope and gradient in straight-line graphs, trigonometry of right-angled triangles)

　　1.5.4　**Linear interpolation and extrapolation**

1.6　Functions, relations, and equations

　　1.6.1　**Patterns, relations, and functions** (number patterns, relations and their properties, functions and their properties, representation of relations and functions, families of functions [graphs and properties], operations on functions, related functions [inverse, derivative, etc.], relationship of functions and equations [e.g., zeros of functions as roots of equations], interpretation of function graphs, functions of several variables, recursion)

　　1.6.2　**Equations and formulas** (representation of numerical situations; informal solution of simple equations; operations with expressions; equivalent expressions [factorization and simplification]; linear equations and their formal [closed] solutions; quadratic equations and their formal [closed] solutions; polynomial equations and their solutions; trigonometrical equations and identities; logarithmic and exponential equations and their solutions; solution of equations reducing to quadratics, radical equations, absolute value equations, etc.; other solution methods for equations [e.g., successive approximation]; inequalities and their graphical representation; systems of equations and their solutions [including matrix solutions]; systems of inequalities; substituting into or rearranging formulas; the general equation of the second degree)

1.7 **Data representation, probability, and statistics**

    1.7.1   **Data representation and analysis** (collecting data from experiments and simple surveys; representing data; interpreting tables, charts, plots, and graphs; kinds of scales [nominal, ordinal, interval, ratio]; measures of central tendency; measures of dispersion; sampling, randomness, and bias; prediction and inferences from data; fitting lines and curves to data; correlations and other measures of relations; use and misuse of statistics)

    1.7.2   **Uncertainty and probability** (informal likelihoods and the vocabulary of likelihoods, numerical probability and probability models, counting principles, mutually exclusive events, conditional probability and independent events, Bayes' Theorem, contingency tables, probability distributions for discrete random variables, probability distributions for continuous random variables, expectation, sampling, estimation of population parameters, hypothesis testing, confidence intervals, bivariate distributions, Markov processes, Monte Carlo methods and computer simulations)

1.8 **Elementary analysis**

    1.8.1   **Infinite processes** (arithmetic and geometric sequences, arithmetic and geometric series, Binomial Theorem, other sequences and series, limits and convergence of series, limits and convergence of functions, continuity)

    1.8.2   **Change** (growth and decay, differentiation, integration, differential equations, partial differentiation)

1.9 **Validation and structure**

    1.9.1   **Validation and justification** (logical connectives, quantifiers ["for all," "there exists"], Boolean algebra and truth tables, conditional statements, equivalence of statements [including converse, contrapositive, and inverse], inference schemes [e.g., modus ponens, modus tollens], direct deductive proofs, indirect proofs and proof by contradiction, proof by induction, consistency and independence of axiom systems)

    1.9.2   **Structuring and abstracting** (sets, set notation, and set combinations; equivalence relations, partitions, and classes; groups; fields; linear [vector] spaces; subgroups, subspaces, etc.; other axiomatic systems [e.g., finite geometries])

1.10 **Other content**

    1.10.1   **Informatics** (operation of computers, flow charts, learning a programming language, programs, algorithms with applications to the computer, complexity; history and nature of mathematics; special applications of mathematics [kinematics, Newtonian mechanics, population growth —discrete or continuous models, networks—applications of graph theory, linear programming, critical path analysis, examples from economics]; problem-solving heuristics; non-mathematical science content; non-mathematical content other than science)

Performance Expectations

2.1 **Knowing**

    2.1.1   **Representing** (demonstrating knowledge of a nonverbal mathematical representation of a mathematical object or procedure either by selection or by construction, either formal or informal; representations might be concrete, pictorial, graphical, algebraic, etc.)

    2.1.2   **Recognizing equivalents** (selecting or constructing mathematically equivalent objects [e.g., equivalent common and decimal fractions; equivalent trigonometric functions and power series; equivalent representation of concepts—e.g., place value; equivalent axiomatic systems; etc.])

    2.1.3   **Recalling mathematical objects and properties** (fitting given conditions)

2.2   **Using routine procedures**

    2.2.1   **Using equipment** (using instruments, using calculators and computers)

    2.2.2   **Performing routine procedures** (counting and routine computations; graphing; transforming one mathematical object into another by some formal process, e.g., multiplying by a matrix; measuring)

    2.2.3   **Using more complex procedures** (estimating to arrive at an approximate answer to a question; collecting, organizing, displaying, or otherwise using quantitative data; comparing and contrasting two mathematical objects, quantities, representations, etc.; classifying objects or working with the properties underlying a classification system)

2.3   **Investigating and problem solving**

    2.3.1   **Formulating and clarifying problems and situations** (formulate or clarify a problem related to a real-world or other concrete situation)

    2.3.2   **Developing strategy** (develop a problem-solving strategy or data-gathering experiment and discuss that strategy or experiment [not just applying the strategy or carrying out the experiment])

    2.3.3   **Solving** (execute some known or ad hoc solution strategy)

    2.3.4   **Predicting** (specify an outcome [number, pattern, etc.] that will result from some operation or experiment before it is actually performed)

    2.3.5   **Verifying** (determine the correctness of the result of problem solving ; interpret results in terms of an initial problem situation to evaluate how sensible the results are, etc.)

2.4   **Mathematical reasoning**

    2.4.1   **Developing notation and vocabulary** (develop new notation and vocabulary to record the actions and results of dealing with real-world and other problem situations)

    2.4.2   **Developing algorithms** (develop a formal algorithmic procedure for performing a computation or solving a problem of a certain type)

    2.4.3   **Generalizing** (extend the domain to which the result of mathematical thinking and problem solving is applicable by restating results in more general and more widely applicable terms)

    2.4.4   **Conjecturing** (make appropriate conjectures and conclusions while investigating patterns, discussing ideas, working with an axiomatic system, etc.)

    2.4.5   **Justifying and proving** (provide evidence for the validity of an action or the truth of a statement by an appeal to mathematical results and properties, or by an appeal to logic)

    2.4.6   **Axiomatizing** (explore a formal axiomatic system by relating subsystems, properties, or proposition in the system; consider new axioms and their consequences; examine the consistency of axiom systems, etc.)

2.5   **Communicating**

    2.5.1   **Using vocabulary and notation** (demonstrate the correct use of specialized mathematical terminology and notation)

    2.5.2   **Relating representations** (work with relationships and related mathematical representations to show the linkages between related mathematical ideas or related mathematical objects)

2.5.3   **Describing/discussing** (discuss a mathematical object, concept, pattern, relationship, algorithm, result, or display from a calculator or computer)

2.5.4   **Critiquing** (discuss and critically evaluate a mathematical idea, conjecture, problem solution, method of problem solving, proof, etc.)

Perspectives

3.1   **Attitudes towards science, mathematics, and technology** (curriculum encourages positive attitudes towards science, mathematics, and technology)

3.2   **Careers involving science, mathematics and technology**
  3.2.1   **Promoting careers in science, mathematics, and technology**
  3.2.2   **Promoting the importance of science, mathematics, and technology in non-technical careers**

3.3   **Participation in science and mathematics by underrepresented groups** (curriculum encourages all types of students to study and use science, mathematics and technology; examples of groups that could be targeted: women, racial, and ethnic minorities)

3.4   **Science, mathematics, and technology to increase interest** (curriculum promotes interest and increasing understanding of topics in science, mathematics, and technology by using experiences that are common to students or popular or intriguing information; examples include using sports, news, celebrities, history, literature, and interesting data)

3.5   **Scientific and mathematical habits of mind** (curriculum encourages ways of scientific and mathematical thinking such as openness, objectivity, tolerance of uncertainty, inventiveness, and curiosity)

DETAILED SCIENCE FRAMEWORK CATEGORIES

Content

1.1 **Earth sciences**
   1.1.1   **Earth features**
      1.1.1.1 Composition (earth's crust, mantle, and core; distribution of metals, minerals)
      1.1.1.2 Landforms (mountains, valleys, continents)
      1.1.1.3 Bodies of water (oceans, lakes, ponds, bottom of ocean, rivers)
      1.1.1.4 Atmosphere (layers of atmosphere [ionosphere, stratosphere, etc.])
      1.1.1.5 Rocks, soil (soil types, soil formation, pH of soil, classes of rocks, specific rocks and their uses)
      1.1.1.6 Ice forms (glaciers, icebergs, Antarctic)
   1.1.2   **Earth processes**
      1.1.2.1 Weather and climate (weather maps, weather forecasts, hurricanes, seasons of the year)
      1.1.2.2 Physical cycles (rock cycle, water cycle)
      1.1.2.3 Building and breaking (plate tectonics, earthquakes, volcanoes)
      1.1.2.4 Earth's history (geologic timetable, formation of fossils, fossil fuels, and mineral resources)
   1.1.3   **Earth in the universe**
      1.1.3.1 Earth in the solar system (earth/sun/moon system, night/day, tides, north/south hemisphere, seasons)
      1.1.3.2 Planets in the solar system (planets' features, order of planets in solar system)
      1.1.3.3 Beyond the solar system (galaxies, black holes, quasars, types of stars, constellations of stars)
      1.1.3.4 Evolution of the universe (origin/history/future of universe)

1.2 **Life sciences**
   1.2.1   **Diversity, organization, structure of living things**
      1.2.1.1 Plants, fungi (types of plants, fungi)
      1.2.1.2 Animals (types of animals)
      1.2.1.3 Other organisms (types of microorganisms)
      1.2.1.4 Organs, tissues (circulatory systems, plant leaf, systems for movement, eyes, ears)
      1.2.1.5 Cells (cell membranes, nucleus, mitochondria, vacuoles)
   1.2.2   **Life processes and systems enabling life functions**
      1.2.2.1 Energy handling (energy capture, storage, transformation—photosynthesis, respiration, biosynthesis [protein, carbohydrate, fat, etc.], digestion, excretion)
      1.2.2.2 Sensing and responding (biofeedback in systems, homeostatis, sensory systems, responses to stimuli [e.g., nervous system and brain])
      1.2.2.3 Biochemical processes in cells (regulation of cell functions, translation, protein synthesis, enzymes)
   1.2.3   **Life spirals, genetic continuity, diversity**
      1.2.3.1 Life cycles (life cycles of plants, insects, etc.: growth, development, reproduction, dispersal, aging, death; cell division, cell differentiation)
      1.2.3.2 Reproduction (plant/animal reproduction, asexual/sexual reproduction)

1.2.3.3 Variation and inheritance (Mendelian/non-Mendelian genetics, quantitative inheritance, population genetics)

1.2.3.4 Evolution, speciation, diversity (evidence for evolution, effects of evolution, processes of evolution [e.g., adaptation, natural selection], nature of a species, domestication, importance of diversity)

1.2.3.5 Biochemistry of genetics (concept of the gene, DNA/RNA, gene expression, genetic engineering)

1.2.4 Interactions of living things

1.2.4.1 Biomes and ecosystems (tundra, rain forest, savannah, wetlands, tide pools)

1.2.4.2 Habitats and niches (habitats of endangered species, niches of species)

1.2.4.3 Interdependence of life (food webs/chains, symbiotic relationships, impact of humans)

1.2.4.4 Animal behaviour (migration of birds, mate selection, rearing of young, social groupings of animals [e.g., beehives, elephant herds])

1.2.5 Human biology and health. Note: Many human biology topics will involve double coding. For example, studying the human digestive system (1.2.1.4 and 1.2.5), human impact on the environment (1.2.5 and 1.6), human reproduction (1.2.5 and 1.2.3.2).

1.2.5.1 Nutrition (vitamins and minerals in diet)

1.2.5.2 Disease (disease types, causes, prevention)

1.3 Physical sciences

1.3.1 Matter

1.3.1.1 Classification of matter (homogeneous and heterogeneous materials, elements, compounds, mixtures, solutions)

1.3.1.2 Physical properties (weight, mass, states of matter, malleability of metals, hardness, shape)

1.3.1.3 Chemical properties (periodic table, acidity, reactivity, atomic spectra, organic/inorganic)

1.3.2 Structure of matter

1.3.2.1 Atoms, ions, molecules (atoms, ions, molecules as the basis for different substances)

1.3.2.2 Macromolecules, crystals (polymers, shape/function of biological molecules, crystal structure)

1.3.2.3 Subatomic particles (electrons, protons, neutrons)

1.3.3 Energy and physical processes

1.3.3.1 Energy types, sources, conversions (potential and kinetic; chemical, nuclear, fossil fuels; hydroelectric power; changing one form of energy to another; energy and work, efficiency)

1.3.3.2 Heat and temperature (temperature scales, heat as a form of energy, heat versus temperature)

1.3.3.3 Wave phenomena (wave properties, types [e.g., IR, UV], wave interactions)

1.3.3.4 Sound and vibration (transmission of sound, acoustics, harmonics)

1.3.3.5 Light (nature of light, optics, luminosity, reflection, refraction)

1.3.3.6 Electricity (static electricity, electrical fields, alternating/direct current, electrical circuits)

1.3.3.7 Magnetism (magnets and their magnetic fields, magnetic properties). Note: Electromagnetism topics should be double-coded 1.3.3.6 and 1.3.3.7.

1.3.4 Physical transformations

1.3.4.1 Physical changes (gas laws, changes in states of matter, mixing)

1.3.4.2 Explanations of physical changes (general explanations for boiling, freezing, dissolving, etc.)

1.3.4.3 Kinetic theory (kinetic molecular theory)

1.3.4.4 Quantum theory and fundamental particles (quantum nature of light, photoelectric effect)

1.3.5 **Chemical transformations**

1.3.5.1 Chemical changes (definition of chemical change, types of reactions [e.g., displacement, acid-base, oxidation-reduction, etc.])

1.3.5.2 Explanations of chemical changes (ionic/covalent bonding, electron configurations, electronegativity)

1.3.5.3 Rate of change and equilibria (reagent concentrations, reaction conditions, dynamic equilibrium)

1.3.5.4 Energy and chemical change (activation energy, exothermic and endothermic reactions)

1.3.5.5 Organic and biochemical changes (types of organic compounds, organic reactions, biochemistry)

1.3.5.6 Nuclear chemistry (fission, fusion, isotopes, half-life, mass/energy conversion)

1.3.5.7 Electrochemistry (electrochemical cells/batteries, electrolysis, oxidation-reduction reactions)

1.3.6 **Forces and motion**

1.3.6.1 Types of forces (gravitational force, friction, centripetal force)

1.3.6.2 Time, space, and motion (measurement of time, types of motion [linear/rotational], describing motion [constant velocity, acceleration, momentum], reference frames for motion)

1.3.6.3 Dynamics of motion (balanced and unbalanced forces, action/reaction, momentum and collisions)

1.3.6.4 Relativity theory (mass/energy/velocity relationship, explaining the velocity of light, time frames while traveling at the speed of light)

1.3.6.5 Fluid behaviour (hydraulics, Bernoulli principle, pneumatics)

1.4 **Science, technology, and mathematics**

1.4.1 **Nature or conceptions of technology** (identifying needs and opportunities, generating a design, planning and making, evaluating)

1.4.2 **Interactions of science, mathematics, and technology**

1.4.2.1 Influence of mathematics, technology in science (information about contributions of mathematics and technology to development of scientific thought and the practice of science, e.g., new mathematics and technology make it possible for science to investigate new questions or to analyze data in new ways)

1.4.2.2 Applications of science in mathematics, technology (information about contributions of science to development and practice of mathematics and technology, e.g., development of calculus and classical mechanics, industrial processes, types of simple machines, measuring devices—thermometer, Geiger counter)

1.4.3 **Interactions of science, technology, and society**

1.4.3.1 Influence of science, technology on society (social, economic, ethical impacts of scientific and technological advances, e.g., influence of scientific ideas on social thought, such as social Darwinism; effects of computers on lifestyles)

1.4.3.2 Influence of society on science, technology (information about influence of society on the directions and progress of science and technology, e.g., controversies over research in genetic engineering, use of animals in research)

1.5 **History of science and technology** (famous scientists, classic experiments, historical development of scientific ideas, industrial revolution, classic inventions)

1.6 **Environmental and resource issues related to science**
  1.6.1 **Pollution** (acid rain, thermal pollution, global warming)
  1.6.2 **Conservation of land, water, and sea resources** (rain forest, old growth forests, water supplies)
  1.6.3 **Conservation of material and energy resources** (fossil fuels versus alternative energy sources, recycling aluminum)
  1.6.4 **World population** (population statistics, trends; effects of increasing world population, e.g., world hunger, epidemic diseases)
  1.6.5 **Food production, storage** (agricultural methods, food supply and demand, distribution methods)
  1.6.6 **Effects of natural disasters** (environmental damages of hurricanes/typhoons, volcanoes, drought)

1.7 **Nature of science.** Note: There is a difference between categories 1.7.1 and 3.6.
  1.7.1 **Nature of scientific knowledge** (scientific methods, knowledge subject to verification, knowledge subject to change)
  1.7.2 **The scientific enterprise** (canons of ethics and decision making, professional communication, the scientific community, personnel and processes in large-scale research efforts)

1.8 **Science and other disciplines**
  1.8.1 **Science and mathematics** (explicit mathematics instruction in the science curriculum)
  1.8.2 **Science and other disciplines** (science curriculum incorporated with language arts, social studies, or the arts; examples include chemistry of painting, using art or music to represent or illustrate science concepts, studying the role of science in other cultures, writing stories as metaphors that illustrate science concepts)

Performance Expectations

2.1 **Understanding.** Note: The performance expectation is that students will understand the kinds of information in this category. In some materials, the difference between simple, complex, and thematic information may be difficult to distinguish.
  2.1.1 **Simple information** (information such as vocabulary, facts, equations, simple concepts; examples include defining, describing, naming, quoting, reciting, etc.; specific examples are defining scientific terms [boiling point, niche], knowing symbols [abbreviations for units, chemical symbols], describing simple concepts [materials expand when heated, characteristics of animals])
  2.1.2 **Complex information** (information involving the integration of bits of simple information; examples include differentiating, comparing, contrasting, synthesizing; specific examples are understanding how increased external pressure raises boiling point of liquids, how fire is a part of the life cycle of pine trees)

2.1.3   **Thematic information.** Note: This category should not be coded if students are merely expected to name or describe thematic concepts. (information about concepts with broad applicability that organize and structure knowledge within a discipline or among disciplines; examples include energy, evolution, patterns, change, systems, etc.; a specific example of performance that could indicate understanding of thematic information is using themes to synthesize science knowledge and experiences)

2.2   **Theorizing, analyzing, and solving problems**

2.2.1   **Abstracting and deducing scientific principles** (when presented with facts or scientific data, deducing a scientific principle [e.g., when presented with spectra of several stars, deducing the stars' relative temperatures; when presented with data on plant growth, deducing that light is required])

2.2.2   **Applying scientific principles to solve quantitative problems** (using physical laws such as $f=ma$ to solve quantitative problems: when given acceleration [a] and mass [m], calculating force [f]; writing and balancing chemical equations; using balanced chemical equations to answer questions about chemical systems, e.g., stoichiometry problems)

2.2.3   **Applying scientific principles to develop explanations** (using gas laws to explain changes in gas temperature, pressure, and volume; using ecological principles to predict effect of reducing a population's habitat)

2.2.4   **Constructing, interpreting, and applying models** (using or creating models that represent systems, objects, events, or ideas: drawing a model of the solar system; making an analogy between human thinking and computer logic)

2.2.5   **Making decisions** (using scientific skills and knowledge to make decisions regarding personal, local, or societal issues; examples of issues are water purification, nutrition and resource utilization, air quality and energy production; decision making may include defining the decision to be made, identifying alternative choices, weighing advantages and disadvantages of each choice, and committing to action on a particular choice)

2.3   **Using tools, routine procedures, and science processes**

2.3.1   **Using apparatus, equipment, and computers** (calibrating an eye dropper, reading a meniscus, using pH paper, folding filter paper, preparing a microscope slide, operating a computer, running a computer program)

2.3.2   **Conducting routine experimental operations** (measuring the volume of an irregular-shaped solid by displacement of water, conducting a titration, culturing bacteria)

2.3.3   **Gathering data** (observing, measuring, etc.: perceiving characteristics, similarities, differences, and changes through use of the senses; comparing objects or events to standards of length, area, volume, mass, temperature, force, or time)

2.3.4   **Organizing and representing data** (classifying, constructing graphs, tables, and diagrams; organizing materials, events, and phenomena into local groupings, making graphs of data)

2.3.5   **Interpreting data** (extrapolating or interpolating data from a table or graph, identifying patterns or trends in data)

2.4   **Investigating the natural world**

2.4.1   **Identifying questions to investigate** (observing water droplets on outside surface of a drinking glass and forming questions about where the liquid came from; reading about fish dying in local lakes and forming questions about the cause)

2.4.2   **Designing investigations** (developing hypotheses, developing or choosing procedures, selecting materials and equipment)

2.4.3   **Conducting investigations** (executing procedures and recording data) Note: Students are often given prescribed procedures and told the expected results and conclusions. Such investigations are sometimes called "cookbook" experiments, since following procedures is similar to following a recipe in cooking.

2.4.4   **Interpreting investigational data** (organizing data, analyzing data, using data to address investigation's hypotheses or questions)

2.4.5   **Formulating conclusions from investigational data** (using data to make conclusions about the questions or hypotheses of the investigation)

2.5   **Communicating**

2.5.1   **Accessing and processing information** (finding information, using a library, listening to others for information)

2.5.2   **Sharing information** (reporting work to others, in written or oral form; communicating in a group to solve a scientific problem)

Perspectives

3.1   **Attitudes towards science, mathematics, and technology**

3.1.1   **Positive attitudes towards science, mathematics, and technology** (curriculum encourages positive attitudes towards science, mathematics, technology, and/or the study of them)

3.1.2   **Sceptical attitudes towards use of science and technology** (curriculum encourages students to evaluate disadvantages of the use of science and technology in society)

3.2   **Careers in science, mathematics, and technology**

3.2.1   **Promoting careers in science, mathematics, and technology** (curriculum materials describe or promote careers in science, mathematics, or technology)

3.2.2   **Promoting importance of science, mathematics, and technology in non-technical careers** (curriculum shows that science, mathematics, and technology are important in automobile repair, accounting, flying airplanes, etc.)

3.3   **Participation in science and mathematics by underrepresented groups** (curriculum materials encourage all types of students to study and use science or mathematics; examples of groups that could be targeted: women, racial, and ethnic minorities; students in certain regions of a country)

3.4   **Science, mathematics, and technology to increase interest** (curriculum uses experiences that are common to children as a way of increasing understanding of topics and/or increasing student interest in topics; popular or intriguing information is used to increase student interest in topics: examples include noting science/math aspects of sports and news, noting celebrities interested in science/math)

3.5   **Safety in science performance** (curriculum materials describe safe use of materials and equipment, safe procedures)

3.6   **Scientific habits of mind** (curriculum encourages ways of scientific thinking such as openness, scepticism, objectivity, tolerance of uncertainty, and curiosity)